Emerging Technologies in Meat Processing

Emerging Technologies in Meat Processing

Production, Processing and Technology

EDITED BY

Enda J. Cummins

UCD School of Biosystems and Food Engineering, University College Dublin, Ireland

James G. Lyng

UCD School of Agriculture and Food Science, University College Dublin, Ireland

WILEY Blackwell

This edition first published 2017
This edition first published 2017 © 2017 by John Wiley & Sons, Ltd

Registered office: John Wiley & Sons, Ltd, The Atrium, Southern Gate, Chichester, West Sussex, PO19 8SQ, UK

Editorial offices: 9600 Garsington Road, Oxford, OX4 2DQ, UK
The Atrium, Southern Gate, Chichester, West Sussex, PO19 8SQ, UK

For details of our global editorial offices, for customer services and for information about how to apply for permission to reuse the copyright material in this book please see our website at www.wiley.com/wiley-blackwell.

Library of Congress Cataloging-in-Publication Data

Names: Cummins, Enda John, editor. | Lyng, G. James, editor.
Title: Emerging technologies in meat processing : production, processing and technology / [edited] by Enda John Cummins, James G. Lyng.
Description: Chichester, UK ; Hoboken, NJ : John Wiley & Sons, 2017. |
Includes bibliographical references and index.
Identifiers: LCCN 2016025830| ISBN 9781118350683 (cloth) | ISBN 9781118350775 (epub)
Subjects: LCSH: Meat industry and trade–Technological innovations. |
Meat–Quality. | Food–Risk assessment.
Classification: LCC TS1960 .E53 2017 | DDC 664/.9–dc23 LC record available at https://lccn.loc.gov/2016025830

A catalogue record for this book is available from the British Library.

Wiley also publishes its books in a variety of electronic formats. Some content that appears in print may not be available in electronic books.

Cover image: © sturti/Gettyimages

Set in 9.5/13pt, MeridienLTStd by SPi Global, Chennai, India.

1 2017

Contents

About the IFST Advances in Food Science Book Series

The Institute of Food Science and Technology (IFST) is the leading qualifying body for food professionals in Europe and the only professional organisation in the UK concerned with all aspects of food science and technology. Its qualifications are internationally recognised as a sign of proficiency and integrity in the industry. Competence, integrity, and serving the public benefit lie at the heart of the IFST philosophy. IFST values the many elements that contribute to the efficient and responsible supply, manufacture and distribution of safe, wholesome, nutritious and affordable foods, with due regard for the environment, animal welfare and the rights of consumers.

IFST Advances in Food Science is a series of books dedicated to the most important and popular topics in food science and technology, highlighting major developments across all sectors of the global food industry. Each volume is a detailed and in-depth edited work, featuring contributions by recognized international experts, and which focuses on new developments in the field. Taken together, the series forms a comprehensive library of the latest food science research and practice, and provides valuable insights into the food processing techniques that are essential to the understanding and development of this rapidly evolving industry.

The IFST Advances series is edited by Dr Brijesh Tiwari, who is Senior Research Officer at Teagasc Food Research Centre in Ireland.

Forthcoming titles in the IFST series

Tropical Roots and Tubers: Production, Processing and Technology, edited by Harish K. Sharma, Nicolas Y. Njintang, Rekha S. Singhal, Pragati Kaushal

Ultrasound in Food Processing: Recent Advances, edited by Mar Villamiel, Jose Vicente Garcia-Perez, Antonia Montilla, Juan Andrés Cárcel and Jose Benedito

Herbs and Spices: Processing Technology and Health Benefits, edited by Mohammad B. Hossain, Nigel P. Brunton and Dilip K Rai

List of contributors

Dong U. Ahn Department of Animal Science, Iowa State University, Ames, IA, USA

Cristina Arroyo Institute of Food and Health, UCD, Belfield, Dublin, Ireland

James R. Claus Department of Animal Sciences, Meat Science and Muscle Biology Laboratory, University of Wisconsin-Madison, USA

Malco Cruz-Romero Food Packaging Group, School of Food and Nutritional Sciences, University College Cork, Cork, Ireland

Patrick J. Cullen School of Food Science & Environmental Health, Dublin Institute of Technology, Dublin, Ireland, School of Chemical Engineering, New South Wales University, Sydney, Australia

Enda J. Cummins UCD School of Biosystems and Food Engineering, Belfield, Dublin, Ireland

Maeve Cushen UCD School of Biosystems and Food Engineering, Agriculture and Food Science Centre, Belfield, Dublin, Ireland.

Agapi Doulgeraki Laboratory of Microbiology and Biotechnology of Foods, Department of Food Science and Human Nutrition, Agricultural University of Athens, Athens, Greece

Michel Federighi INRA UMR1014 SECALIM, Nantes, France; LUNAM Université, Oniris, Université de Nantes, Nantes, France

Jesús M. Frías School of Food Science and Environmental Health, Dublin Institute of Technology, Dublin, Ireland

Sandrine Guillou INRA UMR1014 SECALIM, Nantes, France; LUNAM Université, Oniris, Université de Nantes, Nantes, France

Cheorun Jo Department of Animal Biotechnology, Seoul National University, Seoul, Republic of Korea; Department of Agricultural Biotechnology, Seoul National University, Seoul, Korea

Kompal Joshi School of Food Science and Environmental Health, Dublin Institute of Technology, Dublin, Ireland

Joseph P. Kerry School of Food and Nutritional Sciences, University College Cork, Co Cork, Ireland

Muhammad Issa Khan National Institute of Food Science and Technology, University of Agriculture, Faisalabad, Pakistan; Researcher, Department of Animal Biotechnology, Seoul National University, Seoul

Tatiana Koutchma Guelph Food Research Centre, Agriculture and Agri-Food Canada, Guelph, ON, Canada

Fiona Lalor School of Public Health, Physiotherapy and Population Science, University College Dublin, Dublin, Ireland

Marion Lerasle INRA UMR1014 SECALIM, Nantes, France; LUNAM Université, Oniris, Université de Nantes, Nantes, France

James G. Lyng Institute of Food and Health, UCD, Dublin, Ireland

N.N. Misra School of Food Science & Environmental Health, Dublin Institute of Technology, Dublin, Ireland

Ki Chang Nam Department of Animal Science and Technology, Sunchon National University, Sunchon, Korea

Tomas Norton Agricultural Engineering Department, Harper-Adams University College, Newport, UK

George-John Nychas Laboratory of Microbiology and Biotechnology of Food, Department of Food Science and Human Nutrition, Agricultural University of Athens, Athens, Greece

Efstathios Panagou Laboratory of Microbiology and Biotechnology of Foods, Department of Food Science and Human Nutrition, Agricultural University of Athens, Athens, Greece

Kumari Shikha Ojha Teagasc Food Research Centre, Ashtown, Dublin, Ireland

Hélène Simonin UMR Procédés Alimentaires et Microbiologiques, équipe PMB, AgroSup Dijon, France; Université de Bourgogne, Dijon, France

Da-Wen Sun Food Refrigeration and Computerised Food Technology (FRCFT), School of Biosystems Engineering, University College Dublin, National University of Ireland, Agriculture & Food Science Centre, Belfield, Dublin, Ireland.

Brijesh K. Tiwari Department of Food Biosciences, Teagasc Food Research Centre, Dublin, Ireland; Manchester Food Research Centre, Manchester Metropolitan University, Manchester, UK; Teagasc Food Research Centre, Ashtown, Dublin, Ireland

Ubaid-ur-Rahman National Institute of Food Science and Technology, University of Agriculture, Faisalabad, Pakistan

Patrick Wall School of Public Health, Physiotherapy and Population Science, University College Dublin, Dublin, Ireland

Di Wu Food Refrigeration and Computerised Food Technology (FRCFT), School of Biosystems Engineering, University College Dublin, National University of Ireland, Agriculture and Food Science Centre, Belfield, Dublin, Ireland

Yan Zhao Institute of Quality Standard & Testing Technology for Agro Products, Key Laboratory of Agro product Quality and Safety, Chinese Academy of Agriculture Sciences, Beijing, China; Key Laboratory of Agro-product Quality and Safety, Ministry of Agriculture, Beijing, China

CHAPTER 1

Emerging technologies in meat processing

Enda J. Cummins[1] & James G. Lyng[2]
[1] *Biosystems and Food Engineering, UCD, Dublin, Ireland*
[2] *Institute of Food and Health, UCD, Dublin, Ireland*

1.1 Context and challenges

Meat is a global product, which is traded between regions, countries and continents. The onus is on producers, manufacturers, transporters and retailers to ensure an ever-demanding consumer receives a top-quality product that is free from contamination. With such a dynamic product and market place, new innovative ways to process, package and assess meat products are being developed. In some instances, industry uptake of new technologies is stifled by a lack of knowledge about these new technologies and their impact on product quality and safety. With ever-increasing competition and tighter cost margins, industry has shown willingness to engage in seeking novel innovative ways of processing, packaging and assessing meat products while maintaining quality and safety attributes. Several new technologies have emerged with regard to meat processing, packaging and quality assessment, which have the potential to improve production efficiency while maintaining meat safety and quality. A number of novel thermal and non-thermal technologies designed to achieve microbial safety while minimising the effects on its nutritional and quality attributes have also become available.

Minimising changes in quality and safety during processing is a considerable challenge for food processors and technologists. Thus, there is a requirement for detailed industrially relevant information concerning emerging technologies in meat product manufacture. In addition, industrial adoption of novel processing techniques is in its infancy. Applications of new and innovative technologies and resulting effects to those food products either individually or in combination are always of great interest to academic, industrial, nutrition and health professionals.

Emerging Technologies in Meat Processing: Production, Processing and Technology, First Edition.
Edited by Enda J. Cummins and James G. Lyng.

1.2 Book objective

The primary objective of this book on *Emerging Technologies in Meat Processing* is to provide a comprehensive overview of the application of novel processing techniques as applied to the meat industry. The book evaluates recent advances on how meat is produced, processed and stored and is a benchmark reference book on novel processing, packaging and assessment methods of meat and meat products.

1.3 Book structure

Meat processors have a major responsibility to consumers when it comes to producing quality, nutritious and particularly safe foods. Conventional methods of meat processing and preservation (e.g. heat processing, low-temperature preservation or dehydration) have been used for hundreds of years. However, the last century has witnessed a dramatic increase in the development of new technologies, which have, in many cases, been hyped as replacements for conventional methods. However, in spite of much excitement relating to their discovery and potential, the anticipated uptake by industry has not occurred. In many cases, alternative technologies are still expensive in terms of capital outlay and are therefore not attractive options for processors, although they are generally becoming cheaper as time progresses. The reason for the lack of uptake most likely runs deeper than financial, as in many cases the alternatives are more economic or produce a higher quality product than conventional methods so that processors could recoup the initial capital outlay in reduced running costs or by charging higher prices for a premium quality product. It is most likely that the biggest obstacle these technologies face is a lack of basic understanding of their potential and, more importantly, when it comes to preservation, an unwillingness to trust the alternative methods compared to the tried and tested conventional methods. This book does not set out to try and convince food processors to drop conventional methods and replace them with alternatives. Instead, in Part 1, it sets to review alternative or novel processing techniques reinforcing the positive aspects of each operation and also discussing areas of weakness. Part 2 sets out an overview of alternative packaging solutions and meat functionality, clearly listing advantages and disadvantages and providing the reader with case studies where these technologies have been used. Part 3 reviews advances in assessment techniques for improved meat quality and safety.

Part 1 (processing techniques) consists of a number of chapters on novel processing techniques for the meat industry. Recent developments in irradiation, high-pressure processing, electroprocessing, light-based technologies, ultrasound, robotics and other emerging technologies are discussed with emphasis on operational principles and inherent strengths and weaknesses of the technologies. In Chapter 2, the various sources of ionising radiation are described and distinguished. The mode of action is described and the advantages and disadvantages of irradiation are considered. The chapter finishes with a section outlining the author's view of the future for irradiation. Chapter 3 reviews the history of high preservation, and typical pressures used for meat preservation is put in context. The mode of action of high pressure in meat preservation is discussed, as are its advantages and disadvantages. While a lot has been published, more work needs to be done (e.g. pressure resistance problems, which can be overcome by combining pressure with either mild heat or cold) and the future for high pressure is considered in the final section of this chapter.

Electroprocessing has seen many technological developments in recent years. Chapter 4 begins with the classic categorisation of the different forms of electroheating in terms of the electromagnetic spectrum and then goes on to clearly describe and distinguish the heating mechanism of each. A central portion of the chapter is the presentation of case studies outlining situations where each of the electroheating technologies has been used to preserve products commercially or has undergone research and development to a form, which is suitable for commercial application. Chapter 5 focuses on the application of infrared and light-based technologies to meat and meat products. It has been suggested that magnetic UV, IR and high-intensity light pulses all have potential in meat preservation. Some forms are not always suitable for direct application but still have an important role to play in preservation as they can be used for applications such as sterilising packaging, contact surfaces or air within packaging equipment. These forms of electromagnetic radiation can be used in a number of forms (e.g. near vs far infrared) and the identification of where the various forms fit into the electromagnetic spectrum is achieved using a standardised electromagnetic spectrum diagram. This chapter explores the application, interactions and equipment associated with these light-based technologies in addition to illustrating practical case studies.

Chapter 6 begins where the fundamentals of ultrasonics are outlined and high-intensity versus low-intensity forms of ultrasound are distinguished. This is followed by a section in which ultrasonic equipment and specific industry-relevant case studies are discussed. The use of ultrasound for the decontamination of meat forms a central part of this chapter. It finishes with conclusions regarding the possible future for ultrasonics in meat preservation. Chapter 7 introduces the operational principles of emerging technologies such as the hydrodynamic shock wave, with particular emphasis on applications, mode of operation, advantages and disadvantages of the technology. The chapter concludes with some developmental advances in the technology. Part 1 of the book concludes with Chapter 8 which provides an overview of the use and application of robotics in meat processing. The chapter provides details for product handling and processing with emphasis on inherent strengths and weaknesses. The chapter is illustrated by relevant case studies and provides a reference for currently available robotic equipment. The chapter finishes by providing a synopsis of the likely future role for robotics in meat processing.

Part 2 of the book deals with novel packaging and meat functionality. Recently the area of meat packaging has seen many new developments. This section reports on these developments and implications for shelf life, meat safety and quality. In particular, developments in novel packaging systems and smart packaging of meats are evaluated. Chapter 9 considers novel packaging solutions for meat products including the use of case-ready packaging with emphasis on modified atmospheric packaging and oxygen scavenging systems. The operational principles are detailed along with advantage and disadvantages of the technologies. The chapter concludes with a synopsis of the likely future role that novel packaging will play in the preservation of meat products. Packaging in the future is likely to be more than just a physical container that provides food with protection from the surrounding environment. Chapter 10 analyses the theory, mode of action and role of smart packaging systems in today's meat industry. The recent developments of nanotechnology in smart packaging systems are also discussed. In Chapter 11, the authors look at functionality in the meat product itself, with a focus on probiotics for meat products.

Rapid detection of pathogens and microbial contaminants is essential for ensuring meat quality and safety. Part 3 of this book looks at developments in rapid methods for microbial analysis. In addition, carcass evaluation technology and assessment of meat quality characteristics using computer vision and spectral techniques are evaluated. The section finishes with an assessment of meat authenticity.

Rapid detection of pathogens and microbial contaminants is essential for ensuring meat quality and safety and forms the basis of Chapter 12. Traditional detection methods have relied on time-consuming media culture methods with isolation. There have been a number of new innovations in methods for the microbiological analysis of meat. An array of rapid methods has been developed to make detection and identification faster, more convenient, more sensitive and more specific than conventional assays. This chapter assesses developments in this field and provide a synopsis of rapid methods of assessment. Chapter 13 focus on the use of hyperspectral techniques in evaluating quality and safety of meat and meat products. Spectral imaging techniques have emerged as techniques capable of detecting microbes in a non-destructive and rapid way. Case studies are reviewed and details on advantages and disadvantages of the technology are discussed.

Chapter 14 looks at carcass evaluation techniques with particular emphasis on in vivo methods (ultrasound, X-ray computed tomography (CT) and nuclear magnetic resonance (NMR)/magnetic resonance imaging (MRI)). Methods available for the prediction of body and carcass composition are evaluated. Methods for predicting composition of carcasses including video image analysis

(VIA), total body electrical conductivity (TOBEC) and bioelectrical impedance are discussed. Chapter 15 addresses the issue of meat authenticity. With an ever-expanding, open and globalised market place, meat products can be freely transported around the world. As illustrated in previous public health scares (e.g. dioxins in pork), consumer confidence in the meat industry is reliant on effective safety and authenticity. This chapter looks at recent developments in this field including the use of different authenticity techniques. The book concludes with Chapter 16 which provides an overview of the current role of food regulation practices within the European Union (EU) and internationally. International trade law, with emphasis on international food safety systems and food safety regulation within the EU, is discussed in addition to issues surrounding food marketing. The chapter concludes with a perspective on global trends and marketing challenges.

1.4 Conclusion

Emerging technologies do play an important role and have advantages for both processors and consumers. However, any likely uptake in the short term will be as part of a hurdle or minimal processing strategy in conjunction with conventional methods. The long-term success and uptake of emerging technologies depends on practicing food professionals receiving continued exposure to technological possibilities coupled with the education of new graduates of their potential. This text will serve as a comprehensive reference book for students, educators, researchers and food processors providing an up-to-date insight into emerging technologies for meat manufacture. The range of processes covered provides engineers and scientists working in the meat and food industries with a valuable resource for their work. Given the emphasis on novel technologies, the text is expected to have broad and significant appeal. This book can be a valuable reference book for companies, research institutions and universities active in the areas of meat processing, safety and quality evaluation.

PART I
Novel processing techniques

CHAPTER 2

Irradiation of meat and meat products

Ki Chang Nam[1], Cheorun Jo[2] & Dong U. Ahn[3]
[1]*Department of Animal Science and Technology, Sunchon National University, Sunchon, Korea*
[2]*Department of Agricultural Biotechnology, Seoul National University, Seoul, Korea*
[3]*Department of Animal Science, Iowa State University, Ames, IA, USA*

2.1 Summary

The various sources of ionizing radiation are described and distinguished (i.e., isotopes, electron beams, and X-ray radiation). Identification of where the various forms fit into the electromagnetic spectrum is achieved using a standardized electromagnetic spectrum diagram. The mode of actions, advantages, and disadvantages of irradiation and companies that manufacture irradiation equipment are discussed. The distinction between dosage levels (i.e., low, up to 1 kGy; medium, 1–10 kGy; and high, 10–50 kGy) in terms of their effect on meat quality is described and the extent to which irradiation has been accepted by consumers and approved for food use is discussed. The chapter also includes a section outlining the authors' view of the future for irradiation.

2.2 Theory of irradiation of foods

2.2.1 Forms of irradiation

Radiation energies are classified into three categories of electromagnetic radiation (gamma ray, X-ray), charged particle radiation (alpha ray, beta ray, electron beam, photons), and uncharged particles (neutron). Among them, two types of ionizing radiation are basically used for food safety: one is the radiation energy generated from a radionuclide of radioactive source and the other is produced from an accelerator or a nuclear reactor. Gamma rays and X-rays have relatively short wavelengths (high energy) among electromagnetic spectrum including radio waves, microwaves, visible light, ultraviolet, and so on (Satin, 1892). Accelerated electron is a charged particle with high energy. Thus, radiation

Emerging Technologies in Meat Processing: Production, Processing and Technology, First Edition.
Edited by Enda J. Cummins and James G. Lyng.

types that can be applied to meat and meat products are gamma ray, X-ray, and accelerated electron.

The "ionizing" radiation has a power to dislodge electrons from molecules and convert them into electrically charged ions. Gamma ray has a strong enough power to ionize molecules located in a deep position of targeted food and is from a radioactive isotope. Therefore, it should be managed safely. On the other hand, accelerated electrons and X-ray are generated by a machine process. Electron beam is directed to only target food, and the energy efficiency is higher than that of gamma ray. The most predominantly available form of food irradiation is gamma ray or accelerated electrons. Use of X-ray for food irradiation has been tested for commercial utilization as well as in research, but its efficiency is 70–80% of gamma ray and <30% of accelerated electrons (Olson, 1998).

Regardless of radiation sources, the amount of ionizing energy absorbed in target materials is measured as gray (Gy); 1 Gy equals to 1 J/kg. Thus, a dosage of 1 kGy indicates that the irradiated food receives 1000 J/kg of food mass. Generally, minimum dosage is applied to a food to achieve irradiation purpose and to maintain the quality of treated foods (Cleland, 2006).

The comparative characteristics of ionizing radiation sources used for food irradiation are shown in Table 2.1. The effectiveness of irradiation varies based on the type of radiation sources used, the radiation intensity, and the targeted microbes (Kwon, 2010). Following are the relative advantages and disadvantages of the three forms of food irradiation.

2.2.1.1 Gamma ray

Gamma ray is the most widely used form for food irradiation and is normally emitted from the spontaneous disintegration of radionuclide, that is, radioactive isotope. Gamma ray is classified as photons and does not have mass despite its very high energy levels (Satin, 1892). Gamma ray is higher frequency photons than either ultraviolet or X-ray and can penetrate into a target food to a depth of 60–80 cm (Olson, 1998). Thus, gamma ray is appropriate for "post-sterilization" of packaged foods, and the concern about recontamination of final products is minimal (Loaharanu, Kava, and Choi, 2007). The approved source of

Table 2.1 The characteristics of ionizing radiation sources.

	Gamma ray	Electron beam	X-ray
Energy type	Electromagnetic	Charged particle	Electromagnetic
Energy volt (MeV)	1.17 + 1.33	~10	~5
Energy efficiency	Low (~30%)	High (~85%)	Low (~10%)
Penetration capability	Deep (60–80 cm)	Low (8–10 cm)	Deep
Source control	Continuous	Switch (on/off)	Switch (on/off)

Source: From Kwon (2010). Adopted with permission by Korea Food Safety Research Institute.

gamma ray for food irradiation is cobalt-60 and cesium-137 by the US FDA and by the International Standards for Food Irradiation (Olson, 1998). Co-60 emits two gamma rays simultaneously with energies of 1.17 and 1.33 MeV and has a half-life of 5.26 years with activity decay by 12.35% per year (Kwon, 2010). Co-60 is mainly used to sterilize various medical devices and to control pathogens in foods. Co-60 is encapsulated in a thin stainless steel cylinder (called a pencil). Since the Co-60 in the pencils will not contact with the irradiated food, it does not become radioactive. However, Cs-137 is seldom used because large amounts of source are not readily available (Cleland, 2006).

2.2.1.2 Electron beam

The electron beam is a high-energy stream of electrons generated by an electron accelerator that has a similar structure to television tubes. Electron beam has totally different mechanisms from the gamma ray. Electrons can be accelerated up to 10 MeV, which is about eight times higher than the energy level of gamma ray (Olson, 1998). The machine to produce electron beam can be easily controlled by its on/off switch system because it does not use any radioactive sources. However, electron beam has similar effectiveness to eradicate microorganisms in irradiated meats (Kwon, 2010). Electron beam has advantages in terms of ease of process control, irradiation speed, accuracy, energy efficiency, and consumer acceptance compared with the gamma ray, and the application of electron beam is eagerly tried in many developed countries (WHO, 1988). The only disadvantage of electron beam is its limited penetration capability. As electrons have a small mass and slow down quickly as they enter a product, electrons (10 MeV) can penetrate to a depth of only 3.81 cm in meat. Thus, it can be applied for the surface sterilization of meat or thin meat products (WHO, 1988). To overcome the short penetration capability, simultaneous use of two beams positioned oppositely is normally used.

2.2.1.3 X-ray

X-ray radiation facility can be regarded as a more powerful version of the machines that can be easily found in hospitals. X-ray is produced by collision of the high-energy electrons with a metal (tungsten) target without using any radioactive materials (Brynjolfsson, 1989). X-ray is developed to overcome the low penetration capability of electron beam. The penetration capability of X-ray, however, is lower than the gamma ray, and the facility can be freely switched on and off. X-ray is considered as a new technology with advantages of gamma ray and electron beam, but X-ray has relatively low energy efficiency.

Selection of an irradiation source can be determined by treatment goal (pasteurization, sterilization, insecticide, or growth control), characteristics of target food (thickness and density of target material, contamination degree, moisture content, deteriorated rate, density, packaging, or surface sterilization), energy characteristics (penetration capability, energy efficiency, or source control), minimum dose, dose uniformity, processing rate, and economics (Cleland, 2006).

2.2.2 Mode of action

The microorganisms in foods are highly susceptible to irradiation. When ionizing energy passes through a food, some of the atoms or molecules in the food absorb the energy and become reactive ions, free radicals, or damaged (Woods and Pikaev, 1994). Free radicals are highly reactive and destroy cellular components (Olson, 1998). This type of radiation is called ionizing radiation and is used to destroy insects, pathogenic bacteria, and parasites in meats.

The most direct target of ionization energy is DNA molecules. An exposure of bacterial cells to 0.1 kGy irradiation resulted in 2.8% DNA damage, whereas 0.14% of the enzymes and 0.005% of amino acids were altered with the same dose (Diehl, 1995). The loss of replication ability of cells is caused by damaged DNA. Even small changes in the DNA of a bacterial cell can result in the death of the bacteria. Breaking bonds in the DNA results in the loss of a cell's ability to replicate. The biological mechanisms of ionizing radiation can be explained by a direct theory and an indirect theory (Grecz, Rowley, and Matsuyama, 1983). Radiation destroys microorganisms by inactivation of genetic material in living cells either by its direct effects on DNA or through the production of radicals or ions that attack DNA indirectly (WHO, 1994).

According to the direct theory, some molecules in cells or food components are more sensitive to the action of the ionizing energy than others. DNA is the most critical target of irradiation although other cellular components may also be affected. The direct effect of irradiation on nucleic acid is either ionization or excitation. DNA molecules, especially the base part of DNA, are highly susceptible to ionizing radiation resulting in cleavage of phosphodiester bonds of DNA double helix. The damaged DNA causes the loss of a cell's ability to replicate and eventually lead the cell to death, but some can be repaired by DNA polymerase or ligase (Kwon, 2010). Depending on the repairing capacity, radiation susceptibility of each microorganism is also different.

Indirect theory is that radiolytic products (ions or free radicals) of water molecules induce chemical changes in essential compounds or the structure necessary for maintaining life (Smith and Pillai, 2004). Especially, hydroxyl radical (·OH) is known to have 90% damage rate to DNA molecules. Ionizing radiation can also affect cell membrane, resulting in an additional impact on the resistance and susceptibility of cells to irradiation (Kwon, 2010). The susceptibility of bacteria to irradiation is influenced by environmental factors such as temperature, atmosphere gas, water activity, pH, food components, and growth step of microorganisms. Indirect effects of irradiation on DNA include excitation of water molecules, which then diffuse to the medium and contact with chromosomal materials (Moseley, 1989).

Depending on the irradiation dose, food can be either pasteurized to reduce or eliminate pathogens or sterilized except for some viruses (Crawford and Ruff, 1996). Less than 10 kGy of irradiation can kill insects and larvae and destroy pathogenic bacteria and parasites. Very low doses (up to 1 kGy) of radiation can kill at least 99.9% of *Salmonella* in poultry and an even higher percentage of *Escherichia coli* O157:H7 in ground beef (Olson, 1998). Spore-forming bacteria can be controlled by the combination of high dose of irradiation and heat treatment. Irradiation is not effective to control virus that is resistant to irradiation (Dickson, 2001).

Irradiation can dramatically improve the safety of meat products by killing pathogenic bacteria contaminated. Relatively low doses (<10 kGy) of irradiation can increase the shelf life of meat products in cold chain and is called "radurization" or "radiopasteurization" (Table 2.2). Most of the meat and meat products can be irradiated using the effect of radurization because cold chain has been well equipped in meat industry and meat products are stored in refrigerated temperature. Radicidation is killing asporogenous bacteria and foodborne bacteria contaminated in meat, and radappertization is used to eradicate microorganisms except virus completely using more than 10 kGy of radiation doses (WHO, 1994). Depending on the contamination degree, high doses of 10–50 kGy are required (Loaharanu, Kava, and Choi, 2007).

2.2.3 Advantages and disadvantages of irradiation

Meat irradiation is a process that provides several important benefits for consumers as well as industry. It improves the safety of fresh meats by reducing or eliminating foodborne pathogens and extends the shelf life of the products (Thayer, 1994; Murano, 1995). Consumers and regulatory agencies have become increasingly aware of the dangers of pathogenic microorganisms to human health, and this has led to a greater willingness to consider irradiation treatment of meat. FDA and USDA had approved the use of irradiation on red

Table 2.2 Control effect of food microorganisms by irradiation dose level.

Irradiation effect	Dose (kGy)	Purpose	Examples
Radurization	1–5	Decrease in putrefaction bacteria, Increase in the shelf life of cold chain foods	Fresh meat, processed meat products, poultry meat
Radicidation	2–10	Killing nonspore pathogenic bacteria, food-poisoning bacteria	Meat, poultry meat
Radappertization	1–50	Killing all bacteria including spore-forming bacteria (*Bacillus* and *Clostridium*)	Canned, germ-free for patient, astronaut food

Source: From Kwon (2010). Adopted with permission by Korea Food Safety Research Institute.

meats at the maximum absorbed dose of 4.5 and 7.0 kGy for refrigerated and frozen meats, respectively, in 1999 (USDA, 1999). Poultry meat was approved in 1990 at a dosage between 1.5 and 3.0 kGy to control pathogenic bacteria. Irradiation can also provide numerous advantages to meat processors including increased hygienic quality, extended shelf life, and reduction in chemical and toxic residues by reducing the use of nitrite or other chemical preservatives (Lebepe *et al.*, 1990; Radomyski *et al.*, 1994; Murano *et al.*, 1995).

The radiolytic products in irradiated foods are neither unique nor toxicologically significant in the quantities found (Thayer, 1994). However, the interactions of radiolytic products and meat components can deteriorate meat quality. Water is the major component of meat, and the content ranges from 60% to 75% in meat and meat products. When pure water is irradiated, a number of radiolytic products are formed (Swallow, 1984).

$$H_2O \rightarrow \cdot OH\text{ (hydroxyl radical)} + \cdot H\text{ (hydrogen atom)} + H_2$$
$$+ H_2O_2\text{(hydrogen peroxide)} + H_3O^+\text{(hydrated proton)} + e^-_{aq}\text{(hydrated electron)}$$

Among the radiolytic products, hydroxyl radical is a powerful oxidizing agent whereas the hydrated electron is a strong reducing agent (Diehl, 1995). When the primary electrons from a radiation source collide with an electron in the medium, the energy is transferred to the electron in the medium. This causes the ejection of secondary electron and generates a charged radical. Therefore, despite many advantages on meat safety, irradiation can accelerate oxidative chemical changes in meats.

Water-soluble vitamins react with the radiolytic products of water, whereas fat-soluble vitamins react with free radicals formed by the radiolysis of lipids. Thiamine is the most radiation-sensitive but other B vitamins are resistant to irradiation-induced destruction (Woods and Pikaev, 1994). Vitamin C acts as an antioxidant against irradiation-induced changes and is converted to dehydroascorbic acid by ionizing radiation, but its reactivity remains the same as the native form. Vitamins A and E are the only fat-soluble vitamins affected by irradiation (WHO, 1994). α-Tocopherol is the most radiation-sensitive fat-soluble vitamin (Olson, 1998), but can act as an antioxidant against irradiation-induced changes.

Irradiation accelerates lipid oxidation, especially in aerobically packaged meats, and produces characteristic irradiation off-odors. In the presence of oxygen, irradiation of unsaturated fatty acids accelerates autoxidative changes and produces a number of oxygen-containing products such as hydroperoxides and carbonyl compounds (Nawar *et al.*, 1996). Irradiation at 1.5–10 kGy dosage increased 2-thiobarbituric acid reactive substances (TBARS) values in turkey breast muscles when aerobically packaged in oxygen-permeable bags (Hampson *et al.*, 1996; Ahn *et al.*, 2001). Lipid oxidation in meat, however, was not influenced by irradiation with vacuum packaging (Ahn *et al.*, 1999). The lipid oxidation of frozen-stored irradiated meat was different from those of refrigerated storage, which can be explained by the limited mobility of free radicals in frozen state (Nam *et al.*, 2002).

Several off-odor volatile compounds were newly generated or increased in meat after irradiation. In general, aromatic and sulfur-containing amino acids are the most susceptible amino acids to irradiation (Patterson and Stevenson 1995; Ahn *et al.*, 2001). Jo, Lee, and Ahn (1999) showed that 1-heptene content in volatiles was positively related to irradiation dose. Du *et al.* (2001) showed that the production of alkenes and alkanes is associated with irradiation. Irradiation-induced fatty acid degradation may be similar to lipid oxidation, especially when oxygen is available. Several sulfur compounds were newly generated or increased in meat after irradiation: dimethyl sulfide, dimethyl disulfide, and dimethyl trisulfide were among the most prominent sulfur compounds responsible for irradiation off-odor in meat (Ahn *et al.*, 2000). Sulfur compounds formed by radiolysis of sulfur-containing amino acids might be the major contributors for irradiation odor since sulfur compounds have very low threshold for odor detection (Ahn, 2002; Ahn and Lee, 2009).

Gamma irradiation converted the brown metmyoglobin to a red pigment, which is similar but not identical to oxymyoglobin (Satterlee, Wilhelm, and Barnhart, 1971). Irradiated chicken breasts had a definite change from the usual brown or purple color to a more vivid pink or red color in oxygen-permeable film (Millar *et al.*, 1995). Irradiated pork loin muscles had increased redness, and the increased pink color was very stable during refrigerated storage in even aerobic packaging conditions (Ahn *et al.*, 2000). Irradiation generated carbon monoxide gas, which can form a sixth ligand on the heme of myoglobin (carbon monoxide myoglobin), and lowered oxidation–reduction potential of meat. Thus, the pink color formed in irradiated raw and cooked turkey breast was characterized as carbon monoxide myoglobin (Nam and Ahn, 2002a, 2002b). Contrary to the increased redness of irradiated poultry breast and pork loin, the irradiated beef was discolored to brown under aerobic conditions (Nam *et al.*, 2008).

2.3 Irradiation equipment

2.3.1 Case studies – irradiation of foods

Acceptance/rejection of consumer toward innovative food technologies is the result of a complex decision-making process, which involves an assessment of the perceived risks and benefits associated with the new technology and existing alternatives. This process is not simply related to the characteristics of the process itself but also related to the needs, beliefs, and attitudes of consumers and the nature of economic, political, and social environment in which food choices take place (Henson, 1995). Most studies on public acceptance of food irradiation had been conducted in the 1980s and 1990s. The majority of studies on irradiated meats and meat products were done in the United States, Brazil, and Turkey

(Frewer *et al.*, 2011). The main consumer concerns were the perceived "carcinogenicity" of irradiated foods, impaired food quality, risks to factory workers, and the environmental problems during production (Frewer *et al.*, 2011). Some consumers also believed that the benefits accrue differentially to manufacturers compared to consumers. The negative associations that consumers have with the irradiated foods have restricted widespread implementation of the technology by industry (Henson, 1995).

Any irradiated foods or irradiated food ingredients must be labeled with the words "irradiated" or "treated with ionizing radiation" (Directive 1999/2/EC) (European Commission, 2007). One-third of the respondents stated that they would consider the label as a warning and so would try to avoid the product (He, Fletcher, and Rimal, 2005). However, recent analysis suggested that information on the nature and benefits of food irradiation led consumers' perceptions and buying decisions to positive directions (Nayga, Aiew, and Nichols, 2005). Thompson and Knight (2006) found that the belief of community nutrition educators on food safety and their understanding of food irradiation were critical for consumer acceptance of irradiated foods (Thompson and Knight, 2006).

The US Army Surgeon General concluded in 1965 that irradiated foods were safe for consumption at levels up to 56 kGy. Fifteen years later, the Joint Expert Committee on Food Irradiation, convened by the Food and Agricultural Organization (FAO), the International Atomic Energy Agency (IAEA), and the World Health Organization (WHO), concluded that irradiation of any food commodity up to an overall average dose of 10 kGy presents no toxicological hazard; hence, technological testing of food irradiation is no longer required. It was also found that irradiation up to 10 kGy introduced no special nutritional or microbiological problems in foods (WHO, 1981). A majority of participants in a representative survey conducted in the United States were in favor of eating irradiated beef patties and chicken, and many of the respondents were prepared to pay as much as 30 cents per pound for this opportunity (Hayes, 1995).

In China, ready-to-eat meat products, by cooking or boiling, constitute a large portion of meat products. However, the products can be contaminated by pathogenic or spoilage microorganism during storage, transportation, selling, cutting, and packaging. High temperatures or pressures cannot be used for these kinds of Chinese meat products due to their characteristic color, flavor, and taste. A vacuum-packed pickled chicken or pork with grain stillage, which was irradiated with 6 kGy and stored at less than 10 °C, lasted for 10 days without changes in color and flavor by sensory panels while the samples irradiated at 8 kGy and stored at <10 °C did not change in color, flavor, and taste for 15 days. The irradiated products showed a significant improvement in shelf life and sensory and microbial safety (Xu, Feng, and Jiang, 1998).

In Nigeria, the effect of low-dose irradiation, up to 6 kGy, on quality, shelf life, and consumer acceptance of three traditional Nigerian meat and fish products was investigated (Aworh, Okparanta, and Oyedokun, 2002). Irradiation inhibited microbial growth in "suya" and "kilishi" with a substantial reduction in total aerobic counts, yeasts and molds, and *Staphylococcus aureus*. Non-irradiated smoked dried catfish had a shelf life of less than 1 week at tropical ambient temperature (21–31 °C) due to insect infestation. Irradiated "kilishi" and smoked dried catfish packed in sealed bags were shelf stable for a period of 4–6 months and remained free from molds and insect infestation and were considered acceptable in sensory quality by consumer panels (Aworh, Okparanta, and Oyedokun, 2002).

To improve food safety in schools, purchasing and distributing irradiated foods, such as ground beef, have been recommended by National School Lunch Program and was required by the law in the United States (Fan, 2006). For successful implementation, studies have been conducted to evaluate the sensory attributes of irradiated ground beef during 12 months of storage at −18 °C. Results demonstrated that irradiation at 1.35 and 3.0 kGy, as specified by AMS, did not have significant influence on sensory quality of ground beef evaluated either immediately after irradiation or after 6 and 12 months of storage at −18 °C (Fan, 2006). Ahn and Lee (2009) reviewed the quality changes of irradiated meat including color and flavor and the possible preventive methods to minimize the changes by natural additives, packaging methods, and their combinations. Some potential future applications of food irradiation technology such as development of traditional fermented foods, reduction of undesirable or toxic compounds (nitrosamines, residual nitrites, phytic acid, etc.), and allergenicity in foods were also studied (Byun, Jo, and Lee, 2006).

2.3.2 Brief overview of available irradiation equipments

The parameters associated with food irradiation will have an effect on the irradiator design and, therefore, on the economics of food irradiation (Kunstadt, 2001). Applied doses vary significantly according to the effects required. For example, spout inhibition doses are usually below 0.15 kGy, whereas disinfestation doses need to be up to 0.75 kGy and spice sterilization doses sometimes surpass 10 kGy. Often, dose requirements for the same product vary according to its geographic origin. Packing densities of foods are relatively high. This presents challenges to equipment manufacturers in structural design and radiation physics. The throughput capacity requirement is an important determinant for the choice of the irradiation facility that will be best suited for the specific application. Handling and distribution conditions of foods will dictate the physical form in which the product will be irradiated. Many products are perishable and require a short residence or hold-up time at the irradiation facility. Many are susceptible to physical damage and, therefore, irradiator designs need to minimize the effect of product handling. Preferably, the irradiation facility should accommodate the products without repackaging or restacking. In addition, many food products have a maximum tolerance to radiation. In many cases, there are legislative restrictions on maximum doses. Also, the equipment must ensure good dose uniformity; it must be able to deliver the minimum effective dose while not exceeding regulatory limits or product tolerance doses (Kunstadt, 2001).

2.3.2.1 Gamma irradiation equipment

Cobalt-60 is a man-made radionuclide (Cleland, 2006). It can be activated by placing the metallic slugs of stable Co-59 in nuclear power reactor. Production of radioactive cobalt starts with natural cobalt (metal), which is an element with 100% abundance of the stable isotope Co-59. Cobalt-rich ore is rare, and this metal makes up only about 0.0001% of the earth's crust. Slugs (small cylinder) or pellets (Figures 2.1 and 2.2) made of 99.9% pure cobalt sintered powder and welded in Zircaloy capsules are placed in a nuclear power reactor, where they stay for a limited period (18–24 months) depending on the neutron flux at the location (IAEA, 2005). The absorption of a neutron, which was released by the fission of U-235, changes Co-59 to Co-60. Cobalt-60 ($^{60}Co_{27}$) decays into a stable nonradioactive nickel isotope ($^{60}Ni_{28}$) principally emitting one negative beta particle of maximum energy (0.313 MeV) (Figure 2.3). The nickel-60 produced is an excited state, and it immediately emits two photons of energy 1.17 and 1.33 MeV in succession to reach its stable state. These two gamma-ray photons are responsible for radiation processing in Co-60 gamma irradiators (IAEA, 2005). Cobalt-60 has a half-life of 5.26 years, so the activity decays by 12.35% per year. The total activity in the irradiation facility is replenished by adding new sources every year. Older sources are usually kept in the facility for up to 4 half-lives or about 20 years before being replaced (Cleland, 2006). Generally,

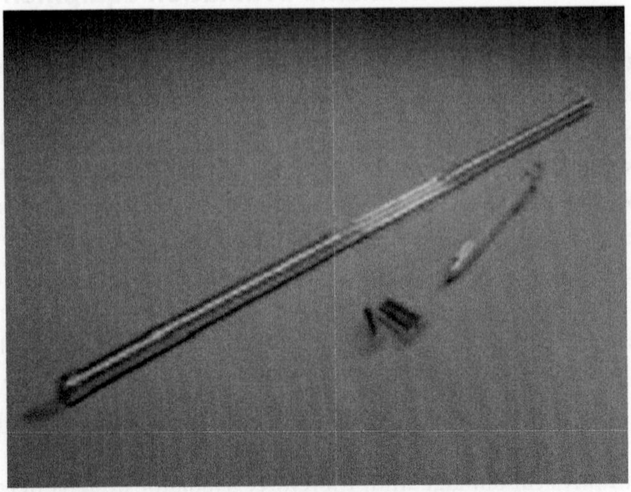

Figure 2.1 Slugs (small cylinders) of cobalt-60, which are the building blocks of the radiation source rack. (Reproduced with permission of Nordion, Canada.)

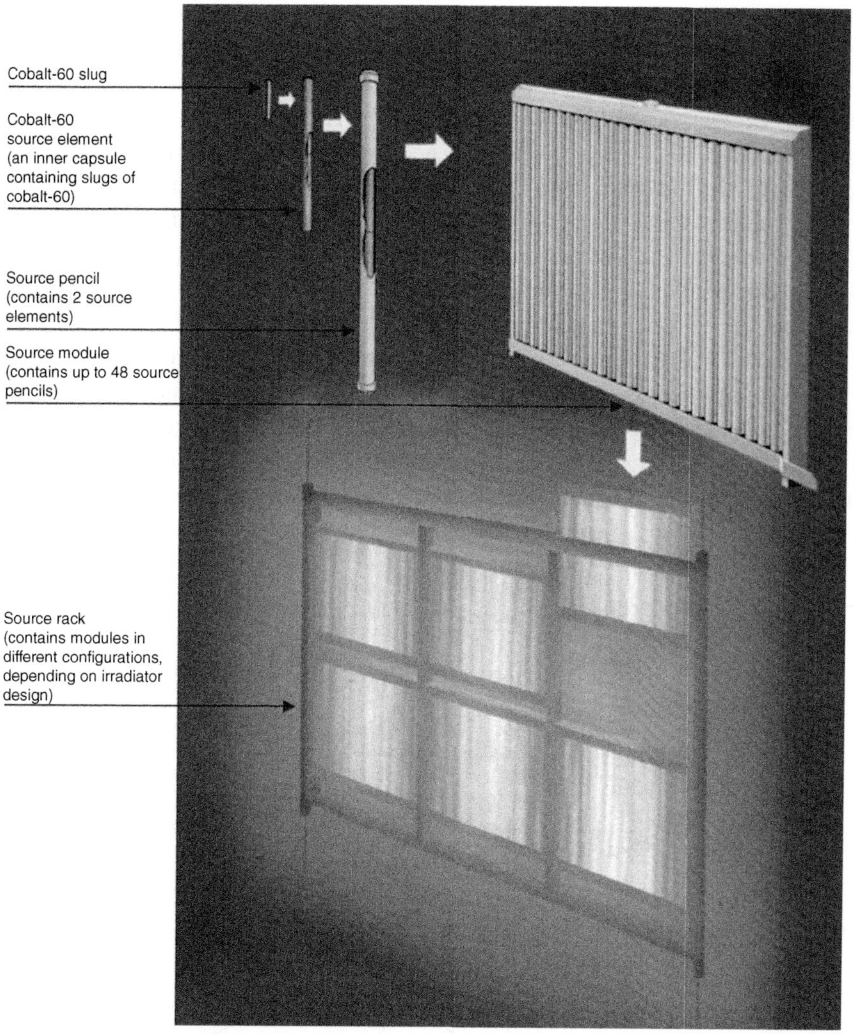

Figure 2.2 Buildup of a typical cobalt source rack from slugs, pencil, and modules. (Reproduced with permission of Nordion, Canada.)

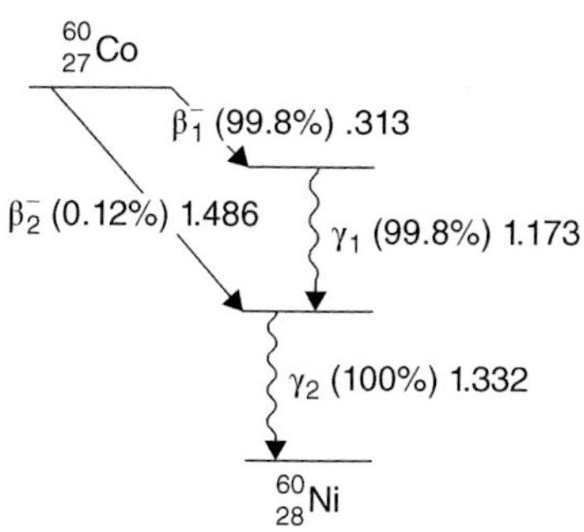

Figure 2.3 Decay scheme of radionuclide cobalt-60. (Reproduced with permission of IAEA.)

they are returned to the supplier for reuse, recycling, or disposal. In about 50 years, 99.9% of Co-60 would decay into nonradioactive nickel (IAEA, 2005).

During an irradiation process, gamma radiation interacts with the product through several types of atomic interactions, such as Compton scattering, photoelectric effects, and pair production (IAEA, 2005). Through these and subsequent interactions, the radiation source (Co-60) imparts energy to the product. As radiation proceeds through the product, its intensity decreases, resulting in a decrease of dose with depth. This is referred to as depth–dose distribution (Figure 2.4). The rate of decrease depends on the composition and density of the product as well as on the energy of the gamma radiation.

In a large irradiation facility, irradiation room where the product is treated with radiation is the focal point of facility (Figure 2.5). Other major components of an industrial facility include the following: (1) shielded storage room for radiation source rack, (2) source hoist mechanism, (3) radiation shield surrounding the irradiation room, (4) control console (room), (5) product containers (totes),

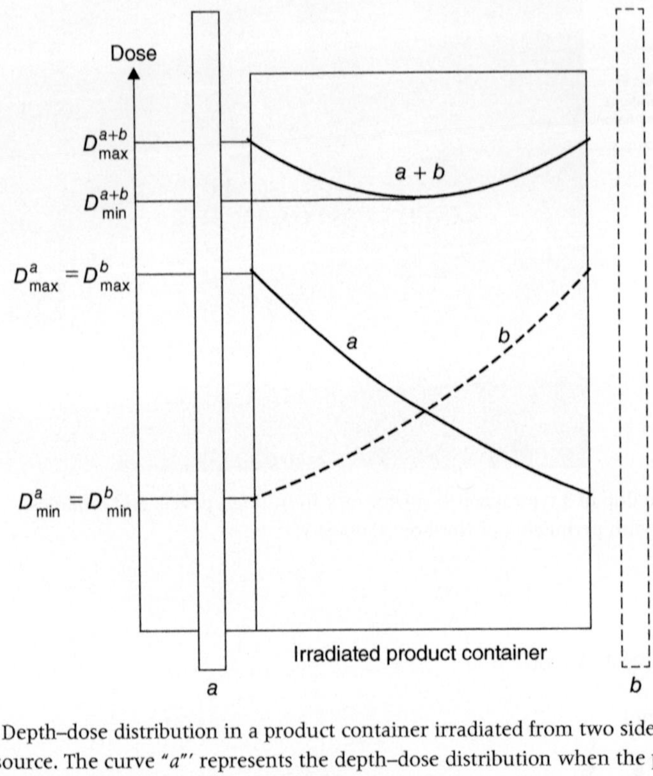

Figure 2.4 Depth–dose distribution in a product container irradiated from two sides with a cobalt-60 source. The curve "*a*"' represents the depth–dose distribution when the product is irradiated from one side only (source is at position "*a*"). Similarly, when the source is at position "*b*", the dose distribution is represented by the curve "*b*". The total dose due to irradiation from two sides is then shown as the curve "*a* + *b*". Notice that this total dose is much more uniform than that of single-side irradiation (curves "*a*" and "*b*"). (Reproduced with permission of IAEA.)

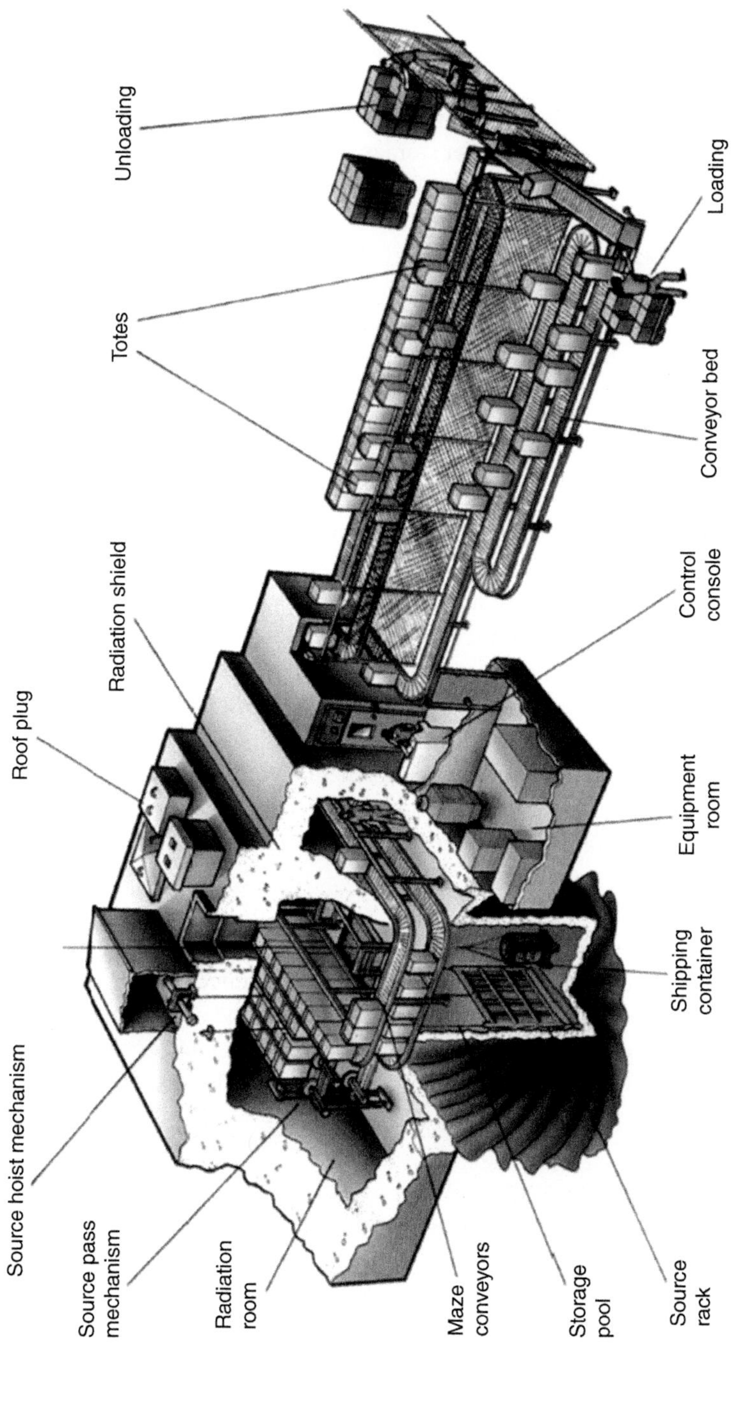

Figure 2.5 Schematic diagram of a typical panoramic, wet-storage gamma irradiation facility. (Reproduced with permission of Nordion, Canada.)

(6) product transport system through the shielding maze, (7) control and safety interlock system, (8) areas for loading and unloading of products, and (9) supporting service equipment (IAEA, 2005).

The radiation source is either in the irradiation room (during irradiation of the product) or in its shielded storage room (generally located under the irradiation room), which could be dry or wet. There is enough shielding provided by solid wall (dry storage) or water (wet storage) so that the personnel can work in the irradiation room for maintenance when the source is in the storage room. Water has several desirable characteristics as a shielding material; it is an easily available liquid that is convenient to circulate for heat transfer and is transparent. For a wet-storage facility, nearly all materials used to construct source rack, guide system, and source containers are stainless steel to eliminate galvanic corrosion. Surrounding the irradiation room is the radiation shield, which is also referred to as biological shield, generally consisting of a concrete wall thick enough (normally 2 m in thickness) to attenuate the radiation emanating from the source, so as to maintain the radiation level at the location of the control console close to natural background. The concrete wall is constructed as a maze (labyrinth) so as to permit movement of the product and yet significantly reduce the scattered radiation reaching the control console from where the operator can control or monitor the movement of the source and the product. For continuous irradiation, the product containers are moved around the radiation source on a conveyor bed that passes through the maze. The irradiation facility also provides areas for the storage of the unprocessed product and the processed product. It is a regulatory requirement that the design of the facility is such that these two types of product cannot be mixed inadvertently. All facilities must have laboratories suitable for carrying out dosimetry measurements. Some facilities also have a microbiology laboratory or a materials testing laboratory (IAEA, 2005).

Over the years, the manufacturers and suppliers of gamma irradiators have put efforts in response to the needs of the industry. The main elements that have been the focus of continuous attention include cost-effectiveness of radiation process, dose uniformity in product, turnaround time, and operational reliability. There are two types of commercially available gamma irradiators: self-contained irradiators (IAEA categories I and III) and panoramic irradiators (IAEA categories II and IV) (IAEA, 2002).

2.3.2.2 Electron beam equipment

High-current electron beam (EB) accelerators are used in various industries to enhance the physical and chemical properties of materials and to reduce undesirable contaminants such as pathogens or toxic by-products (Chmielewski and Berejka, 2008). The international standards for food irradiation allow up to 10 MeV of energy (Codex, 2003) to avoid inducing radioactive nuclides in foods (WHO, 1981).

All EB accelerators have some common features: (1) electrons are emitted from heated cathodes; (2) electrons are focused into a beam with an extraction electrode; and (3) electrons are accelerated within an evacuated space with a strong electric field (IAEA, 2011). Electrons pass into the air through a thin titanium foil window. Accelerators differ in how they attain the final electron energy, which is determined by the electronic charge times the voltage in direct current (DC) accelerators. For microwave linear accelerators (Linac), the energy is determined by the electronic charge times the forward electric field integrated over the path length (IAEA, 2011). Figure 2.6 illustrates the principles for electron accelerators used in DC EB equipment, which are similar to those used in television picture tubes or a computer monitor, except that the voltage is much higher (Cleland, 2006).

The percentage depth–dose distributions in water for electron energies are shown in Figure 2.7. For treatment from one side with 10 MeV electrons, the thickness where the exit dose equals the entrance dose is about 3.7 cm after subtracting the equivalent thicknesses of the electron beam window (40 μ of titanium) and the air space (15 cm) between the window and the water (Cleland, 2006). For treatment from both sides with 10 MeV electrons, the thickness can be increased to about 8.6 cm. Then the dose in the middle would be the same as the entrance and exit doses. This thickness is enough to irradiate most retail packages of fresh meat, except for whole turkey. The depth–dose curves for irradiation on both sides with 5 MeV electrons are shown in Figure 2.8. Other types of EB accelerators for producing high energy and high power such as constant potential accelerators and radio frequency accelerators are also available (Cleland, 2006).

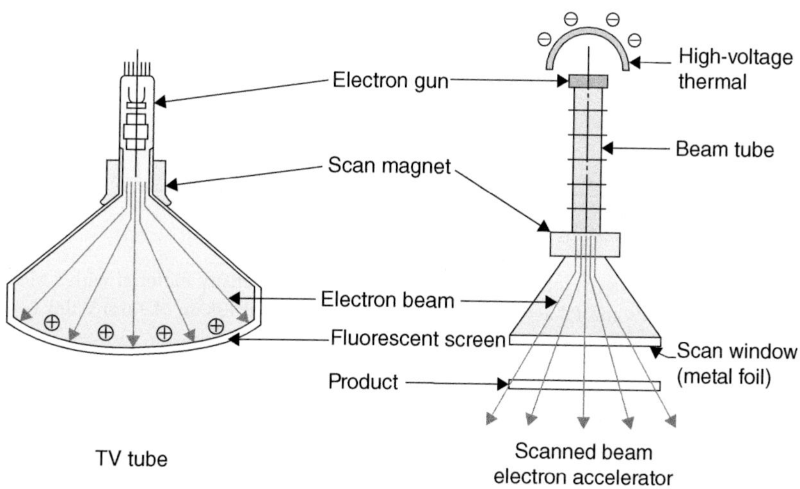

Figure 2.6 Direct current electron beam operating principles. (From IAEA (2011). Reproduced with permission of IAEA.)

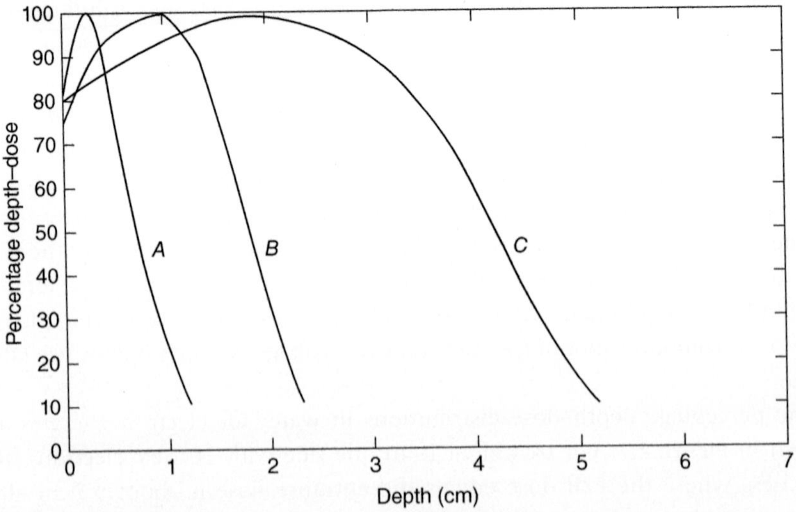

Figure 2.7 Percentage depth–dose curves for a single-side electron irradiation of water. *A*, 1.8 MeV; *B*, 4.7 MeV; *C*, 10.6 MeV. (From Woods and Pikaev (1994). Reproduced with permission of John Wiley & Sons Ltd.)

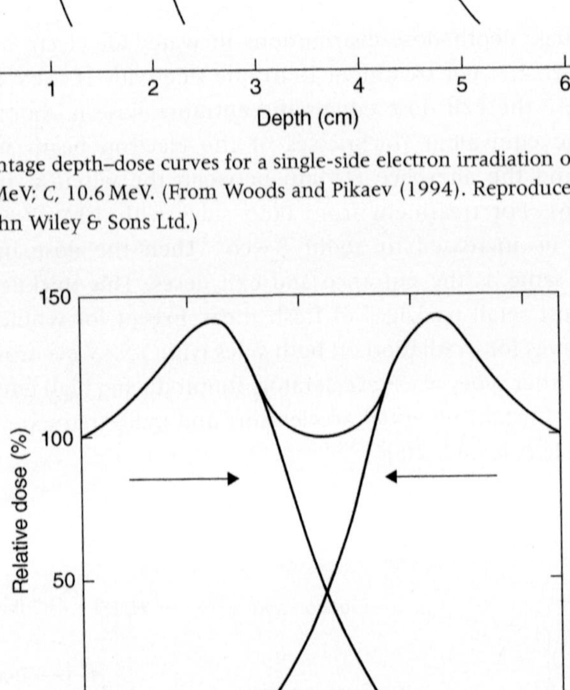

Figure 2.8 Depth–dose curves for two-side irradiation of a unit-density material with 5 MeV electrons. (From Woods and Pikaev (1994). Reproduced with permission of John Wiley & Sons Ltd)

2.3.2.3 X-ray equipment

When an accelerated electron impinges upon any materials, it generates X-rays. Characteristic monoenergetic X-ray photons are produced when accelerated electrons interact with orbital electrons; bremsstrahlung photons are produced by the interaction of accelerated electrons with the nucleus of an atom (IAEA,

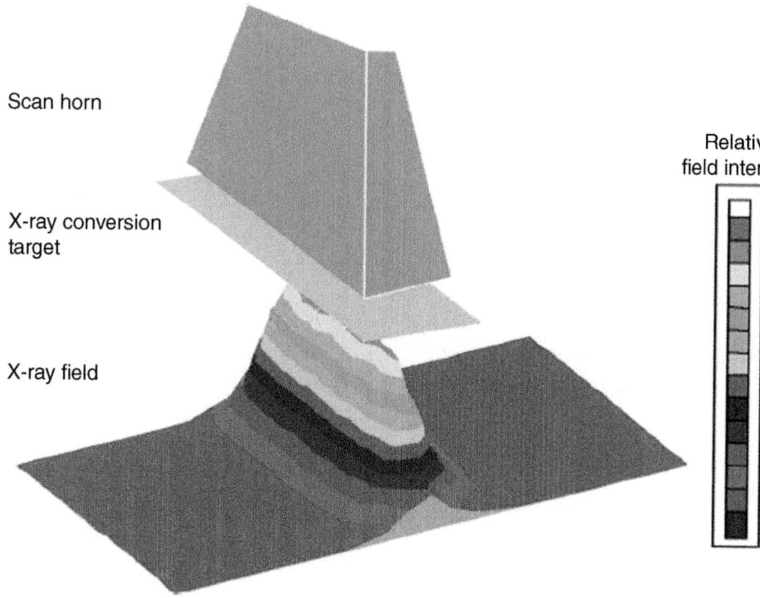

Figure 2.9 Forward peaked emission of 5.0 MeV X-rays. (From IAEA (2011). Reproduced with permission of IAEA.)

2011). The intensities of X-ray from high-power, high-energy industrial X-ray generators far exceed that of common medical X-ray equipment. As a result, thick shielding is needed to prevent worker exposure to high levels of X-ray. Shielding thickness increases with maximum energy of the X-rays. High-energy beams for food use are usually shielded using concrete (IAEA, 2011).

With the advent of very high current and mid-energy and high-energy accelerators, X-ray can now be used as an alternative to the use of radioactive isotopes in such areas as medical device sterilization and food treatments. X-rays are produced by interposing a metal target between the EB and the product to be treated. To enhance electron-to-photon conversion, these X-ray targets are made of high-atomic-number metals (high Z). Water-cooled tantalum is preferred for large-area targets. The conversion efficiency of electron to X-ray depends on accelerator energy and the target Z-value; efficiencies ranging between 8% and 12% have been shown for electron energies between 5.0 and 7.5 MeV. X-ray has a forward peaked emission and the rate at which a material receives X-ray photons, the dose rate, can be controlled by a combination of the distance from the target, the beam current, and under-beam transport speed (IAEA, 2011). Figure 2.9 illustrates the forward peaked emission of X-rays, which is significantly different from the panoramic emission of gamma-ray sources. This property facilitates the treatment of single pallet loads of product. Figure 2.10 shows that X-ray penetration is much greater than the highest energy industrial EB systems, 10 MeV, and is comparable to gamma-ray penetration (IAEA, 2011).

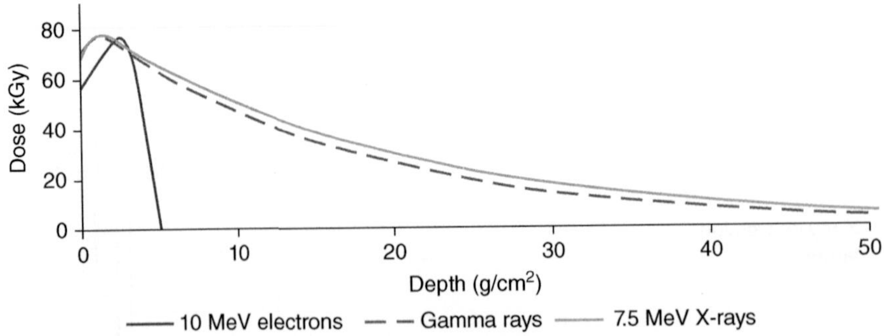

Figure 2.10 Electron beam, gamma ray, and X-ray penetration. (From IAEA (2011). Reproduced with permission of IAEA.) (*See color plate section for the color representation of this figure.*)

The high-power, high-electron energy accelerators can compensate the inefficiencies of X-ray conversion efficiency (16.2%, 13.3%, and 8.2% at 10, 7.5, and 5 MeV, respectively; Cleland, 2006). Table 2.3 shows the example of the cost-effectiveness of X-ray facility at different energy and power (IAEA, 1995). To attain more uniform dose distributions in large pallet loads of packaged products, a turntable can be used to rotate the pallet in front of a tall, collimated X-ray target (Figure 2.11; IAEA, 2011).

Table 2.4 summarizes the properties of EB, X-ray, and gamma-ray sources. The dose rates of X-ray are at least one order of magnitude higher than gamma rays but significantly less than EB. Product throughput rates for X-ray processing based on high-power EB equipment can be used for food irradiation (Table 2.5).

2.3.3 Manufacturers of irradiation equipment

More than 200 gamma-ray facilities are being used for various industrial applications, mainly for sterilizing medical devices and food irradiation (Cleland, 2006). Gamma rays from Co-60 and Cs-137 are allowed by international standards

Table 2.3 Examples of the effect of increasing the energy, increasing the power, and reducing the dose on economics of X-ray irradiation of foods.

Energy (MeV)	Beam power (kW)	Absorbed dose (kGy)	Cost/tone (US$) at 50% utilization	Cost/ton (US$) at 100% utilization
5	100	2.5	77.0	52.5
7.5	100	2.5	51.6	35.0
7.5	200	2.5	28.3	20.0
7.5	200	1.0	11.3	8.0

Source: Adopted from IAEA (1995) with permission by the IAEA.

Figure 2.11 Rotation of product on a Palletron™ in front of X-ray target. (From IAEA (2011). Reproduced with permission of IAEA.)

Table 2.4 Ionizing radiation sources for food use.

	Electron beam	X-rays	Gamma rays
Power source	Electricity	Electricity	Radioactive isotope (mainly Co-60)
Power activity	Electrical on/off	Electrical on/off	5.27 years half-life
Property	Electrons mass = 9.1×10^{-31} kg	Photons $\lambda = 4.1 \times 10^{-3}$ nm	Photons (1.25 MeV) $\lambda = 1.0 \times 10^{-3}$ nm
Charge	1.6×10^{-19} C	None	None
Emission	Undirectional (can be scanned and bent by magnets)	Forward peaked	Isotropic
Penetration	Finite range	Exponential attenuation	Exponential attenuation
Dose rate	360,000 kGy/h 100 kGy/s	100 kGy/h 2.7×10^{-2} kGy/s	10 kGy/h 2.8×10^{-3} kGy/s

Source: Adopted from IAEA (2011).

Table 2.5 X-ray processing throughput potential*.

Source	Equal-entrance, equal-exit penetration in unit density (mm)	Emitted power (kW)	Potential throughput (kg/h)
10 MeV linac EB mode	38	20	720
10 MeV Rhodotron™ EB mode	38	200	7200
10 MeV linac 20 kW X-ray mode	480	3.2	109
7 MeV Rhodotron 700 kW X-ray mode	450	77	2772
5 MeV Dynamitron™ 300 kW X-ray mode	385	24	864

*Assuming an underbeam process efficiency of 25% (typical of cart systems) and a dose of 25 kGy and X-ray conversion efficiencies of 16% at 10 MeV, 11% at 7 MeV, and 8% at 5 MeV.
Source: Adopted from IAEA (2011) with permission by author Berjeka.

for food irradiation (Codex, 2003). However, Cs-137 is seldom used because large Cs-137 sources are not readily available. The main commercial suppliers of encapsulated Co-60 sources are located in Canada (MDS Nordion) and the United Kingdom (PURIDEC). There are various manufacturers of gamma irradiation units but following ones are more prominent: MDS Nordion, Inc., Canada (including its predecessor Atomic Energy of Canada Ltd.); Sterigenics International, USA; Institute of Isotopes, Hungary; H.S. Marsh, UK; Shanghai Institute of Nuclear Research, China; Techsnabexport Co. Ltd., Russia; Board of Radiation & Isotope Technology, India; Sulzer, Switzerland; Gammaster, The Netherlands; Radiation Technology Inc., USA, and Isotron, UK (IAEA, 2005).

The major manufacturers of electron beam accelerator and X-ray conversion are Wasik Associates, USA; Budker Institute of Nuclear Physics, Russia; Nissin High Voltage, Japan; IBA E-beam X-ray Solutions, USA; Mevex Corporation, Canada; Linac Technologies, France; Titan Scan, USA; Iotron Industries, Canada; MDS Nordion, Canada (Cleland, 2006). The major manufacturers of irradiation equipment is summarized in Table 2.6. Tables 2.7 and 2.8 summarize gamma and electron beam (including X-ray) service units in the world.

Table 2.6 Major irradiation equipment manufacturers.

Name	Location	Product
Aerial CRT	Illkirch, France	Dosimeters
Australian	Kirrawee, Australia	Dosimeters
BINE High-Tech. Co., Ltd.	Beijing, China	Gamma irradiator Distributor for MDS Nordion sources of Co-60 in China
Best Theratronics	Springfield, VI, USA	Self-contained gamma irradiators
Board of Radiation & Isotope Technology	Mumbai, India	Gamma irradiation facility
Bruce Power	Tiverton, Ontario, Canada	Co-60 for MDS Nordion
E-beam Service Inc.	Lebanon, OH, USA	Electron beam facility
EB-Tech	Daejeon, South Korea	Electron beam facility
ETIGAM bv	Apeldoorn, The Netherlands	Sterilization indicators
Far West Technology, Inc.	Goleta, CA, USA	Radiation detection system
Gamma-Service Recycling GmbH	Leipzig, Germany	Handling and shipment of radioactive materials
Getinge Linac Technologies	Villebon, France	Electron beam systems
GEX Corp.	Centennial, CO, USA	Dosimetry systems
GRAY*STAR, Inc	Arilington, NJ, USA	Self-contained irradiators
Harwell Dosimeters Ltd	Oxfordshire, UK	Dosimeters
Hungaroster Co., Ltd	Budapest, Hungary	Gamma irradiation facility
IBA Industrial	Louvain-La-Neuve, Belgium	Electron beam, X-ray, and Proton accelerators
Jiangsu Dasheng Electron Accelerator Co., Ltd	Jiangsu, China	Electron beam facility
Jiangsu Zhongke Hi-wits Technol. Develop. Co., Ltd	Jiangsu, China	Electron beam equipment
MEVEX Co.	Stittsville, ON, Canada	Electron beam facility
NHV Cooperation	Methuen, MA, USA	Electron processing equipment
Nordion Inc.	Ottawa, ON, Canada	Gamma irradiators and electron beam facility
REVISS Services (UK) Ltd	Buckinghamshire, UK	Gamma irradiators and services
RISO	Frederiksborgvej, Denmark	Dosimetry system
Tongxing Nuclear Technology Co., Ltd	Beijing, China	Co-60 radioactive sources
Wuxi El Pont	Wuxi, China	Electron accelerators

Source: Modified from IMRP (2011) and iiA (2013) with permission by authors Berejka and Cleland.

Table 2.7 Gamma irradiation processing service facilities in the world.

Region (number of countries)	Countries (number of gamma irradiation processing units)
Africa (3)	Egypt (1); Ghana (1); South Africa (3) Subtotal: 5
East Asia and the Pacific (12)	Australia (2); Bangladesh (2); China (21); India (3); Indonesia (1); Japan (2); South Korea (1); Malaysia (4); Philippines (1); Taiwan (2); Thailand (4) Vietnam (1) Subtotal: 44
Europe (18)	Austria (1); Belgium (2); Bulgaria (1); Croatia (1); Germany (3); Greece (1); Hungary (3); Ireland (1); Italy (2); Portugal (1); Romania (1); Serbia and Montenegro (1); Sweden (1); Switzerland (1); Netherlands (3); Turkey (2); Ukraine (1), United Kingdom (5) Subtotal: 31
Latin America (5)	Argentina (1); Brazil (4), Chile (1), Mexico (2), Peru (1) Subtotal: 9
North America (2)	Canada (1); USA (28) Subtotal: 29
West Asia (5)	Iran (1): Israel (1); Jordan (1); Saudi Arabia (1); Syria (1) Subtotal: 5

Source: Adopted from IAEA (2005) with permission by authors Berjeka and Cleland.

2.4 Future role for irradiation in the preservation of foods

Food irradiation has a 60-year history of scientific research and testing, including toxicological and microbiological evaluation as well as testing for wholesomeness (WHO, 1994), and has been approved in about 60 countries and approximately 71 commercial irradiation facilities are operating in the world (Kume and others, 2009). Irradiation is the best known physical decontamination technologies for inactivating foodborne pathogens as well as nonpathogenic bacteria, and improving the safety of meats and other foods. Irradiation of foods, therefore, can increase not only the safety of foods by killing pathogen but also indigenous microflora, thereby extending the shelf life of foods. The major advantages of irradiating foods are as follows: it is a nonthermal processing, maintains the integrity of products, and leaves no chemical residues. Also, the products can be treated after final packaging, which prevents further cross-contamination during postprocessing handling. Therefore, irradiation may play a key role in improving the safety and preservation of foods in the future. However, current use of irradiation in foods is minimal except for certain food items such as herbs, spices, and tropical fruits for export (Table 2.9). For irradiation to be used as a major

Table 2.8 Electron beam processing service facilities in the world (selected for beam energy at 5 MeV or higher).

Facility	Location	Beam energy (MeV)
BeamOne	Denver, CO; Lima, OH, USA	10
E-Beam Services	Cranbury, NJ, USA	10
	Cranbury, NJ, USA	5
	Lebanon, OH, USA	5
NeoBeam	Middlebury, OH, USA	5
Sadex Corporation	Sioux City, IA, USA	10
Sterigenics	Bridgeport, NJ, USA	10*
Steris	Libertyville, IL, USA	5
Acsion	Pinawa, Manitoba, Canada	10
Iotron	Port Coquitlam, BC	10
BeamOne	Coyol Park, Costa Rica	10
Ionisos	Orsay, France	10
Beta Gamma Service	Bruchsal, Germany	10*
	Bruchsal, Germany	5
	Saal, Germany	10
Bioster	Padua, Italy Seriate, Italy	10 10
Sintez	Kurgan, Russia	5–8
NIP	Lytarkino, Russia	10
Federal Medical Biological Center Burzanyan	Moscow, Russia	10
Institute of Physical Chemistry	Moscow, Russia	5–8
MIRTI	Moscow, Russia	10
Tori	Moscow, Russia	10
Budker Institute of Nuclear Physics	Novosibirsk, Russia	5
NIIFCI	Obninsk, Russia	10
RAD	Saint Petersburg, Russia	10
Ionisos	Chaumesnil, Spain	10
Leoni Studer Hard	Daniken, Switzerland	5–10
	Daniken, Switzerland	X-ray only*
Isotron	Daventry, UK	10
	Harwell, UK	10–12
EB-Tech	Daejeon, South Korea	10
Korea Atomic Energy Research Institute	Jeongeup, South Korea	10*
West-Japan Irradiation Service Co. Ltd	Osaka, Japan	5*
Nuclear Fuel Industries, Ltd	Osaka, Japan	10*
Radia Industries	Takasaki, Japan	5*
Tsukuba EB Center	Tsukuba, Japan	5*

*X-ray capability.

Source: Adopted from IAEA (2011) with permission by authors Berejka and Cleland.

Table 2.9 Approval of food irradiation in the United States.

Date	Products	Dose (kGy)	Purpose
1964, 1965	Potatoes	0.05–0.15	Inhibit sprouting
1983	Spices and dry seasonings	<30	Disinfestation and decontamination
1985	Pork	0.3–1.0	Control of *Trichinella spiralis*
1985, 1986	Dehydrated enzymes	<10	Control insects and microbes
1986	Fruits and vegetables	<1	Delay maturation and disinfection
1986	Herb, spices, and seasonings	<30	Control of microorganisms
1990	Poultry, fresh, and frozen	<3.0	Control of microorganisms
1995	Meat, frozen, and packaged	>44	Sterilization only for NASA
1997, 1999	Red meat, chilled	<4.5	Control of microorganisms
	Red meat, frozen	<7.5	
2000	Shell eggs	<3.0	Control of *S. Enteritidis*
2000	Sprouts	<8.0	Control of pathogens in seeds
2005	Fresh or frozen molluscan and other shellfish	<5.5	Control of *Vibrio* species and foodborne pathogens
2008	Iceberg lettuce and spinach	<4.0	Control of foodborne pathogens and extension of shelf life

technology to improve the safety and extend the shelf life of foods, the following should be done.

2.4.1 Consumer education

A large proportion of consumers believe that irradiated foods pose a health risk and viewed food irradiation as moderately or highly risky (AMIF, 1993). Another issue related to consumer acceptance of irradiated foods is quality changes by irradiation. The acceptance of irradiated food was mainly associated with consumers' knowledge about food irradiation (Lusk, Fox, and McIlvain, 1999). Consumer attitudes toward irradiated foods can be changed through education: the more they know about the safety benefits of food irradiation, the higher the positive attitudes toward irradiation indicating the importance of consumer education (Bruhn, 1995; Resurreccion and Galvez, 1999).

2.4.2 Use of irradiation as a part of hurdle technology

The efficacy of irradiation against pathogenic as well as nonpathogenic microorganisms in foods has been proven for decades (Farkas, 1987; Thayer, 1993; Olson, 1998). The microcidal effects of irradiation in foods are affected by many factors including irradiation dose, food composition, temperature, gaseous atmosphere, and microbial factors. The reduction of microbial populations in foods is proportional to radiation doses; however, use of high-dose irradiation can have negative effects on the desirable sensory attributes of foods. Most of the processed foods are added with various additives, spices, and herbs with antimicrobial effects. Combined use of irradiation with other intervention technologies such as antimicrobial food preservatives, heat, high hydrostatic pressure, and so on can increase the microcidal efficacy of irradiation. Therefore, even with low doses of irradiation, the microbial safety of foods can be improved while the desirable sensory attributes are maintained.

2.4.3 Research on the mechanisms and prevention of quality changes in irradiated foods

Odor and color are important quality parameters that determine consumer acceptance of irradiated raw meat, but flavor and taste changes are important issues for cooked meat. Extensive work has been done to elucidate the mechanisms involved in quality changes and their prevention in irradiated raw meat. However, only limited information on the mechanisms and causes of taste/flavor changes in irradiated further processed or precooked ready-to-eat meat products is available because irradiation of meats with additives are not permitted. Therefore, additional research to elucidate the causes and mechanisms of flavor and taste changes, the roles of spices and additives on taste/flavor and microorganisms, and the development of methods that can control taste/flavor changes in irradiated further processed meat products is necessary. In addition, regulatory approval for the use of irradiation in processed meats and other foods would be the key for the success of food irradiation in the future.

References

iiA. (2013) Member directory. *International Irradiation Association*, http://www.iiaglobal.com/index.php?page=members (accessed 15 May 2013).

Ahn, D.U. (2002) Production of volatiles from amino acid homopolymers by irradiation. *Journal of Food Science*, **67** (7), 2565–2570.

Ahn, D.U. and Lee, E.J. (2009) Mechanisms and prevention of quality changes in meat by irradiation, in *Food Irradiation Research and Technology*, 2nd edn (eds C. Sommers and X. Fan), IFT Symposium Book Series.

Ahn, D.U., Jo, C. and Olson, D.G. (1999) Volatile profiles of raw and cooked turkey thigh as affected by purge temperature and holding time before purge. *Journal of Food Science*, **64**, 230–233.

Ahn, D.U., Jo, C., Du, M. *et al.* (2000) Quality characteristics of pork patties irradiated and stored in different packaging and storage conditions. *Meat Science*, **56**, 203–209.

Ahn, D.U., Nam, K.C., Du, M. and Jo, C. (2001) Effect of irradiation and packaging conditions after cooking on the formation of cholesterol and lipid oxidation products in meats during storage. *Meat Science*, **57**, 413–418.

American Meat Institute Foundation (1993) *Consumer Awareness, Knowledge, and Acceptance of Food Irradiation*, American Meat Institute Foundation, Washington, DC.

Aworh, O.C., Okparanta, R.N. and Oyedokun, E.O. (2002) *Effect of Irradiation on Quality, Shelf Life and Consumer Acceptance of Traditional Nigerian Meat and Fresh Products. Study of the Impact of Food Irradiation on Preventing Losses: Experience in Africa*. IAEA-TECDOC-1291.

Bruhn, C.M. (1995) Consumer attitudes and market response to irradiated food. *Journal of Food Protection*, **58** (2), 175–181.

Brynjolfsson, A. (1989) Future radiation sources and identification of irradiated foods. *Food Technology*, **43**, 84–89.

Byun, M.W., Jo, C. and Lee, J.W. (2006) Potential application of ionizing radiation, in *Food Irradiation Research and Technology* (eds C.H. Sommers and X. Fan), Blackwell Publishing, Oxford, UK, pp. 249–262.

Chmielewski, A.G. and Berejka, A.J. (2008) *Radiation Sterilization Centers World-Wide. Trends in Radiation Sterilization of Health Care Products*, International Atomic Energy Agency, Vienna.

Cleland, M.R. (2006) Advances in gamma ray, electron beam, and X-ray technologies for food irradiation, in *Food Irradiation Research and Technology* (eds C.H. Sommers and X. Fan), Blackwell Publishing, Ames, IA, pp. 11–36.

Codex. (2003) Codex General Standard for Irradiated Foods. Codex Alimentarius Commission. Rome, Italy.

Crawford, L.M. and Ruff, E.H. (1996) A review of the safety of cold pasteurization through irradiation. *Food Control*, **7**, 87–97.

Dickson, J.S. (2001) Radiation inactivation of microorganisms, in *Food Irradiation: Principles and Applications* (ed. R.A. Molins), Wiley-Interscience, NY, pp. 23–35.

Diehl, J.F. (1995) Safety of irradiated foods, in *Revised and Expanded*, 2nd edn, Marcel Dekker, Inc., NY.

Du, M., Ahn, D.U., Nam, K.C. and Sell, J.L. (2001) Volatile profiles and lipid oxidation of irradiated cooked chicken meat from laying hens fed with diets containing conjugated linoleic acid. *Poultry Science*, **80**, 235–241.

Fan, X. (2006) Irradiated ground beef for national school lunch program, in *Food Irradiation Research and Technology* (eds C.H. Sommers and X. Fan), Blackwell Publishing, Oxford, UK, pp. 237–248.

Farkas, J. (1987) Contamination, including parasite control, of dried, chilled and frozen foods by irradiation. *Acta Alimentaria*, **16**, 351–358.

Frewer, L.J., Bergmann, K., Brennan, M. *et al.* (2011) Consumer response to novel agri-food technologies: implication for predicting consumer acceptance of emerging food technologies. *Trends in Food Science and Technology*, **22**, 442–456.

Grecz, N., Rowley, D.B. and Matsuyama, A. (1983) The action of radiation on bacteria and viruses, in *Preservation of Foods by Ionizing Radiation* (eds E.S. Josephson and M.S. Peterson), CRC Press, Boca Raton, FL.

Hampson, J.W., Fox, J.B. Jr.,, Lakritz, L. and Thayer, D.W. (1996) Effect of low dose gamma irradiation on lipids in five different meats. *Meat Science*, **42**, 271–276.

Hayes, D.J. (1995) The economics of marketing irradiated foods, in *Food Irradiation, A source book* (ed. E.A. Murano), Iowa State University Press, Ames, IA, USA, pp. 111–126.

He, S., Fletcher, S. and Rimal, A. (2005) Unwillingness to consume irradiated beef. *Journal of Food Distribution Research*, **36**, 71–78.

Henson, S. (1995) Demand-side constraints on the introduction of new food technologies: the case of food irradiation. *Food Policy*, **20**, 111–127.

IAEA. (1995) The development of X-ray machines for food irradiation: proceedings of a consultants' meeting. International Atomic Energy Agency, Vienna, Austria.

IAEA. (2002) Dosimetry for food irradiation. TRS No. 409. International Atomic Energy Agency, Vienna, Austria.

IAEA. (2005) Gamma irradiators for radiation processing. International Atomic Energy Agency, Vienna, Austria.

IAEA. (2011) Industrial radiation processing with electron beams and X-rays. International Atomic Energy Agency, Vienna, Austria.

IMRP. (2011) Participant list of International Meeting on Radiation Processing, Montreal, Canada.

Jo, C., Lee, J.I. and Ahn, D.U. (1999) Lipid oxidation, color changes and volatiles production in irradiated pork sausage with different fat content and packaging during storage. *Meat Science*, **51**, 355–361.

Kume, T., Furuta, M. and Todoriki, S. (2009) Quantity and economic scale of food irradiation in the world. *Radioisotopes*, **58**, 25–35.

Kunstadt, P. (2001) Economic and technical considerations in food irradiation, in *Food Irradiation: Principles and Application* (ed. R.A. Molins), John Wiley & Sons, Inc., New York, pp. 415–442.

Kwon, J.H. (2010). Safety and understanding of irradiated food. S. Y. Yoo and K. W. Lee Korea Food Safety Research Institute. Seoul, Korea.

Lebepe, S., Molins, R.A., Chawen, S.P. *et al.* (1990) Change in microflora and other characteristics of vacuum-packaged pork loins irradiated at 3.0 kGy. J. *Food Science*, **55**, 918–924.

Loaharanu, P., Kava, R., and Choi, E.H. 2007. *Irradiated Foods*. 6th Ed., American Council on Science and Health, http://www.acsh.org (accessed 15 May 2013).

Lusk, J.L., Fox, J.A. and McIlvain, C.L. (1999) Consumer acceptance of irradiated meat. *Food Technology*, **53**, 56–59.

Millar, S.J., Moss, B.W., MacDougall, D.B. and Stevenson, M.H. (1995) The effect of ionizing radiation on the CIELAB color co-ordinates of chicken breast meat as measured by different instruments. *International Journal of Food Science and Technology*, **30**, 663–674.

Moseley, B.E.B. (1989) Ionizing radiation: action and repair, in *Mechanisms of Action of Food Preservation Procedures* (ed. G.W. Gould), Elsevier Applied Science, NY, pp. 43–70.

Murano, P.S. (1995) Quality of irradiated foods, in *Food Irradiation: A Source Book* (ed. E.A. Murano), Iowa State University Press, Ames, IA, pp. 89–126.

Nam, K.C. and Ahn, D.U. (2002a) Carbon monoxide-heme pigment complexes are responsible for the pink color in irradiated raw turkey breast meat. *Meat Science*, **60**, 25–33.

Nam, K.C. and Ahn, D.U. (2002b) Mechanisms of pink color formation in irradiated precooked turkey breast. *Journal of Food Science*, **67**, 600–607.

Nam, K.C., Hur, S.J., Ismail, H. and Ahn, D.U. (2002) Lipid oxidation, volatiles, and color changes in irradiated turkey breast during frozen storage. *Journal of Food Science*, **67**, 2061–2066.

Nam, K.C., Min, B.R., Ko, K.Y. *et al.* (2008) Packaging determines color and odor of irradiated ground beef, in *Radiation Physics Research Progress* (ed. A.N. Camilleri), Nova Publisher, pp. 287–300.

Nawar, W.W., Zhu, Z., Wan, H. *et al.* (1996) Progress in the detection of irradiated food by measurement of lipid derived volatiles, in *Detection Methods for Irradiated Foods* (eds C.H. McMurray, E.M. Stewart, R. Gray and J. Pearce), Royal Society of Chemistry, Cambridge, UK, pp. 241–249.

Nayga, R.M., Aiew, W. and Nichols, J.P. (2005) Information effects on consumers' willingness to purchase irradiated food products. *Review of Agricultural Economics*, **27**, 37–48.

Olson, D.G. (1998) Irradiation of food. *Food Technology*, **52**, 56–62.

Patterson, R.L.S. and Stevenson, M.H. (1995) Irradiation induced off-odor in chicken and its possible control. *British Poultry Science*, **36**, 425–441.

Radomyski, T., Murano, E.A. and Olson, D.G. (1994) Elimination of pathogens of significance in food by low-dose irradiation: a review. *Journal of Food Protection*, **57**, 73–86.

Resurreccion, A.V.A. and Galvez, F.C.F. (1999) Will consumers buy irradiated beef? *Food Technology*, **53** (3), 52–55.

Satin, M. (1892) *Food Irradiation: A Guidebook*, Technomic Publishing Co. Inc., Lancaster, Basel, pp. 3–26.

Satterlee, L.D., Wilhelm, M.S. and Barnhart, H.M. (1971) Low-dose gamma irradiation of bovine metmyoglobin. *Journal of Food Science*, **36**, 549–551.

Smith, J.S. and Pillai, S. (2004) Irradiation and food safety. *Food Technology*, **58**, 48–55.

Swallow, A.J. (1984) Fundamental radiation chemistry of food components, in *Recent Advances in the Chemistry of Meat* (ed. A.J. Bailey), The Royal Soc. Chem. Burlington House, London, UK, pp. 165–177.

Thayer, D.W. (1993) Extending the shelf-life of poultry and red meat by irradiation processing. *Journal of Food Protection*, **56** (831–833), 846.

Thayer, D.W. (1994) Wholesomeness of irradiated foods. *Food Technology*, **48**, 132–135.

Thompson, B.M. and Knight, S.L. (2006) Determining the food irradiation beliefs of community nutrition educators: do beliefs influence educational outreach? *Journal of Nutrition Education and Behavior*, **38**, 50–55.

USDA. (1999) Backgrournders: USDA issues final rule on meat and poultry irradiation. FSIS USDA, Washington, D.C.

WHO. (1981) Wholesomeness of irradiated food. WHO Technical Report 659. Joint FAO/IAEA/WHO Expert Committee.

WHO (1988) Food Irradiation: a technique for preservation and improving the safety of food. World Health Organization, Geneva, pp. 24–43.

WHO (1994) Food irradiation, in *Safety and Nutritional Adequacy of Irradiated Food*, WHO, Geneva, pp. 5–13.

Woods, R.J. and Pikaev, A.K. (1994) Selected topics in radiation chemistry, in *Applied Radiation Chemistry: Radiation Processing* (eds R.J. Woods and A.K. Pikaev), John Wiley & Sons, Inc., NY, pp. 165–210.

Xu, Z., Feng, Z. and Jiang, P. (1998) *Test marketing and consumer acceptance of irradiated meat products. Consumer acceptance and market development of irradiated food in Asia and the Pacific.* IAEA-TECDOC-1219.

CHAPTER 3

High-pressure processing of meat and meat products

Sandrine Guillou[1], Marion Lerasle[1], Hélène Simonin[2] & Michel Federighi[1]
[1] *UMR1014 SECALIM, INRA, Oniris, 44307, Nantes, France*
[2] *UMR Procédés Alimentaires et Microbiologiques, équipe PMB, Université de Bourgogne, AgroSup Dijon, 21000, Dijon, France*

3.1 Introduction

High pressure (HP) was first introduced at the end of the nineteenth century in the field of materials chemistry. The first application of high pressure in the food industry was proposed by Hite in 1899 to pasteurize milk and fruit products (Hite, 1899). Later, research has been performed on the resistance of bacterial spores to high pressure (Basset and Macheboeuf, 1932), and a Nobel Prize in Physics was awarded in 1946 to the American physicist P.W. Bridgman for his invention of a hermetic vessel especially designed for conducting high-pressure treatments (Federighi *et al.*, 2001). This technique was first applied in the food industry in Japan in the late 1980s. Also named pascalization, it consists of maintaining the product at very high hydrostatic pressure (HHP, 100–1000 MPa) for a definite time and temperature. The HHP treatment is considered as a nonthermal process, alternative to heat treatments, as it enables inactivation of pathogenic and spoilage microorganisms in foods. Since its first application in Japan, this novel technology has gained significant development with regard to technological aspects and the range of pressurized products commercially available. Today, more than 160 industrial vessels are in use throughout the world. Moreover, treatment of meat products by high pressure is a fast-growing sector as attested by the numerous patents deposited in the last few years.

Emerging Technologies in Meat Processing: Production, Processing and Technology, First Edition.
Edited by Enda J. Cummins and James G. Lyng.

3.2 Theory of high-pressure preservation and decontamination of foods

3.2.1 What is the mode of action when used for preservation and decontamination?

High-pressure processing is a technique that involves subjecting food to a high level of pressure between 100 and 600 MPa at a specific temperature for a few minutes (Farkas and Hoover, 2000). This technology is based on two principles that determine the behavior of foods under pressure (Hugas, Garriga and Monfort, 2002). Firstly, an important principle that underlies the effects on reaction equilibria is known as the principle of Le Chatelier according to which any phenomenon accompanied by a decrease in volume will be enhanced by an increase in pressure and vice versa (Balasubramaniam, Farr and Turek, 2008). Therefore, application of pressure leads to physical and chemical changes (chemical reactivity and reaction, change in molecular configuration, phase transition). Physically, an increase in pressure contributes to a decrease in the average intermolecular distance between molecules (Perrier-Cornet, Moussa and Gervais, 2009). Physical compression during high-pressure treatment also results in an increase in temperature through adiabatic heating of approximately 3 °C per 100 MPa depending on food composition (Rastogi *et al.*, 2007). Chemical changes are minimal. Only noncovalent bonds such as hydrogen and ionic bonds and hydrophobic interactions in food are disrupted by pressure (Mújica-Paz *et al.*, 2011). Pressure significantly affects microbial behavior. Microbial inactivation contributing to food preservation indeed results from pressure-induced damage to cell membranes, protein denaturation, and changes in enzyme activity and conformation of ribosome (Torres and Velazquez, 2005). However, pressure has also effects on food components. Hence, protein denaturation can modify texture with positive or negative effects on food quality. Since covalent bonds are not affected by pressure, vitamins and flavor compounds are not modified (Mújica-Paz *et al.*, 2011). As a consequence, microorganisms are selectively targeted by high-pressure treatment while nutritional and sensorial properties of foods are mainly preserved compared to a conventional heat treatment.

The second principle associated with high-pressure effects is the isostatic principle, which states that pressure is instantaneously and uniformly transmitted throughout the entire product with no gradient of pressure (Torres and Velazquez, 2005; Smelt, 1998). Thus, pressure is characterized as isostatic pressure. The pressure is applied independently of size and geometry of food because transmission of pressure to the core is not mass/time dependent (Farkas and Hoover, 2000). Then the food returns to its original shape as soon as the pressure is released.

The high-pressure equipment consists of a vessel into which pressure is transmitted to the product through the use of a pressurization fluid, usually water,

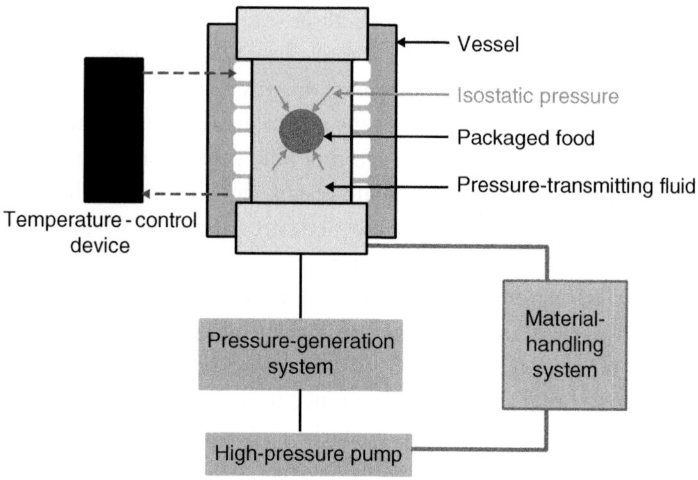

Figure 3.1 Diagram of the principle of high-pressure equipment (adapted from Tonello (1998)).

pressure generation system, material-handling system, and temperature control device (Figure 3.1) (Tonello, 1998). The product packaging must be both flexible to withstand the pressure and water resistant. Packaged food is loaded into the pressure vessel. The vessel is sealed and pressure is increased by either reducing the volume (direct compression) or injecting the pressure-transmitting fluid into the vessel through a high-pressure pump (indirect compression). The fluid displaces any air around the food until the desired pressure is reached. Then, the pump is stopped and the desired pressure is maintained for a given time. When the process is completed, the vessel is decompressed by releasing the pressure-transmitting fluid that returns to atmospheric pressure. The vessel is opened and the packaged food is removed (Yordanov and Angelova, 2010).

3.2.2 Influence of high pressure on microbial contamination of meat

3.2.2.1 General aspects

- Factors influencing the efficiency of high pressure
  - Type of microorganism

    The influence of high pressure on microorganisms has been extensively studied (Smelt, 1998; Simonin, Duranton and de Lamballerie, 2012; Cheftel, 1995; Cheftel and Culioli, 1997; Rendueles *et al.*, 2011; Patterson, 2005). It is well known that the extent of microbial inactivation depends on the nature of microorganisms. Generally speaking, the resistance of microorganisms is highly variable and depends on the type of organism and the food matrix (Rendueles *et al.*, 2011). HHP usually has a higher destructive effect in organisms with a greater degree of organization and structural complexity. Prokaryotes such as bacteria are usually more resistant compared to eukaryotes. Among eukaryotes, protozoa and foodborne parasites (*Toxoplasma gondii, Cryptosporidium parvum, Anisakis simplex, Trichinella spiralis, Ascaris*) are considered to be more sensitive to pressure than molds and yeasts (Rendueles *et al.*, 2011). Bacteria are generally more resistant than yeast and molds, the latter having intermediate resistance to pressure. However, mold spores particularly resistant to high pressures have been reported (Chapman *et al.*, 2007). Gram-positive bacteria are generally more resistant to high pressure than Gram-negative bacteria. However, some exceptions to the rule have been observed. Indeed, Wuytack, Diels and Michiels (2002) found that the groups represented by Gram-positive bacteria and Gram-negative bacteria overlapped regarding their resistance to high-pressure treatments from 200 to 400 MPa at 25 °C. Finally, bacterial endospores can exhibit extremely high resistance to high pressure, just as they are resistant to other physical treatments such as irradiation and heat (Patterson, 2005).

High-pressure treatments, in general, are able to inactivate most vegetative bacteria (more than $4\log_{10}$) under pressure in the range 200–600 MPa at room temperature (Rendueles *et al.*, 2011).

Apart from the type of microorganism treated, it has been shown that within the same species, strong variations in resistance to high pressure could be observed. Alpas *et al.* (1999) reported that resistance to high pressure at 25 °C was highly dependent on strain, particularly for *Listeria monocytogenes, Salmonella, Staphylococcus aureus*, and *Escherichia coli* O157:H7. This variation in strain resistance is illustrated in Figure 3.2. To take into account natural variability in sensitivity to high pressure, it is suggested to use a cocktail of microorganisms in validation or challenge studies, that is, validation of kinetic models when using food as medium (Balasubramaniam *et al.*, 2004).

The physiological state of microorganisms also plays an important role in the treatment efficiency. It is well known that microbial cells in the stationary phase of growth are more resistant to high pressure than in exponential phase (Manas and Mackey, 2004; Pagan and Mackey, 2000; Linton, McClements and Patterson, 2001).

- Pressure kinetics

  Apart from the pressure level, the pressure kinetics (i.e., pressurization rate, pressure-holding time, and depressurization rate) affects the efficacy of microbial inactivation. Chapleau *et al.* (2006) demonstrated that an increase of pressurization and depressurization and increase in the holding time favored higher inactivation.

- Influence of temperature

  Many other factors can influence the efficacy of high-pressure treatments. They have been extensively reviewed (Patterson and Linton, 2008). The temperature at which microbial cells are submitted to high pressure also plays an important role. It has been shown that an optimal temperature of

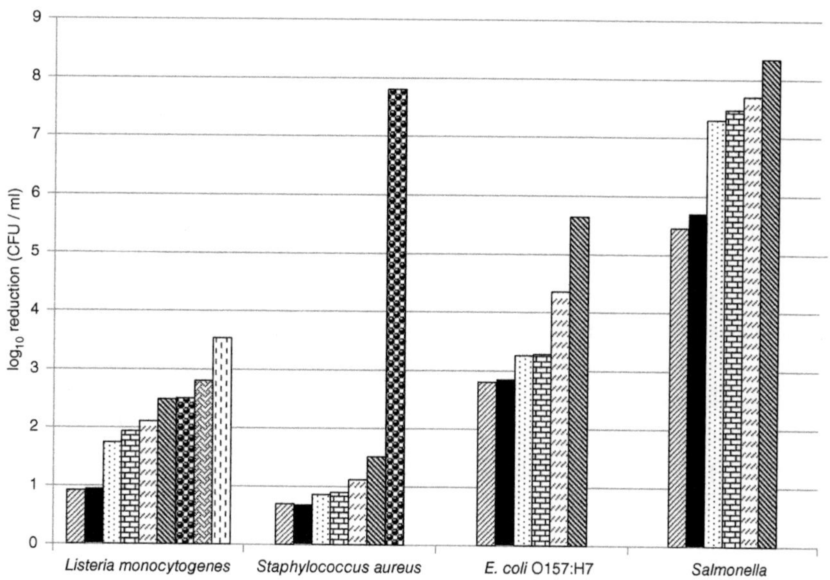

Figure 3.2 Variation in resistance of various strains of foodborne pathogens (adapted from Alpas *et al.* (1999)).

inactivation by high pressure can be assigned to each microorganism (Ritz and Federighi, 2001). Inactivation of vegetative microorganisms is enhanced at temperatures of 50–70 °C and perhaps below 0 °C (Cheftel and Culioli, 1997). This was observed for *L. monocytogenes, S. aureus, E. coli* O157: H7, and *Salmonella* as temperature treatment was increased from 25 to 50 °C (Alpas *et al.*, 2000). Even increasing temperature from 25 to 35 °C could produce an additional microbial reduction of 5.48 and 3.3 $\log_{10}$ in *E. coli* O157:H7 and *Salmonella Enteritidis*, respectively, when 345 MPa was applied during 10 min (Alpas *et al.*, 2000). The combined use of high pressure and moderate temperature of 50–60 °C could be an efficient means of inactivation of pathogens and spoilage microorganisms in food products (Lerasle *et al.*, 2012). There are also studies showing that treatments performed at temperatures below 20 °C or even at subzero temperature are more efficient in killing cells than at 20 °C (Patterson and Linton, 2008). High pressure applied at 4 °C was shown to reduce to a greater extent the microbial population of *Citrobacter freundii, Pseudomonas aeruginosa,* and *Listeria innocua* inoculated into fresh minced beef muscle than a treatment at 20 °C (Carlez *et al.*, 1993). Ritz *et al.* (2008) showed that the maximum log reduction (4.2 $\log_{10}$) in *L. monocytogenes* was reached when a 200-MPa pressure treatment was applied at −18 °C compared to a similar treatment performed at −7 and −14 °C, whereas no inactivation could be achieved when the same pressure level was applied at 20 °C.

- Influence of the medium treatment
  The composition of the medium was also shown to influence the efficiency of high pressure in inactivating microorganisms. Resistance of microorganisms to pressure is enhanced in more complex media, that is, foods compared to buffered solutions (Cheftel and Culioli, 1997). Moreover, the presence in the food of chemical substances considered as protectors such as fats, proteins, minerals, and sugars, affects the microbial susceptibility to pressure (Rendueles *et al.*, 2011; Molina-Hoppner *et al.*, 2004; Black *et al.*, 2007; Hauben, Bernaerts and Michiels, 1998). As high-pressure treatment induces sublethal stress and injury, it is recommended to enumerate viable organisms using microbiological nonselective media (Balasubramaniam and Farkas, 2008).
- Influence of pH and a_w
  Physical parameters such as pH and a_w significantly determine the microbial lethality induced by high pressure. It is generally accepted that acidic pH (Alpas *et al.*, 2000; Wouters, Glaasker and Smelt, 1998) and a_w above 0.9 (Rendueles *et al.*, 2011; Black *et al.*, 2007; Hayman *et al.*, 2008; Patterson *et al.*, 1995) enhance microbial inactivation by high pressure. Low pH would also impair recovery of injured microbial cells from the high-pressure stress (Smelt, 1998).

3.2.2.2 Use of high pressure to improve the food safety of meat products

Meat and meat products represent most sources of human infections with *Salmonella* spp., *Campylobacter jejuni/coli, Yersinia enterocolitica,* Verotoxigenic *Escherichia coli* (VTEC), and, to some extent, *L. monocytogenes*. Harbored by the gastrointestinal tract of healthy animals, they are transferred to foods through transformation, handling, and cross-contamination along the food chain. Occurrences of *Salmonella* spp., *C. jejuni/coli, Y. enterocolitica,* and VTEC in fresh red meat are variable, depending on a range of factors including the organism, geographical factors, farming, and/or meat production practices (Nørrung and Buncic, 2008). The two most frequently reported zoonotic diseases in humans in the European Union in 2010 were *Campylobacter* and *Salmonella* infections with EU notification rate for confirmed cases of 48.6 and 21.5 cases per 100,000 population, respectively (EFSA, 2012). The EU notification rate for VTEC and *L. monocytogenes* was 0.83 and 1.2 per 100,000 population, respectively (EFSA, 2012). Meat and meat products thereof comprise the most important single food that contributes to "strong evidence" outbreak-related human salmonellosis (23.30%) in the EU in 2010 while mixed buffet meals and eggs and egg products contributes to 15.34% and 15.15%, respectively (Figure 3.3). Considering "strong evidence" outbreaks related to human campylobacteriosis, broiler meats contribute to 56.67% in the European Union in 2010 (Figure 3.4).

Studies on the influence of HHP on the inactivation of foodborne pathogens have been gathered by Simonin, Duranton and de Lamballerie (2012). *E. coli,*

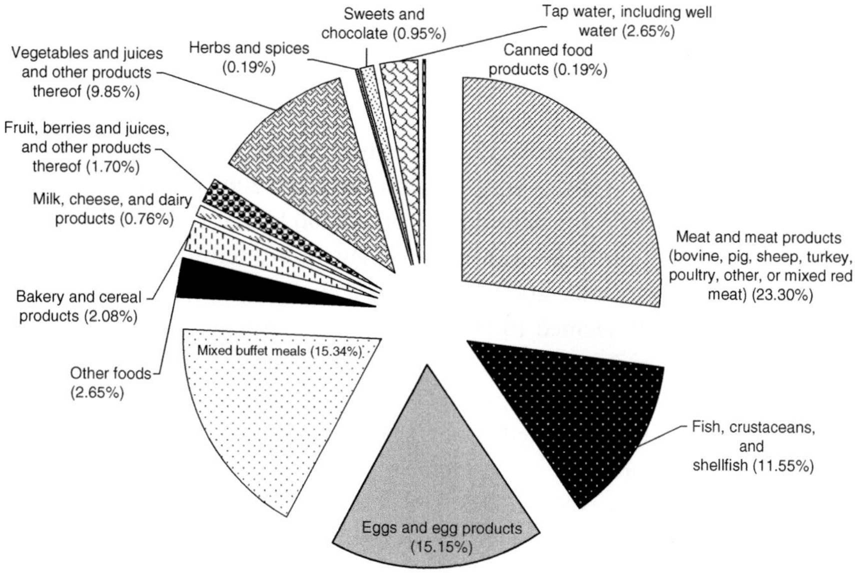

Figure 3.3 Salmonella outbreaks with "strong evidence" and their foodstuff identified origin in the European Union in 2010 (adapted from EFSA (2012)).

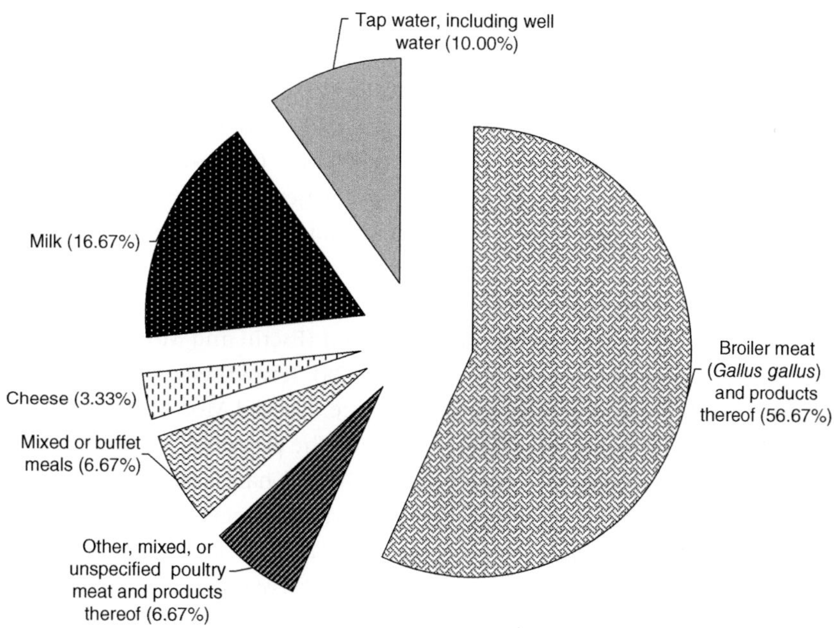

Figure 3.4 *Campylobacter* outbreaks with "strong evidence" and their foodstuff identified origin in the European Union in 2010 (adapted from EFSA (2012)).

considered as a hygiene indicator in most cases, comprises some very pathogenic serovars. Among VTEC, *E. coli* O157:H7 can pose a serious public health risk as it can induce serious complications such as kidney failure. Most studies have shown that *E. coli* was sensitive to high pressure as it could be totally inactivated by 600 MPa during 6 min in marinated beef loin at 16 °C and cooked and dry-cured ham at 6 °C (Garriga *et al.*, 2004; Jofré *et al.*, 2009a). *E. coli* O157:H7 was also completely inactivated by a treatment of 700 MPa at 20 °C during 5 min and 483 MPa at 19 °C during 5 min, respectively, in dry fermented salami and raw minced meat (Gola *et al.*, 2000; Porto-Fett *et al.*, 2010). High-pressure treatments below 500 MPa seemed to fail in inactivating completely *E. coli* and *E. coli* O157:H7. Treatments performed at 400 MPa on model meat system (Garriga, Aymerich and Hugas, 2002a) and ground beef patties (Morales *et al.*, 2009) led to inactivation of 4.5 (of 8 inoculated) and 2.45 $\log_{10}$ (of 7 inoculated), respectively.

L. monocytogenes represents a considerable public health concern. Therefore, many studies have dealt with the influence of high pressure on inactivation of *L. monocytogenes* in meat products (Porto-Fett *et al.*, 2010; Garriga, Aymerich and Hugas, 2002a; Aymerich *et al.*, 2005; Marcos *et al.*, 2008; Simpson and Gilmour, 1997; Chen, 2007; Koseki, Mizuno and Yamamoto, 2007; Jofré, Garriga and Aymerich, 2008). As the composition of the food matrix influences the treatment efficiency, variable decontamination by high pressure was observed. Treatments above 400 MPa during more than 5 min generally enable a significant reduction of *L. monocytogenes* from 1.8 $\log_{10}$ in 25-day-fermented salami treated at 483 MPa (Porto-Fett *et al.*, 2010) to complete inactivation in sliced cooked ham treated for 10 min at 400 and 500 MPa (Aymerich *et al.*, 2005; Marcos *et al.*, 2008) and 35-day-fermented salami treated for 5 min at 600 MPa (Porto-Fett *et al.*, 2010).

It should be noted that the ability of *L. monocytogenes* to grow at refrigerated temperature requires that HHP treatments would be efficient enough not to enable surviving cells to regrow during chilled storage.

Salmonella, a Gram-negative bacteria, was generally significantly inactivated by HHP treatments from 400 MPa in minced chicken (Escriu and Mor-Mur, 2009), meat model (Garriga, Aymerich and Hugas, 2002a), fermented (Jofré, Aymerich and Garriga, 2009b), and poultry sausages (Yuste, Pla and Mor-Mur, 2000a).

Jofré, Garriga and Aymerich (2008) examined the potential use of high pressure to improve the safety of cooked and dry-cured ham and marinated beef loin by performing challenge tests with foodborne pathogens *L. monocytogenes, Salmonella enterica, S. aureus, Y. enterocolitica,* and *Campylobacter jejuni*, a hygiene indicator (*E. coli*) and spoilage microorganisms (*Debaryomyces hansenii* and LAB) (Jofré *et al.*, 2009a). The treatment at 600 MPa enabled efficient inactivation of *E. coli*, yeasts, and the foodborne pathogens *L. monocytogenes, Salmonella, Y. enterocolitica, and C. jejuni* inoculated at 3.5 log CFU/g. *C. jejuni* and *Y. enterocolitica* were shown to be the most inactivated, becoming undetectable in 10 g of product. Concentrations of *L. monocytogenes* and *Salmonella* below 10 CFU/g could be obtained for all products except for marinated beef loin for which complete inactivation could be achieved. However, *S. aureus* was by far the most resistant to the HHP treatment among the other pathogenic inoculated species. The highest inactivation of 2.7 $\log_{10}$ CFU/g was obtained for marinated beef loin (Jofré *et al.*, 2009a). High resistance of *S. aureus* to high pressure has been observed by other authors. No significant inactivation was even reached in model meat system and pork Marengo treated at 400 MPa during 10 and 30 min, respectively (Garriga, Aymerich and Hugas, 2002a; Moerman, 2005).

3.2.2.3 Use of high pressure to extend the shelf life of meat products

High pressure could represent a convenient means of inactivating microorganisms in ready-to-eat (RTE) products, intended to be eaten without cooking at which shelf life is highly dependent on good hygiene/manufacturing practices, particularly during slicing and packaging operations. Therefore, many studies deal with the use of HHP treatment to extend the shelf life of specialty sausages (Yuste, Pla and Mor-Mur, 2000a; Yuste *et al.*, 2000b), dry-cured ham (Garriga *et al.*, 2004; Jofré *et al.*, 2009a) and sausages (Ruiz-Capillas *et al.*, 2007a; Rubio *et al.*, 2007a), cooked ham (Garriga *et al.*, 2004; Aymerich *et al.*, 2005; Lopez-Caballero, Carballo and Jimenez-Colmenero, 1999; López-Caballero, Carballo and Jiménez-Colmenero, 2002a; Karłowski *et al.*, 2002) and cooked sausages (Diez *et al.*, 2008a; Ruiz-Capillas, Carballo and Jiménez-Colmenero, 2007b), etc. Simonin, Duranton and de Lamballerie (2012) gathered studies aiming at extending the shelf life of meat products by the use of HHP treatments. Studies have been distinguished according to the classification introduced by Garriga *et al.* (2004) as a function of the nature of meat products, that is, raw (including marinated with high a_w), dry cured, and cooked meat products.

Spoilage microorganisms in meat are mainly caused by enterobacteria and lactic acid bacteria (LAB) whereas low-temperature alterations are mainly due to psychrotrophs. Psychrotrophs dominated by *Pseudomonas* species in meat are considered to be more sensitive to high pressure than mesophiles (Simonin, Duranton and de Lamballerie, 2012). Hence, treatments performed at 300 and 450 MPa were shown to completely inactivate indigenous *Pseudomonas* (López-Caballero, Carballo and Jiménez-Colmenero, 2002a; Carlez *et al.*, 1994) although regrowth was observed to occur following HHP treatment after a period of time in some studies (Carlez *et al.*, 1994; López-Caballero *et al.*, 2002b). It appeared that high pressure applied at 600 MPa during 6 min at 31 °C was efficient to completely inactivate yeasts and enterobacteria in dry cured and cooked ham throughout the storage duration of 120 days at 4 °C (Garriga *et al.*, 2004; Garriga, Aymerich and Hugas, 2002a). Thus, HHP treatment was efficient in reducing potential capacity of these two flora to produce off-flavors and gas. However, the efficiency of HHP treatment in inactivating LAB was shown to depend on the nature of meat products. Although 400 MPa for 20 min at 7 °C enabled a significant reduction in the LAB population in sliced cooked ham (Lopez-Caballero, Carballo and Jimenez-Colmenero, 1999), a higher pressure treatment of 600 MPa for 10 min at 15 °C performed on blood sausages had almost no effect on LAB counts (Diez *et al.*, 2008b). Indeed it enabled complete inactivation of LAB in dry cured and marinated food products, but regrowth was observed in cooked meat products from 60 days of storage, leading to a final microbial load similar to nontreated and pressurized cooked meat products (Garriga *et al.*, 2004). In the study of Jofré *et al.* (2009a), HHP treatment enabled higher inactivation of LAB in cooked ham than in dry-cured ham (4.6 vs 1.6 $\log_{10}$). However, presumably because of the low a_w of these products, no regrowth of LAB occurred throughout the chilled storage in dry-cured products, whereas the population of LAB at the end of the 120-day storage at 4 °C reached the same level in pressurized and nontreated cooked products (Jofré *et al.*, 2009a). Spoilage or antimicrobial properties of LAB have to be appreciated to determine whether the well-documented baroresistance of LAB has to be considered as positive or negative, depending on its impact on the product shelf life.

Interestingly, the application of high pressure does not always lead to immediate microbial inactivation but to delayed reduction of the population. This was observed after treatment of blood sausages by 300–600 MPa for 10 min at 15 °C for *Pseudomonas* and enterobacteria, which led to no apparent microbial reduction immediately after HHP, whereas bacteria became undetectable the following day (Diez *et al.*, 2008b). On the other hand, high pressure above 400 MPa was shown to affect recovery of bacteria and led to delayed bacterial regrowth (Garriga *et al.*, 2004; Jofré *et al.*, 2009a; Yuste *et al.*, 2000b). However, in some cases, bacterial regrowth after HHP negates the benefits of the immediate microbial decrease obtained after high pressure. This was observed by Garriga, Aymerich and Hugas (2002a) as the population of LAB and *E. coli* of nontreated cooked ham homogenized with water was rejoined by that of the pressurized product (400 MPa for 10 min at 17 °C) after less than 20 days of chilled storage. Since recovery was shown to depend on food matrix and particularly on the presence of exudates (Simpson and Gilmour, 1997) or fat (Rubio *et al.*, 2007a), on product a_w (Jofré *et al.*, 2009a)), and on the presence a cooking step before HHP treatment (Garriga *et al.*, 2004; Garriga, Aymerich and Hugas, 2002a; Garriga and Aymerich, 2009), it is an important aspect to consider to control microbial growth following HHP treatments.

3.2.2.4 Modeling of microbial inactivation by high pressure

- Primary modeling

  First-order model

  The kinetics of microbial inactivation by heat is traditionally described by a first-order decay derived from the equation:

$$\log\left(\frac{N_0}{N}\right) = \frac{t}{D} \quad (t \geq 0) \tag{3.1}$$

where N_0 is the initial number of cells (CFU/ml), N is the number of survivors (CFU/ml), t is the exposure time (min), and D is the decimal reduction time. The first-order kinetic model is commonly used in inactivation studies and has also been applied to model high-pressure inactivation for *L. innocua* in liquid UHT dairy cream (Raffalli *et al.*, 1994); *E. coli*, *L. innocua*, and *S.* Enteritidis in liquid whole egg (Ponce *et al.*, 1998); *E. coli* in a physiological saline solution (Smelt and Rijke, 1992); *E. coli* in broth, milk, peach, and orange juice (Erkmen and Doğan, 2004); and *L. monocytogenes* in raw milk (Mussa, Ramaswamy and Smith, 1998). However, deviations from the first-order kinetics have often been observed (Figure 3.5). Hence, it is common to observe tails and shoulders in high-pressure inactivation curves. This may be due to the presence of subpopulations of microorganisms unequally resistant to pressure (Smelt, 1998). Then the use of non-first-order model types is necessary to describe inactivation by HHP. Three non-first-order models are commonly used: biphasic, Weibull, and modified Gompertz models.

Biphasic model

This model considers that the microbial population is composed of two subpopulations each following a first-order kinetics (Xiong *et al.*, 1999). The equation corresponding to this model is as follows:

$$\log\left(\frac{N_0}{N}\right) = \log\left(\frac{N_0}{N}\right)_0 - \log\left[\alpha \cdot 10^{-\frac{t}{D_1}} + (1-\alpha)\cdot 10^{-\frac{t}{D_2}}\right] \qquad (3.2)$$

and

$$\alpha \frac{N_1(t=0)}{N_1(t=0) + N_2(t=0)}$$

where D_1 and D_2 represent decimal reduction times of the two subpopulations (D_1 for the most resistant population N_1 and D_2 for the least one, N_2).

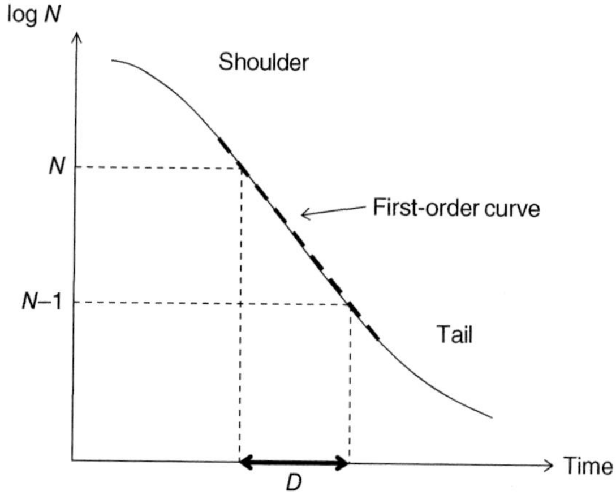

Figure 3.5 Typical inactivation curve obtained following heat or high-pressure treatment.

This model was applied to analyze inactivation by pressure of *E. coli* in Hepes buffer and in carrot juice (Van Opstal *et al.*, 2005). This model could be considered as a suitable alternative to the first-order model for fitting inactivation curves with tails (Diels *et al.*, 2007).

Weibull model

The Weibull model has been extensively used to model thermal and nonthermal microbial inactivation (Couvert *et al.*, 2005). It takes into account biological variation within a population of cells (Diels *et al.*, 2007) and follows the equation:

$$\log(N) = \log(N_0) - \left(\frac{t}{\delta}\right)^p \tag{3.3}$$

where δ (scale parameter) corresponds to the first reduction time that leads to a 10-fold reduction of the surviving population and p is the shape parameter. Its value depends on the shape of the survival curves; $p < 1$ for concave upwards survival curves, $p = 1$ for linear survival curves, and $p > 1$ for concave downwards survival curves (Figure 3.6).

The Weibull model has been applied to modeling inactivation of *L. monocytogenes* and *L. innocua* by HHP and mild heat (Chen and Hoover, 2003a; Buzrul and Alpas, 2004) and HHP alone (Chen and Hoover, 2004). Inactivation by combined use of temperature and high pressure has also been modeled for *E. coli* in Hepes buffer pH 7.0 (Van Opstal *et al.*, 2005) and *Alicyclobacillus acidoterrestris* in culture medium pH 3.0 (Buzrul, Alpas and Bozoglu, 2005) and *Y. enterocolitica* in phosphate buffer pH 7.0 and UHT whole milk (Chen and Hoover, 2003b). The main advantage of this model is that it is simple to use with only two parameters to adjust. However, the most important drawback would be the lack of biological significance of these parameters (Diels *et al.*, 2007).

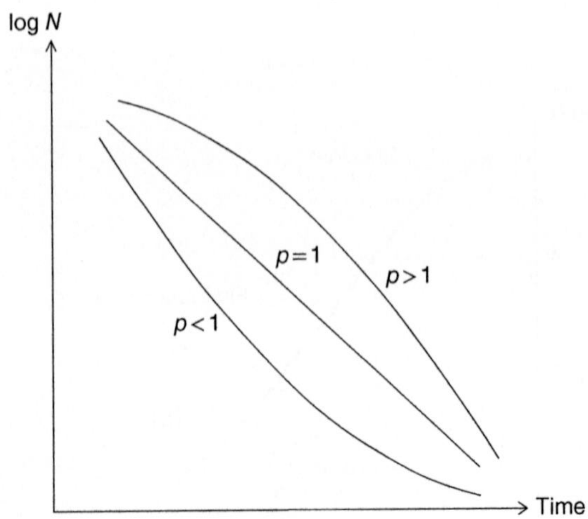

Figure 3.6 Microbial inactivation curves modeled using Weibull equation: $p < 1$ concave upwards curve; $p > 1$ concave downwards curve; $p = 1$ linear curve with $\delta = D$ of the first-order model.

Modified Gompertz model

Originally implemented by Gibson, Bratchell and Roberts (1988) to model microbial growth, it has been subsequently applied to modeling thermal and pressure inactivation curves (Xiong *et al.*, 1999; Linton *et al.*, 1995; Erkmen, 2001a). The modified Gompertz model is derived from the equation as follows:

$$\log\left(\frac{N_0}{N}\right) = \left(\frac{N_0}{N}\right)_0 - A\exp\left\{-\exp\left[\frac{\mu e}{A}(\lambda - t) + 1\right]\right\} \tag{3.4}$$

where $A\log\left(\frac{N}{N_0}\right)$ is the lower asymptote value, μ is the inactivation rate, λ is the time for complete inactivation, t is the time, and e is a mathematical parameter. It has been used to model inactivation of *L. monocytogenes* (Chen and Hoover, 2003a) and *Y. enterocolitica* (Chen and Hoover, 2003b) by high pressure, with a poorer fit than that obtained with the Weibull and log logistic models according to Diels *et al.* (2007).

Log logistic model

The equation of this model proposed by Cole *et al.* (1993) is as follows:

$$\log\left(\frac{N_0}{N}\right) = \log N_0 - \log N = \frac{\omega - \alpha}{1 + e^{4\sigma(\tau-6)(\omega-\alpha)}} - \frac{\omega - \alpha}{1 + e^{4\sigma(\tau-\log t)/(\omega-\alpha)}} \tag{3.5}$$

where α and ω are, respectively, the upper and lower asymptotes, σ is the maximum rate of inactivation, and τ is the logarithmic time to the maximum rate of inactivation. A small time, that is, 10^{-6}, but enabling the calculation of $\log t$ was used, which is as follows:

$$\log N_0 = \alpha + \frac{\omega - \alpha}{1 + e^{4\sigma(\tau-6)(\omega-\alpha)}} \tag{3.6}$$

However, the parameters of this model would not have any physical significance (Diels *et al.*, 2007).

The choice of the model is based on its ability to fit experimental data. Fitting performance of the model increases with decreasing MSE (mean square error) (Neter *et al.*, 1996) and $R^2{}_{\text{adjusted}}$ (Ross, 1996) and A_f (accuracy factor) values tending to 1. MSE provides a measure of the average difference between observed and fitted values (Van Opstal *et al.*, 2005) and derives from

$$\text{MSE} = \frac{\text{SSE}}{n - p} \tag{3.7}$$

and

$$R^2_{\text{adjusted}} = 1 - \frac{(n - i)(1 - R^2)}{(n - p)} \tag{3.8}$$

where SSE is the sum of squares of errors, n is the number of observations, and p is the number of parameters.

As a conclusion, the model chosen should be as simple as possible. However as microbial inactivation by high pressure is influenced by many factors, the choice of the model will depend upon data sets (Diels *et al.*, 2007). In the study by van Asselt and Zwietering performed on 4000 thermal D-values collected in the literature, it appeared that the first-order model was finally the best choice even if it made shoulders and tails disappear in the analysis (van Asselt and Zwietering, 2006). Diels *et al.* (2007) suggested a similar recommendation to analyze HHP inactivation curves as soon as there is sufficient data.

- Secondary modeling

  Once the primary parameters are expressed in the equation of the primary model, secondary modeling will consist of expressing primary parameters as a function of other factors such as temperature and high-pressure kinetic parameters. Polynomial models are the most frequently used secondary models (Diels *et al.*, 2007). If a first-order model is used as the primary model, secondary modeling will consist of expressing *D*-values as a function of temperature and pressure. The associated representation of such an equation is iso-*D* curves with an example described in Figure 3.7. This type of curve established for a microorganism enables the determination of the right combinations of temperature and pressure as a function of the targeted inactivation rate. It has been applied to high-pressure inactivation of *Saccharomyces cerevisiae, Lactobacillus plantarum* (Perrier-Cornet *et al.*, 2005), *Zygosaccharomyces bailii* (Reyns *et al.*, 2000), and *E. coli* (Van Opstal *et al.*, 2005).

  These curves illustrate the feature that log *D* as a function of temperature is not a linear relationship as is the heat inactivation with

$$\log D = -\frac{T}{Z} + c^{te} \tag{3.9}$$

- Model validation in real foods

  Once chosen, the model has to be validated by using experimental data obtained in real food. The nature and composition of the food have been found to greatly influence inactivation induced by high pressure. Consequently, the shape of iso-*D* contours may be affected by the media in which

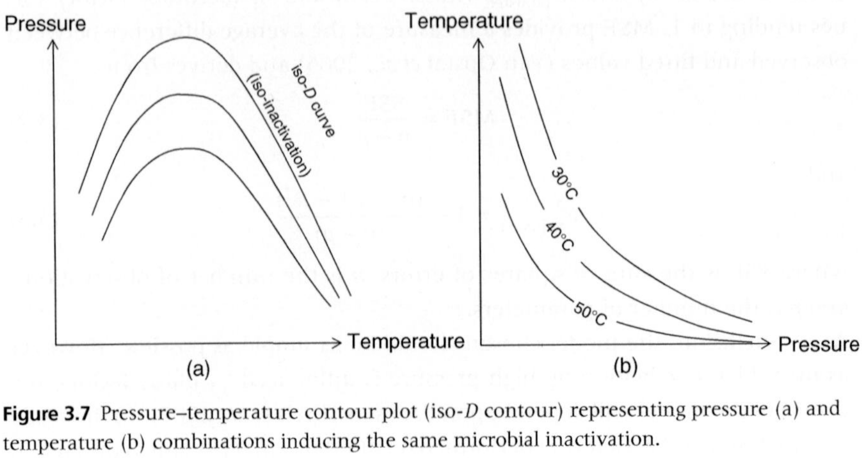

Figure 3.7 Pressure–temperature contour plot (iso-*D* contour) representing pressure (a) and temperature (b) combinations inducing the same microbial inactivation.

the organism is HHP-treated. This has been shown for *E. coli* treated in two different media with straight-shaped iso-*D* curves in carrot juice and elliptical-shaped iso-*D* contours in Hepes buffer (Van Opstal *et al.*, 2005). Hence, the model built for one matrix is not necessarily transposable to another matrix. This observation has been done by Reyns *et al.* (2000) who noted that the biphasic model developed for inactivation of *Z. bailii* in a buffer at pH 6.5 was applicable in a low-pH buffer but not in apple and orange juice.

- Integrating food quality in inactivation models
  High pressure is a nonthermal process that is known to affect nutritional and organoleptic qualities of food. However, the impact of high pressure on both food safety and food quality has never been modeled to our knowledge. Attempts have been done to consider together enzyme inactivation and microbial inactivation (Ludikhuyze *et al.*, 2002; Valdramidis *et al.*, 2007). It appeared that enzyme denaturation by high pressure was more critical than microbial inactivation in the optimization of the process combining pressure and temperature.

3.2.3 Modification of meat induced by high pressure

3.2.3.1 Effect of high pressure on meat texture

A wide range of textural modifications following high pressure has been reported, depending on the processing conditions, meat composition, and preparation (raw, cooked, dry, ...) (Jung, Tonello Samson and De Lamballerie, 2011). The influence of HHP on the texture of meat and meat products has been recently reviewed by Sun and Holley (2010). It is generally accepted that high pressure at mild temperature can tenderize meat when applied in *pre-rigor*, that is, a few hours after death (Jung, Tonello Samson and De Lamballerie, 2011; Bouton *et al.*, 1977). However, applied in *post-rigor*, the reverse effect of high pressure on tenderness is often reported. By way of conclusion, Sun and Holley (2010) summarized the two main effects of high pressure as a function of its time of application in relation to muscle maturation, that is, *rigor mortis*, and temperature. Hence, what has to be retained is a tenderization effect of low-pressure (<200 MPa) treatment when applied on *pre-rigor* meat, whereas higher temperature (40–80 °C) has to be combined to high-pressure treatment in order to achieve tenderization of *post-rigor* meat. Tenderization of *pre-rigor* meat is indeed one of the researched applications of high pressure to meat products. This application has been dealt with in section "Meat tenderization". Therefore, in this section, only the influence of high pressure on texture of *post-rigor* meat will be highlighted. High-pressure-induced increase in toughness of *post-rigor* beef muscle (Ma and Ledward, 2004) and chicken muscle treated at 400 MPa (Del Olmo *et al.*, 2010; Zamri, Ledward and Frazier, 2006) have been observed. Similarly, a treatment of 520 MPa led to enhanced hardness of beef muscle (Jung, Ghoul and de Lamballerie-Anton, 2000a). On the contrary, an increase in beef meat tenderness was observed following high-pressure treatment at 500 MPa at 8 °C (Ichinoseki, Nishiumi and Suzuki, 2006). Injection of papain in bovine muscle before high-pressure treatment from 100 to 300 MPa has been found to prevent an increase in meat toughness (Schenková *et al.*, 2007).

High-pressure-induced hardness has also been observed for processed meat products such as ham. A treatment of 500 MPa at 15 °C during 6 min induced a reduction in tenderness from the HHP treatment throughout the storage duration of 50 days (Clariana and Garcia-Regueiro, 2011). High-pressure-treated cooked sausages at 500 MPa and 65 °C lost firmness and gained cohesiveness with lower weight loss compared to heat-pasteurized sausages cooked at 80–85 °C for 40 min (Mor-Mur and Yuste, 2003). Studies performed on meat batters have shown that high pressure, when performed before cooking, tended to increase gel elasticity compared to cooked-only batters (Iwasaki *et al.*, 2006) and reduced cooking losses (Sikes, Tobin and Tume, 2009). Combining high pressure and thermal treatments resulted in a variable modification of meat texture as heat-induced protein denaturation. Generally, an increase in temperature above 70 °C during high-pressure treatment resulted in lower hardness and apparent elasticity (Fernández-Martín *et al.*, 1997) but higher cohesiveness (Yuste *et al.*, 1999).The gel of raw ostrich sausages became stronger as the temperature increased from 40 to 60 °C, and the amount of released and expressible water decreased with increasing severity of the treatment (Supavititpatana and Apichartsrangkoon, 2007; Chattong and Apichartsrangkoon, 2009). Treating meat batters at 200 or 400 MPa and at 70–80 °C favored better fat and water retention than was found in cooked-only samples (Jimenez-Colmenero *et al.*, 1998). Product formulation and, for instance, salt content were shown to influence texture and water retention in meat batters (Iwasaki *et al.*, 2006; Sikes, Tobin and Tume, 2009). Tripolyphosphates act synergistically with high pressure (400 MPa for 30 min at 70 °C) in increasing water-holding capacity in pressure- and heat-treated pork batters (Fernandez-Martin *et al.*, 2002). Simonin, Duranton and de Lamballerie (2012) indicated that the addition to meat products of ingredients having gelation properties under high pressure would affect the resulting texture of the product. For instance, dried egg white (Trespalacios and Pla, 2009), starch (Fernandez *et al.*, 1998; Fernández-Martín *et al.*, 2000a), and carrageenan (Fernandez *et al.*, 1998) generally increased the toughness of high-pressure-treated meat batters. High pressure could also be used to create meat-to-meat binding in cold-set restructured meat products (Hong *et al.*, 2008a; Hong *et al.*, 2008b). Ingredients such as NaCl, tripolyphosphate, glucono-delta-lactone, and κ-carrageenan (Hong *et al.*, 2008b) or transaminase (Trespalacios and Pla, 2007a; Trespalacios and Pla, 2007b) could be added wisely in order to consolidate the protein network of meat to be restructured. They represent promising potentialities of high pressure for the creation of new texturized products (Simonin, Duranton and de Lamballerie, 2012).

The mechanisms underlying alteration of meat texture by the application of high pressure involve depolymerization, solubilization, denaturation, and aggregation in myofibrillar proteins (Cheftel and Culioli, 1997; Chapleau *et al.*, 2004). These have been thoroughly reviewed by Sun and Holley (2010).

3.2.3.2 Effect of high pressure on oxidation of meat

Many studies have shown that lipid oxidation could be observed following high-pressure treatments applied in poultry meat and meat products immediately (Tuboly *et al.*, 2003) or during subsequent refrigeration storage (Dissing, Bruun-Jensen and Skibsted, 1997; Orlien, Hansen and Skibsted, 2000; Beltran *et al.*, 2003; Beltran *et al.*, 2004a). Lipid oxidation is generally associated with undesirable production of rancidity and off-flavors. Therefore, much attention has been paid on this potentially detrimental characteristic of high-pressure processing. Generally, high pressure has little effect on lipid oxidation below 300 MPa (Simonin, Duranton and de Lamballerie, 2012). However, lipid oxidation was shown to be increased with the pressure level with the production of secondary oxidation products. Pressure above 600 MPa was shown to induce lipid oxidation after 10 days of chilled storage in vacuum-packaged chicken breast muscle (Orlien, Hansen and Skibsted, 2000). In the study of Beltran *et al.* (2004a), vacuum packaging was supposed to help in delaying lipid oxidation in raw minced chicken thighs treated at 500 MPa and 50 °C for 30 or 60 min. Indeed one strategy proposed to minimize rancidity in high-pressure-treated products was to use packaging under modified atmosphere with N_2 and CO_2 (Andres *et al.*, 2006) or vacuum packaging (Campus *et al.*, 2008). Oxygen concentration in the package would indeed affect lipid oxidation when lowered during the refrigerated storage, whereas it seems to have no influence during the high-pressure treatment itself (Cheah and Ledward, 1996).

Conflicting results have been obtained concerning the influence of meat origin on its susceptibility to lipid oxidation after high-pressure treatment. Ma *et al.* (2007) found that chicken was less susceptible to lipid oxidation when submitted to pressure of 600 and 800 MPa than beef muscle. On the other hand, the reverse observation was noted by Schindler *et al.* (2010) who found that beef was more stable than chicken to high-pressure treatments at 400 or 600 MPa. Contact with oxygen is favored by operations such as homogenization, grinding, mincing, and slicing, so that oxidation is generally enhanced in RTE products (Beltran *et al.*, 2004a; Fuentes *et al.*, 2010). Product formulation also plays an important role in acceleration or prevention of high-pressure-induced lipid oxidation. For instance, salt added at 5% in chicken slurries has been shown to have a pro-oxidant effect (Beltran *et al.*, 2003). On the other hand, some additives with antioxidant properties may minimize lipid oxidation. In this respect, rosemary and sage extract have successively been added to comminuted chicken treated by high pressure (300–800 MPa) (Beltran *et al.*, 2004b; Bragagnolo, Danielsen and Skibsted, 2007; Mariutti *et al.*, 2008). Other antioxidants, such as EDTA, for its chelating role of iron, protected chicken meat slurries against lipid oxidation following high-pressure treatment (300 or 500 MPa) and subsequent chilled storage (Beltran *et al.*, 2004b).

Multiple explanations of oxidation acceleration in meat after high pressure have been mentioned: release of iron ions that catalyzes lipid oxidation (Cheah and Ledward, 1997; Angsupanich and Ledward, 1998), membrane damage in raw chicken breast muscle (Orlien, Hansen and Skibsted, 2000), or rupture of adipocytes in beef patties (Carballo *et al.*, 1997).

3.2.3.3 Effect of high pressure on color and sensory quality of meat

Meat color is an important quality trait influencing the purchase decisions of consumers at the point of sale. However, high pressure is known to modify meat color, especially the color of raw red meat. The Lightness (L^*) component in the L^*, a, b color scale, has been found to increase, respectively, in chicken, pork, and beef meat, following pressures above 200 MPa (Del Olmo *et al.*, 2010; Beltran *et al.*, 2004a; Korzeniowski, Jankowska and Kwiatkowska, 1999; Marcos, Kerry and Mullen, 2010). This whitening effect can appear after high-pressure treatments with durations as short as 1 min (Del Olmo *et al.*, 2010; Jung, Ghoul and de Lamballerie-Anton, 2003). It has been related to either (1) protein coagulation with a resulting loss of solubility of sarcoplasmic and/or myofibrillar proteins that affect structure and surface properties (Goutefongea *et al.*, 1995) or (2) globin denaturation and heme group displacement or release (Carlez, Veciana-Nogues and Cheftel, 1995). The a^* component, correlating to the red index, has also been shown to decrease following pressure treatment of beef meat above 350–400 MPa (Marcos, Kerry and Mullen, 2010; Jung, Ghoul and de Lamballerie-Anton, 2003; Carlez, Veciana-Nogues and Cheftel, 1995). This decrease has been attributed to oxidation of ferrous myoglobin to ferric metmyoglobin (Jung, Ghoul and de Lamballerie-Anton, 2003; Carlez, Veciana-Nogues and Cheftel, 1995; Cava *et al.*, 2009) and possibly to further denatured myoglobin ferric species, responsible for the brownish color (Wackerbarth *et al.*, 2009). High fat content (20–25%) would significantly enhance meat color alteration, especially of beef patties (Carballo *et al.*, 1997) and pork sausages (Colmenero *et al.*, 1997). However, since pressurized raw meat products are intended, for the most part, to be further cooked, detrimental effects of high pressure on raw meat color can be generally erased (Gola *et al.*, 2000; Mor-Mur and Yuste, 2003). Hence, these aspects only remain negative for products that are intended to be sold as "raw-like" products (Beltran *et al.*, 2004a; Jung, Ghoul and de Lamballerie-Anton, 2003). For RTE meat products, such as cured products, high pressure has been shown to modify meat color to a lesser extent (Karłowski *et al.*, 2002; Rubio *et al.*, 2007b). Moreover, high-pressure-induced alteration of meat color sometimes disappears during storage (Cava *et al.*, 2009). Preliminary cooking cured products before pressurization can also be an interesting means of reducing color modification (Goutefongea *et al.*, 1995).

Concerning the sensory quality of pressurized meat products, a study performed on the aroma profiles of raw meat (chicken and beef) after pressure treatment did not show any significant change in the aroma of vacuum-packed beef and chicken treated below 600 MPa compared to fresh untreated meat (Schindler *et al.*, 2010). It has to be noted that differences between aroma profiles are sometimes in favor of pressurized products as they are no longer being spoiled by contaminating bacteria. Interestingly, the perception of salt, that is, saltiness, in dry-cured ham has been found to be intensified by high-pressure treatment (Saccani *et al.*, 2004; Fulladosa *et al.*, 2009; Clariana *et al.*, 2011), possibly due to weakening of the interaction between Na^+ and proteins (Clariana *et al.*, 2011).

Color, texture, aroma, and taste modifications induced by high pressure drives sensory acceptance of high-pressure-treated meat products. In this respect, pressurized raw meat products raise problems mainly in relation to color alteration (Simonin, Duranton and de Lamballerie, 2012). On the other hand, cooked and cured meat products are generally less affected by high pressure, and it results in more commercial applications of these types of products.

3.2.4 Advantages and disadvantages of high-pressure preservation and decontamination

Kadam *et al.* (2012) listed advantages and limitations of high-pressure technology. Table 3.1 lists some of their items in addition to others that we identified.

3.3 High-pressure applications

3.3.1 HP pasteurization

HP pasteurization is the more common use of high-pressure technology in food preservation. It consists of applying high pressure (from 300 to 600 MPa) at a temperature close to ambient temperature or below. It can be used as thermal pasteurization replacement, the main purpose being to extend shelf life and ensuring food safety of products by inactivating foodborne pathogens, while maintaining nutritional and sensorial qualities. Some applications of high pressure, especially for RTE meats or meals, enable the removal of preservatives in favor of "clean label" products. Some commercially available products treated by HP pasteurization are listed in Table 3.2.

Table 3.1 Advantages and limitations of high hydrostatic pressure technology in preserving and decontaminating food products.

Advantages	Limitations
Products	
Suitable for products with high water content	Lower efficiency of HHP on low a_w products
Flavor, texture, and appearance closer to that of a raw product than that of a cooked product	But meat texture is affected
	Oxidation of meat may be accelerated by high pressure
	Discoloration of meat, particularly red meat, may occur
	Inappropriate to solid food containing air such as bread and cakes, mousses
Little or no heating	
Isostatic process, so pressure is transmitted rapidly and uniformly throughout both the pressure medium and the food	
Effect independent of shape and size	
Process suitable for products for which existing decontamination technologies revealed to be unsatisfactory	Process generally not dedicated to high-value products due to the capital investment required
Packaging	
	Packaging has to be adapted to high-pressure technology
	Completely rigid packaging is unsuitable (canned foods, glass packaging)
Availability of product packaging able to accommodate a change up to 15% in volume followed by a return to its original shape without losing seal integrity or barrier properties	Packaging has to be designed not to deform the treated product
	Transfer of chemical substances from the packaging to the product has to be evaluated
Food safety and shelf life	
Extension of shelf life by reducing spoilage microorganisms and inactivating enzymes	Lactic acid bacteria are the most resistant group of meat endogenous flora
Ensuring food safety by reducing foodborne pathogens	However, HHP is known to fail in inactivating bacterial spores
	Decontamination efficiency highly depends on matrix, so that HHP has to be examined product by product
Equipment	
Batch processing and semicontinuous processing available	Rate-limiting step in the process
	High cost for a commercial scale
	Complexity in the HHP equipment requiring extremely high precision in its construction, use, and maintenance
Regulation	
	The European "Novel Foods" Directive has introduced regulatory problems and slowed the introduction of new pressurized products

Table 3.2 Main but nonexhaustive commercially available HP-treated products.

Product category	Company (country)	Website
Fruit and vegetables		
Guacamole	Verfruco (USA)	http://www.verfruco.com/
	Fresherized Foods (USA)	http://www.fresherizedfoods.com/
Pineapple salsa dip	Fresherized Foods (USA)	http://www.fresherizedfoods.com/
Dips and salsas	Fresh Garden Gourmet (USA)	http://www.gardenfreshsalsa.com/
Pasta sauce and salsa	Simply Fresco (USA)	http://www.simplyfresco.com/
Avocado pulp	Verfruco (USA)	http://www.verfruco.com/
Broccoli and apple juice	Beskyd (Czech republic)	http://www.beskyd.cz/
Grapefruit juice	Pokka (Japan)	
Orange, lemon, and grapefruit juice	Ulti (France)	http://ulti.fr/
Various fruit juices	Donnyboy (Australia)	http://www.preshafruit.com.au/
	ReBootYourLife (Australia)	http://www.rebootyourlife.com.au/
	FruityLine (The Netherlands)	http://www.fruity-line.nl/
	Austchilli (Australia)	http://www.austchilli.com.au/
	Orchard House (UK)	http://www.ohf.co.uk/
Fruit jam, fruit sauces, jellies	Meidi-Ya (Japan)	
	CJ Cheil Jedang (Japan)	http://www.cjthekitchen.co.kr/
Smoothies	Invo (Spain)	http://www.invo.es/
	FruityLine (The Netherlands)	http://www.fruity-line.nl/
Revitalizing drinks	Villa de Patos (Spain)	http://villadepatos.com/
Herbs	Austchilli (Australia)	http://www.austchilli.com.au/
Ready-to-eat vegetable meals	The Ian Group (Spain)	http://www.grupoian.com/
Meat products		
Sliced and cooked products	Zwanenberg (The Netherlands)	http://www.zwanenberg.nl/
	Perdue (USA)	http:// www.perdue.com/
Sliced ham, turkey or chicken cuts, ready-to-eat products, whole pieces of cured ham, tapas	Espũna (Spain)	http://www.espuna-charcuterie.fr/
	Abraham (Germany)	http://www.abraham.de/
	Itoham (Japan)	http://www.itoham.co.jp/
	Foster Farms (USA)	http://www.fosterfarms.com/
	Santa Maria Foods (Canada)	http://www.sharemastro.com/
Dry-cured meats	Campofrio (Spain)	http://www.campofrio.es/
	Ferrarini (Italy)	http://www.ferrarini.it/
	Rovagnati (Italy)	http://www.rovagnati.it/
Seafood products		
Crustacean and mollusk meat extraction	5 degrés ouest (France)	http://www.5do.fr/
Oyster sauce	Motivatit (USA)	http://www.motivatit.com/

(*continued overleaf*)

Table 3.2 (*Continued*)

Product category	Company (country)	Website
Fish products	Campofrio (Spain)	http://www.campofrio.es/
Mollusks and crustaceans	Clearwater (Canada)	http://www.clearwater.ca/
Mussel powder	Maclab (New Zealand)	http://www.maclab.net/
Lobster meat extraction	Ocean Choice (Canada)	http://www.oceanchoice.com/
Dairy products		
Colostrum beverage	Col + (New Zealand)	http://colplus.com/
Cheese	Deli24 (UK)	http://deli24.co.uk/
Yoghurts	Fonterra (New Zealand)	http://www.fonterra.com/
Ready-to-eat meals		
Snack foods	Creta Farm (Greece)	http://www.cretafarms.gr/
	Deli24 (UK)	http://deli24.co.uk/
	Moira Mac's (Australia)	http://www.moiramacs.com.au/
Sandwich meat	Hormel (USA)	http://www.hormelnatural.com/

3.3.2 Sterilization – pressure-assisted thermal processing (PATP) or pressure-assisted thermal sterilization (PATS)

The vocabulary used to qualify the category of high-pressure treatments derives from that traditionally used to qualify heat treatments. Hence, pasteurization by high pressure aims to inactivate pathogenic vegetative microorganisms, whereas sterilization by high pressure targets all microorganisms, including vegetative cells and spores, and enzymes. In general, HP sterilization is performed with a start temperature higher than for pasteurization.

Temperatures ranging from 60 to 90 °C and adiabatic compression for rapid heating are components of HP sterilization (Matser *et al.*, 2004). Recent studies have evaluated the impact of the pressurization rate and pressure pulsing on the lethality of *Bacillus amyloliquefaciens* spores. Pressure of 600 MPa in association with high temperature of 105 °C has been tested at two pressurization rates (18.06 and 3.75 MPa/s) and five holding times (0.5, 1, 2, 3, and 5 min) (Ratphitagsanti *et al.*, 2009). Slow pressurization rates are revealed to be more efficient in inactivating spores than fast pressurization rates, especially for short holding times. Pressure pulsing considerably increased the efficacy of pressure-assisted thermal processing (PATP) treatment against bacterial spores (Ratphitagsanti *et al.*, 2009). In another study, these researchers evaluated the efficiency of PATP (700 MPa – 105 °C) on the same spores in association with organic acids (100 mM acetic, citric, and lactic acids) as a function of holding time. They compared inactivation induced by PATP with that obtained following thermal treatment (105 °C) and high-pressure treatment (700 MPa) alone. They conclude that neither thermal treatment nor HHP treatment alone succeeded in inactivating spores, whereas there was an association of both reduced spores to undetectable levels with or without organic acids after 3 min PATP holding time. Moreover, the addition of some organic acids provided significant lethality enhancement ($p < 0.05$) during PATP treatments and suppressed spore recovery in treated carrot puree (Ratphitagsanti *et al.*, 2010).

In 2009, the United States Food and Drug Administration approved a petition for the commercial use of PATS for application in the production of low-acid foods. Although this new process is promising particularly for soup, stew, and mashed potatoes, which are severely affected by conventional thermal treatments, there are currently no PATS-processed shelf-stable products commercially available in the market (Koutchma, 2011). High-temperature conditions are likely to produce toxic substances that impact on human health, such as acrylamide, polycyclic aromatic hydrocarbons (PAHs), heterocyclic amines (HCAs), and N-nitroso compounds (NOCs). Segovia Bravo *et al.* (2012) reviewed specific chemical safety risk that could be associated with PATP foods. This review comes to the conclusion that there is a need to expand research on chemical changes in PATP foods focusing on toxicity risks for an effective and safe implementation of this emerging technology (Segovia Bravo *et al.*, 2012).

3.3.3 Texturization

As high pressure has important effects on structure and functionality of proteins, it also affects meat texture. In some applications this characteristic of high pressure, considered in some cases as a major drawback, can be highlighted in the processing of raw meat. Indeed high pressure has been shown to tenderize pre-rigor meat (Jung, De Lamballerie-Anton and Ghoul, 2000b). In other studies, high-pressure technology has been used as a stabilizing and texturizing technique for meat sausage, by enhancing protein gelation properties and cohesiveness (Mor-Mur and Yuste, 2003; Supavititpatana and Apichartsrangkoon, 2007).

3.3.4 Freezing – thawing

Figure 3.8 illustrates the phase diagram of water as a function of pressure. The liquid–solid state change of water occurs at 0 °C under atmospheric pressure (additional pressure of 0 MPa). The pressure increase from 0 to 200 MPa, enables the lowering of the temperature at which liquid-solid state change occurs. Above 200 MPa, the phenomenon is reversed. This property can be used for instantaneous freezing of meat products at −20 °C by releasing pressure from 200 MPa to atmospheric pressure (De Lamballerie, Taylor and Culioli, 2002). This rapid freezing has the advantage to produce very small ice crystals that do not damage meat tissue.

Concerning the thawing process, the application of a pressure of 200 MPa would enable the thawing of 2 kg of frozen beef meat much faster (80 min instead of 7 h) than conventional methods (Pothakamury *et al.*, 1995). Thawing

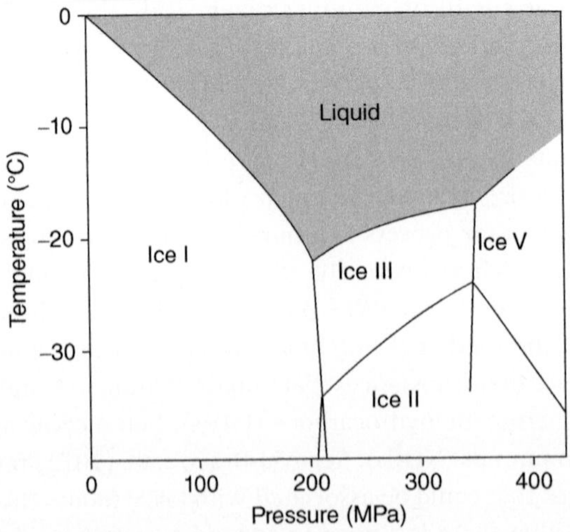

Figure 3.8 Phase diagram of water under pressure (adapted from Kalichevsky, Knorr and Lillford (1995)).

of frozen ground beef could be performed at temperatures as low as −24 °C when applying pressure from 210 to 280 MPa. However, surface discoloration could occur at pressures above 210 MPa (Zhao, Flores and Olson, 1998). Thawing products by the use of high pressure was the purpose of SAFE ICE, a project supported by the European Community. The project is currently developing prototypes, processes, and product concepts for industry (Urrutia *et al.*, 2007).

High-pressure freezing and thawing processes that have been modeled include subzero cooling at high pressure, pressure-shift freezing (PSF), pressure-assisted freezing (PAF), pressure-assisted thawing, and pressure induced thawing (PIT) (Norton and Sun, 2008). Key findings in the area of PSF and PAF for all categories of food products have been summarized by Rastogi *et al.* (2007).

In Table 3.3, the studies on PSF and PAF relating exclusively to meat products are presented.

3.4 High-pressure equipment

3.4.1 Case studies – commercial application of HHP-treated meat products

The first commercial food products treated by high pressure were jellies and jams. They appeared in Japan in the 1990s and were commercialized by Meidi-Ya company. Then, Ulti, a French company, was the first European company developing this technology for freshly squeezed citrus juice. In 1997, a US company Avomex, which became Fresherized Foods, began the first industrial production of HP-pasteurized avocado products, such as Guacamole (Tonello, 2011).

Table 3.3 Key findings in the area of pressure-shift freezing and pressure-assisted freezing applied to meat products.

Products	HP treatment	Main findings	References
Pork and beef muscle	200 MPa, −20 °C With or without water freezing	Considerable and even complete cold denaturation was observed for myosin and actin, respectively. Pressurization alone and pressure-shift freezing induced deleterious effects on color and textural properties.	Fernández-Martín *et al.* (2000b)
Pork muscle	62, 115, 157, and 199 MPa −5 to 20 °C	The ice crystal ratio in pure water and pork muscle was determined using a calorimeter. HP calorimetry seemed an effective method to evaluate the ratio of ice instantaneously formed after depressurization in pressure-shift freezing of pure water and pork muscle.	Zhu, Ramaswamy and Bail (2005)
	100, 150, and 200 MPa −11, −16.5, and −21.5 °C	Cylindrical specimens of fresh pork muscle were frozen by air blaster freezing (ABF), liquid immersion freezing (LIF), and pressure-shift freezing (PSF). The size and shape of ice crystals differed along the radial direction. Ice crystal formation was affected by supercooling during/after depressurization and subsequent freezing.	Zhu *et al.* (2004a)
	100, 150 and 200 MPa	Cylindrical specimens of fresh pork muscle (boneless rib portions) were frozen by pressure-shift freezing (PSF), air blast freezing (ABF), and liquid immersion freezing (LIF). PSF at 150 and 200 MPa resulted in considerable denaturation of myofibrillar proteins of pork muscle and increase in the pork muscle toughness.	Zhu *et al.* (2004b)

Adapted from Rastogi *et al.* (2007).

At present, 60 companies worldwide are using high-pressure technology with marketing of more than 250 HP-treated products (Tonello, 2011). This technology is generally adopted due to the minimal changes it induces in the "fresh" characteristics of foods in comparison with thermal treatments. Indeed, it maintains the nutritional value, vitamins, and the majority of small substances responsible for product flavors (Aymerich, Picouet and Monfort, 2008). Moreover, it ensures food safety and extends refrigerated product shelf life. Fruit and vegetable products and meat products represented more than two-thirds of the HHP food global production in 2009 (Figure 3.9). Meat products indeed represent 27% of the installed HHP machines (Tonello, 2012b).

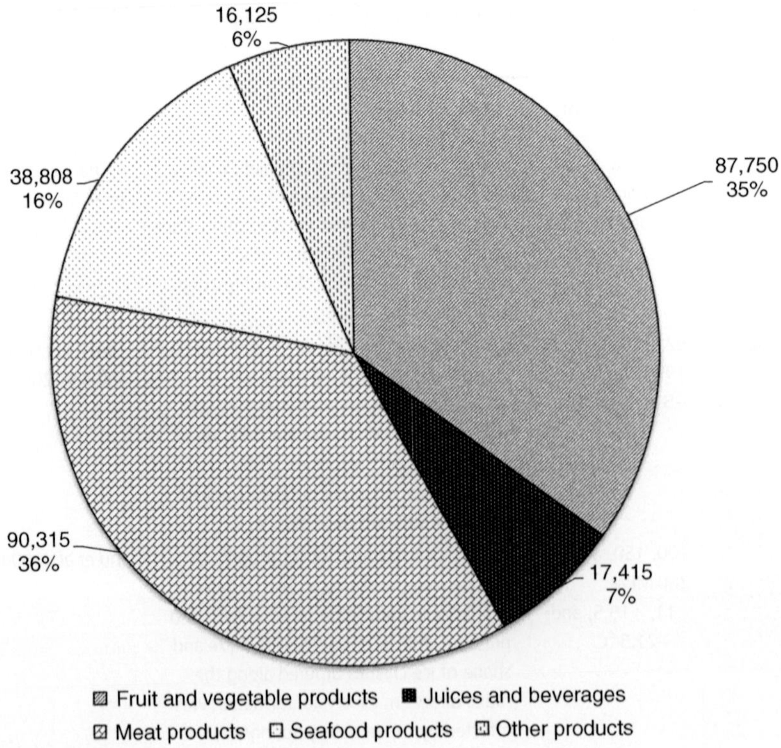

Figure 3.9 Estimation of HPP food global production in 2009 (tons/year) (adapted from Tonello (2012a)).

HHP-treated meat products commercially available belong mainly to the RTE category. They include the following:

- Minced beef (carpaccio)
- Sliced and diced cooked products including sliced cooked ham
- Marinated meats
- RTE meals (turkey and delicatessen, sausage tapas, poultry strips)
- Dry-cured meats including dry-cured ham, Parma ham, mortadella, bacon, salami, and other smoked or nonsmoked sausages.

Commercially available HHP-treated meat products come from countries all around the world (Japan, the United States, Italy, Spain, Germany, and Australia). According to Tonello (2011), pressurized meat products represent the fastest growing sector in the last 5 years. Indeed as illustrated by Table 3.4, many patents have been lodged in the last few years.

In the meat sector, two Spanish meat companies using HHP equipment emerge (Esteban Espuña, SA and Campofrio Alimentacion, SA). In the United States, several meat companies have made this methodology available (e.g., Hormel Foods and Purdue Farms). Companies using high pressure and the associated

Table 3.4 Main patents deposited in which high pressure is applied to meat products.

Year	Inventor	Title	Applicant	Patent number	Application
1994	N. Masatoshi	Production of high-pressure processed ham	Mutsutsutaa Ham KK	JP6319490	Processed raw pork is salted, smoked, and subjected to a high-pressure treatment under 200–250 MPa for a prescribed period to obtain the high-pressure processed ham. The ham produced by this process exhibits appetizing red meat color and has soft texture, moderately seasoned taste with controlled salt content, delicious taste, refreshing appearance, and unique palatability.
1998 1999	F. Kortschack	Method for producing stable meat product and sausages, produced in the presence of microorganisms	Kortschack Fritz	WO/1998/026667 EP 97954402	A method for fixing pH values in raw sausages, sausages suitable for storing, raw ham, and similar products, making the durable stabilization of these values possible. For this, the sausages and/or meat products are enclosed in impermeable packing and exposed to high pressure (approx. 400–600 MPa) for several minutes.
1999	R. S. Meyer	Ultra-high-pressure and high-temperature food preservation process	Meyer Richard S.	WO/1999/029187	Methods for the commercial sterilization of foods that involve preheating the food to a temperature of about 43–71 °C, and thereafter pressurizing the food to 138–827 MPa for a short period of time to achieve sterilization in a few minutes or less without exceeding the taste transition temperature of the food, thus retaining a fresh flavor. These methods are especially advantageous in preserving foods having a $pH \leq 4.5$.

(continued overleaf)

Table 3.4 (*Continued*)

Year	Inventor	Title	Applicant	Patent number	Application
2000				WO/2000/015053	Commercial sterilization of foods particularly advantageous for foods having a pH $\geq$ 4.5 involving concomitant treatment of the food to two or more cycles of high heat, high pressure, with a brief pause between cycles.
2002	H. Tanaka	Low-temperature and high-pressure processing for animal meat and fish meat mixed and kneaded product	Hokurei Co., Ltd. (Hokkaido, JP)	US6440484	Novel foods having texture such as raw ham or vinegar-pickled mackerel and containing less bacteria can be obtained from fish meat and/or a kneaded fish meat product as a raw material.
2000	I. Fernandets	Method for producing sausages with low content of salt and/or phosphate, and/or fat, and/or fat, and/or high water content	Sos'ete De Prodjui Nestle SA (CH)	RU2147404	The method involves preparing meat farce; according to receipt, adding ice, nitrite salt, phosphate and/or fat; cutting meat farce; injecting cut farce into casing; treating farce at hydrostatic pressure of 50–400 MPa for 60 min, preferably for 10 min, at temperature of −15 °C to +100 °C; cooking. Such method provides dense structure of sausage, increased cooking yield, and improved quality of product.
2000	E. A. Voisin	A process of elimination of bacteria in shellfish and of shucking shellfish	Voisin, Ernest A.	WO/2000/065923	Process for eliminating pathogenic organisms and other bacteria from raw food products, such as raw shellfish, whereby the molluskan shellfish is exposed to hydrostatic pressure between 138 and 1034 MPa for 1–15 min in the range of between ambient temperature and 65 °C, leaving the raw shellfish in its desired raw state, such that the pathogenic organisms are destroyed, while sensory characteristics of the raw shellfish remain high. The same process can be used for shucking bivalve mollusks.

2002	S. N. Hong	Production of seasoned meat	Korean Food Development Institute	KR1020000034835	Minced fresh meat is mixed with conventional spices containing additionally 1–5% by weight of a grapefruit seed extract and subjected to vacuum packing and high-pressure treatment at 203–709 MPa for 10–30 min. The process improves the quality of the product by increasing tenderness and palatability thereof.
2003	A. J. M. Garwood	Pressurized decontamination of meat using carbonic acid	SafeFresh Technologies LLC	WO/2003/101210	The high pressure favors the formation of carbonic acid, which reduces the pH to a low level sufficient to kill microorganisms residing on the meat.
2003	N.L. Kudrjashov	Smoked product preparing method	Kemerovskij tekhnologicheskij institut pishchevoj promyshlennosti	RU2207024	Processing raw meat with salt brine comprising smoking preparation "Liquid smoke plus" to provide for combined salting and smoking processes; carrying out the said process in cyclic processing conditions initially under high pressure of 400–600 MPa for 3 min with following processing at atmospheric pressure for 5 min, with cycle being repeated three times until complete readiness of product. EFFECT: Intensified preparing of smoked product and provision for obtaining ecologically clean meat product.

(continued overleaf)

Table 3.4 (*Continued*)

Year	Inventor	Title	Applicant	Patent number	Application
2003	G. H. Nam	Instant cooking braised chicken and production thereof	Youjin Food Bank Co. Ltd.	KR1020030073019	A dressed chicken is cut to a suitable size, soaked in sauce, heated, and then allowed to stand to remove fat. The boiled chicken is packed with sauce and then sterilized at high temperature and high pressure. For example, a chicken is cut to about 4–10 cm, aged with sauce-containing red pepper powder, say sauce, sugar, black pepper, garlic, ginger, etc., for 12–72 h and then heated for 5–30 min at 300–800 °C.
2003	P. Van Ysseldijk	Seafood product and process	Navital (UK) Ltd.	WO/2003/013279	A process for preparing ready-to-cook seafood is described. The seafood is sealed in a container under partial, substantial, or full vacuum, in which it undergoes ultra-high hydrostatic pressure. Thus, the seafood remains "raw", i.e., uncooked, but is at least partially sterilized from infectious agents. The seafood also retains its spores and enzymes, hence its taste and quality, with a significantly extended shelf life, from several days to possibly several weeks.
2003	M. J. Wilson	High-temperature/ultra-high pressure sterilization of low-acid foods	Kal Kan Foods	EP1295537	Method for sterilizing low-acid foods using both ultra-high pressures and high temperatures. The process involves heating a low-acid food to a pre-pressurized temperature, subjecting the food to ultra-high pressure, which instantaneously raises the temperature of the food, and then releasing the pressure so that the temperature returns to the original pre-pressurized temperature.

2003	T. Yamaguchi	Bacon ham with skin	Minamo Kyushu Chikusan Kogyo KK	JP2003189822	The production process of the bacon ham with skin comprises a combination of a pre-treating process, salting out process, smoking process, softening process, drying process, finishing process, shipping process, etc. The softening process comprises a boiling, temporarily packing, and high-temperature and high-pressure treating processes.
2003	J. T. C. Yuan	Method of using low-temperature and high-/low-pressure processing to preserve food products	L'air liquide – Société anonyme à directoire et conseil de surveillance pour l'étude et l'exploitation des procédés Georges Claude	WO/2003/094637	The method includes exposing the food product to a gas and/or injecting a gas into a container containing food product, and then subjecting the food product to a temperature of less than about 50 °C and, concurrently, to more than one pressure treatment cycle above about 69 MPa. The food product is generally contacted with the gas under pressure conditions for a time sufficient to substantially sanitize or disinfect the food or food product following depressurization.
2004	H. L. M. Lelieveld	High-pressure preservation of food	Unilever	WO/2004/057970 US2004126480	Use of high-pressure preservation avoids or minimizes the need for refrigeration of the food, thus reducing costs, sparing resources, and lowering space requirements. Elimination or decreasing refrigeration requirements decreases preparation and serving time. High-pressure preservation refers to subjecting the food to a pressure of at least 10 MPa.

(continued overleaf)

Table 3.4 (*Continued*)

Year	Inventor	Title	Applicant	Patent number	Application
2005	A. J. M. Garwood	Decontamination methods for meat using carbonic acid at high pressures	Safe Fresh Technologies, LLC	US2005/0260311	A method for separating lean meat from lean meat-containing material includes combining a particulate with fluid, subcritical carbon dioxide at a pH of about 7 or less and a pressure of about 4.1 MPa.
2006	T. J. Carroll	High-pressure processing of bioactive compositions	Fonterra Cooperative Group Ltd.	WO/2006/096074	The present invention relates to a method of pressure treating a bioactive composition comprising at least one bioactive component (protein hydrolysates, lipid hydrolysates, carbohydrates, and probiotic factor) to prevent the growth of at least one unwanted microorganism while retaining a desired level activity of at least one bioactive component. The pressure treatment is at a predetermined pressure from about 350–1000 MPa.
2006 2007	F. Kortschack	Method for treating foodstuffs, particularly for preserving fresh sausage or meat products	K. Fritz	WO/2006/097248 EP1858354	The invention relates to a method for treating foodstuffs, particularly for preserving fresh sausage or meat products involving the time-limited influence of temperature and/or pressure or high pressure and for subsequently bringing the products to market.
2006	S. McKenna	Meat tenderizing apparatus and a method of operation thereof	Mac Food Ltd.	WO/2006/043085	Meat tenderizing apparatus comprising a pressurizing means and an irradiating means whereby, in use, the pressurizing means is adapted to apply pressure to a meat product and the irradiating means is adapted to apply radiation to the meat product, wherein the resulting meat product remains uncooked. Additionally, there is a described method of using such apparatus and a tenderized meat product resulting therefrom to produce uncooked and tenderized meat products.

2006	E. J. Turek	Method for preserving food products	Kraft Food Holding Inc.	ZA2005/05930	An improved method is provided for inactivating microorganisms in food products by introducing an edible phenolic compound in a food product and subjecting the resulting food product to high-pressure conditions. This particular combination of physical and chemical treatments on the food products inactivates undesirable microorganisms in a food product to a significantly greater degree possible than that achieved with any one of the treatments by itself, and the combined antimicrobial effect of these different types of treatments is synergistically greater than any additive effect that might be expected from the individual treatments.
2007	A. Kazuhico	High-protein-containing food and method for producing the same	Miyagi Prefecture	JP2007244266	To provide high-protein-containing food excellent in chewability/swallowing property without leaving raw material meat fibers in the mouth, by heating high-protein food material or subjecting the material to ultra-high-pressure treatment in advance so as to make the protein degenerate.
2007	A. Lebail	Method of manufacturing a food preparation that is heterogeneous and complex, fresh, and packaged	Marie	EP1844663	The manufacturing of a heterogeneous, complex, fresh, and conditioned food preparation comprises assembling and dressing of food preparation on a support making inner packaging, packing of the food preparation by vacuum draping using a flexible thermoplastic film, applying a treatment cycle with high pressure of 400–700 MPa for 3 min at 30–35 °C, instantaneously relaxing, and cooling. The fresh cooked dishes are fish and meat dishes with sauce and mixed salads. The support making inner packing is rigid or semirigid packing, namely trays, plates, and dish.

(continued overleaf)

Table 3.4 (*Continued*)

Year	Inventor	Title	Applicant	Patent number	Application
2008	R. S. Meyer	High-pressure frozen sterilization process	JCR Technologies LLC	WO/2009/003040	Methods of sterilizing a temperature-sensitive material, such as an acidified or nonacidified food product, pharmaceutical product, or cosmetic product, are disclosed. The methods comprise freezing the temperature-sensitive material to an initial temperature $\leq -2\,°C$ and then pressurizing to a first elevated pressure above 250 MPa for 90 s or 3 min, releasing pressure for a predetermined time, then pressurizing to a second elevated pressure above 250 MPa for a predetermined second period of at least 90 s.
2009	V. Heinz 2010	Method for the automatable production of raw sausage	Deutsches Inst. Lebensmittel Tec.	WO/2009/074446 EP 2219463	The invention relates to a process for the automatable production of raw sausage by mincing and mixing parts or pieces of meat and adding starter cultures, spices, or similar materials. The material or sausage meat remains exposed to a pressure treatment within a water-permeable film until the isoelectric point is reached in the material. In addition, the sausage meat in the film is subjected to a tempering treatment during the maturing phase, in order to control the activity of the microorganisms responsible for the maturing process. The pre-shaped sausage meat preforms are exposed to a high-pressure treatment in order to inactivate the microorganisms and then removed from the enveloping film, subjected to a drying step and then processed further.

2009	McKenna	High-pressure meat product processing	Cargill, Inc.	US20090232959	Processes that utilize one or more pressurization and depressurization treatments to tenderize fresh cuts of meat are described
2010 2011	F. Kortschack	Method for controlling product characteristics in the manufacture of meat products	Triton Gmbh	WO/2010/026171 EP2320755 US223285	The invention relates to a method for controlling product characteristics in the manufacture of foods belonging to the product groups of raw sausage, cooked sausage, pates, cooked meats, or the like, in particular for maintaining spreadability or sliceability. A high-pressure treatment is carried out for the purposes of protein coagulation, the said treatment leading to the deactivation of microorganisms at the same time, wherein the mechanically separated meat raw materials remain separated.
2010 2011	N. R. Smit	Use of high-pressure processing to improve muscle quality by inhibiting postmortem glycolysis	Hormel Food Corp.	WO/2010/022305 EP2317863	The method includes subjecting the meat (pork or turkey) to *high-pressure processing at a pressure of at least* 175 MPa *for a* time of 20 min or less, wherein postmortem glycolysis is inhibited. For beef, a pressure of at least 200 MPa and a hold time from 30 s to 20 minutes is maintained.
2011	C. J. Jo	Producing method of high-quality noncrushed meat product using the super-high-pressure technique, and the noncrushed meat product	The industry and academic cooperation in Chungnam National University (IAC)	KR1020110099440	A producing method of high-quality noncrushed meat product (pork, beef, lamb, chicken breast, goad, duck, turkey, horse, and dog meat) is provided to super-high-pressure process the meat product after vacuum packing after mixing noncrushed meat with vegetable or animal fat, or seasoning sauce.

(*continued overleaf*)

Table 3.4 (*Continued*)

Year	Inventor	Title	Applicant	Patent number	Application
2011	Y. J. Kim	Method for processing meat and processed food product	Niigata Industrial Creation Organization	JP2011083228	The method for processing meat includes subjecting raw meat to a sodium bicarbonate treatment where the raw meat is soaked in a sodium bicarbonate liquid having a concentration of 0.1–0.5 M (mol/l) for 30–60 min, a seasoning liquid treatment where the meat is soaked in a soaking seasoning liquid mainly containing soy sauce and Mirin (Japanese sweet cooking wine), and a high-pressure processing where pressure at 100–400 MPa is applied to the meat.
2011	R. Lanzrath	High-pressure pasteurizing of whole meat products	Cargill Inc.	WO/2011/149900	A process for preserving fresh meat that includes the steps of (a) packaging fresh meat in a sealed package; (b) placing the packaged fresh meat in a pressurization vessel; (c) pressurizing the packaged fresh meat to at least 344 MPa pressure; (d) maintaining the elevated pressure on the packaged fresh meat for 1–300 s and at a temperature of −2 °C to 8 °C; (e) reducing the pressure on to ambient pressure. Pathogens are effectively killed using this process, providing manufacturing efficiencies and longer product shelf life as compared to other meat-handling procedures.
2011	J. W. Summerfield	Use of high-pressure processing to aid in the hair or feather removal from animal carcasses/hides	Hormel Foods Corp.	WO/2011/040935	High-pressure processing is utilized to remove the external carcass portions from an animal carcass. The animal carcass is subjected to pressures of 50 MPa or greater to loosen the external carcass portions. In addition, high-pressure processing is utilized to strengthen the hides of animal carcasses by subjecting them to pressures of 10–50 MPa.

2011	A. Yamazaki	Method for retorting edible organ meat and method for producing edible organ meat-containing retort food	Echigo Seika Co. Ltd.	JP2011036141	The method for retorting the edible organ meat includes adding a proper amount of salt to raw edible organ meat, rubbing the organ meat with salt, washing salt away with water, subjecting the washed edible organ meat to high-pressure treatment at 100–800 MPa, boiling the edible organ meat subjected to high-pressure treatment and cooling, and sealing the edible organ meat in a heat-resistant retort pouch followed by retorting.
2011 2012	J. D. Parker	High-pressure pasteurizing of frozen ground meats	Cargill Inc.	WO/2011/038237 EP2490546	A process for preserving frozen ground meat comprises the steps of (a) packaging frozen ground meat in a sealed package; (b) placing the packaged frozen ground meat in a pressurization vessel; (c) pressurizing the packaged frozen ground meat to at least about 344 MPa pressure; (d) maintaining the elevated pressure for 1 s to about 300 s and at a temperature from −45 °C to about 7 °C; (e) reducing the pressure to ambient pressure. Pathogens are effectively killed using this process, providing manufacturing efficiencies and longer product shelf life as compared to other ground meat-handling procedures.
2011 2013	V. Heinz	Method for controlling product characteristics in the production of foodstuffs	Triton GMBH	WO/2011/107335 EP2542102	The invention relates to a method for controlling product characteristics in the production of foodstuffs belonging to the product groups of raw and cooked sausage, pates, cooked meat products, or the like, in particular in order to preserve spreadability or sliceability.

applications are summarized in Table 3.5. Some equipment suppliers are as follows: Flow International, ACB Pressure systems in France, Hiperbaric (Nicolas Correa group) in Spain, and Avure technologies AB, a Swedish-based company with branches in Germany, the United Kingdom, France, Italy, and Spain.

A treatment of 600 MPa during 2–10 min is currently used to process meat products (Aymerich, Picouet and Monfort, 2008). Despite a high initial investment cost, the processing cost approximated to 0.14 €/kg of product treated at 600 MPa, including investment and operation costs (Anonymous, 2002), remains moderate. Moreover, a three-country (France, Germany, and the United Kingdom) survey on the consumer perception of high-pressure processing reported 67% acceptability (Baron *et al.*, 1999). Concerning USDA regulation, high pressure has been included in preservation techniques acceptable to control *Listeria* in RTE refrigerated meat products (Tonello, 2011). The use of high pressure for processing of various food products has received approval from the FDA and Health Canada on the basis of "internationally accepted principles for establishing the safety of novel Foods" (Duranton, 2012). No specific labeling is required, as no chemical substances are added to the pressurized product (Yordanov and Angelova, 2010). Hence, new applications of high pressure highlight less additives with a "clean label" (Tonello, 2011). Experimentation for the industrial application of high pressure in the meat sector aims to (1) determine the best treatment conditions in every product and (2) optimize commercial presentation by combining, for example, new packaging systems (Hugas, Garriga and Monfort, 2002).

3.4.2 High-pressure carbon dioxide decontamination (HPCD)

High-pressure carbon dioxide decontamination has some fundamental advantages over other pathogen inactivation methods for the food industry, which is nontoxic, nonflammable, inexpensive, and physiologically safe with low-level residues in foods (Bae *et al.*, 2011). The use of high-pressure carbon dioxide (HPCD) for food decontamination has been reviewed by Garcia-Gonzalez *et al.* (2007). In this technique, food is in contact with pressurized sub- or supercritical CO_2 for a certain time in batch or semicontinuous conditions. Supercritical CO_2 refers to the CO_2 state in the phase diagram as temperature and pressure are above its critical point (*c*) values ($T_c = 31.1\,°C$, $P_c = 7.38\,MPa$) (Figure 3.10). At the supercritical phase, CO_2 possesses properties of both liquid and gaseous CO_2 (Figure 3.10). On the other hand, subcritical CO_2 is CO_2 under its critical point; it can be in either liquid or gaseous phase depending on the pressure level. The main advantage of HPCD over HP alone is the significantly lower pressure needed to reach microbial decontamination. Indeed pressures as low as 20 MPa can be applied in preservation purposes (Garcia-Gonzalez *et al.*, 2007). The hypothesized mode of action of pressurized CO_2 in bacterial inactivation was

Table 3.5 Some available commercial meat products.

Country	Company	Product	Pressure treatment conditions (MPa)	Shelf life (days)	Source
Romania	Cris-Tim	Fermented sausages (salami)			Cris-Tim (2013)
Netherlands	Zwanenberg	Filet American Steak Tartar Carpaccio			Hiperbaric (2013) Zwanenberg (2010a) Zwanenberg (2010b)
USA	Cargill	Hamburger (Fressure®)	345	42	Cargill (2010)
Greece	Creta Farm	Deli meats infused with extra virgin olive oil			Hiperbaric (2012)
USA	Perdue Farms	Poultry strips	600 without preservatives		Ikeuchi (2011)
Greece	Ifantis	Sliced processed meat (mortadella, ham, and salami)			Ifantis (2009)
USA	Tyson Food	Oven roasted chicken	600	45	Ikeuchi (2011)
UK	Deli 24	Meat and cheese snack products			Watson (2011)
USA	Hormel	Natural Choice Product range® (ham, poultry strips)	600 without preservatives	Until 110	Hormel (2012)
USA	Columbus Foods	Deli meat and fermented meats			Foodprocessing-Technology (2011)
Spain	Campofrio	Thick sliced ham	500 (lower salt level)	56	Ikeuchi (2011)
		Marinated poultry	600	>30	
Spain	Espuña	Tapas	400 to 600	67	Tapas Minute (2012)
Spain	Espuña	Sliced ham	400	60	Espuna (2012)
Spain	Espuña	Sliced ham	600	270	Espuna (2012)
Germany	Abraham	Sliced prosciutto	600	240	Abraham (2005)
Italy	Ferrarini	Sliced prosciutto	600		Abraham (2005)
Canada	Santa Maria Foods	Sliced dry-cured meat and cooked meats			Foods (2013)
Canada	DuBreton	Organic sliced ham	Without nitrites	90	DuBreton (2013)

Adapted from Bajovic, Bolumar and Heinz (2012), and Duranton (2012).

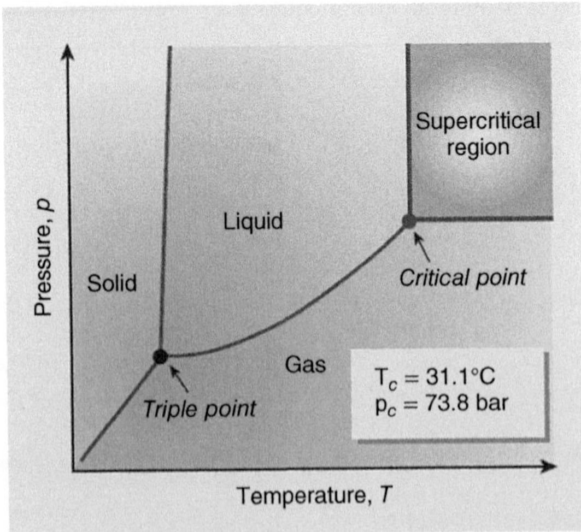

Figure 3.10 Carbon dioxide temperature–pressure phase diagram (adapted from Leitner (2000)).

schematized (Garcia-Gonzalez *et al.*, 2007). Although several phenomena would simultaneously occur in a complex manner, it appears that several steps could be distinguished in the decontamination process induced by HPCD.

3.4.2.1 Mode of action of HPCD

The mechanisms involved in decontamination by HPCD have been reviewed and illustrated by Garcia-Gonzalez (Figure 3.11).

1. Solubilization of pressurized CO_2 in the external liquid phase of foods
   This first step leads to transformation of CO_2 into bicarbonic acid and its associated bases HCO_3^- and CO_3^{2-} and to concomitant acidification. This acidification could be partly responsible for the lethal effect of pressurized CO_2. Cell membrane permeability to CO_2 would be enhanced by lowered extracellular pH, which would have a preparing effect for step 2.
2. Cell membrane modification
   Accumulation of CO_2 in phospholipids would functionally and structurally disorder cell membrane and enhance its fluidity.
3. Intracellular pH (pHi) decrease
   Following easy penetration of CO_2 into bacterial cells through CO_2-induced membrane fluidization, equilibrium between CO_2 and its dissociated carbonated form occurs as a consequence of pH homeostasis to maintain pHi constant. As the influx of CO_2 becomes too high, H^+-ATPase pomp efflux fails in expelling protons from the cytoplasm in sufficient quantity to maintain pHi, and the latter begins to decrease with a serious impact on cell viability.

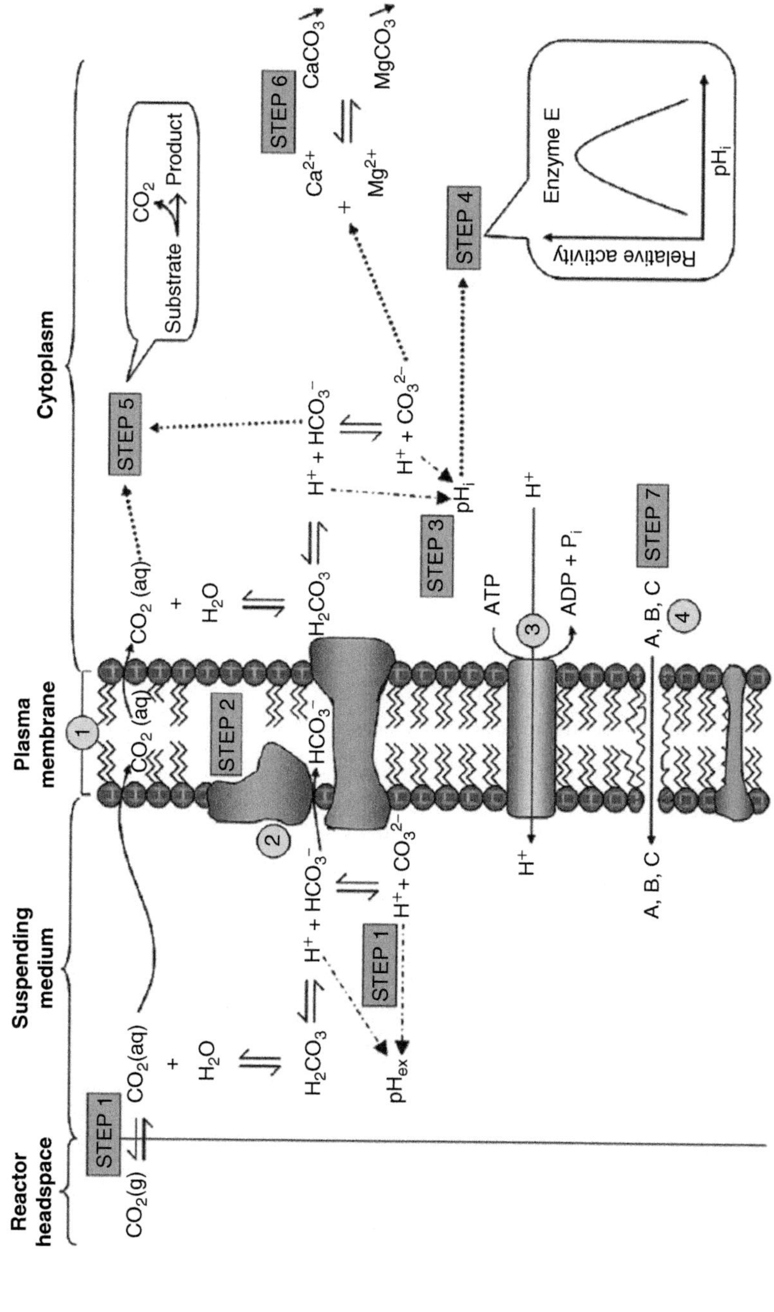

Figure 3.11 Hypothesized mode of action of carbon dioxide in microbial decontamination by HPCD (adapted from Garcia-Gonzalez *et al.* (2007)).

4 Key enzyme inactivation/cellular metabolism inhibition due to pHi lowering
As pHi is lowered, activities of cytosolic enzymes, under pH away from the optimum, will be reduced and even stopped with a great impact on metabolic and regulation processes.
5 Direct (inhibitory) effect of molecular CO_2 and HCO_3^- on metabolism
As CO_2 and HCO_3^- were shown to influence carboxylation and decarboxylation reactions, the HPCD will affect microbial metabolism, especially in those cells for which these reactions play an important role.
6 Disordering of the intracellular electrolyte balance
Intracellular inorganic electrolytes such as Ca^{2+} and Mg^{2+} may precipitate in the presence of high quantities of CO_3^{2-}. Since these ions are involved in osmotic processes and antiports driven by the proton motive force, this could impact cellular volume and pH homeostasis.
7 Removal of vital constituents from cells and cell membranes
Due to its high solvating power, CO_2 could extract cellular constituents such as phospholipids and hydrophobic compounds from the cell or cell membranes, especially as the applied pressure is suddenly released. This would lead to deformation of cell walls and damage of cell membranes.

3.4.2.2 Factors influencing microbial inactivation by HPCD

- Process parameters (pressure, temperature, CO_2 state, depressurization rate, agitation)

  Above all other parameters, pressure and temperature have been shown to be the most influential on microbial inactivation. CO_2 pressure and temperature are indeed directly linked to the characteristics of CO_2 mass transfer (Garcia-Gonzalez *et al.*, 2007). As long as the saturation solubility of CO_2 is not reached in the suspending medium, the higher CO_2 pressure the higher microbial inactivation is (Hong, Park and Pyun, 1997; Hong, Park and Pyun, 1999; Lin, Cao and Chen, 1994; Lin, Yang and Fu, 1993). Increase in temperature also favors higher microbial inactivation by enhancing diffusivity of CO_2 (Hong, Park and Pyun, 1997; Hong, Park and Pyun, 1999; Lin, Yang and Fu, 1993). However, this technique cannot operate at a temperature far above the CO_2 critical point and would lose its beneficial effect on food products if the temperature was too high. Supercritical CO_2 was shown to be more effective in inactivating microbial cells than CO_2 under subcritical conditions (Gunes, Blum and Hotchkiss, 2006). This could be attributed to the high solvating power of the liquid-like density of this physical state in between gas and liquid (Garcia-Gonzalez *et al.*, 2007).

  Among the other process parameters influencing decontamination, agitation was shown to improve microbial inactivation efficiency by facilitating CO_2 dissolution (Hong, Park and Pyun, 1997; Lin, Cao and Chen, 1994; Lin, Yang and Fu, 1993; Lin, Yang and Chen, 1992a; Dillow *et al.*, 1999). Concerning the depressurization rate, controversial results have been reported, questioning whether mechanical rupture due to explosive decompression is the predominating inactivation mechanism during an HPCD treatment. Indeed, while some investigators have shown that high depressurization rate enhanced disruption of microbial cells (Lin, Yang and Chen, 1992a; Lin, Yang and Chen, 1992b; Lin *et al.*, 1991; Nakamura *et al.*, 1994), others argued that this parameter had no significant effect on microbial inactivation (Hong, Park and Pyun, 1997; Debs-Louka *et al.*, 1999; Kumagai, Hata and Nakamura, 1997; Arreola *et al.*, 2007).

- Microorganisms (susceptibility, stage of growth, initial concentration)
  Although it is generally accepted that Gram-positive bacteria are more resistant to HPCD than Gram-negative bacteria (Garcia-Gonzalez *et al.*, 2007), no difference in sensitivity could be observed in the recent screening study of Garcia-Gonzalez *et al.* (2009) performed on foodborne and spoilage bacteria. Microbial susceptibility toward HPCD treatments in the pressure and temperature conditions tested followed the sequence: Gram negative ≈ Gram positive > yeasts > fungal and bacterial spores. It appeared to be related to the acid resistance of the organisms as *A. acidoterrestris*, *Enterobacter faecium*, and yeast, in particular *Z. bailii*, known as being highly tolerant to pH, exhibited a very low HPCD susceptibility (Garcia-Gonzalez *et al.*, 2009). Hong and Pyun (1999) confirmed the belief of higher resistance of microorganisms in the stationary phase of growth than in the exponential phase for HPCD treatments applied to *L. plantarum*.
  Unlike many physical inactivation processes, efficiency of decontamination by HPCD was shown to be strongly affected by the initial bacterial load. The higher the initial population, the longer the exposure time to HPCD treatments has to be (Erkmen, 2000a,b,c).
- Suspending medium
  Some characteristics of the medium have the same impact on the HPCD efficiency as for the HP treatment itself. Indeed, the composition of the suspending medium has been shown to influence the microbial inactivation by HPCD with an enhanced efficiency in laboratory media compared to food matrixes. Low a_w was also shown to decrease the treatment efficiency (Garcia-Gonzalez *et al.*, 2007). However, at similar a_w but below 0.95, the kind of solute used to adjust a_w has been shown to be influential. Fat and different constituents in solutions would have protective effects (Lin, Cao and Chen, 1994). Some other characteristics of the medium related to the mode of action of supercritical or subcritical CO_2, which specifically influence microbial inactivation by HPCD. For example, the buffering capacity of the suspending medium has been mentioned to prevent the extracellular pH from decreasing following HPCD (Erkmen, 2000a,b,c; Erkmen and Karaman, 2001; Erkmen, 2001b,c). On the other hand, low pH, by contributing to an increased permeability of cells to CO_2, has been reported to enhance inactivation by HPCD (Lin, Cao and Chen, 1994; Lin, Yang and Fu, 1993; Garcia-Gonzalez *et al.*, 2009; Hong and Pyun,

1999). Concentrations of salt above 2% have been reported to have a stimulating effect on decontamination by HPCD, whereas addition of sunflower oil (from 10 to 30%) has been shown to induce the reverse effect on inactivation of *Pseudomonas fluorescens* (Garcia-Gonzalez *et al.*, 2009). A protective effect of high viscosity was also shown by Garcia-Gonzalez *et al.* (2009) regardless of solute used, to artificially increase medium viscosity. a_w lowered by sucrose and even more by glycerol resulted in reduced inactivation of *P. fluorescens* and *S. aureus*, whereas a_w lowered by NaCl addition induced an accelerated decontamination (Garcia-Gonzalez *et al.*, 2009).

3.4.2.3 Studies performed on meat products

The use of supercritical carbon dioxide as an alternative to thermal treatments was shown to be efficient in inactivating microorganisms and enzymes not only in culture media (Hong and Pyun, 1999; Erkmen, 2000a) or liquid foods (Ramirez-Rodrigues *et al.*, 2013; Spilimbergo and Ciola, 2010) but also in fresh-cut fruits (Valverde, Marín-Iniesta and Calvo, 2010; Ferrentino *et al.*, 2012a; Ferrentino, Balzan and Spilimbergo, 2012b; Ferrentino, Balzan and Spilimbergo, 2013a) and vegetables (Zhong *et al.*, 2008; Bi *et al.*, 2011). However, few studies have been performed on meat products (Bae *et al.*, 2011; Ferrentino, Balzan and Spilimbergo, 2013a; Wei *et al.*, 1991; Sirisee, Hsieh and Huff, 1998; Erkmen, 2000d; Ferrentino, Balzan and Spilimbergo, 2013b; Choi *et al.*, 2009). Chicken and shrimp meat inoculated, respectively, with *Salmonella* and *L. monocytogenes* could be efficiently decontaminated (from 95 to 99%) by 13.7 MPa and 35 °C for 2 h (Wei *et al.*, 1991). Sirisee, Hsieh and Huff (1998) examined the tolerance of *E. coli* and *S. aureus* to supercritical CO_2 treatment in ground beef systems. Inactivation of both microorganisms was shown to be clearly less efficient in ground beef than in phosphate buffer. While applying the same treatment (31.03 MPa at 42.5 °C), it took almost 10 times longer to inactivate *S. aureus* in ground beef than in phosphate buffer. Moreover, ground beef processed with a supercritical treatment looked like cooked color meat. The inactivation effects of 6.05 MPa CO_2 pressure at 45 °C were studied by Erkmen (2000d) on a spoilage bacterium, *Brochothrix thermosphacta,* in skinned and minced meat as a function of the treatment duration. It appeared that *B. thermosphacta* in skinned meat was below the detection threshold after 150 min of treatment, whereas the population of *B. thermosphacta* was still high ($7.3.10^4$ CFU/g) in minced meat. Recent studies of Ferrentino, Balzan and Spilimbergo (2013a, 2013b) examined the feasibility of HPCD to pasteurize ham-type meat products. The first study aimed at evaluating the potential use of this technique to inactivate *L. monocytogenes* inoculated on the surface of dry-cured ham (Ferrentino, Balzan and Spilimbergo, 2013a). Total inactivation of a population of 10^7 CFU/g of *L. monocytogenes* could be achieved by a treatment of 12 MPa performed at 50 °C during 15 min. In the second study, the objective was to optimize the HPCD treatment that enables the best inactivation efficiency of natural microbial flora (mesophilic aerobic, psychrophilic and lactic acid bacteria, yeast and molds, and coliforms) while preserving color, pH, and texture characteristics of cooked ham cubes (Ferrentino, Balzan and Spilimbergo, 2013b). The response of a reduction of mesophilic aerobic bacteria, lactic acid bacteria, and psychrophilic bacteria could be modeled in a linear polynomial model with regression coefficients following the equation:

$$Y = \beta_0 + \beta_1 A + \beta_2 B + \beta_3 C \tag{3.10}$$

where Y is the predicted response of microbial reduction; β_0 is the intercept; β_1, β_2, and β_3 are the linear coefficients; and A, B, and C are the pressure, temperature, and treatment time variables.

Response surface analyses indicated that pressure and temperature significantly influenced microbial inactivation while treatment time had a slight effect. Significant interaction between pressure and temperature could only be observed for mesophilic aerobic bacteria. The optimal process conditions in terms of microbial counts were found to be 12 MPa, 50 °C, and 5 min at which 3.0, 2.5, and 1.6 log (CFU/g) reductions were achieved for mesophilic aerobic bacteria, lactic acid bacteria, and psychrophilic bacteria, respectively.

Supercritical carbon dioxide high-pressure treatments from 10 to 14 MPa at 40 and 45 °C were assayed for their ability to inactivate mesophilic aerobic flora in ground pork inoculated with swine feces (Bae *et al.*, 2011). Inactivation of mesophilic aerobic flora was shown to significantly occur from 14 MPa at 45 °C while in ground pork inoculated with *L. monocytogenes* and *S.* Typhimurium, the treatment temperature was shown to have more effect than the pressure level itself. The authors suggested that this technology could be helpful for improving the microbiological safety of comminuted meat products.

In the study of Choi *et al.* (2009), the combined use of HPCD (12 MPa and 35 °C for 30 min) and addition of organic acids (lactic and acetic acids) was investigated for the reduction of nonpathogenic *E. coli* and three pathogenic bacteria (*L. monocytogenes, S.* Typhimurium, and *E. coli* O157:H7) in fresh pork. For all bacteria, almost one log reduction more could be achieved by the combined treatment of 3% organic acid and HPCD in comparison with the HPCD treatment alone. However, except for nonpathogenic *E. coli*, lactic acid addition induced the same additional microbial reduction when combined with HPCD as acetic acid addition.

Among the numerous patents dealing with treatment of meat products by high pressure, only two patents related to meat decontamination by HPCD were deposited by Garwood (Garwood, 2003; Garwood, 2005). Their purpose is decontamination of Lean Fine Textured Beef (LFTB), harvested from discarded fat from carcasses. Meat is immersed in high-pH alkaline solution than in low-acid solution, such as carbon dioxide and water, respectively, at HHP. Pathogenic bacteria such as *L. monocytogenes, Salmonella,* and *E. coli* O157:H7 have to be killed by this technique.

3.4.3 Brief overview of available high-pressure equipment

The first systems allowing one to reach isostatic pressures of more than 100 MPa were constructed at the end of the nineteenth century for experimental purposes. They comprised the two essential elements found in the systems used today: a vessel permitting maintenance of the pressure in a finite volume and a pump permitting generation of this high pressure. At this time, the pressure was generated by means of a large flywheel turned by hand. The greatest difficulty at this time, and up until World War II, was to obtain materials having a high resistance, as the mechanical stresses are very high in steel undergoing cycles of compression–decompression. The mechanical characteristics of the most resistant materials were not as good as today; thus, it happened that the vessels developed cracks or parts of the pumps broke. An example of the high-pressure treatment of foodstuffs was given by Hite (1899). One of these vessels burst under a pressure of 900 MPa while filled with milk contaminated by *Salmonella* Typhi. A member of the research team was contaminated by splashes of pressurized milk and developed violent typhoid fever, which put an end to the experiments on pathogenic bacteria conducted by these researchers. The evolution of the process was limited by technical problems. Hence, it was not conceivable at this stage to construct large-scale equipment to process food on an industrial scale. It was in the 1960s that industrial installations began to appear for the molding of metallic, plastic, or ceramic powders and carbon/graphite alloys. The pressure ranges from 50 to 600 MPa in these applications outside the food industry, but lies, in most cases, between 100 and 300 MPa. The products or materials are pressurized in oil, solvents, or gases for periods of a few seconds to several weeks at temperatures ranging from ambient to 2200 °C (Federighi *et al.*, 2001).

Today, the systems employed in the food-processing industry are derived from these installations but operate in general, at higher pressures, between 350 and 800 MPa and, at low temperatures, between −20 and +90 °C. Water is the pressurization fluid communally used for practical, sanitary, and safety reasons. Oxidizable materials are therefore excluded for the construction of the parts, which are in direct contact with the fluid. The ease of cleaning of the equipment is also a much more important parameter in the food industry than in chemistry or metallurgy. The duration of the cycles is of the order of 5 min to some tens of minutes. Moreover, as food products have an added market value less than that of materials or chemicals treated by high pressure, the design of the equipment has had to be improved to permit very short phases of opening, closing, loading, unloading, and pressure increase. Thus, the productivity of the equipment is improved and the processing costs are lower (Federighi *et al.*, 2001). The installations therefore have to be designed to withstand a large number of cycles. In fact, the material fatigue is created more by the deformation stresses, at each increase or decrease in pressure, than by the maintenance of a constant high pressure. In 2010, the number of industrial high-pressure processing equipment is 156, mainly (53%) in North America (Figure 3.12)

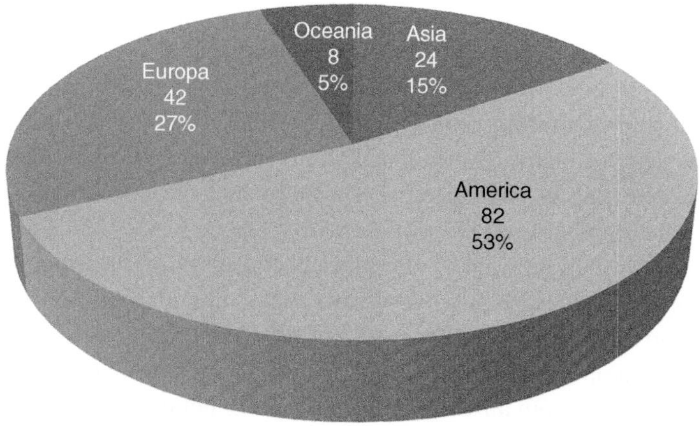

Figure 3.12 Distribution of industrial high-pressure processing equipment (number of installations) around the world (adapted from Duranton (2012)).

The equipment used in the food-processing sector consists of a vessel, a compression pump, high-pressure circuits, a control unit, and eventually a heating-cooling unit.

3.4.4 The principal components of high-pressure equipment

3.4.4.1 Vessel

The body of the vessel is always cylindrical as this is the form most resistant to pressure. To increase its mechanical resistance and life span, various techniques allow one to introduce a compressive stress at the level of the internal wall of the vessel, which goes from the exterior to the interior, to minimize the effect of the pressure, which itself acts from the interior to the exterior. Two types of stoppers are used in industrial vessels. If the diameter of the vessel is small, the stoppers can be screwed into the ends of the high-pressure vessel and it is the continuous or discontinuous screw threads inside the vessel that absorb the axial force. But the large majority of horizontal installations are equipped with a mesh and smooth plugs. During pressurization, the axial force transmitted to the plugs is absorbed by the mesh against which the plugs press. The mesh is composed of a pile of steel sheets linked to one another by straps. The mesh and/or the upper plug or both plugs are removed for loading and unloading of the products of the installation, thus improving the productivity.

3.4.4.2 Compression pump

Since pressure is a force applied on a surface, all high-pressure pumps function according to the same principle. A motor fluid under low or medium pressure (between 0.7 and 30 MPa) pushes a piston with a large surface area, which itself pushes a piston having a smaller surface area, which compresses the fluid to be pressurized. The same force acting on a smaller piston surface creates a higher pressure. The ratio of the surface areas of the pistons determines the factor by which the pressure is multiplied.

3.4.4.3 Heating or cooling unit

The temperature of high-pressure installations has to be controlled for certain applications. Heating or cooling can be ensured through circulation of a heat transfer fluid either between two walls of the vessel or inside the vessel (but its useful volume is then reduced by the space occupied by the heating or cooling circuit). More simply, the equipment can be heated by placing electric heating collars around the vessel. Or, if the treatment has to be done under refrigeration, the equipment supplied with water at the required temperature (4 °C, 10 °C, ...) can be installed in a room air conditioned at this temperature.

3.4.5 Compression systems

Two types of compression system exist, according to whether the high-pressure pump compresses the pressurization fluid and injects it into a vessel of fixed internal volume, or whether it compresses a transmission fluid that pushes a piston which reduces the volume of the vessel.

3.4.5.1 Direct compression

When the pressure is generated by a piston in contact with the medium to be pressurized that reduces the volume of the vessel, the compression is termed "direct." This system permits a high speed of compression (which may reach 1500 MPa/min), but it can be limited in practice to vessels of quite small diameter due to leaks between the piston and the wall of the vessel. The products may be processed, packaged or, if they are liquid, injected directly into the vessel. In this case, the treatment process can be semicontinuous when the equipment is composed of a series of several vessels operating successively (e.g., filling of one being carried out during pressurization of a second and emptying of a third). The pressurized liquid is then bottled aseptically. Today, less than 5% of the installations existing in the world use this compression system.

3.4.5.2 Indirect compression

The compression system called "indirect" uses a high-pressure pump, which injects a pressurization fluid into a vessel of constant volume. This is the method most often employed in industrial applications. The products that must be packaged are placed in the vessel and it is closed. The compression pump then injects the fluid under pressure into the interior of the vessel through resistant metallic pipes. In the first step, this fluid replaces the air around the products, whereas in the second step it allows an increase in the pressure inside the vessel. Closing a valve placed in the circuit permits maintenance of the pressure within the vessel for the required time. Opening this valve causes the pressure to drop. Then one

Table 3.6 Main manufacturers of industrial high-pressure processing equipment.

Name of the company	Target industries	Potential pressure treatment (MPa)
Avure technologies Inc. (http://avure.com)	Food industries	600
NC Hyperbaric (http://www.nchyperbaric.com)	Food industries	600
Elmhurst Research Inc. (http://elmhurstresearch.com)	Food industries	689
Mitsubishi Heavy Industries (http://www.mhi.co.jp)	Several industries	686
Engineered Pressure Systems Inc. (http://www.epsi-highpressure.com)	Several industries	100–900
Kobelco (http://www.kobelco.co.jp)	Several industries	98–686

Adapted from Norton and Sun (2008).

has to open the vessel and take out the treated products. This therefore implies a discontinuous process of loading and unloading of the packaged foodstuffs in the vessel whose useful volume is reduced to about 50–80% of the internal volume, depending on the form of packaging. Positioning the vessel horizontally has enabled an important increase in the treatment and productivity yields. Loading is carried out from one side and unloading from the other with respect to the forward progression of the products and avoids any confusion between processed and nonprocessed products.

3.4.6 Manufacturers of high-pressure equipment

The main manufacturers of high-pressure processing equipment are presented in Table 3.6. They are in correspondence with the world repartition of the equipment, main manufacturers for food industries in North America, Europe, and Asia (Table 3.6). It is known now that the investment for a HHP machine is placed in a range of 500–2000 k€ (US$ 750,000–3,000,000).

3.5 Future role for high pressure in the preservation and decontamination of foods

3.5.1 Substitution of thermal treatments for high value-added and new functional products

According to Purroy, Val Aparicio and Tonello (2012), substitution of thermal treatments and meat tenderization are likely to become commercial applications in a few years.

Indeed they examined emerging trends of high-pressure processing. They reported that extensive research had been carried out in Germany, New Zealand, and the United States over the last few years. From this research, new opportunities have been adopted by food industries in the manufacturing of spreadable sausages, liver sausages, and cooked-cured loin products. With the substitution of traditional thermal treatments in these products, improvement of texture and retention of sensorial and nutritional properties could be achieved (Purroy, Val Aparicio and Tonello, 2012).

High pressure could represent an interesting alternative method of preservation, particularly for high value-added and heat-sensitive products (Torres and Velazquez, 2005). These products include new functional foods formulated with biologically active compounds (specialty supplements, functional foods, nutraceuticals, and natural personal care products). HHP could meet consumer and food safety requirements by providing products containing ingredients with preserved biological properties and free of pathogens.

Among new functional products, juice and dairy industries play an important role. Some products are already available such as pomegranate and broccoli juice (Purroy, Val Aparicio and Tonello, 2012). HHP at room temperature has been shown to improve the quality of pomegranate juice, getting closer to the color of the fresh juice and preserving the content of nutraceutical components such as natural anthocyanins and polyphenols (Ferrari, Maresca and Ciccarone, 2010). On the other hand, chemoprotective effects of broccoli juice that is known to contain vitamins, polyphenolic compounds, glucosinolates, and other constituent substances have been shown not to be suppressed by a high-pressure treatment of 500 MPa (Mandelová and Totušek, 2007).

In the dairy industry, high pressure proved to be a good replacement for thermal treatment in the case of dairy products containing probiotics, particularly sensitive to heat treatments (Purroy, Val Aparicio and Tonello, 2012). Innovation in this area led to the commercialization by Fonterra, a manufacturer from New Zealand, of a colostrum drink called Col+ (http://www.colplus.com).

3.5.2 Meat tenderization

The second potential emerging tendency of HHP considered by Purroy-Balda, Val Aparicio and Tonello (2012) is meat tenderization. This process developed by Australian and North American scientists consists of submitting meat cuts to high pressure (200 MPa, a few minutes), after slaughter in *pre-rigor* conditions. To withstand increase in hardness induced by the application of high pressure in *post-rigor* meats, HHP treatments have to be performed at moderate temperature (60 °C) (Sikes, Tornberg and Tume, 2010). Meat tenderness is related to cathepsins and calpains that were shown to be affected by the high-pressure treatment (Sikes, Tornberg and Tume, 2010). Meat tenderization could represent an interesting means for the food industry of adding value to poor value beef and lamb cuts (Purroy-Balda, Val Aparicio and Tonello, 2012).

New high-pressure technologies have emerged in the recent years. Hydrodynamic pressure (HDP) or shock waves have been developed to meet tenderness and meat quality requirements not addressed by existing technologies (Bajovic, Bolumar and Heinz, 2012). This concept was first patented by Godfrey (1970). HDP generates shock wave pressure front at velocities exceeding the speed of sound from underwater detonations of a high-energy explosive (Solomon *et al.*, 2010). Shock wave pressure fronts from 70 to 100 MPa occur in fractions of milliseconds with virtually no increase in meat temperature. HDP induces meat tenderization by the physical disruption of the sarcomere and the breaking of the myofibrils, and presumably an additional indirect tenderizing effect on meat (Bajovic, Bolumar and Heinz, 2012). As tenderness problems mainly affect red meats, this technology has been directed at beef and pork tenderization (Solomon *et al.*, 2010). However, some benefits have been observed on using HDP on turkey meat to improve brine absorption, texture, and color characteristics (Bowker, Callahan and Solomon, 2010).

3.5.3 Combination of HPP and other technologies

Considering the inability of HPP to inactivate spores, the development of a derived technique of HPP to fill this gap is still challenging. The combination of HPP and heat – high-pressure sterilization (HPS) or pressure-assisted temperature sterilization (PATS) is one of the strategies that can be used for this purpose (see section "Sterilization – pressure-assisted thermal processing (PATP) or pressure-assisted thermal sterilization (PATS)"). For example, high-pressure destruction kinetics of *Clostridium sporogenes* spores at elevated temperature (80–100 °C) in extra-lean ground beef was examined by Zhu *et al.* (2008). They concluded that *Clostridium* spores could be destroyed with a shorter time with this method than with conventional thermal processing.

The combination of high pressure and the addition of antibacterial or bacteriostatic substances has been extensively studied. Many studies explored, for example, the combined use of bacteriocins and high pressure (Aymerich *et al.*, 2005; Marcos *et al.*, 2008; Jofré, Garriga and Aymerich, 2008; Yuste *et al.*, 2002; Garriga *et al.*, 2002b; Hauben *et al.*, 1996; Masschalck *et al.*, 2001). Among bacteriocins, nisin is currently the only bacteriocin that has been recognized as a food preservative by FAO/WHO, and it explains why it has been the focus of so many studies. Considering the synergistic mechanism underlying high pressure and nisin, it is suggested that high pressure would favor entry of antibacterial proteins by altering outer membrane permeability (Aymerich *et al.*, 2005; Marcos *et al.*, 2008; Hauben *et al.*, 1996; Masschalck *et al.*, 2001; Ogihara *et al.*, 2009).

Garriga *et al.* (2002b) examined the behavior of several foodborne bacteria inoculated in a meat model system with added bacteriocins (enterocins A and B, sakacin K, pediocin AcH, or nisin) after pressurization (400 MPa, 10 min, 17 °C). Nisin was generally the most efficient in inactivating *E. coli* and *Staphylococcus carnosus* during 60 days of storage at 4 °C, except for *L. monocytogenes* where the population reached the same level as the control at the end of the storage. The combination of nisin (200 ppm) and high pressure (450 MPa) also proved to be efficient in decontaminating mechanically recovered poultry meat (Yuste *et al.*, 2002).

Ogihara *et al.* (2009) studied the synergistic effect of high pressure and 1% concentration of food additives on the inactivation of *S.* Enteritidis. Among 27 food additives, 11 exhibited a significant synergism with high-pressure treatment of 250 MPa for 30 min. In some cases, the synergistic effect could be attributable to the lowering of the pH following the addition of the food additives. Those additives (glycerin mono-caprylic acid ester (C8), glycerin mono-capric acid ester (C10), and protamine sulfate) were not associated with a pH below 5. Complete inactivation of *S.* Enteritidis could be achieved with 0.25% of C8, 0.25% of C10, and 0.06 of protamine sulfate. In another study, the synergistic effect of lactate more specifically and high pressure was investigated (Aymerich *et al.*, 2005). A concentration of 1.8% potassium lactate in combination with a high-pressure treatment at 400 MPa for 10 min at 17 °C was shown to be efficient to inhibit complete recovery of *L. monocytogenes* during chilled storage at 6 °C. A mixture of 1.4% lactate and 0.1% diacetate was also essayed for its potential ability to prevent regrowth of *L. monocytogenes* after a treatment at 400 MPa during 10 min in sliced cooked ham (Marcos *et al.*, 2008). The authors demonstrated that the addition of additives enabled the population of *L. monocytogenes* to be maintained at the level reached after HHP, during the entire storage duration (60 days) at 1 °C.

References

Hite, B.H. (1899) The effect of pressure in the preservation of milk. *West Virginia University Agricultural Experiment Station Bulletin*, 15–35.

Basset, J. and Macheboeuf, M.A. (1932) Study of the biological effects of ultra-pressures: Resistances of bacteria, diastases and toxins to very high pressures. *Comptes Rendus Hebdomadaires des Sceances de l'Academie des Sciences*, **195**, 1431–1433.

Federighi, M., Tonello, C., De Lamballerie, M. and Ritz, M. (2001) Les traitements hautes pressions des aliments, in *Traitements ionisants et hautes pressions des aliments* (eds M. Federighi and J.L. Tholozan), Economica, Paris, pp. 151–227.

Farkas, D.F. and Hoover, D.G. (2000) High Pressure Processing *Journal of Food Science*, **65**, s8, 47–64.

Hugas, M., Garriga, M. and Monfort, J.M. (2002) New mild technologies in meat processing: High pressure as a model technology. *Meat Science*, **62**, 359–371.

Balasubramaniam, V.M., Farr, S.B. and Turek, J. (2008) Preserving foods through high pressure processing. *Food Technology*, **62**, 32–38.

Perrier-Cornet, J.-M., Moussa, M., Gervais, P. (2009) Applications des hautes pressions hydrostatiques en agroalimentaire. Techniques de l'ingénieur base documentaire: TIB430DUO.

Rastogi, N.K., Raghavarao, K.S., Balasubramaniam, V.M. *et al.* (2007) Opportunities and challenges in high pressure processing of foods. *Critical Reviews in Food Science and Nutrition*, **47**, 69–112.

Mújica-Paz, H., Valdez-Fragoso, A., Samson, C. *et al.* (2011) High-pressure processing technologies for the pasteurization and sterilization of foods. *Food and Bioprocess Technology*, **4**, 969–985.

Torres, J.A. and Velazquez, G. (2005) Commercial opportunities and research challenges in the high pressure processing of foods. *Journal of Food Engineering*, **67**, 95–112.

Smelt, J.P.P.M. (1998) Recent advances in the microbiology of high pressure processing. *Trends in Food Science and Technology*, **9**, 152–158.

Tonello, C. (1998) Applications des hautes pressions en agroalimentaire. *Techniques de l' Ingenieur*, F2, F3220.

Yordanov, D.G. and Angelova, G.V. (2010) High pressure processing for foods preserving. *Biotechnology & Biotechnological Equipment*, **24**, 1940–1945.

Simonin, H., Duranton, F. and de Lamballerie, M. (2012) New insights into the high-pressure processing of meat and meat products. *Comprehensive Reviews in Food Science and Food Safety*, **11**, 285–306.

Cheftel, J.C. (1995) Review: High pressure microbial inactivation and food preservation. *Food Science and Technology*, **1**, 75–90.

Cheftel, J.C. and Culioli, J. (1997) Effects of high pressure on meat: A review. *Meat Science*, **46**, 211–236.

Rendueles, E., Omer, M.K., Alvseike, O. *et al.* (2011) Microbiological food safety assessment of high hydrostatic pressure processing: A review. *Lebensmittel-Wissenschaft und -Technologie*, **44**, 1251–1260.

Patterson, M.F. (2005) Microbiology of pressure-treated foods. *Journal of Applied Microbiology*, **98**, 1400–1409.

Chapman, B., Winley, E., Fong, A.S.W. *et al.* (2007) Ascospore inactivation and germination by high pressure processing is affected by ascospore age. *Innovative Food Science & Emerging Technologies*, **8**, 531–534.

Wuytack, E.Y., Diels, A.M. and Michiels, C.W. (2002) Bacterial inactivation by high-pressure homogenisation and high hydrostatic pressure. *International Journal of Food Microbiology*, **77**, 205–212.

Alpas, H., Kalchayanand, N., Bozoglu, F. *et al.* (1999) Variation in resistance to hydrostatic pressure among strains of food-borne pathogens. *Applied and Environmental Microbiology*, **65**, 4248–4251.

Balasubramaniam, V.M., Ting, E.Y., Stewart, C.M. and Robbins, J.A. (2004) Recommended laboratory practices for conducting high-pressure microbial inactivation experiments. *Innovative Food Science & Emerging Technologies* **5**, 299–306.

Manas, P. and Mackey, B.M. (2004) Morphological and physiological changes induced by high hydrostatic pressure in exponential- and stationary-phase cells of *Escherichia coli*: Relationship with cell death. *Applied and Environmental Microbiology*, **70**, 1545–1554.

Pagan, R. and Mackey, B. (2000) Relationship between membrane damage and cell death in pressure-treated *Escherichia coli* cells: Differences between exponential- and stationary-phase cells and variation among strains. *Applied and Environmental Microbiology*, **66**, 2829–2834.

Linton, M., McClements, J.M.J. and Patterson, M.F. (2001) Inactivation of pathogenic *Escherichia coli* in skimmed milk using high hydrostatic pressure. *Innovative Food Science & Emerging Technologies*, **2**, 99–104.

Chapleau, N., Ritz, M., Delepine, S. *et al.* (2006) Influence of kinetic parameters of high pressure processing on bacterial inactivation in a buffer system. *International Journal of Food Microbiology*, **106**, 324–330.

Patterson, M.F. and Linton, M. (2008) Factors affecting inactivation of food-borne bacteria by high pressure, in *High pressure microbiology* (eds C. Michiels, D.H. Bartlett and A. Aertsen), ASM Press, Washington, DC, pp. 181–193.

Ritz, M. and Federighi, M. (2001) Effets des hautes pressions sur les microorganismes, in *Traitements ionisants et hautes pressions des aliments* (eds J.L. Tholozan and M. Federighi), Polytechnica, pp. 171–204.

Alpas, H., Kalchayanand, N., Bozoglu, F. and Ray, B. (2000) Interactions of high hydrostatic pressure, pressurization temperature and pH on death and injury of pressure-resistant and pressure-sensitive strains of foodborne pathogens. *International Journal of Food Microbiology*, **60**, 33–42.

Lerasle, M., Duranton, F., Simonin, H. *et al.* (2012) Traitements par hautes pressions hydrostatiques des denrées alimentaires: état de l'art. *Revue de Médecine Vétérinaire*, **163**, 595–614.

Carlez, A., Rosec, J.-P., Richard, N. and Cheftel, J.-C. (1993) High Pressure Inactivation of *Citrobacter freundii, Pseudomonas fluorescens* and *Listeria innocua* in Inoculated Minced Beef Muscle. *Lebensmittel-Wissenschaft und -Technologie*, **26**, 357–363.

Ritz, M., Jugiau, F., Federighi, M. *et al.* (2008) Effects of high pressure, subzero temperature, and pH on survival of *Listeria monocytogenes* in buffer and smoked salmon. *Journal of Food Protection*, **71**, 1612–1618.

Molina-Hoppner, A., Doster, W., Vogel, R.F. and Ganzle, M.G. (2004) Protective effect of sucrose and sodium chloride for *Lactococcus lactis* during sublethal and lethal high-pressure treatments. *Applied and Environmental Microbiology*, **70**, 2013–2020.

Black, E.P., Huppertz, T., Fitzgerald, G.F. and Kelly, A.L. (2007) Baroprotection of vegetative bacteria by milk constituents: A study of *Listeria innocua*. *International Dairy Journal*, **17**, 104–110.

Hauben, K.J., Bernaerts, K. and Michiels, C.W. (1998) Protective effect of calcium on inactivation of *Escherichia coli* by high hydrostatic pressure. *Journal of Applied Microbiology*, **85**, 678–684.

Balasubramaniam, V.M. and Farkas, D. (2008) High-pressure Food Processing. *Food Science and Technology International*, **14**, 413–418.

Wouters, P.C., Glaasker, E. and Smelt, J.P.P.M. (1998) Effects of High Pressure on Inactivation Kinetics and Events Related to Proton Efflux in *Lactobacillus plantarum*. *Applied and Environmental Microbiology*, **64**, 509–514.

Hayman, M.M., Kouassi, G.K., Anantheswaran, R.C. *et al.* (2008) Effect of water activity on inactivation of Listeria monocytogenes and lactate dehydrogenase during high pressure processing. *International Journal of Food Microbiology*, **124**, 21–26.

Patterson, M.F., Quinn, M., Simpson, R. and Gilmour, A. (1995) Sensitivity of vegetative pathogens to high hydrostatic pressure treatment in phosphate-buffered saline and foods. *Journal of Food Protection*, **58**, 524–529.

Nørrung, B. and Buncic, S. (2008) Microbial safety of meat in the European Union. *Meat Science*, **78**, 14–24.

EFSA (2012) The European union summary report on trends and sources of zoonoses, zoonotic agents and food-borne outbreaks in 2010. *EFSA Journal*, **10**, 2597.

Garriga, M., Grèbol, N., Aymerich, M.T. *et al.* (2004) Microbial inactivation after high-pressure processing at 600 MPa in commercial meat products over its shelf life. *Innovative Food Science & Emerging Technologies*, **5**, 451–457.

Jofré, A., Aymerich, T., Grèbol, N. and Garriga, M. (2009a) Efficiency of high hydrostatic pressure at 600 MPa against food-borne microorganisms by challenge tests on convenience meat products. *Lebensmittel-Wissenschaft und -Technologie*, **42**, 924–928.

Gola, S., Mutti, P., Manganelli, E. *et al.* (2000) Behaviour of *E. coli* 0157:H7 strains in model system and in raw meat by hpp: Microbial and technological aspects. *High Pressure Research*, **19**, 91–97.

Porto-Fett, A.C.S., Call, J.E., Shoyer, B.E. *et al.* (2010) Evaluation of fermentation, drying, and/or high pressure processing on viability of *Listeria monocytogenes, Escherichia coli* O157:H7, *Salmonella* spp., and *Trichinella spiralis* in raw pork and Genoa salami. *International Journal of Food Microbiology*, **140**, 61–75.

Garriga, M., Aymerich, T. and Hugas, M. (2002a) Effect of High Pressure Processing on the Microbiology of Skin Vacuum Packaged Sliced Meat Products: Cooked Pork Ham, Dry Cured Pork Ham and Marinated Beef Loin. *Report from the Food Microbiology and Biotechnology Unit of IRTA (Institute for Food and Agricultural Research and Technology)*, Monells, 17 pp.

Morales, P., Calzada, J., Rodríguez, B. *et al.* (2009) Inactivation of *Salmonella* Enteritidis in chicken breast fillets by single-cycle and multiple-cycle high pressure treatments. *Foodborne Pathogens and Disease*, **6**, 577–581.

Aymerich, T., Jofré, A., Garriga, M. and Hugas, M. (2005) Inhibition of Listeria monocytogenes and Salmonella by Natural Antimicrobials and High Hydrostatic Pressure in Sliced Cooked Ham. *Journal of Food Protection*, **68**, 173–177.

Marcos, B., Jofré, A., Aymerich, T. *et al.* (2008) Combined effect of natural antimicrobials and high pressure processing to prevent Listeria monocytogenes growth after a cold chain break during storage of cooked ham. *Food Control*, **19**, 76–81.

Simpson, R.K. and Gilmour, A. (1997) The resistance of *Listeria monocytogenes* to high hydrostatic pressure in foods. *Food Microbiology*, **14**, 567–573.

Chen, H. (2007) Temperature-assisted pressure inactivation of *Listeria monocytogenes* in Turkey breast meat. *International Journal of Food Microbiology*, **117**, 55–60.

Koseki, S., Mizuno, Y. and Yamamoto, K. (2007) Predictive modelling of the recovery of *Listeria monocytogenes* on sliced cooked ham after high pressure processing. *International Journal of Food Microbiology*, **119**, 300–307.

Jofré, A., Garriga, M. and Aymerich, T. (2008) Inhibition of *Salmonella* sp. *Listeria monocytogenes* and *Staphylococcus aureus* in cooked ham by combining antimicrobials, high hydrostatic pressure and refrigeration. *Meat Science*, **78**, 53–59.

Escriu, R. and Mor-Mur, M. (2009) Role of quantity and quality of fat in meat models inoculated with *Listeria innocua* or *Salmonella* Typhimurium treated by high pressure and refrigerated stored. *Food Microbiology*, **26**, 834–840.

Jofré, A., Aymerich, T. and Garriga, M. (2009b) Improvement of the food safety of low acid fermented sausages by enterocins A and B and high pressure. *Food Control*, **20**, 179–184.

Yuste, J., Pla, R. and Mor-Mur, M. (2000a) *Salmonella enteritidis* and aerobic mesophiles in inoculated poultry sausages manufactured with high-pressure processing. *Letters in Applied Microbiology*, **31**, 374–377.

Moerman, F. (2005) High hydrostatic pressure inactivation of vegetative microorganisms, aerobic and anaerobic spores in pork Marengo, a low acidic particulate food product. *Meat Science*, **69**, 225–232.

Yuste, J., Pla, R., Capellas, M. *et al.* (2000b) High-Pressure Processing Applied to Cooked Sausages: Bacterial Populations during Chilled Storage. *Journal of Food Protection*, **63**, 1093–1099.

Ruiz-Capillas, C., Jiménez Colmenero, F., Carrascosa, A.V. and Muñoz, R. (2007a) Biogenic amine production in Spanish dry-cured "chorizo" sausage treated with high-pressure and kept in chilled storage. *Meat Science*, **77**, 365–371.

Rubio, B., Martínez, B., García-Cachán, M.D. *et al.* (2007a) The effects of high pressure treatment and of storage periods on the quality of vacuum-packed "salchichón" made of raw material enriched in monounsaturated and polyunsaturated fatty acids. *Innovative Food Science & Emerging Technologies*, **8**, 180–187.

Lopez-Caballero, M.E., Carballo, J. and Jimenez-Colmenero, F. (1999) Microbiological changes in pressurized, prepackaged sliced cooked ham. *Journal of Food Protection*, **62**, 1411–1415.

López-Caballero, M.E., Carballo, J. and Jiménez-Colmenero, R. (2002a) Microbial Inactivation in Meat Products by Pressure/Temperature Processing. *Journal of Food Science*, **67**, 797–801.

Karłowski, K., Windyga, B., Fonberg-Broczek, M. *et al.* (2002) Effects of high pressure treatment on the microbiological quality, texture and colour of vacuum packed pork meat products. *High Pressure Research*, **22**, 725–732.

Diez, A.M., Urso, R., Rantsiou, K. *et al.* (2008a) Spoilage of blood sausages morcilla de Burgos treated with high hydrostatic pressure. *International Journal of Food Microbiology*, **123**, 246–253.

Ruiz-Capillas, C., Carballo, J. and Jiménez-Colmenero, F. (2007b) Consequences of high-pressure processing of vacuum-packaged frankfurters on the formation of polyamines: Effect of chilled storage. *Food Chemistry*, **104**, 202–208.

Carlez, A., Rosec, J.-P., Richard, N. and Cheftel, J.-C. (1994) Bacterial growth during chilled storage of pressure-treated minced meat. *Lebensmittel-Wissenschaft und -Technologie*, **27**, 48–54.

López-Caballero, M., Carballo, J., Solas, M. and Jiménez-Colmenero, F. (2002b) Responses of *Pseudomonas fluorescens* to combined high pressure/temperature treatments. *European Food Research and Technology*, **214**, 511–515.

Diez, A.M., Santos, E.M., Jaime, I. and Rovira, J. (2008b) Application of organic acid salts and high-pressure treatments to improve the preservation of blood sausage. *Food Microbiology*, **25**, 154–161.

Garriga, M. and Aymerich, T. (2009) Advanced decontamination technologies: High hydrostatic pressure on meat products, in *Safety of meat and processed meat* (ed. F. Toldra), Springer, New York, pp. 183–208.

Raffalli, J., Rosec, J.P., Carlez, A. *et al.* (1994) High pressure stress of *Listeria innocua* in inoculated dairy cream. *Sciences des Aliments*, **14**, 349–358.

Ponce, E., Pla, R., Mor-Mur, M. *et al.* (1998) Inactivation of *Listeria innocua* inoculated in liquid whole egg by high hydrostatic pressure. *Journal of Food Protection*, **61**, 119–122.

Smelt, J.P.P.M. and Rijke, A.G.F. (1992) High pressure as a tool for pasteurization of foods, in *High pressure and Biotechnology* (eds C. Balny, R. Hayashi, K. Heremans and P. Masson), John Libby and Company, Ltd, London, pp. 361–364.

Erkmen, O. and Doğan, C. (2004) Kinetic analysis of *Escherichia coli* inactivation by high hydrostatic pressure in broth and foods. *Food Microbiology*, **21**, 181–185.

Mussa, D.M., Ramaswamy, H.S. and Smith, J.P. (1998) High pressure (HP) destruction kinetics of *Listeria monocytogenes* Scott A in raw milk. *Food Research International*, **31**, 343–350.

Xiong, R., Xie, G., Edmondson, A.E. and Sheard, M.A. (1999) A mathematical model for bacterial inactivation. *International Journal of Food Microbiology*, **46**, 45–55.

Van Opstal, I., Vanmuysen, S.C., Wuytack, E.Y. *et al.* (2005) Inactivation of *Escherichia coli* by high hydrostatic pressure at different temperatures in buffer and carrot juice. *International Journal of Food Microbiology*, **98**, 179–191.

Diels, A., Van Opstal, I., Masschalck, B. and Michiels, C.W. (2007) Modelling of high-pressure inactivation of microorganisms in foods, in *Modelling microorganisms in food* (eds S. Brul, S. Van Gerwen and M. Zwietering), Woodhead Publishing Limited and CRC Press LLC, Washington, DC, pp. 161–197.

Couvert, O., Gaillard, S., Savy, N. *et al.* (2005) Survival curves of heated bacterial spores: Effect of environmental factors on Weibull parameters. *International Journal of Food Microbiology*, **101**, 73–81.

Chen, H. and Hoover, D.G. (2003a) Modeling the combined effect of high hydrostatic pressure and mild heat on the inactivation kinetics of *Listeria monocytogenes* Scott A in whole milk. *Innovative Food Science & Emerging Technologies*, **4**, 25–34.

Buzrul, S. and Alpas, H. (2004) Modeling the synergistic effect of high pressure and heat on inactivation kinetics of *Listeria innocua*: A preliminary study. *FEMS Microbiology Letters*, **238**, 29–36.

Chen, H. and Hoover, D.G. (2004) Use of Weibull model to describe and predict pressure inactivation of *Listeria monocytogenes* Scott A in whole milk. *Innovative Food Science & Emerging Technologies*, **5**, 269–276.

Buzrul, S., Alpas, H. and Bozoglu, F. (2005) Use of Weibull frequency distribution model to describe the inactivation of *Alicyclobacillus acidoterrestris* by high pressure at different temperatures. *Food Research International*, **38**, 151–157.

Chen, H. and Hoover, D.G. (2003b) Pressure inactivation kinetics of *Yersinia enterocolitica* ATCC 35669. *International Journal of Food Microbiology*, **87**, 161–171.

Gibson, A.M., Bratchell, N. and Roberts, T.A. (1988) Predicting microbial growth: Growth responses of *salmonellae* in a laboratory medium as affected by pH, sodium chloride and storage temperature. *International Journal of Food Microbiology*, **6**, 155–178.

Linton, R.H., Carter, W.H., Pierson, M.D. and Hackney, C.R. (1995) Use of a Modified Gompertz Equation to Model Nonlinear Survival Curves for *Listeria monocytogenes* Scott A. *Journal of Food Protection*, **58**, 946–954.

Erkmen, O. (2001a) Mathematical modeling of *Escherichia coli* inactivation under high-pressure carbon dioxide. *Journal of Bioscience and Bioengineering*, **92**, 39–43.

Cole, M.B., Davies, K.W., Munro, G. *et al.* (1993) A vitalistic model to describe the thermal inactivation of *Listeria monocytogenes*. *Journal of Industrial Microbiology*, **12**, 232–239.

Neter, J., Kutner, M., Nachtsheim, C. and Wasserman, W. (1996) *Applied Linear Statistical Models*, McGraw-Hill, New York.

Ross, T. (1996) Indices for performance evaluation of predictive models in food microbiology. *The Journal of Applied Bacteriology*, **81**, 501–508.

van Asselt, E.D. and Zwietering, M.H. (2006) A systematic approach to determine global thermal inactivation parameters for various food pathogens. *International Journal of Food Microbiology*, **107**, 73–82.

Perrier-Cornet, J.M., Tapin, S., Gaeta, S. and Gervais, P. (2005) High-pressure inactivation of *Saccharomyces cerevisiae* and *Lactobacillus plantarum* at subzero temperatures. *Journal of Biotechnology*, **115**, 405–412.

Reyns, K.M., Soontjens, C.C., Cornelis, K. *et al.* (2000) Kinetic analysis and modelling of combined high-pressure-temperature inactivation of the yeast *Zygosaccharomyces bailii*. *International Journal of Food Microbiology*, **56**, 199–210.

Ludikhuyze, L., Van Loey, A., Indrawati, I. *et al.* (2002) Effects of high pressure on enzymes related to food quality, in *Ultra High Pressure Treatments* (eds M. Hendrickx and D. Knorr), Kluwer Academic/Plenum Publishers, New York, pp. 115–160.

Valdramidis, V.P., Geeraerd, A.H., Poschet, F. *et al.* (2007) Model based process design of the combined high pressure and mild heat treatment ensuring safety and quality of a carrot simulant system. *Journal of Food Engineering*, **78**, 1010–1021.

Jung, S., Tonello Samson, C. and De Lamballerie, M. (2011) High hydrostatic pressure food processing, in *Alternative to conventional food processing, vol. Chapter 6* (ed. A. Proctor), RSC Publishing, pp. 254–306.

Sun, X.D. and Holley, R.A. (2010) High hydrostatic pressure effects on the texture of meat and meat products. *Journal of Food Science*, **75**, R17–R23.

Bouton, P.E., Ford, A.L., Harris, P.V. *et al.* (1977) Pressure-heat treatment of postrigor muscle: Effects on tenderness. *Journal of Food Science*, **42**, 132–135.

Ma, H.J. and Ledward, D.A. (2004) High pressure/thermal treatment effects on the texture of beef muscle. *Meat Science*, **68**, 347–355.

Del Olmo, A., Morales, P., Ávila, M. *et al.* (2010) Effect of single-cycle and multiple-cycle high-pressure treatments on the colour and texture of chicken breast fillets. *Innovative Food Science & Emerging Technologies*, **11**, 441–444.

Zamri, A.I., Ledward, D.A. and Frazier, R.A. (2006) Effect of combined heat and high-pressure treatments on the texture of chicken breast muscle (*Pectoralis Fundus*). *Journal of Agricultural and Food Chemistry*, **54**, 2992–2996.

Jung, S., Ghoul, M. and de Lamballerie-Anton, M. (2000a) Changes in lysosomal enzyme activities and shear values of high pressure treated meat during ageing. *Meat Science*, **56**, 239–246.

Ichinoseki, S., Nishiumi, T. and Suzuki, A. (2006) Tenderizing effect of high hydrostatic pressure on bovine intramuscular connective tissue. *Journal of Food Science*, **71**, E276–E281.

Schenková, N., Šikulová, M., Jeleníková, J. *et al.* (2007) Influence of high isostatic pressure and papain treatment on the quality of beef meat. *High Pressure Research*, **27**, 163–168.

Clariana, M. and Garcia-Regueiro, J.A. (2011) Effect of high pressure processing on cholesterol oxidation products in vacuum packaged sliced dry-cured ham. *Food and Chemical Toxicology*, **49**, 1468–1471.

Mor-Mur, M. and Yuste, J. (2003) High pressure processing applied to cooked sausage manufacture: Physical properties and sensory analysis. *Meat Science*, **65**, 1187–1191.

Iwasaki, T., Noshiroya, K., Saitoh, N. *et al.* (2006) Studies of the effect of hydrostatic pressure pretreatment on thermal gelation of chicken myofibrils and pork meat patty. *Food Chemistry*, **95**, 474–483.

Sikes, A.L., Tobin, A.B. and Tume, R.K. (2009) Use of high pressure to reduce cook loss and improve texture of low-salt beef sausage batters. *Innovative Food Science & Emerging Technologies*, **10**, 405–412.

Fernández-Martín, F., Fernández, P., Carballo, J. and Jiménez Colmenero, F. (1997) Pressure/heat combinations on pork meat batters: Protein thermal behavior and product rheological properties. *Journal of Agricultural and Food Chemistry*, **45**, 4440–4445.

Yuste, J., Mor-Mur, M., Capellas, M. *et al.* (1999) Mechanically recovered poultry meat sausages manufactured with high hydrostatic pressure. *Poultry Science*, **78**, 914–921.

Supavititpatana, T. and Apichartsrangkoon, A. (2007) Combination effects of ultra-high pressure and temperature on the physical and thermal properties of ostrich meat sausage (yor). *Meat Science*, **76**, 555–560.

Chattong, U. and Apichartsrangkoon, A. (2009) Dynamic viscoelastic characterisation of ostrich-meat yor (Thai sausage) following pressure, temperature and holding time regimes. *Meat Science*, **81**, 426–432.

Jimenez-Colmenero, F., Fernandez, P., Carballo, J. and Fernandez-Martin, F. (1998) High-pressure-cooked low-fat pork and chicken batters as affected by salt levels and cooking temperature. *Journal of Food Science*, **63**, 656–659.

Fernandez-Martin, F., Cofrades, S., Carballo, J. and Jimenez-Colmenero, F. (2002) Salt and phosphate effects on the gelling process of pressure/heat treated pork batters. *Meat Science*, **61**, 15–23.

Trespalacios, P. and Pla, R. (2009) Development of low-fat chicken meat and dried egg white gels by high pressure. *High Pressure Research*, **29**, 150–161.

Fernandez, P., Cofrades, S., Solas, M.T. *et al.* (1998) High pressure-cooking of chicken meat batters with starch, egg white, and iota carrageenan. *Journal of Food Science*, **63**, 267–271.

Fernández-Martín, F., Guerra, M.A., López, E. *et al.* (2000a) Characteristics of pressurised pork meat batters as affected by addition of plasma proteins, apple fibre and potato starch. *Journal of the Science of Food and Agriculture*, **80**, 1230–1236.

Hong, G.-P., Min, S.-G., Ko, S.-H. and Choi, M.-J. (2008a) Effect of high pressure treatments combined with various levels of k-carrageenan on cold-set binding in restructured pork. *International Journal of Food Science and Technology*, **43**, 1484–1491.

Hong, G.-P., Ko, S.-H., Choi, M.-J. and Min, S.-G. (2008b) Effect of glucono-δ-lactone and k-carrageenan combined with high pressure treatment on the physico-chemical properties of restructured pork. *Meat Science*, **79**, 236–243.

Trespalacios, P. and Pla, R. (2007a) Simultaneous application of transglutaminase and high pressure to improve functional properties of chicken meat gels. *Food Chemistry*, **100**, 264–272.

Trespalacios, P. and Pla, R. (2007b) Synergistic action of transglutaminase and high pressure on chicken meat and egg gels in absence of phosphates. *Food Chemistry*, **104**, 1718–1727.

Chapleau, N., Mangavel, C., Compoint, J.-P. and de Lamballerie-Anton, M. (2004) Effect of high-pressure processing on myofibrillar protein structure. *Journal of the Science of Food and Agriculture*, **84**, 66–74.

Tuboly, E., Lebovics, V.K., Gaál, Ö. *et al.* (2003) Microbiological and lipid oxidation studies on mechanically deboned turkey meat treated by high hydrostatic pressure. *Journal of Food Engineering*, **56**, 241–244.

Dissing, J., Bruun-Jensen, L. and Skibsted, L.H. (1997) Effect of high-pressure treatment on lipid oxidation in turkey thigh muscle during chill storage. *Zeitschrift für Lebensmittel-Untersuchung und-Forschung*, **205**, 11–13.

Orlien, V., Hansen, E. and Skibsted, L.H. (2000) Lipid oxidation in high-pressure processed chicken breast muscle during chill storage: Critical working pressure in relation to oxidation mechanism. *European Food Research and Technology*, **211**, 99–104.

Beltran, E., Pla, R., Yuste, J. and Mor-Mur, M. (2003) Lipid oxidation of pressurized and cooked chicken: Role of sodium chloride and mechanical processing on TBARS and hexanal values. *Meat Science*, **64**, 19–25.

Beltran, E., Pla, R., Capellas, M. *et al.* (2004a) Lipid oxidation and colour in pressure- and heat-treated minced chicken thighs. *Journal of the Science of Food and Agriculture*, **84**, 1285–1289.

Andres, A., Adamsen, C., Moller, J. *et al.* (2006) High-pressure treatment of dry-cured Iberian ham. Effect on colour and oxidative stability during chill storage packed in modified atmosphere. *European Food Research and Technology*, **222**, 486–491.

Campus, M., Flores, M., Martinez, A. and Toldrá, F. (2008) Effect of high pressure treatment on colour, microbial and chemical characteristics of dry cured loin. *Meat Science*, **80**, 1174–1181.

Cheah, P.B. and Ledward, D.A. (1996) High pressure effects on lipid oxidation in minced pork. *Meat Science*, **43**, 123–134.

Ma, H.J., Ledward, D.A., Zamri, A.I. *et al.* (2007) Effects of high pressure/thermal treatment on lipid oxidation in beef and chicken muscle. *Food Chemistry*, **104**, 1575–1579.

Schindler, S., Krings, U., Berger, R.G. and Orlien, V. (2010) Aroma development in high pressure treated beef and chicken meat compared to raw and heat treated. *Meat Science*, **86**, 317–323.

Fuentes, V., Ventanas, J., Morcuende, D. *et al.* (2010) Lipid and protein oxidation and sensory properties of vacuum-packaged dry-cured ham subjected to high hydrostatic pressure. *Meat Science*, **85**, 506–514.

Beltran, E., Pla, R., Yuste, J. and Mor-Mur, M. (2004b) Use of antioxidants to minimize rancidity in pressurized and cooked chicken slurries. *Meat Science*, **66**, 719–725.

Bragagnolo, N., Danielsen, B. and Skibsted, L.H. (2007) Rosemary as antioxidant in pressure processed chicken during subsequent cooking as evaluated by electron spin resonance spectroscopy. *Innovative Food Science & Emerging Technologies*, **8**, 24–29.

Mariutti, L.B., Orlien, V., Bragagnolo, N. and Skibsted, L. (2008) Effect of sage and garlic on lipid oxidation in high-pressure processed chicken meat. *European Food Research and Technology*, **227**, 337–344.

Cheah, P.B. and Ledward, D.A. (1997) Inhibition of metmyoglobin formation in fresh beef by pressure treatment. *Meat Science*, **45**, 411–418.

Angsupanich, K. and Ledward, D.A. (1998) High pressure treatment effects on cod (*Gadus morhua*) muscle. *Food Chemistry*, **63**, 39–50.

Carballo, J., Fernandez, P., Carrascosa, A.V. *et al.* (1997) Characteristics of low-and high-fat beef patties: Effect of high hydrostatic pressure. *Journal of Food Protection*, **60**, 48–53.

Korzeniowski, W., Jankowska, B. and Kwiatkowska, A. (1999) The effect of high pressure on some technological properties of pork. *Electronic Journal of Polish Agricultural Universities*, **2**, 1–7.

Marcos, B., Kerry, J.P. and Mullen, A.M. (2010) High pressure induced changes on sarcoplasmic protein fraction and quality indicators. *Meat Science*, **85**, 115–120.

Jung, S., Ghoul, M. and de Lamballerie-Anton, M. (2003) Influence of high pressure on the color and microbial quality of beef meat. *Lebensmittel-Wissenschaft und -Technologie*, **36**, 625–631.

Goutefongea, R., Rampon, V., Nicolas, J. and Dumont, J. (1995) *Meat Color Changes Under High Pressure Treatment*. Proceeding of the 41st International Congress of Meat Science and Technology, San Antonio, Texas.

Carlez, A., Veciana-Nogues, T. and Cheftel, J.-C. (1995) Changes in colour and myoglobin of minced beef meat due to high pressure processing. *Lebensmittel-Wissenschaft und -Technologie*, **28**, 528–538.

Cava, R., Ladero, L., Gonzalez, S. *et al.* (2009) Effect of pressure and holding time on colour, protein and lipid oxidation of sliced dry-cured Iberian ham and loin during refrigerated storage. *Innovative Food Science & Emerging Technologies*, **10**, 76–81.

Wackerbarth, H., Kuhlmann, U., Tintchev, F. *et al.* (2009) Structural changes of myoglobin in pressure-treated pork meat probed by resonance Raman spectroscopy. *Food Chemistry*, **115**, 1194–1198.

Colmenero, F.J., Carballo, J., Fernández, P. *et al.* (1997) High-pressure-induced changes in the characteristics of low-fat and high-fat sausages. *Journal of the Science of Food and Agriculture*, **75**, 61–66.

Rubio, B., Martínez, B., García-Cachán, M.D. *et al.* (2007b) Effect of high pressure preservation on the quality of dry cured beef "Cecina de Leon". *Innovative Food Science & Emerging Technologies*, **8**, 102–110.

Saccani, G., Parolari, G., Tanzi, E. and Rabbuti, S. (2004). Sensory and Microbiological Properties of Dried Hams Treated with High Hydrostatic Pressure. Proceedings of the 50th international congress of meat science and technology, Helsinki, Finland.

Fulladosa, E., Serra, X., Gou, P. and Arnau, J. (2009) Effects of potassium lactate and high pressure on transglutaminase restructured dry-cured hams with reduced salt content. *Meat Science*, **82**, 213–218.

Clariana, M., Guerrero, L., Sárraga, C. *et al.* (2011) Influence of high pressure application on the nutritional, sensory and microbiological characteristics of sliced skin vacuum packed dry-cured ham. Effects along the storage period. *Innovative Food Science & Emerging Technologies*, **12**, 456–465.

Kadam, P.S., Jadhav, B.A., Salve, R.V. and Machewad, G.M. (2012) Review on the high pressure technology (HPT) for food preservation. *Journal of Food Processing & Technology*, **3**, 135.

Matser, A.M., Krebbers, B., van den Berg, R.W. and Bartels, P.V. (2004) Advantages of high pressure sterilisation on quality of food products. *Trends in Food Science and Technology*, **15**, 79–85.

Ratphitagsanti, W., Ahn, J., Balasubramaniam, V.M. and Yousef, A.E. (2009) Influence of pressurization rate and pressure pulsing on the inactivation of *Bacillus amyloliquefaciens* spores during pressure-assisted thermal processing. *Journal of Food Protection*, **72**, 775–782.

Ratphitagsanti, W., De Lamo-Castellvi, S., Balasubramaniam, V.M. and Yousef, A.E. (2010) Efficacy of pressure-assisted thermal processing, in combination with organic acids, against *Bacillus amyloliquefaciens* spores suspended in deionized water and carrot puree. *Journal of Food Science*, **75**, M46–M52.

Koutchma, T. (2011) Regulatory status of novel sterilization and pasteurization, in *Handbook of Food Safety Engineering* vol. Chapter 14 (ed. D.-W. Sun), Wiley-Blackwell, Oxford, UK.

Segovia Bravo, K., Ramírez, R., Durst, R. *et al.* (2012) Formation risk of toxic and other unwanted compounds in pressure-assisted thermally processed foods. *Journal of Food Science*, **77**, R1–R10.

Jung, S., De Lamballerie-Anton, M. and Ghoul, M. (2000b) Textural changes in bovine meat treated with high pressure. *High Pressure Research*, **19**, 69–74.

Kalichevsky, M.T., Knorr, D. and Lillford, P.J. (1995) Potential food applications of high-pressure effects on ice-water transitions. *Trends in Food Science and Technology*, **6**, 253–259.

De Lamballerie, M., Taylor, R.G. and Culioli, J. (2002) High pressure processing of meat, in *Meat processing: Improving quality* vol. Chapter 16 (eds J. Kerry, J. Kerry and D. Ledward), Woodhead Publishing Limited/CRC Press, pp. 320–330.

Pothakamury, U.R., Barbosa-Canovas, G.V., Swanson, B.G. and Meyer, R.S. (1995) The pressure builds for better food processing. *Chemical Engineering Progress*, **91**, 45–53.

Zhao, Y., Flores, R.A. and Olson, D.G. (1998) High hydrostatic pressure effects on rapid thawing of frozen beef. *Journal of Food Science*, **63**, 272–275.

Urrutia, G., Arabas, J., Autio, K. *et al.* (2007) SAFE ICE: Low-temperature pressure processing of foods: Safety and quality aspects, process parameters and consumer acceptance. *Journal of Food Engineering*, **83**, 293–315.

Norton, T. and Sun, D.-W. (2008) Recent advances in the use of high pressure as an effective processing technique in the food industry. *Food and Bioprocess Technology*, **1**, 2–34.

Fernández-Martín, F., Otero, L., Solas, M.T. and Sanz, P.D. (2000b) Protein denaturation and structural damage during high-pressure-shift freezing of porcine and bovine muscle. *Journal of Food Science*, **65**, 1002–1008.

Zhu, S., Ramaswamy, H.S. and Bail, A.L. (2005) High-pressure calorimetric evaluation of ice crystal ratio formed by rapid depressurization during pressure-shift freezing of water and pork muscle. *Food Research International*, **38**, 193–201.

Zhu, S., le Bail, A., Ramaswamy, H.S. and Chapleau, N. (2004a) Characterization of ice crystals in pork muscle formed by pressure-shift freezing as compared with classical freezing methods. *Journal of Food Science*, **69**, FEP190–FEP197.

Zhu, S., Le Bail, A., Chapleau, N. *et al.* (2004b) Pressure shift freezing of pork muscle: Effect on color, drip loss, texture, and protein stability. *Biotechnology Progress*, **20**, 939–945.

Tonello, S.C. (2012a) Estimation of HPP Food Global Production in 2009. Personal communication.

Tonello, C. (2011) Case studies on high pressure processing of foods, in *Nonthermal Processing Technologies for Food* vol. Chapter 4 (eds H.Q. Zhang, G.V. Barbosa-Canovas, V.M. Balasubramaniam *et al.*), Wiley-Blackwell, pp. 36–50.

Aymerich, T., Picouet, P.A. and Monfort, J.M. (2008) Decontamination technologies for meat products. *Meat Science*, **78**, 114–129.

Tonello, S.C. (2012b) *Personal communication*. October 12, 2012

Bajovic, B., Bolumar, T. and Heinz, V. (2012) Quality considerations with high pressure processing of fresh and value added meat products. *Meat Science*, **92**, 280–289.

Duranton, F. (2012) Potential use of High Pressure for sustainable formulation of pork-based meat products. PhD Thesis. University of Nantes.

Cris-Tim (2013) *Tehnolologiei de Protectie prin Inhalta Presiune*. Available at: http://www.cristim.ro/en/about-us/high-pressure-technology.html (accessed: 25 February 2013).

Hiperbaric (2013) *Applications. Meat Products*. Available at: http://www.hiperbaric.com/en/meat (accessed: 28 February 2013).

Zwanenberg (2010a) *Steak Tartare Improved Thanks to Unique High Pressure Technology*. Available at: http://www.zwanenberg.nl/en/334/steak-tartare-improved-thanks-to-unique-high-pressure-technology/ (accessed: 1 March 2013).

Zwanenberg (2010b) *Carpaccio Tastier and 'silky smooth' Thanks to HPP Technology*. Available at: http://www.zwanenberg.nl/en/343/carpaccio-tastier-and-silky-smooth-thanks-to-hpp-technology/ (accessed: 1 March 2013).

Cargill, I. (2010) *Fressure™ Ground Beef Patties: Freshness without Compromise™*. Available at: http://www.cargill.com/wcm/groups/internal/@ccom/documents/document/na3040085.pdf (accessed: 1 March 2013).

Hiperbaric (2012) Creta Farm. Available at: http://www.hiperbaric.com/en/creta-farm (accessed: 1 March 2013).

Ikeuchi, Y. (2011) Recent advances in the application of high pressure technology to processed meat products, in *Processed Meats-Improving Safety, Nutrition and Quality* (eds J.P. Kerry and K.J.F. Kerry), Woodhead Publishing, Cambridge, pp. 590–609.

Ifantis (2009) *Technology freshpress - Ultimate Food Protection*. Available at: http://www.freshpress.gr/index.cfm?action=home&lang=EN, (accessed: 1 March 2013).

Watson E (2011). Deli 24 gains radical shelf-life boost with novel process technology. Available at: http://www.foodmanufacture.co.uk/Manufacturing/Deli-24-gains-radical-shelf-life-boost-with-novel-processing-technology, (accessed: 1 March 2013).

Hormel (2012) *About High Pressure Processing*. Available at: http://www.hormelnatural.com/ASSETS/E89897EB5D8D4AEBBC29E9D32C2A3ECA/High_Pressure_Processing_Fact_Sheet.pdf, (accessed: 1 March 2013).

Foodprocessing-Technology (2011) *Columbus Foods Slicing and Packaging Facility, United States of America*. Available at: http://www.foodprocessing-technology.com/projects/columbus-foods/ (accessed: 1 March 2013).

Espuna (2012) *Tapas Minute*. Available at: http://www.espuna-charcuterie.fr/fr/produits-/tapas-minute--snacks.aspx (accessed: 1 March 2013).

Espuna (2012) Charcuterie en tranches. Available at: http://www.espuna-charcuterie.fr/fr/produits-/charcuterie-en-tranches.aspx (accessed: 1 March 2013).

Abraham (2005) Abraham leads the market with its innovations. Available at: http://www.abraham.de/fileadmin/user_upload/en/pdf/abraham-leads-the-market.pdf (Accessed: 1 March 2013).

Foods, S.M. (2013) *Discover Our Products*. Available at: http://www.sharemastro.com/discover-our-products/ (accessed: 1 March 2013).

DuBreton 2013. Products. Available at: http://www.dubreton.com/en/products (accessed: 1 March 2013).

Anonymous (2002) Screentec inspection and systems. *Voedingsmiddelentechnologie*, **35**, 6.

Baron A, Bayer O, Butz P, Geisel B, Gupta B, Oltersdorf U, Tauscher, B. (1999) *Consumer Perception of High Pressure Processing: A Three Country Survey*. Proceedings European Conference on Emerging Food Science and Technology, Tampere, Finland.

Bae, Y.Y., Kim, N.H., Kim, K.H. *et al.* (2011) Supercritical carbon dioxide as a potential intervention for ground pork decontamination. *Journal of Food Safety*, **31**, 48–53.

Garcia-Gonzalez, L., Geeraerd, A.H., Spilimbergo, S. *et al.* (2007) High pressure carbon dioxide inactivation of microorganisms in foods: The past, the present and the future. *International Journal of Food Microbiology*, **117**, 1–28.

Leitner, W. (2000) Green chemistry: Designed to dissolve. *Nature*, **405**, 129–130.

Hong, S.-I., Park, W.-S. and Pyun, Y.-R. (1997) Inactivation of *Lactobacillus* sp. from Kimchi by High Pressure Carbon Dioxide. *Lebensmittel-Wissenschaft und -Technologie*, **30**, 681–685.

Hong, S.-I., Park, W.-S. and Pyun, Y.-R. (1999) Non-thermal inactivation of *Lactobacillus plantarum* as influenced by pressure and temperature of pressurized carbon dioxide. *International Journal of Food Science and Technology*, **34**, 125–130.

Lin, H.-M., Cao, N. and Chen, L.-F. (1994) Antimicrobial effect of pressurized carbon dioxide on *Listeria monocytogenes*. *Journal of Food Science*, **59**, 657–659.

Lin, H.-m., Yang, Z. and Fu, C.L. (1993) Inactivation of *Leuconostoc dextranicum* with carbon dioxide under pressure. *Chemical Engineering Journal*, **52**, B29–B34.

Gunes, G., Blum, L.K. and Hotchkiss, J.H. (2006) Inactivation of *Escherichia coli* (ATCC 4157) in diluted apple cider by dense-phase carbon dioxide. *Journal of Food Protection*, **69**, 12–16.

Lin, H.M., Yang, Z. and Chen, L.F. (1992a) Inactivation of *Saccharomyces cerevisiae* by supercritical and subcritical carbon dioxide. *Biotechnology Progress*, **8**, 458–461.

Dillow, A.K., Dehghani, F., Hrkach, J.S. *et al.* (1999) Bacterial inactivation by using near- and supercritical carbon dioxide. *Proceedings of the National Academy of Sciences of the United States of America*, **96**, 10344–10348.

Lin, H.M., Yang, Z. and Chen, L.F. (1992b) An improved method for disruption of microbial cells with pressurized carbon dioxide. *Biotechnology Progress*, **8**, 165–166.

Lin, H.-M., Chan, E.-C., Chen, C. and Chen, L.-F. (1991) Disintegration of yeast cells by pressurized carbon dioxide. *Biotechnology Progress*, **7**, 201–204.

Nakamura, K., Enomoto, A., Fukushima, H. *et al.* (1994) Disruption of microbial cells by the flash discharge of high-pressure carbon dioxide. *Bioscience Biotechnology and Biochemistry*, **58**, 1297–1301.

Debs-Louka, E., Louka, N., Abraham, G. *et al.* (1999) Effect of compressed carbon dioxide on microbial cell viability. *Applied and Environmental Microbiology*, **65**, 626–631.

Kumagai, H., Hata, C. and Nakamura, K. (1997) CO_2 sorption by microbial cells and sterilization by high-pressure CO_2. *Bioscience Biotechnology and Biochemistry*, **61**, 931–935.

Arreola, A., Balaban, M., Wei, C. *et al.* (2007) Effect of supercritical carbon dioxide on microbial populations in single strength orange juice. *Journal of Food Quality*, **14**, 275–284.

Garcia-Gonzalez, L., Geeraerd, A.H., Elst, K. *et al.* (2009) Influence of type of microorganism, food ingredients and food properties on high-pressure carbon dioxide inactivation of microorganisms. *International Journal of Food Microbiology*, **129**, 253–263.

Hong, S.I. and Pyun, Y.R. (1999) Inactivation kinetics of *Lactobacillus plantarum* by high pressure carbon dioxide. *Journal of Food Science*, **64**, 728–733.

Erkmen, O. (2000a) Inactivation of *Salmonella typhimurium* by high pressure carbon dioxide. *Food Microbiology*, **17**, 225–232.

Erkmen, O. (2000b) Antimicrobial effect of pressurised carbon dioxide on *Enterococcus faecalis* in physiological saline and foods. *Journal of the Science of Food and Agriculture*, **80**, 465–470.

Erkmen, O. (2000c) Effect of carbon dioxide pressure on *Listeria monocytogenes* in physiological saline and foods. *Food Microbiology*, **17**, 589–596.

Erkmen, O. and Karaman, H. (2001) Kinetic studies on the high pressure carbon dioxide inactivation of *Salmonella typhimurium*. *Journal of Food Engineering*, **50**, 25–28.

Erkmen, O. (2001b) Note. Antimicrobial effect of pressurized carbon dioxide on *Yersinia enterocolitica* in broth and foods. *Food Science and Technology International*, **7**, 245–250.

Erkmen, O. (2001c) Kinetic analysis of Listeria monocytogenes inactivation by high pressure carbon dioxide. *Journal of Food Engineering*, **47**, 7–10.

Ramirez-Rodrigues, M.M., Plaza, M.L., Ferrentino, G. *et al.* (2013) Effect of dense phase carbon dioxide processing on microbial stability and physicochemical attributes of Hibiscus sabdariffa beverage. *Journal of Food Process Engineering*, **36**, 125–133.

Spilimbergo, S. and Ciola, L. (2010) Supercritical CO_2 and N_2O pasteurization of peach and kiwi juice. *International Journal of Food Science and Technology*, **45**, 1619–1625.

Valverde, M.T., Marín-Iniesta, F. and Calvo, L. (2010) Inactivation of *Saccharomyces cerevisiae* in conference pear with high pressure carbon dioxide and effects on pear quality. *Journal of Food Engineering*, **98**, 421–428.

Ferrentino, G., Balzan, S., Dorigato, A. *et al.* (2012a) Effect of supercritical carbon dioxide pasteurization on natural microbiota, texture, and microstructure of fresh-cut coconut. *Journal of Food Science*, **77**, E137–E143.

Ferrentino, G., Balzan, S. and Spilimbergo, S. (2012b) On-line color monitoring of solid foods during supercritical CO_2 pasteurization. *Journal of Food Engineering*, **110**, 80–85.

Ferrentino, G., Balzan, S. and Spilimbergo, S. (2013a) Supercritical carbon dioxide processing of dry cured ham spiked with *Listeria monocytogenes*: Inactivation kinetics, color, and sensory evaluations. *Food and Bioprocess Technology, Food and Bioprocess Technology* **6**(5): 1164–1174.

Zhong, Q., Black, D.G., Davidson, P.M. and Golden, D.A. (2008) Nonthermal inactivation of *Escherichia coli* K-12 on spinach leaves, using dense phase carbon dioxide. *Journal of Food Protection*, **71**, 1015–1017.

Bi, X., Wu, J., Zhang, Y. *et al.* (2011) High pressure carbon dioxide treatment for fresh-cut carrot slices. *Innovative Food Science & Emerging Technologies*, **12**, 298–304.

Wei, C.I., Balaban, M.O., Fernando, S.Y. and Peplow, A.J. (1991) Bacterial effect of high pressure CO_2 treatment on foods spiked with Listeria or Salmonella. *Journal of Food Protection*, **54**, 189–193.

Sirisee, U., Hsieh, F. and Huff, H.E. (1998) Microbial safety of supercritical carbon dioxide processes. *Journal of Food Processing and Preservation*, **22**, 387–403.

Erkmen, O. (2000d) Antimicrobial effects of pressurised carbon dioxide on *Brochothrix thermosphacta* in broth and foods. *Journal of the Science of Food and Agriculture*, **80**, 1365–1370.

Ferrentino, G., Balzan, S. and Spilimbergo, S. (2013b) Optimization of supercritical carbon dioxide treatment for the inactivation of the natural microbial flora in cubed cooked ham. *International Journal of Food Microbiology*, **161**, 189–196.

Choi, Y.M., Kim, O.Y., Kim, K.H. *et al.* (2009) Combined effect of organic acids and supercritical carbon dioxide treatments against nonpathogenic *Escherichia coli*, *Listeria monocytogenes*, *Salmonella typhimurium* and *E. coli* O157:H7 in fresh pork. *Letters in Applied Microbiology*, **49**, 510–515.

Garwood A.J.M. (2003) Pressurized decontamination of meat using carbonic acid. Patent WO 2003101210 A1

Garwood A.J.M. (2005) Decontamination methods for meat using carbonic acid at high pressures. US Patent US9131707

Purroy-Balda, F., Val Aparicio, B. and Tonello, S.C. (2012) Industrial high pressure processing of foods: Review of evolution and emerging trends. *Journal of Food Science and Engineering*, **2**, 543–549.

Ferrari, G., Maresca, P. and Ciccarone, R. (2010) The application of high hydrostatic pressure for the stabilization of functional foods: Pomegranate juice. *Journal of Food Engineering*, **100**, 245–253.

Mandelová, L. and Totušek, J. (2007) Broccoli juice treated by high pressure: Chemoprotective effects of sulforaphane and indole-3-carbinol. *High Pressure Research*, **27**, 151–156.

Sikes, A., Tornberg, E. and Tume, R. (2010) A proposed mechanism of tenderising post-rigor beef using high pressure-heat treatment. *Meat Science*, **84**, 390–399.

Godfrey, C.S. (1970) Apparatus for tenderizing food. US patent 1970-3492688.

Solomon, M.B., Liu, M.N., Patel, J.R. *et al.* (2010) Hydrodynamic Pressure Processing to improve meat quality and safety, in *Advanced Technologies For Meat Processing* vol. Chapter 9 (eds L.M.L. Nollet and F. Toldra), CRC Press, Taylor & Francis, New York, pp. 219–244.

Bowker, B.C., Callahan, J.A. and Solomon, M.B. (2010) Effects of hydrodynamic pressure processing on the marination and meat quality of turkey breasts. *Poultry Science*, **89**, 1744–1749.

Zhu, S., Naim, F., Marcotte, M. *et al.* (2008) High-pressure destruction kinetics of *Clostridium sporogenes* spores in ground beef at elevated temperatures. *International Journal of Food Microbiology*, **126**, 86–92.

Yuste, J., Pla, R., Capellas, M. and Mor-Mur, M. (2002) Application of high-pressure processing and nisin to mechanically recovered poultry meat for microbial decontamination. *Food Control*, **13**, 451–455.

Garriga, M., Aymerich, M.T., Costa, S. *et al.* (2002b) Bactericidal synergism through bacteriocins and high pressure in a meat model system during storage. *Food Microbiology*, **19**, 509–518.

Hauben, K.J.A., Wuytack, E.Y., Soontjens, C.C.F. and Michiels, C.W. (1996) High-pressure transient sensitization of *Escherichia coli* to nisin and lysozyme by disruption of outer-membrane permeability. *Journal of Food Protection*, **59**, 350–355.

Masschalck, B., Van Houdt, R., Van Haver, E.G. and Michiels, C.W. (2001) Inactivation of gram-negative bacteria by lysozyme, denatured lysozyme, and lysozyme-derived peptides under high hydrostatic pressure. *Applied and Environmental Microbiology*, **67**, 339–344.

Ogihara, H., Yatuzuka, M., Horie, N. *et al.* (2009) Synergistic effect of high hydrostatic pressure treatment and food additives on the inactivation of *Salmonella enteritidis*. *Food Control*, **20**, 963–966.

CHAPTER 4

Electroprocessing of meat and meat products

Cristina Arroyo & James G. Lyng
Institute of Food and Health, UCD, Belfield, Dublin, Ireland

4.1 Introduction to electroprocessing technologies

In food processing operations, electrical energy can be applied directly to foodstuffs (e.g. ohmic heating, electrical stimulation or pulsed electric fields) or can be converted into electromagnetic energy (e.g. radio frequency or microwave radiation), which can in turn interact electrically with foods (Table 4.1). This electrical energy generally induces movement at a subcellular level in foods, which can lead to appreciable heating (i.e. electro heating) and can also be applied to foodstuffs to induce structural changes with minimal heating (i.e. non-thermal applications). For the purpose of this chapter, we refer to these unit operations as electroprocessing and focus on the application of these technologies in the processing of meat and meat products.

4.2 Non-thermal electroprocessing of meat

Electrical stimulation (ES) and pulsed electric fields (PEF) are electroprocessing technologies categorised as non-thermal as their main effects are not due to an increase in temperature within the foodstuff. A review of the current scientific literature shows 1022 publications on ES and meat with first reports in the early 1970s, whereas the application of PEF on meat was first mentioned in 2001 and less than 20 studies have been published up to date on this field.

4.2.1 Electrical stimulation

The process of electrical stimulation (ES) was first developed in New Zealand in the 1970s to tackle the toughening in lambs resulting from cold-shortening when frozen rapidly after slaughter to export. Nonetheless, earlier studies on the long-term effect of ES on post-mortem fall of pH in muscles of landrace pigs are reported (Hallund and Bendall, 1965). Nowadays, ES is a well-established

Emerging Technologies in Meat Processing: Production, Processing and Technology, First Edition.
Edited by Enda J. Cummins and James G. Lyng.

Table 4.1 Summary of electroprocessing conditions and applications in food processing.

	Technology	Processing conditions and applications
Non-thermal	Electrical stimulation (ES)	• Low-voltage AC (typically ≤100 V peak) • Medium-voltage AC (300 V peak) • High-voltage AC (up to 1143 V peak) Application: Intervention technique in beef and lamb processing (accelerates rigor onset)
	Pulsed electric fields (PEF)	Extremely high DC voltages (≤100,000 V) applied in short (e.g. 1–50 μs) pulses at frequencies of up to 1000 Hz for short total treatment times (often <1 s). Application: Permeabilisation of cell membranes leading to microbial inactivation and cell disruption, which can be exploited for accelerating mass transfer (e.g. extraction, curing and drying)
Thermal (electro heating)	Ohmic heating (OH)	50–500 V AC applied directly to foodstuffs at a frequency of 50–25,000 Hz. Application: Thermal processing and heating foods
	Radio frequency (RF) heating	Electrical energy converted to RF electromagnetic radiation. ISM frequencies between 13.6 and 40.7 MHz. Application: Thermal processing and heating foods
	Microwave (MW) heating	Electrical energy converted to MW electromagnetic radiation at ISM frequencies between 896 and 2450 MHz. Application: Thermal processing and heating foods

intervention method used primarily in the beef and lamb industries. There are some studies considering ES for pork, although care must be taken to prevent PSE (Pale, Soft and Exudative) meat. Provided ES is used properly and in conjunction with other slaughter and chilling procedures, the main commercial incentive is to accelerate the tenderization of meat cuts and reduce the variability in meat eating quality. Electrical stimulation involves passing an electric alternating current through the carcass of a freshly slaughtered animal to accelerate the process of rigor mortis through two phases. During stimulation, the current causes the muscles to contract vigorously, increasing the rate of glycolysis with a rapid drop in the muscle pH. Following stimulation, a temperature- and voltage-dependent change in the rate of pH decline occurs. The biochemical and physical effects of ES on beef and sheep meat tenderness were extensively reviewed by Hwang *et al.* (2003). The tenderising effect of ES has been attributed to (1) the weakening of muscle fibres as a result of the disruption of the myofibrillar network caused by the severe muscle contractions, (2) the enhanced activity of acid proteases and especially (3) the fact that electrically stimulated carcasses enter rigor faster and at higher temperatures which subsequently accelerates the rate of tenderization and prevents from cold-induced shortening (Simmons *et al.*, 2008). A range of other meat quality attributes are also affected by this carcass intervention including meat colour, colour stability and water-binding properties.

The processing/treatment options for ES have been traditionally classified as low-voltage stimulation (LVS, typically ≤100 V peak) and high-voltage stimulation (HVS, typically up to 1143 V peak) although a new approach has been developed in Australia based on medium voltage stimulation (MVS, 300 V peak) using alternating frequencies (Pearce *et al.*, 2009). Both result in effective stimulation with advantages and disadvantages although it is assumed that higher voltages are required the longer the delay between exsanguination and ES is. A wide range of combinations for the ES parameters (i.e. voltage, electrical current and pulse width, number, polarity, waveform and frequency) are being currently applied in the abattoirs depending on specific market requirements. Other parameters to consider when defining stimulation protocols for desired outcomes are the type of species, the pathway for stimulation (via nerve or direct), the pre-stimulation delay and the chilling rate. For extensive information on meat electrical stimulation, we forward the reader to the following comprehensive reviews: Lawrie and Ledward (2006) and Devine *et al.* (2014).

4.2.2 High-voltage pulsed electric fields

Unlike electrical stimulation, Pulsed Electric Fields (PEF) is not a technology originally developed to address a meat industry need and has no current commercial application on this field. For the fundamentals and applications of PEF in the food industry, we direct the reader to a comprehensive book edited by Raso and Heinz (2006). PEF involves the generation of an electric field between the two electrodes that confine the treatment gap of the PEF chamber where a product is located (in either a static or continuous design). This electric field is directly proportional to the magnitude of potential difference (V) and inversely proportional to the distance (cm) between the electrodes and is responsible for the reversible/irreversible cell membrane permeabilisation. In this case, certainly an electric current is driven across the cross section of the product although this is not the principal effect. In terms of processing parameters, PEF uses high-voltage pulses with durations in the range of microseconds to milliseconds to generate electric fields of different strengths depending on the specific purpose. PEF was first used by the 1940s in food processing for (1) expanding the shelf life of liquid products by the inactivation of pathogen and spoilage microorganisms and (2) enhancing mass transfer processes by increasing the permeability of plant and animal tissues. The well-known permeabilisation effect caused by PEF on microbial and plant cell membranes – when the strength of the applied electric field exceeds a critical value – has recently increased the interest in the potential of PEF application for meat and meat products. Nonetheless, it is important to consider the complexity of the structure and the size of a skeletal muscle cell in comparison with a microbial or a plant cell, the transformation process from muscle to meat as well as the potentially greater variability in the raw material.

Concepts for the application of PEF in meat and meat products focus on the disruption of the cell structure to influence maturing and on the enhancement of mass transfer to (1) accelerate water removal during drying, (2) accelerate the uptake of substances such as curing agents, spices and flavour enhancers in marinating/curing processes and (3) enhance the micro-diffusion of water-binding agents for the modification of water-binding characteristics. It is therefore thought that if PEF is applied to meat and the electric field exceeds the critical threshold, it could affect muscle fibres and consequently modify quality and functional attributes such as tenderness, colour and water-holding capacity. Recent trends also focus on exploring the possibilities of incorporating exogenous proteases from plant or microbial origin that might act synergistically with PEF to affect meat microstructure (Faridnia *et al.*, 2014a). Table 4.2 comprises a comprehensive summary of the published research data on the application of PEF in meat and meat products up to date.

According to Bekhit and Hopkins (2014) and Suwandy *et al.* (2015c), PEF performance on meat seems to be independent of the orientation of the muscle fibre direction with respect to the electric field. Table 4.2 shows that in most of the studies the long axis of the muscle fibre is arranged to be perpendicular to the electric field. However, clinical studies on in vivo mouse muscle electroporation indicate that electroporation threshold values differ depending on the fibre orientation with respect to the electric field due to the anisotropic electric properties of the muscle tissue (Čorović *et al.*, 2010).

After PEF application, an immediate increase in conductivity of meat samples has been observed. Töpfl (2006) reported a trend towards a higher increase in tissue conductivity when increasing the treatment intensity, and also he observed an average 2.5 mS/cm increase in pork fillet and pork shoulder conductivity when the same energy input was applied (10 kJ/kg) at different combinations of electric field strength and pulse number. A similar increase in conductivity was observed with higher intensity treatments in beef *m. semitendinosus* (84 kJ/kg) (O'Dowd *et al.*, 2013), *m. longissimus lumborum* (36 kJ/kg) (Bekhit *et al.*, 2014) and *m. semimembranosus* (70 kJ/kg) (Suwandy *et al.*, 2015b). This increase in conductivity could be then caused either by PEF-induced electroporation or as a result of the temperature increase produced by PEF since the conductivity values were similar to the untreated samples after cooling down the samples (O'Dowd *et al.*, 2013). It is worth mentioning that although PEF is classified as a non-thermal technology, during processing certain amount of energy is transformed into heat within the product. In any case, an increase in conductivity in meat might be a useful indicator of cell membrane integrity as changes in the permeability of the cell membrane enhances the mobility of intracellular constituents. In fact, considering that approximately 85% of muscle moisture is located within the myofibrils, a relatively small weight loss in meat samples following PEF has been

Table 4.2 Summary of studies of PEF application in meat up to date.

E (kV/cm)	No. of pulses	Pulse width (μs)	*F* (Hz)	Energy (kJ/kg)	*T*ª (°C)	Muscle(s)	Fibre orientationª	Equipment used/sample size (weight)	Reference
1.36	40	2	ns	ns	'room'	Chicken breast	ns	ns/ns	Gudmundsson and Hafsteinsson (2001)
3.5	100	ns	20	ns	ns	Beef *triceps brachii* (long and medial heads)	ns	ns/3 cm ø 2 cm thick	Lopp and Weber (2005)
0.5–5	Up to 11111	10–400	Variable	1–25	ns	Pork fillet, shoulder and haunch	ns	Exponential decay technical scale pulse modulator, 40 kV, 8000 J/s, TUB 7/variable sample size	Töpfl (2006); Töpfl and Heinz (2007, 2008)
3.75–15	50–150	10	5	ns	4	Chicken breast	ns	Elcrack-HPV5/1 × 1 × 1 and 1 × 1 × 4 cm	Haughton *et al.* (2012)
1.1–2.8	152–300	20	5–200	12.7–225.8	4	Beef *semitendinosus* (72 h pm)	⊥	Elcrack-HPV5/2 × 2 × 2 cm (~10 g)	O'Dowd *et al.* (2013)
0.27–0.56	Up to 2726	20	20–90	Up to 73.2	ns	Cold-boned beef *longissimus lumborum* (24 h pm) and *semimembranosus* (24 and 72 h pm)	⊥	Elcrack-HPV5/13 × 8 × 5 cm	Bekhit *et al.* (2014)
0.2–0.56	30–1528	20	1–50	0.05–34.33	4	Beef *longissimus thoracis* (48 h pm)	⊥	Elcrack-HPV5/10 × 8 × 4 cm (~350 g)	Faridnia *et al.* (2014a)
1.4	1032	20	50	250	4	Beef *semitendinosus* (24 h pm)	⊥	Elcrack-HPV5/6 × 4 × 6 cm (~70 g)	Faridnia *et al.* (2014b)
1.2–2.3	150–300	20	100–200	ns	4	Pork *longissimus thoracis et lumborum* (48 h pm)	⊥	Elcrack-HPV5/2 × 2 × 6 cm (~30 g)	McDonnell *et al.* (2014)
1.4	300–600	20	10	25–50	4	Beef *longissimus thoracis et lumborum* (48 h pm)	ns	Elcrack-HPV5/2 × 2 × 6 cm (~30 g)	Arroyo *et al.* (2015a)

(continued overleaf)

Table 4.2 (*continued*)

E (kV/cm)	No. of pulses	Pulse width (μs)	*F* (Hz)	Energy (kJ/kg)	T^a (°C)	Muscle(s)	Fibre orientation[a]	Equipment used/sample size (weight)	Reference
1.1–3	100–300	20	5–110	11–194	4 and −18	Raw and frozen turkey breast (24 h pm)	⊥	Elcrack-HPV5/2 × 2 × 6 cm (~30 g)	Arroyo *et al.* (2015b)
5–10 kV (voltage)	ns	ns	20–90	ns	ns	Cold-boned beef *longissimus lumborum* (24 h pm) and *semimembranosus* (24 and 72 h pm)	⊥	Elcrack-HPV5/13 × 8 × 5 cm	Suwandy *et al.* (2015a)
0.28–0.56	607–2724	20	20–90	3.2–70.1	24	Hot-boned beef *longissimus lumborum* and *semimembranosus* (4 h pm)	⊥	Elcrack-HPV5 / 13 × 8 × 5 cm (~365 g)	Suwandy *et al.* (2015b)
0.58–0.73	ns	20	90	16.2–19.3	9.9	Cold-boned beef *longissimus lumborum* (24 h pm)	⊥, =	Elcrack-HPV5 / 13 × 8 × 5 cm (~318 g)	Suwandy *et al.* (2015c)
0.50–0.58	Series of PEF treatments	20	90	ns	ns	Cold-boned beef *longissimus lumborum* and *semimembranosus* (6 h pm)	ns	Elcrack-HPV5 / 13 × 8 × 5 cm	Suwandy *et al.* (2015d)
0.32–0.48	Series of PEF treatments	20	90	17.5–70.8	24.5	Hot-boned beef *longissimus lumborum* and *semimembranosus* (6 h pm)	ns	Elcrack-HPV5 / 13 × 8 × 5 cm (~365 g)	Bekhit *et al.* (2016)

E, electric field strength; *F*, frequency; T^a, initial temperature of the sample; ns, no specified; pm, post-mortem.

[a]Orientation of the long axis of muscle fibres with respect to the electric field. ⊥ indicates perpendicular (i.e. fibres' direction parallel to the electrodes) and = indicates parallel.

observed (Töpfl, 2006; O'Dowd *et al.*, 2013; Bekhit *et al.*, 2014; Faridnia *et al.*, 2014a,b; McDonnell *et al.*, 2014; Arroyo *et al.*, 2015b; Suwandy *et al.*, 2015b), so it is likely for PEF to induce microstructural changes.

In fact, several studies have demonstrated physical changes in meat structure caused by PEF. Firstly, observations under the microscope performed by Gudmundsson and Hafsteinsson (2001) revealed that a PEF treatment at 1.4 kV/cm and 40 pulses of 2 μs caused a reduction in chicken muscle cell size (39% smaller) but without gapping between them or a visible change in the appearance. Cryo-SEM images performed by Faridnia *et al.* (2014a) demonstrated that the structure of PEF-treated beef *m. longissimus thoracis* (0.3–0.6 kV/cm, 8.5–34 kJ/kg) became more porous as the electric field strength increased, whereas the microstructure of untreated samples showed intact cells with no visible gapping. Further TEM images of frozen-thawed beef *m. semitendinosus* showed jagged edges in the muscular cells and myofibril separation from z-lines due to PEF (Faridnia *et al.*, 2014b). Previous studies on the myofibrils size performed by O'Dowd *et al.* (2013) suggested that PEF-treated beef was more fragmented than the control samples, which was correlated with the increase in drip loss immediately post-PEF. Muscle fibre bundles appeared smaller in diameter and muscle fibres with less volume were observed for beef samples exposed to PEF. However, NMR relaxation measurements did not confirm major changes to the cell membrane (O'Dowd *et al.*, 2013). Tissue integrity was also indirectly supported by Töpfl (2006) as the mechanical stress during tumbling of pork shoulder pieces had an impact on sample conductivity and weight loss similar to a PEF treatment although in a different timescale (hour/s vs second/s), suggesting that PEF treatment might provide a potential to reduce tumbling time during cooked ham processing. Further REM images of cooked pork showed a porous, swamp-like structure with an enhanced protein swelling in those samples exposed to PEF before brine and phosphate injection and tumbling (Töpfl, 2006). All these physical changes as a consequence of the permeabilisation of cell membranes have opened the possibility that PEF could affect the tenderness and the water-binding capacity of meat cuts. However, conflicting results are reported regarding the effect of PEF on meat tenderization. Lopp and Weber (2005) reported a significant decrease (averaged 22%) in the maximum shear force for beef *m. triceps brachii* (3.5 kV/cm, 20 Hz, 100 pulses) as well as the tenderness of beef *m. longissimus lumborum* and *m. semimembranosus* benefited (maximum reduction of 19.5% in shear force values) from PEF (0.27–0.56 kV/cm; 3.4–40.7 kJ/kg) (Bekhit *et al.*, 2014). It is noteworthy to mention that texture measurements performed by Bekhit *et al.* (2014) were taken within 1 week of freezing so it could be questioned the effect of the freezing and thawing steps on the results attributed to PEF. A similar significant decrease in the shear force of beef cold-boned (Suwandy *et al.*, 2015a) *m. longissimus lumborum* and both cold-boned (Suwandy *et al.*, 2015a) and hot-boned (Suwandy *et al.*, 2015b) *m. semimembranosus* was found under similar experimental conditions.

Furthermore, protein profiles showed that there was a faster rate of protein degradation in PEF-treated beef cold-boned (Suwandy *et al.*, 2015a) or hot-boned (Suwandy *et al.*, 2015b) *m. longissimus lumborum* than in non-treated muscle, which was reflected by the increased degradation of troponin-T and desmin. Both troponin-T and desmin degradations are used as markers of loss of myofibrillar integrity and consequently meat tenderization (Suwandy *et al.*, 2015a,b). Experiments performed by Faridnia *et al.* (2014b) showed that PEF was an effective method to tenderise frozen-thawed *m. semitendinosus* samples with maximum shear force reductions of 20%. It was hypothesised that thawed meat would have provided a more susceptible environment for endogenous enzymes to interact with their substrates. Although it has been suggested, there is no evidence of PEF enhancing the release of protease enzymes from the lysosomes or a faster release of Ca++ ions for accelerated proteolysis responsible for textural changes. Moreover, the biochemical basis of the tenderization benefit provided by PEF warrants further exploration. On the contrary, other studies have reported no enhancement of meat tenderness due to PEF at different stages during the ageing process. Despite PEF caused an increased myofibrillar fragmentation in beef *m. semitendinosus*, no changes in texture were revealed by shear force measurements (O'Dowd *et al.*, 2013). Similarly, texture of beef *m. longissimus thoracis* (Faridnia *et al.*, 2014a), *m. longissimus thoracis et lumborum* (Arroyo *et al.*, 2015b) hot-boned *m. longissimus lumborum* (Suwandy *et al.*, 2015b), cold-boned *m. longissimus lumborum* (Suwandy *et al.*, 2015c) and turkey breasts (Arroyo *et al.*, 2015a) was not affected by PEF application. The apparent discrepancy may be associated with differences in the experimental conditions such as the use of different processing parameters, muscle types, sample size and sample preparation prior and after PEF treatment. Other parameters such as colour and cook loss have been studied, and it is agreed that PEF has no detrimental effect on the colour of meat samples (O'Dowd *et al.*, 2013; Faridnia *et al.*, 2014a; Arroyo *et al.*, 2015a,b; Suwandy *et al.*, 2015c), whereas cook loss either decreased (Bekhit *et al.*, 2014; Arroyo *et al.*, 2015b), increased (Suwandy *et al.*, 2015b) or remained unaffected (Faridnia *et al.*, 2014a; Suwandy *et al.*, 2015c). According to Faridnia *et al.* (2014b), PEF enhanced lipid oxidation and altered the volatile profile of frozen-thawed beef *m. semitendinosus* during ageing although no alteration of the nutritional value of fatty acids was detected. However, neither Arroyo *et al.* (2015a) found differences in TBARS for raw, cooked and frozen turkey breasts after 5 days of storage post-PEF nor Suwandy *et al.* (2015c) for cold-boned *m. longissimus lumborum* after 9 days of display after PEF. The most recent research (Suwandy *et al.*, 2015d; Bekhit *et al.*, 2016) has been done on the effect of repeated (1x, 2x and 3x) PEF treatments on the quality of beef loins and topsides with variable results.

Removal of water in drying processes accounts for a significant amount of energy costs in the food industry. Cell membrane permeabilisation caused by PEF can be therefore utilised as a method for enhancing drying rates. Töpfl (2006) found an 80% reduction (from 300 to 60 h) of drying times for PEF-treated salted pork shoulder and that the drying rate was dependent on treatment intensity and salting procedure. In addition, a PEF pre-treatment of 3 kV/cm and 5 kJ/kg was better than 4 kV/cm and 20 kJ/kg for cured meat drying, noting that high-energy inputs resulted in a crust formation that could act as a barrier to mass transfer. In another form of mass transfer, PEF could be used to improve the uptake of salt during production of salted products as the diffusive mass transport along a concentration gradient is enhanced after breaking the permeability barrier of the cells. In this regard, Töpfl and Heinz (2007) demonstrated that a PEF treatment (3 kV/cm; 5 kJ/kg) prior to immersion in cover brine (0.08 kg/kg nitrite salt) could improve the diffusion of salt and nitrate of pork haunches. Similarly, McDonnell *et al.* (2014) showed that the NaCl content of pork *m. longissimus thoracis et lumborum* increased by 10% and 13% when applying a 1.2 kV/cm (68 kJ/kg) and 2.3 kV/cm (181 kJ/kg) PEF pre-treatment, respectively. The time required for meat curing and drying is essentially determined by the diffusion rate of curing agents and water within the tissue, and these studies highlight the potential for PEF as a pre-treatment to accelerate this process. Further research has to be carried out on the optimisation of PEF efficiency through parameter combinations which result in the lowest effective total treatment time. The fact that ingredients such as salt will modify tissue has also to be taken into account when setting up the PEF processing parameters. An acceleration of acid lactic fermentation in mincemeat for raw sausage production (salami type) was also observed by improving the availability of intracellular liquid for fermenting cultures. After a 2 kV/cm treatment with an energy input of 10 kJ/kg, the time required to achieve a pH drop to a value of 5.0 was reduced by 30% (Töpfl, 2006).

As mentioned previously, an increase in drip loss following PEF has been described. However, during the production of cooked ham, PEF and phosphate addition provided a synergetic effect on enhancing the water-binding capacity of meat and reducing its toughness after 2 h tumbling (Töpfl, 2006). Cell permeabilisation may have caused an improved access of phosphate to protein filaments resulting in an inclusion of free water after brine injection. Therefore, improvements in water retention properties of meat leading to a better yield during cooking as well as a tender product can be expected when PEF is used as a pre-treatment.

The food industry is poised to adopt new concepts and technologies that offer advantages over conventional systems or just have particular benefits. Current knowledge indicates that PEF may aid in numerous applications within meat processing industry; however, extensive validation based on a deep understanding of the phenomena taking place at a microstructural level is still necessary to justify its adoption.

4.3 Thermal electroprocessing (i.e. electroheating) of meat

Ohmic (OH) heating, microwave (MW) heating and radio frequency (RF) heating are electroprocessing technologies that induce substantial movement at a micromolecular level in a foodstuff, the overwhelming effect of which is the generation of heat due to friction. A review of the current scientific literature shows 1838 publications which associations between these technologies and meat with 3.2%, 12.6% and 84.2% of the publications on OH, RF and MW, respectively. In terms of the number of publications per annum, for MW heating, there were approximately 2–3 publications per annum from 1959 to 1973, 10–20 publications per annum from 1974 to 2006 while from 2007 this number has dramatically increased to 90–110 per annum. Similar trends are evident in OH and RF technologies.

4.3.1 Heating mechanisms

OH is a term used to describe direct electrical resistance heating of food products while MW and RF heating involve the conversion of electrical energy to electromagnetic radiation, which in turn interacts with the product and leads to heat generation. Figure 4.1 illustrates the difference between these technologies. The three rectangles are meant to represent meat products, the product in Figure 4.1a is ohmically heated, the product in Figure 4.1b is heated by RF radiation while the product in Figure 4.1c is heated by MW radiation. The three main points about this diagram are as follows:

1 OH involves the passage of an electrical current through a food product (therefore, direct electrode–product contact is necessary or the product must be

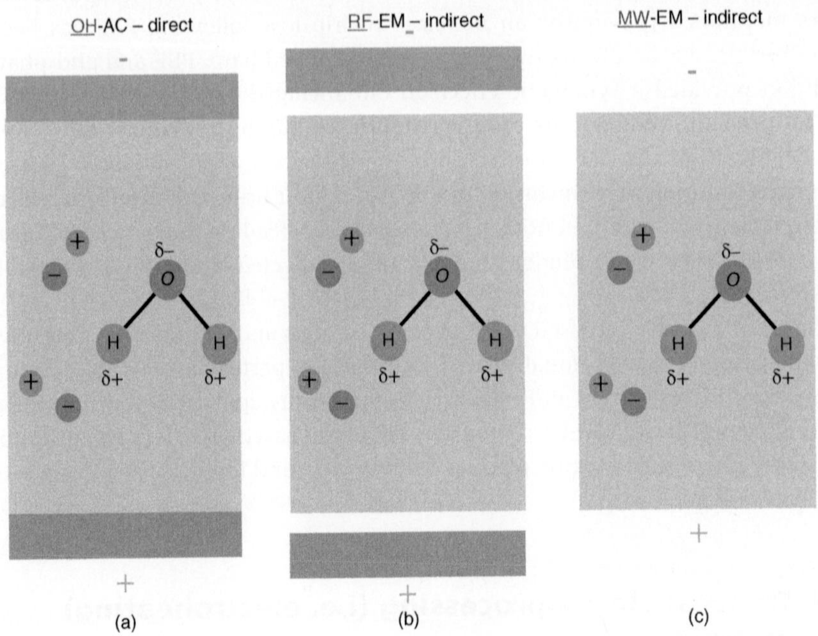

Figure 4.1 An illustration of the application differences between ohmic, radio frequency and microwave heating of materials. (a) Frequency (*f*), 50–25,000 Hz (lowest); (b) *f*, 13.6–40.7 MHz (intermediate); (c) *f*, 896–2450 MHz (highest).

packaged in appropriate conductive packaging to allow the current to pass into the foodstuff). In RF and MW radiation, electrical energy is converted into electromagnetic radiation. This radiation in turn is applied to a product via pair of electrodes not touching the product in the case of RF heating or via an oven cavity in the case of MW heating. RF and MW radiation can penetrate packaging.

2 All three technologies generate electric fields with positive and negative regions that attract/repel dissociated ions and the charged regions of dipoles. It is the friction induced by the movement of dipoles and/or ions that leads to heat generation.

3 The polarity of the electric fields is not static with constant changes in polarity occurring. OH has the lowest frequency (e.g. 50 Hz in European Union or 60 Hz in the United States although frequencies of up to 25,000 Hz are now been used (the latter is used to minimise electrolysis)), RF is intermediate (1–300 MHz although ISM (Industrial Scientific and Medical) frequencies between 13.6 and 40.7 MHz are generally the only frequencies used in food processing) while MW have the highest frequencies (300 MHz–30 GHz although ISM frequencies between 896 and 2450 MHz are generally the frequencies used in food processing). Frequency is the key factor determining whether the majority of heat generation comes from dipoles or ions. Higher frequency fields (e.g. MW) allow less time for ion movement although dipoles are continually oscillating while lower frequency fields (e.g. OH) allow more time for ion movement while dipoles relative to MW frequencies terms may have time to pause between polarity changes in OH.

4.3.2 History of electroprocessing of meat

4.3.2.1 History of OH of meat

As far back as 1882, Fowler (1882) proposed the flow of electrical current through a meat product as a cooking method. Since then several ohmic heating processes and applications have been patented and developed for the food industry (Lyng and McKenna, 2013). However, the commercial application for solid foods (i.e. meat and processed meat) has not been adopted and applications to date are limited to the processing of pumpable liquid-particulate foods. Although OH was used in its early stage mainly to pasteurise milk (Anderson and Finkelstein, 1919; Prescott, 1927), the developments of OH systems waned during 1930–1960 probably due to high electricity cost constraints and the lack of suitable electrode materials and controls (Sastry, 1994). Nevertheless, in the late 1980s, OH regained attention and considerable commercial interest as a potential alternative heating method. This was largely helped by the formation of a consortium of 25 partners in 1992 from all fields (industry, academia and government) to develop products and to evaluate the capabilities of one of the first commercial continuous flow ohmic systems, manufactured by APV Baker (Zoltai and Swearingen, 1996). While this new technology appeared very promising, ohmic commercialisation was not further pursued at that stage due to the business risk associated with start-up costs and unknown market potential (Zoltai and Swearingen, 1996). However, its application to commercial production of high value-added recipe meals was approved by the Advisory Committee on Novel Foods and Processes, and the production was set up in the United States, Europe and Japan (Jamieson and Williamson, 1999). This success in producing liquid and particulate foods has re-ignited interest in ohmic cooking of solid foods (Piette *et al.*, 2001). Strategies were also tested for a NASA project, and a simulation was carried out to ohmically heat soups in reusable pouches with incorporated electrodes (Jun and Sastry, 2007). In the past number of years, a number of more in-depth studies have been conducted on ohmic cooking of meat and meat products (Piette *et al.*, 2000a,b, 2001, 2004; Shirsat *et al.*, 2004a,b,c,d; De Halleux *et al.*, 2005; Zell *et al.*, 2009a,b,c, 2010a,b, 2011, 2012), and an extensive review of the literature relating to the OH of meat products is given by Lyng and McKenna (2013).

4.3.2.2 History of RF heating of meat

The potential application of RF heating to food processing had been recognised in the early 1940s (Piyasena *et al.*, 2003). After Second World War, Sherman (1946) described 'electric heat', how it was produced and its potential applications to the food industry. Potential uses of RF energy over a broad range of possible applications were reported in work performed between 1944 and 1947 (Cathcart *et al.*, 1947; Kinn, 1947; Moyer and Stotz, 1947). These included cooking various processed meat products, heating bread, dehydrating and blanching vegetables and so on. However, the work resulted in very few commercial installations due to the high overall operating costs associated with using RF energy. By the 1960s, studies on the application of RF energy to foods focused on thawing of frozen products, which resulted in several commercial production lines (Jason and Sanders, 1962a,b; Anon, 1992). Pircon *et al.* (1953) were the first to study RF pasteurization of meats (9 MHz, 15 kW oscillator) and described a heat process for boned ham in plastic-walled containers with metal caps as electrodes. Further developments occurred in 1970s when Bengtsson and Green (1970) developed a method for continuous RF pasteurization of cured hams packaged in Cryovac casings at 60 MHz. Houben *et al.* (1990, 1991,1994) described continuous RF pasteurization of sausage emulsion pumped through a tube (2-min treatment to elevate temperature from 15 to 80 °C at a flow rate of 120 kg/h). Microbial analysis showed that RF had a lethal effect on the test organisms at least comparable to a conventional heat treatments at the same pasteurization values, and the final products were well coagulated and had a good appearance and taste. Laycock *et al.* (2003) described the use of a 1.5 kW RF generator for cooking ground, comminuted and whole muscle meat products at 27.12 MHz using a cylindrical RF applicator, which was wrapped around the electrodes.

Instrumental quality measurements indicated that RF cooked samples had lower juice losses and also acceptable colour and water-holding capacity. However, the same workers found that the texture of ground beef was too chewy and elastic, and the whole beef colour was considered inferior. The comminuted meat displayed a large cross-sectional temperature differential, which the authors attributed to an uneven salt distribution. The authors also postulated that well-mixed comminuted and ground meat products appeared to be the most promising for RF cooking. More recently, studies on RF heating of highly comminuted encased pork meat products including small-diameter white puddings and large-diameter luncheon rolls were performed at University College Dublin (Zhang *et al.*, 2004a,b; Brunton *et al.*, 2005). In both cases, the RF heating was carried out in a cell with recirculating hot water so as to avoid an electrical arcing problem which occurred when cooking the products in air. While no significant difference ($P \geq 0.05$) was found between selected quality attributes of RF and conventionally cooked white puddings, instrumental assessments showed that RF-heated luncheon rolls had a greater ability to hold water; were significantly harder, chewier and gummier ($P < 0.001$); and had less cook colour development than their steam-cooked counterparts. Differences were also detected by sensory methods. These results clearly indicated that the effects of RF heating on the quality aspects of comminuted meats are product dependent. Further work by Tang *et al.* (2006) involved assessing the quality of RF-heated encased turkey and beef products. Compared with heating times of 150 min for meat rolls (1–1.2 kg) using a conventional steam oven at 80 °C, cooking times were reduced approximately 3.5-fold for brined turkey breast and beef muscle (*m. biceps femoris*) and 4.5-fold for non-brined rolls. Proximate analysis of macro components and assays of thiamine and riboflavin revealed no significant difference ($P \geq 0.05$) between all meat products cooked by the two heating methods. Although texture profile analysis (TPA) did not reveal significant cooking method related differences ($P \geq 0.05$) between the turkey rolls, instrumental parameters measuring toughness were significantly lower ($P < 0.05$) for RF-cooked non-brined beef rolls and for brined rolls containing water-binding dielectrically inactive additives. While a sensory panel could distinguish between RF and steam cooked turkey rolls, panellists were unable to detect texture differences that had been indicated by instrumental analysis for the beef products.

4.3.2.3 History of MW heating of meat

MW heating dates back to the 1940s although the first widespread applications in food processing only commenced in the 1980s (Mermelstein, 1998). The technology and potential applications have been comprehensively reviewed by Fito, Chiralt and Martin (2005) and Ramaswamy and Tang (2008). MW technology differs from the other two thermal technologies in that it has had considerable success in domestic application with MW ovens now being standard appliances in over 80% of homes in the United Kingdom and Northern Europe, while in Australia, the United States and Japan the saturation level is over 100% (Ohlsson and Bengtsson, 2001). In contrast to its domestic success, MW technology has not received equivalent uptake in an industrial context with economics (in particular, the higher capital cost of the equipment), technology (requirement to purpose-built equipment for specific application) and safety issues (concerns over leakage were an issue in the earlier days and also issues relating to the transfer of contaminants from susceptors and containers to the food) often being viewed as impediments to its commercial uptake (Buffler, 1993; Fleischman, 2001; Orsat *et al.*, 2005). That said the technology has been used for a broad range of food processing applications including cooking, reheating, thawing, drying and, more recently, even for crisping and roasting though many of the successful commercial applications for this technology are niche areas where the technology can be used to product with unique features not otherwise obtained by conventional processing (Orsat *et al.*, 2005).

4.3.3 Physical properties affecting OH, RF and MW heating

The physical properties of meats relevant to heating by either OH, RF or MW means have been reviewed extensively (Shirsat *et al.*, 2004a,c,d; Zhang *et al.*, 2004a,b, 2007; Lyng *et al.*, 2005; Brunton *et al.*, 2006; Zell *et al.*, 2009a).

4.3.3.1 Electrical conductivity (OH)

The electrical conductivity (σ) of a product determines its suitability for OH (De Alwis *et al.*, 1989). Piette (2001) claimed that it is theoretically possible to provide any food with enough ohmic power to generate a target temperature rise. However, in practical terms, it is generally accepted that OH is only possible between σ values of 0.01 and 10 S/m, and that it works optimally in the range of 0.1–5 S/m. In OH of meat, the flow of alternating electrical current by electrolytic conduction is influenced by the level of salt and ionic ingredients (which increase current flow) and fat and non-ionic ingredients (which reduce current flow). A uniform distribution of these ingredients is critical. Therefore, injection and tumbling procedures (where used) must be such that ionic ingredients are distributed uniformly within the product as uneven distribution will lead to current flow through more conductive regions of the product and ultimately uneven heating (Zell *et al.*, 2009a). By contrast, the presence of fat can reduce current flow (Shirsat *et al.*, 2004c) with the form of fat being particularly important. An increase in the level of endogenous fat in an entire meat product tends to reduce current flow to a greater extent than a similar increase in the level of added fat in a comminuted meat (Zell *et al.*, 2009b). In such products, the current can navigate more readily around the fat globules while the endogenous fat can be more problematic for the current especially when there are seams of fat.

4.3.3.2 Dielectric properties (RF and MW)

RF and MW are forms of electromagnetic radiation with RF occupying the frequency range 1–300 MHz while MW occupies a range from 300 MHz to 30 GHz. However, for practical application purposes, the frequencies that can be used for heating foods are limited to the ISM frequencies (Table 4.3).

As forms of electromagnetic radiation, RF and MW can be reflected, transmitted or absorbed by foods. The dielectric properties of a material are key in terms of their influence on its interaction with electromagnetic radiation. The relative dielectric constant (ε') is used to estimate the proportion of electrical energy reflected away from (P_r) or transmitted into (P_t) the product and is also used to estimate the amount of energy stored in the product (Buffler, 1993). Within products, the percentage of reflection/transmission generally does not vary hugely between ISM frequencies, although the impact of this factor on ε' becomes more apparent at higher frequencies (Barba and d'Amore, 2012). Temperature also has an influence on ε' of meats, with a substantial drop in ε' evident below freezing while composition can also have an influence (e.g. lean vs 50:50 mix of lean and fat vs fat) (Farag *et al.*, 2008a). In addition, ε' changes are evident during the cooking of meats and can be associated with phenomenon such as protein denaturation (Brunton *et al.*, 2006). Moisture and content of ionic ingredients can also have a significant influence on ε'. As a general rule, a reduction in moisture content leads to a reduction in ε' (Feng *et al.*, 2002) while an increase in concentration of ionic ingredients tends to decrease ε' although the effects are more pronounced with small changes in ionic ingredients with larger chances in moisture content required in order to observe an effect (Lyng, 2013).

The relative dielectric loss factor (ε'') describes how lossy a product is (i.e. how well a material absorbs energy from the electrical field passing through it) (Buffler, 1993). ε'' is a key property influencing penetration depth (d_p) and

Table 4.3 Industrial scientific and medical (ISM) allocated frequency (f) bands.

Technology	f (MHz)	Tolerance (±) (MHz)	Area permitted
RF	13.56	0.00678	Worldwide
RF	27.12	0.16272	Worldwide
RF	40.68	0.02034	Worldwide
MW	433.92	0.88	Austria, the Netherlands, Portugal, Germany, Switzerland, United Kingdom
MW	896	10	United Kingdom
MW	915	13	North and South America
MW	2375	50	Albania, Bulgaria, Hungary, Romania, Czechoslovakia, Russia
MW	2450	50	Worldwide

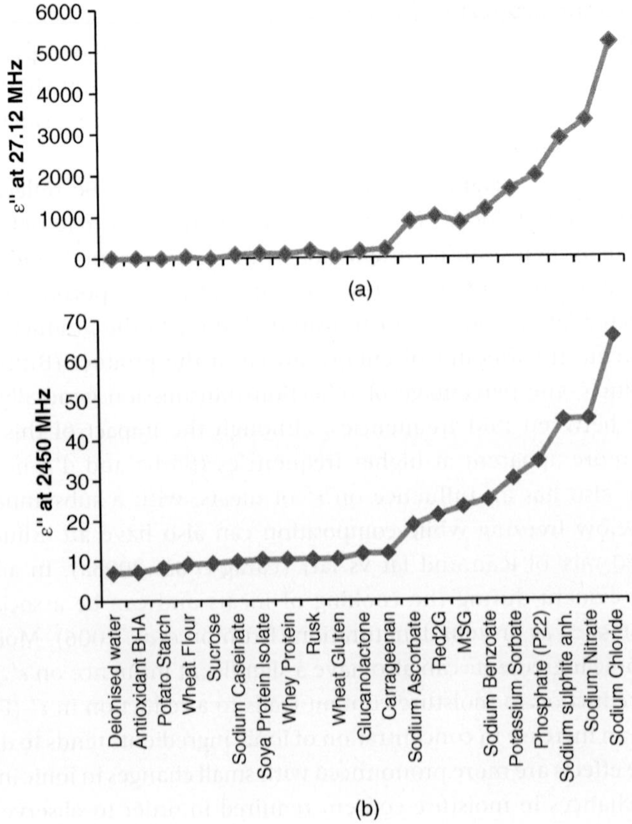

Figure 4.2 Relative dielectric loss factor (E″) of 5% aqueous solutions/suspensions of selected ingredients at (a) 27.12 and (b) 2450 MHz (adapted from Lyng (2005)). BHA, butylated hydroxyanisole; MSG, monosodium glutamate; anh., anhydrous.

power absorption (P_{abs}) with an increase in ε'' decreasing d_p and increasing P_{abs}. In terms of meat, meat products and ingredients, a survey by Lyng *et al.*, (2005) on 5% aqueous solutions of ingredients used in meat product manufacture shows similar trends in ε'' between ISM frequencies but 27.12 MHz tends to have considerably higher ε'' than 2450 MHz (Figure 4.2a,b). Furthermore, a correlation between σ, ε'' 2450 MHz and ε'' 27.12 MHz shows a virtually linear relationship between these three physical properties with ε'' 27.12 MHz considerably being higher than ε'' 2450 MHz, which in turn is considerably higher than σ (Figure 4.3a–c). Temperature also has a significant influence on ε'' with a depression in ε'' particularly evident below freezing (particularly in meats with higher ε'' e.g. lean vs fat) (Farag *et al.*, 2008a). Similar to ε', the effect of temperature on ε'' is less clear above freezing with phenomena such as protein denaturation exerting an influence at certain temperatures (Brunton *et al.*, 2006).

In terms of penetration depth (d_p), air has minimal absorption on RF and MW radiation, but on entering the product, RF or MW energy is absorbed quickly

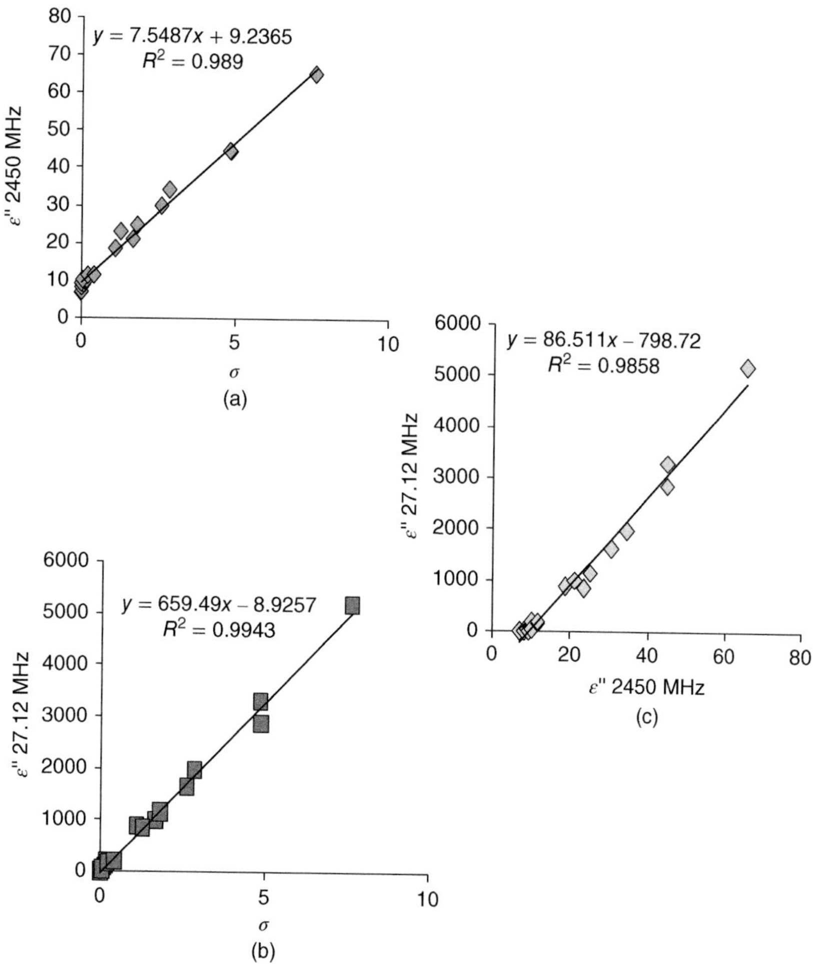

Figure 4.3 (a–c) Correlations between dielectric properties (27.12 and 2450 MHz) and electrical conductivities (S/m) of 5% aqueous solutions/suspensions of selected ingredients (adapted from Lyng *et al.* (2005)).

and the strength of the beam weakens. d_p gives an index of this as it describes distance into product to which power is reduced to approximately one-third of the power at the surface. In terms of frequency, d_p tends to be higher at lower frequencies with high ε'' ingredients decreasing d_p and vice versa (Lyng *et al.*, 2005).

4.3.3.3 Thermal and other physical properties

While σ, ε' and ε'' are the key properties influencing interaction with electrical energy, it is important to consider the impact of other thermal (e.g. specific heat capacity (C_p) (kJ/kg °C), thermal conductivity (k) (W/m °C) and thermal diffusivity (α) (m^2/s)) and mechanical (e.g. viscosity) properties on rates of heating and heat transfer. For example, products with lower C_p values (e.g. fats or oils) will tend to have lower ε'' values. However, while these products will absorb less MW or RF radiation, due to their lower C_p they require less energy for a given temperature increase and will often perform quite well during electroprocessing (Lyng, 2013). Another physical property influencing heat distribution is product viscosity. More viscous products will have poorer convection and are, therefore, less capable of rapidly redistributing absorbed energy and will sometimes display edge overheating during a MW or RF cook cycle (Lyng, 2013).

4.3.4 Applications in meat processing

4.3.4.1 Defrosting of meats

In some sectors of the meat industry (e.g. burger manufacturing industry), processors hold a stock of frozen ingredients including lean and fat from which meat products are manufactured by further processing (e.g. grinding, bowl chopping, slicing and dicing) and combination with other ingredients. In commercial setting, this meat is frozen in large (generally 25 kg) blocks and defrosted as required. The difficulty with complete thawing (i.e. elevating product temperatures from −20 °C to +2 to +4 °C) of these large blocks using RF and particularly MW technology is localised heating at corners and edges. This phenomenon goes back to dielectric properties. As MW energy bounces around an oven cavity, it tends to pass through the corners and edges of the blocks more frequently. As a result, corners and edges tend to heat more rapidly, which leads to the ice melting, which leads to a phenomenon called 'runaway heating' at corners and edges. This runaway heating can be overcome to some extent by reducing the microwave power, extending the thawing period or using pulsed microwaves to allow time for temperature equilibrium. However, complete thawing of meats is generally not practiced commercially.

By contrast, one of the most successful commercial applications for RF and MW heating is tempering of frozen foods (Farag *et al.*, 2008a,b, 2009a,b, 2010, 2011). Tempering is a form of defrosting where the product temperature is raised from frozen storage temperature (e.g. −20 °C) to a temperature just below the freezing point of the food product (−5 to −2 °C). Under these conditions, there is limited phase change (ε'' still low), and as a result overheating problems can be avoided. Lower frequency MW (e.g. 896 or 915 MHz) is generally used for tempering such blocks of meat as the longer wavelengths allow for greater penetration for thicker products. Reported advantages include faster processing (e.g. meat blocks defrosted in 10 min compared to several days in tempering rooms), less storage space required (i.e. no need for conventional tempering rooms) and low drip loss, which increases product yields and reduces nutritional losses. RF or MW tempering also has the advantage of lower labour requirement as the product is defrosted in boxes that improves hygiene and reduces the amount of handling while it has been reported that quality problems such as bacterial growth can be reduced compared to thawing (James and James, 2002). Overall, there is much better control over defrosting conditions and hence improved product quality and greater ability to respond more quickly to product demand. RF and MW systems are available in batch or continuous formats.

4.3.4.2 Pasteurization of meat

Thermal processing of meat dates back to the beginning of the nineteenth century when in 1810 Nicolas Appert found that meat would remain edible if it was heated in a sealed container and the seal maintained until the meat was to be eaten. Heat processing is still viewed as one of the key methods for meat preservation as it kills pathogenic and spoilage microorganisms likely to grow during normal distribution and shelf life of a product while also inactivating enzymes. Conventional pasteurization of meat products generally involves placing a product in a steam oven or immersing it in tanks of hot water. When the product is placed in heating media, the outer surface of the product heats first and then heat transfers from this outer surface to the interior predominantly by conduction that is relatively slow. The net effect of this is that the product must be left in the heating media for a relatively long period of time for the interior to heat to an appropriate temperature. Meanwhile, the outer surface of the product will be at this high temperature for a relatively long period of time (Figure 4.4).

In contrast, OH, RF and MW heat volumetrically, which theoretically means that all parts of the product more or less heat simultaneously and at the same

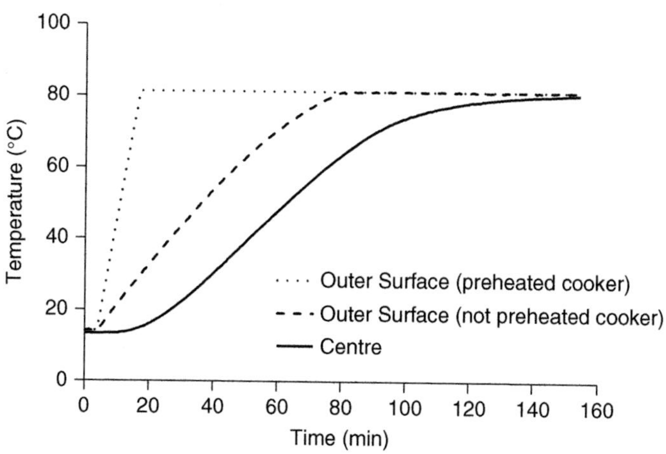

Figure 4.4 An example of a typical temperature profile (surface vs centre) of a meat product placed in a preheated and a non-preheated cooker.

rate. This largely avoids overheating in the outer regions of the product and could potentially increase product quality and nutritional value while reducing cooking times. This in turn could be expected to lead to the production of products with improved nutritional quality. In addition, it also makes continuous processing a reality for such products (Lyng and McKenna, 2013).

There are a number of critical process parameters that need to be considered in applying electro heating technologies to meat products. While RF and MW will penetrate non-conductive packaging such as casings, the use of traditional metal clips may be problematic and these may need to be replaced by plastic clips. Furthermore, a number of studies have reported issues such as arcing and burning of packaging when RF heating of meats in air, although this can be eliminated if the product is heated while immersed in hot water (Lyng *et al.*, 2007). In the case of OH, packaging must be such that it is sufficiently conductive to allow the current into the product. Furthermore, to avoid conductive heat losses from products, OH electrodes should be relatively thin. In all cases (OH, RF and MW) while heat is generated volumetrically, if the product is in a cool air environment during heating, heat losses can occur, leading to the location of cold spots in areas such as product surfaces (Zell *et al.*, 2011). Therefore, for large meat products, it is generally advisable to have the product in a hot air environment during heating.

The volumetric nature of these technologies is such that they have the advantage of allowing products to be elevated to target temperatures that vary rapidly. This can then be followed by a holding time, which in the case of meat pasteurization could involve immersion in hot water or hot air (Lyng and McKenna, 2013). The purpose of a holding time is to ensure sufficient microbial destruction occurs, and this can be quantified using values such as the F_0 (in the case of sterilisation) or PU (Pasteurisation Units, in the case of pasteurisation). However, in the case of meat products, the heat process should be sufficient to develop chemical reactions, which give meat the characteristic flavour, colour and texture of a cooked product. These reactions can be quantified using the cook value (C_s). Conventional heat processing of large-diameter meat products can result in overheating of the outer surface of the product while waiting for the product centre to reach the appropriate temperature. The graph in Figure 4.5 shows the results when the yield of RF or OH meat products is compared to steam-cooked products. In the case of comminuted products, no significant yield difference between RF and steam-cooked products is evident. However, for the non-comminuted products (ham and beef), in all cases the RF cooked had higher yield than the steam cooked, with the difference being about 1–1.5% in the ham and 4–6% in the beef. As products cook, proteins such as myosin and collagen denature and when they denature they lose water. This result may be indicative of less heat-induced protein denaturation (and associated water loss) in the outer regions of the more rapidly OH-/RF-heated cooked products, which manifests itself in products with superior yield to conventionally cooked

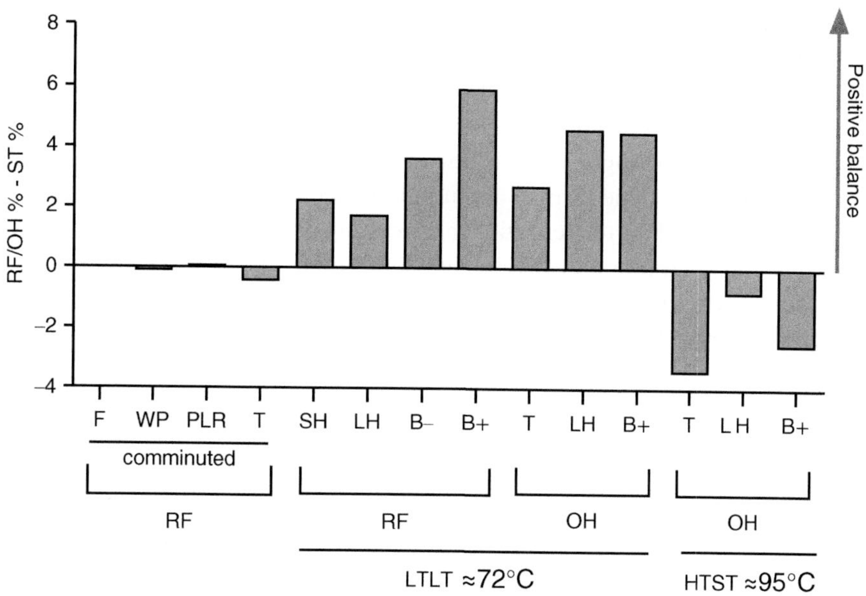

Figure 4.5 A comparison of the yield of conventionally versus electroheated (ohmic or radio frequency heated) meats. F: Frankfurter; WP: White Pudding; PLR: Pork Luncheon Roll; T: Turkey; SH: Shoulder Ham; LH: Leg Ham; B–: Beef (no added ingredients); B+: Beef (added ingredients). ST: Steam; RF: Radio Frequency; OH: Ohmic Heating; LTLT: Low Temperature Long Time; HTST: High Temperature Short Time.

counterparts. A possible cause of this may be less advanced heat-induced protein denaturation with less loss of associated water from myosin and collagen denaturation. Proximate analysis results also corroborate this with reports that volumetrically cooked large-diameter products have higher moisture content and lower protein and fat than their conventional counterparts. The more prolonged heating at the product surface is also evident in the sample colour with darker colours in surface regions of conventionally cooked meats due to prolonged time at higher temperatures. In addition, it can be detected in the variation in product texture across the product diameter with conventional products showing variation in hardness, springiness, gumminess and chewiness from surface to centre, a phenomenon which is not evident in a volumetrically cooked product (e.g. OH).

4.3.4.3 MW cooking of bacon

The two most successful uses of MW energy for food processing, in terms of a number of operating systems, are tempering meat, poultry and fish and precooking of bacon, with processors in North America and the UK precooking about 2,500,000 slices of bacon each hour (Eves, 1996).

4.4 Future of electroprocessing of meat

Electrical stimulation is an established technology in meat processing while microwave heating and radio frequency heating have successful commercial applications in tempering and defrosting of meats and cooking of bacon. However, much remains to be learned about pulsed electric field processing meat in terms of its potential for tenderization, accelerating drying and curing. Similarly, ohmic heating is a technology that is likely to see more future applications in the heat processing of meat as it is lower cost and more energy-efficient alternative to microwave and radio frequency technologies.

4.5 Equipment suppliers

Supplier	Website, email and telephone	ES	PEF	OH	RF	MW
AGL Consultancy Ltd. Cobham, Surrey, England	Website: http://aglconsultancy.com/page3.htm; Tel: +44 (0)844 504 8071	✓				
Sperling Industries Ltd. Sarpy Ave, Bellevue, NE 68147, USA	Website: www.sperlingind.com; Email: sales@boss.omhcoxmail.com; Tel: +1 402 556 4070	✓				
Jarvis Products Corp. 33 Anderson Road, Middletown, CT 06457-4926, USA	Website: http://www.jarviscanada.com/; Email: jarvis.products.corp@snet.net; Tel:+1 (860) 347 7271	✓				
BCH Limited Spring Place, Millfold, Whitworth, Lancashire, England, OL12 8DN	Website: http://www.bchltd.com/; Email: info@bchltd.com; Tel: +44 (0)1706 85 2122			✓	✓	✓
Sairem 12, porte du Grand Lyon, BP 80214, 01702 Neyron Cedex, France	Website: http://www.sairem.com/; Email: welcome@sairem.com; Tel: +33(0)4 72 01 81 60				✓	✓

Supplier	Website, email and telephone	ES	PEF	OH	RF	MW
Ctech Innovation Capenhurst Technology Park, Capenhurst, Chester CH1 6EH, UK	Website: http://www.ctechinnovation.com/; Email: info@ctechinnovation.com; Tel: +44 (0)151 347 2900		✓	✓	✓	✓
Elea Technology Prof.-von-Klitzing-Str., 7 D-49610, Quakenbrück, Germany	Website: http://www.elea-technology.eu/; Email: info@elea-technology.com; Tel: +49 (0)5431 183 442		✓			
Diversified Technologies, Inc. 35 Wiggins Avenue, Bedford, MA 01730, USA	Website: http://www.divtecs.com/; Email: info@divtecs.com; Tel: +1 781-275-9444		✓			
Strayfield Ltd. Units 10/11, Ely Road, Theale, Berkshire, England, RG7 4BQ	Website: http://www.strayfield.co.uk/; Email: info@strayfield.co.uk; Tel: +44 (0)118 932 7760				✓	
STALAM S.p.A. Via dell'Olmo, 7 (Z.I.) - 36055 NOVE (Vicenza), Italy	Website: http://www.stalam.it/; Email: stalam@stalam.com; Tel: +39 0424 597400				✓	
Emmepiemme Srl Via Paolo Burali, 2 Piacenza PC, Italy	Website: http://emmepiemmesrl.it/; Tel: +39 0523 451583			✓		
Pulsemaster Rootven 24 - 5531 MB Bladel – The Netherlands	Website: www.pulsemaster.us; Email: info@pulsemaster.us; Tel: +31 497 820300		✓			

Acknowledgements

This research has been partly funded by grant aid under the Food Institutional Research Measure, which is administered by the Department of Agriculture, Food and The Marine, Ireland.

References

Anderson, A.K. and Finkelstein, R. (1919) A study of the electropure process of treating milk. *Journal of Dairy Science*, **2**, 374–406.

Anon (1992) Radio Waves cut meat defrosting time. *Process Engineering*, **73**, 21.

Arroyo, C., Eslami, S., Brunton, N.P. *et al.* (2015a) An assessment of the impact of pulsed electric fields processing factors on oxidation, color, texture and sensory attributes of turkey breast meat. *Poultry Science*, **94**, 1088–1095. doi: 10.3382/ps/pev097.

Arroyo, C., Lascorz, D., O'Dowd, L. *et al.* (2015b) Effect of pulsed electric field treatments at various stages during conditioning on quality attributes of beef *longissimus thoracis et lumborum* muscle. *Meat Science*, **99**, 52–59. doi: 10.1016/j.meatsci.2014.08.004.

Barba, A.A. and d'Amore, M. (2012) Relevance of dielectric properties in microwave assisted processes, in *Microwave materials characterisation* (ed. S. Costanzo), pp. 91–118. InTech, Croatia. ISBN 978-953-51-0848-1.

Bekhit, A.E.D. and Hopkins, D.L. (2014) *Enhancement of Meat Quality by Pulsed Electric Field Application*, Meat and Livestock Australia. Available at: http://www.ampc.com.au/site/assets/media/reports/2015/A.MQA.0005-Enhancement-of-meat-quality-by-pulse-electric-field-application.pdf (last accessed May 2016).

Bekhit, A. E. D., Suwandy, V., Carne, A., van de Ven, R. and Hopkins, D. L. (2016) Effect of repeated pulsed electric field treatment on the quality of hot-boned beef loins and topsides. *Meat Science*, **111**, 139–146.

Bekhit, A.E.D., van de Ven, R., Suwandy, V. *et al.* (2014) Effect of pulsed electric field treatment on cold-boned muscles of different potential tenderness. *Food and Bioprocess Technology*, **7**, 3136–3146.

Bengtsson, N.E. and Green, W. (1970) Radio frequency pasteurization of cured hams. *Journal of Food Science*, **35**, 681–687.

Brunton, N.P., Lyng, J.G., Li, W.Q. *et al.* (2005) Effect of radio frequency (RF) heating on the texture, colour and sensory properties of a comminuted pork meat product. *Food Research International*, **38**, 337–344. doi: 10.1016/j.foodres.2004.06.016

Brunton, N.P., Lyng, J.G., Zhang, L. and Jacquier, J.C. (2006) The use of dielectric properties and other physical analyses for assessing protein denaturation in beef *biceps femoris* muscle during cooking from 5 to 85 °C. *Meat Science*, **72**, 236–244.

Buffler, C.R. (1993) *Microwave Cooking and Processing: Engineering Fundamentals for the Food Scientist*, Van Nostrand Reinhold, New York.

Cathcart, W.H., Parker, J.J. and Beattie, H.G. (1947) The treatment of packaged bread with high frequency heat. *Food Technology*, **1**, 174–177.

Čorović, S., Županič, A., Kranjc, S. *et al.* (2010) The influence of skeletal muscle anisotropy on electroporation: In vivo study and numerical modelling. *Medical & Biological Engineering & Computing*, **48**, 637–648.

De Alwis, A.A.P., Halden, K. and Fryer, P.J. (1989) Shape and conductivity effects in the ohmic heating of foods. *Chemical Engineering Results Dissemination*, **67**, 159–168.

De Halleux, D., Piette, G., Buteau, M.-L. and Dostie, M. (2005) Ohmic cooking of processed meats: Energy evaluation and food safety considerations. *Canadian Biosystems Engineering*, **47**, 3.41–3.47.

Devine, C.E., Hopkins, D.L., Hwang, I.H. *et al.* (2014) Electrical stimulation, in *Encyclopedia of Meat Sciences*, 2nd edn (eds M. Dikeman and C. Devine), Academic Press, Elsevier, USA, pp. 486–496.

Eves, G. (1996) Industrial microwave food processing. *Microwave World*, **17**, 15–18.

Farag, K.W., Lyng, J.G., Morgan, D.J. and Cronin, D.A. (2008a) Dielectric and thermophysical properties of different beef meat blends over a temperature range of −18 to +10 °C. *Meat Science*, **79**, 740–747. doi: 10.1016/j.meatsci.2007.11.005

Farag, K.W., Lyng, J.G., Morgan, D.J. and Cronin, D.A. (2008b) A comparison of conventional and radio frequency tempering of beef meats: Effects on product temperature distribution. *Meat Science*, **80**, 488–495. doi: 10.1016/j.meatsci.2008.01.015

Farag, K.W., Lyng, J.G., Morgan, D.J. and Cronin, D.A. (2009a) Effect of low temperatures (−18 to +5 °C) on the texture of beef lean. *Meat Science*, **81**, 249–254. doi: 10.1016/j.meatsci.2008.07.028

Farag, K.W., Duggan, E., Morgan, D.J. *et al.* (2009b) A comparison of conventional and radio frequency defrosting of lean beef meats: Effects on water binding characteristics. *Meat Science*, **83**, 278–284. doi: 10.1016/j.meatsci.2009.05.010

Farag, K.W., Marra, F., Lyng, J.G. *et al.* (2010) Temperature changes and power consumption during radio frequency tempering of beef lean/fat formulations. *Food and Bioprocess Technology*, **3**, 732–740. doi: 10.1007/s11947-008-0131-5

Farag, K.W., Lyng, J.G., Morgan, D.J. and Cronin, D.A. (2011) A comparison of conventional and radio frequency thawing of beef meats: Effects on product temperature distribution. *Food and Bioprocess Technology*, **4**, 1128–1136. doi: 10.1007/s11947-009-0205-z

Faridnia, F., Bekhit, A.E.D., Niven, B. and Oey, I. (2014a) Impact of pulsed electric fields and post-mortem vacuum ageing on beef *longissimus thoracis* muscles. *International Journal of Food Science and Technology*, **49**, 2339–2347.

Faridnia, F., Ma, Q.L., Bremer, P.J. *et al.* (2014b) Effect of freezing as pre-treatment prior to pulsed electric field processing on quality traits of beef muscles. *Innovative Food Science & Emerging Technologies*, **29**, 31–40

Feng, H., Tang, J. and Cavalieri, R.P. (2002) Dielectric properties of dehydrated apples as affected by moisture and temperature. *Transactions of the ASAE*, **45**, 129–135.

Fito, P., Chiralt, A. and Martin, M.E. (2005) Current state of microwave applications to food processing, in *Novel Food Processing Technologies Book Series: Food Science and Technology: A Series of Monographs, Textbooks, and Reference Books* vol. 141 (eds G.V. Barbosa Canovas, M.S. Tapia and M.P. Cano), pp. 525–537. CRC Press, Marcel Dekker. ISBN: 0-8247-5333-X.

Fleischman, G.J. (2001) Safety in microwave processing, in *Handbook of Microwave Technology for Food Applications* (eds A.K. Datta and R.C. Anantheswaran), Marcel Dekker, New York, pp. 471–497.

Fowler, A. (1882). Apparatus for preserving meats. U.S. Patent 267,685.

Gudmundsson, M. and Hafsteinsson, H. (2001) Effect of electric field pulses on microstructure of muscle foods and roes. *Trends in Food Science & Technology*, **12**, 122–128.

Hallund, O. and Bendall, J.R. (1965) Long-term effect of electrical stimulation on post-mortem fall of pH in muscles of landrace pigs. *Journal of Food Science*, **30**, 296.

Haughton, P.N., Lyng, J.G., Cronin, D.A. *et al.* (2012) Efficacy of pulsed electric fields for the inactivation of indicator microorganisms and foodborne pathogens in liquids and raw chicken. *Food Control*, **25**, 131–135.

Houben, J., Schoenmakers, L., van Putten, E. *et al.* (1991) Radio frequency pasteurization of sausage emulsions as a continuous process. *Journal of Microwave Power and Electromagnetic Energy*, **26**, 202–205.

Houben, J., van Roon, P. and Krol, B. (1994) Continuous radio frequency pasteurization in sausage manufacturing lines. *Fleischwirtschaft International*, **2**, 35–36.

Houben, J.H., van Roon, P.S., Krol, B. and Jansen, W.J.L. (1990) Radio frequency pasteurization of moving sausage emulsions, in *Processing and Quality of foods*, vol. 1 (eds P. Zuethen *et al.*), Elsevier Science Publishers, London, pp. 171–177.

Hwang, I.H., Devine, C.E. and Hopkins, D.L. (2003) The biochemical and physical effects of electrical stimulation on beef and sheep meat tenderness. *Meat Science*, **65**, 677–691.

James, S.J. and James, C. (2002) Thawing and tempering, in *Meat Refrigeration*, (eds. James, S.J & James, C.) Woodhead publishing, Cambridge, pp. 159–187.

Jamieson, L. and Williamson, P. (1999) The potential of electro-technologies for the processing of foods. *Food Science and Technology*, **13**, 97–101.

Jason, A.C. and Sanders, H.R. (1962a) Dielectric thawing of fish. I. Experiments with frozen herrings. *Food Technology*, **16**, 101–106.

Jason, A.C. and Sanders, H.R. (1962b) Dielectric thawing of fish. II. Experiments with frozen white fish. *Food Technology*, **16**, 107–112.

Jun, S. and Sastry, S.K. (2007) Reusable pouch development for long term space missions: A 3D ohmic model for verification of sterilization efficacy. *Journal of Food Engineering*, **80**, 1199–1205.

Kinn, T.P. (1947) Basic theory and limitations of high frequency heating equipment. *Food Technology*, **1**, 161–173.

Laycock, L., Piyasena, P. and Mittal, G.S. (2003) Radio frequency cooking of ground, comminuted and muscle meat products. *Meat Science*, **65**, 959–965.

Lawrie, R.A. and Ledward, D. (2006) The storage and preservation of meat: I Temperature control, in *Lawrie's Meat Science*, 7th edn (eds R.A. Lawrie and D. Ledward), Woodhead publishing Limited, UK, pp. 189–234.

Lopp, A. and Weber, H. (2005) Untersuchungen zur Optimierung der Zartheit von Rindfleisch: Research into the optimising the tenderness of beef from parts of the forequarter. *Fleischwirtschaft*, **85**, 111–116.

Lyng, J.G. (2013). *Dielectric Properties of Food Materials*. IFT 13 Conference, July 13–16, McCormick Place, South Chicago, IL, USA.

Lyng, J.G., Zhang, L. and Brunton, N.P. (2005) A survey of the dielectric properties of meats and ingredients used in meat product manufacture. *Meat Science*, **69**, 589–602. doi: 10.1016/j.meatsci.2004.09.011

Lyng, J.G. and McKenna, B.M. (2013) Ohmic pasteurization of meat and meat products (Chapter 20), in *Food Machinery Design Handbook: Farming, Processing, and Packaging*, 2nd edn (ed. M. Kutz), Academic Press, San Diego, CA. ISBN: 978-0-12-385881-8.

Lyng, J.G., Cronin, D.A., Brunton, N.P. *et al.* (2007) An examination of factors affecting radio frequency heating of an encased meat emulsion. *Meat Science*, **75**, 470–479. doi: 10.1016/j.meatsci.2006.07.022

McDonnell, C.K., Allen, P., Chardonnereau, F.S. *et al.* (2014) The use of pulsed electric fields for accelerating the salting of pork. *LWT - Food Science and Technology*, **59**, 1054–1060.

Mermelstein, N.H. (1998) Microwave and radio frequency drying. *Food Technology*, **52**, 84–86.

Moyer, J.C. and Stotz, E. (1947) The blanching of vegetables by electromagnetics. *Food Technology*, **1**, 252–257.

O'Dowd, L.P., Arimi, J.M., Noci, F. *et al.* (2013) An assessment of the effect of pulsed electrical fields on tenderness and selected quality attributes of post rigour beef muscle. *Meat Science*, **93**, 303–309.

Ohlsson, T. and Bengtsson, N. (2001) Microwave technology and foods. *Advances in Food and Nutrition Research*, **43**, 65–140.

Orsat, V., Raghavan, V. and Meda, V. (2005) Microwave technology for food processing: an overview, in *The microwave processing of foods* (eds H. Schubert and M. Reigier), Woodhead publishing Limited, Cambridge, pp. 105–118.

Pearce, K.L., Hopkins, D.L., Williams, A. *et al.* (2009) Alternating frequency to increase the response to stimulation from medium voltage electrical stimulation and the effect on objective meat quality. *Meat Science*, **81**, 188–195.

Piette, G., Dostie, M. and Ramaswamy, H. (2001). *Ohmic Cooking of Processed Meats – State of the Art and Prospects*. 47th International Congress of Meat Science and Technology, 62–67.

Piette, G., Buteau, M.L., de Halleux, D. *et al.* (2004) Ohmic cooking of processed meats and its effects on product quality. *Journal of Food Science*, **69**, FEP71–FEP78.

Piette, G., de Halleux, D., Dostie, M., Raymond, Y. and Buteau M.-L. (2000a). *Energetic Aspects of Ohmic Cooking for Processed Meats*. 46th International Congress of Meat Science and Technology, pp. 244–245.

Piette, G., Ramaswamy, H., de Halleux, D., Dostie, M., Raymond, Y., Buteau M.-L. and Vachon, L. (2000b). *Ohmic Heating Behaviour of Processed Meats*. 46th International Congress of Meat Science and Technology, pp. 246–247.

Pircon, L.J., Loquercio, P. and Doty, D.M. (1953) High frequency heating as a unit operation in meat processing. *Journal of Agriculture and Food Chemistry*, **1**, 844–847.

Piyasena, P., Ramaswamy, H.S., Awuah, G.B. and Defelice, C. (2003) Dielectric properties of starch solutions as influenced by temperature, concentration, frequency and salt. *Journal of Food Process Engineering*, **26**, 93–119.

Prescott, S.C. (1927) The treatment of milk by an electrical method. *American Journal of Public Health*, **17**, 221–223.

Ramaswamy, H. and Tang, J. (2008) Microwave and radio frequency heating. *Food Science and Technology International*, **14**, 423–427.

Raso, J. and Heinz, V. (eds) (2006) *Pulsed Electric Fields Technology for the Food Industry – Fundamentals and Applications*, Springer, US.

Sherman, V.W. (1946) Electronic heat in the food industries. *Food Industry*, **18**, 506–509, 628–630.

Sastry, S.K. (1994) Ohmic heating, in *Minimal Processing of Foods and Process Optimization: An Interface* (eds R.P. Singh and F.A.R. Oliveira), CRC Press, pp. 17–33.

Shirsat, N., Brunton, N.P., Lyng, J.G. and McKenna, B. (2004a) Water holding capacity, dielectric properties and light microscopy of conventionally and ohmically cooked meat emulsion batter. *European Food Research and Technology*, **219**, 1–5. doi: 10.1007/s00217-004-0920-7

Shirsat, N., Brunton, N.P., Lyng, J.G. *et al.* (2004b) Texture, colour and sensory evaluation of a conventionally and ohmically cooked meat emulsion batter. *Journal of the Science of Food and Agriculture*, **84**, 1861–1870. doi: 10.1002/jsfa.1869

Shirsat, N., Lyng, J.G., Brunton, N.P. and Mckenna, B.M. (2004c) Conductivities and ohmic heating of meat emulsion batters. *Journal of Muscle Foods*, **15**, 121–137. doi: 10.1111/j.1745-4573.2004.tb00716.x

Shirsat, N., Lyng, J.G., Brunton, N.P. and McKenna, B. (2004d) Ohmic processing: Electrical conductivities of pork cuts. *Meat Science*, **67**, 507–514. doi: 10.1016/j.meatsci.2003.12.003

Simmons, N.J., Daly, C.C., Cummings, T.L. *et al.* (2008) Reassessing the principles of electrical stimulation. *Meat Science*, **80**, 110–122.

Suwandy, V., Carne, A., van de Ven, R. *et al.* (2015a) Effect of pulsed electric field on the proteolysis of cold boned beef *M. Longissimus lumborum* and *M. Semimembranosus*. *Meat Science*, **100**, 222–226.

Suwandy, V., Carne, A., van de Ven, R., Bekhit, A. E. D. and Hopkins, D. L. (2015b). Effect of pulsed electric field treatment on hot-boned muscles of different potential tenderness. *Meat Science*, **105**, 25–31.

Suwandy, V., Carne, A., van de Ven, R., Bekhit, A. E. D. and Hopkins, D. L. (2015c). Effect of pulsed electric field treatment on the eating and keeping qualities of cold-boned beef loins: impact of initial pH and fibre orientation. *Food and Bioprocess Technology*, **8**, 1355–1365.

Suwandy, V., Carne, A., van de Ven, R., Bekhit, A. E. D. and Hopkins, D. L. (2015d). Effect of repeated pulsed electric field treatment on the quality of cold-boned beef loins and topsides. *Food and Bioprocess Technology*, **8**, 1218–1228.

Tang, X.Y., Lyng, J.G., Cronin, D.A. and Durand, C. (2006) Radio frequency heating of beef rolls from biceps femoris muscle. *Meat Science*, **72**, 467–474. doi: 10.1016/j.meatsci.2005.08.011

Töpfl, S. (2006). Pulsed electric fields (PEF) for permeabilization of cell membranes in food- and bioprocessing – Applications, process and equipment design and cost analysis. PhD Thesis, Technische Universität Berlin, Germany.

Töpfl, S. and Heinz, V. (2007) Application of pulsed electric fields to improve mass transfer in dry cured meat products. *Fleischwirtschaft International*, **22**, 62–64.

Töpfl, S. and Heinz, V. (2008) The stability of uncooked cured products. Use of pulsed electric fields to accelerate mass transport operations in uncooked cured products. *Fleischwirtschaft*, **88**, 127–130.

Zell, M., Lyng, J.G., Cronin, D.A. and Morgan, D.J. (2009a) Ohmic heating of meats: Electrical conductivities of whole meats and processed meat ingredients. *Meat Science*, **83**, 563–570. doi: 10.1016/j.meatsci.2009.07.005

Zell, M., Lyng, J.G., Cronin, D.A. and Morgan, D.J. (2009b) Ohmic cooking of whole beef muscle-Optimisation of meat preparation. *Meat Science*, **81**, 693–698. doi: 10.1016/j.meatsci.2008.11.012

Zell, M., Lyng, J.G., Morgan, D.J. and Cronin, D.A. (2009c) Development of rapid response thermocouple probes for use in a batch ohmic heating system. *Journal of Food Engineering*, **93**, 344–347. doi: 10.1016/j.jfoodeng.2009.01.039

Zell, M., Lyng, J.G., Morgan, D.J. and Cronin, D.A. (2011) Minimising heat losses during batch ohmic heating of solid food. *Food and Bioproducts Processing*, **89**, 128–134. doi: 10.1016/j.fbp.2010.04.003

Zell, M., Lyng, J.G., Cronin, D.A. and Morgan, D.J. (2010a) Ohmic cooking of whole turkey meat-effect of rapid ohmic heating on selected product parameters. *Food Chemistry*, **120**, 724–729. doi: 10.1016/j.foodchem.2009.10.069

Zell, M., Lyng, J.G., Cronin, D.A. and Morgan, D.J. (2010b) Ohmic cooking of whole beef muscle-Evaluation of the impact of a novel rapid ohmic cooking method on product quality. *Meat Science*, **86**, 258–263. doi: 10.1016/j.meatsci.2010.04.007

Zell, M., Lyng, J.G., Morgan, D.J. and Cronin, D.A. (2012) Quality evaluation of an ohmically cooked ham product. *Food and Bioprocess Technology*, **5**, 265–272.

Zoltai, P. and Swearingen, P. (1996) Product development considerations for ohmic processing. *Food Technology*, **50**, 263–266.

Zhang, L., Lyng, J.G. and Brunton, N.P. (2007) The effect of fat, water and salt on the thermal and dielectric properties of meat batter and its temperature following microwave or radio frequency heating. *Journal of Food Engineering*, **80**, 142–151. doi: 10.1016/j.jfoodeng.2006.05.016

Zhang, L., Lyng, J.G. and Brunton, N.P. (2004a) Effect of radio frequency cooking on the texture, colour and sensory properties of a large diameter comminuted meat product. *Meat Science*, **68**, 257–268. doi: 10.1016/j.meatsci.2004.03.011

Zhang, L., Lyng, J.G., Brunton, N. *et al.* (2004b) Dielectric and thermophysical properties of meat batters over a temperature range of 5–85°C. *Meat Science*, **68**, 173–184. doi: 10.1016/j.meatsci.2004.02.009

CHAPTER 5

Application of infrared and light-based technologies to meat and meat products

Tatiana Koutchma
Guelph Food Research Centre, Agriculture and Agri-Food Canada, Guelph, ON, Canada N1C 5G9

5.1 Introduction

There are three innovative food processing techniques–ultraviolet (UV) and pulsed light, and infrared irradiation that found applications in meat operations for treatment of raw and finished products and improving safety of meat facilities by disinfecting air and non-food contact and food contact surfaces.

For finished ready-to eat-products (RTE) these technologies can be used as post-lethality treatments. Post-lethality treatments are defined as lethality treatment that are applied or are effective after post-lethality exposure. They are applied to the final product or sealed package of product in order to reduce or eliminate the level of pathogens resulting from contamination from post-lethality exposure. With the increase in national and wholesale distribution systems, which require that products have longer refrigerated shelf lives, the opportunity for *Listeria monocytogenes* to survive and grow in vacuum-packaged, RTE meat products had increased. According to the USDA-FSIS guidelines antimicrobial interventions and post-lethality treatments can be used for Listeria growth inhibition and control. Post-packaging lethality treatments offer the advantage of destroying the surface of *L. monocytogenes* contamination that may have occurred after the original cooking process. Traditional post-packaging processing methods that can be used in place of antimicrobial ingredients include freezing, high salt levels, acidification, bacteriocins, and drying.

Because of the non-thermal and mild thermal character, the application of light-based technologies also have been extended beyond RTE meats to raw meat and poultry products to combat a health risk of food poisoning associated with food borne illnesses. The terms "raw" and "fresh" when used on the label or in labeling of a food suggests or implies that the food itself and all ingredients are unprocessed, meaning that the food is in its natural untreated state and has not

Emerging Technologies in Meat Processing: Production, Processing and Technology, First Edition.
Edited by Enda J. Cummins and James G. Lyng.

been frozen or subjected to any form of thermal processing or any other form of preservation.

This chapter explores of how light technologies, ultraviolet (UV) and pulsed, and infrared irradiation can be effectively used in meat industry to improve safety of RTE and raw meats and poultry including an overview of principles and mode of operation, manufacturers of available equipment and reported case studies.

5.2 Theory of UV, IR, and high-intensity light pulse preservation of foods

There are several innovative food processing techniques that are based on electromagnetic energy and have found applications in meat operations for treatment of raw and finished products and improvement of the safety of meat facilities by disinfecting air and nonfood contact and food contact surfaces. The types of electromagnetic radiation are broadly classified into the gamma radiation, X-ray radiation, ultraviolet (UV), visible, infrared, microwave radiation, and radio waves. The classification is based on the increasing order of wavelengths. The technical effects depend on the distinction in photon energies, corresponding depth of penetration and different sources. This chapter will focus on nonthermal and thermal applications of light-based electromagnetic technologies using ultraviolet (UV), pulsed light (PL), and infrared (IR) radiation for the processing and treatment of meat and meat products.

The wavelength of UV rays is shorter than the violet end of the visible spectrum but longer than the X-ray, that is, in the range between 400 and 10 nm, corresponding to photon energies from 3 to 124 eV. After the first application of UV light for disinfection of water more than hundred years ago, UV light technology has expanded as a food safety technique in the last few decades. UV has been taking its niche in food production as a nonchemical physical treatment to control and enhance safety of processing plants and storage facilities.

Pulsed light (PL) is another light technique that has been developed and tested for surface treatments. PL has a broader range of the wavelengths of 170–1000 nm and often combines visible and UV photons at energies of 0.01–50 J/cm^2. PL is delivered in several flashes of light per second, allowing fast throughput of product and low energy usage. The intensity of pulsed light is about 20,000 times the intensity of UV light.

Infrared (IR) light is an electromagnetic radiation with longer wavelengths than those of visible light and covers the range from approximately 300 GHz (1 mm) to 400 THz (750 nm). Unlike conventional systems where heat energy is transferred from a hot medium to a cooler product resulting in large temperature gradients, IR or thermal radiation is one of the principal mechanisms of heat transfer. IR light is emitted or absorbed by molecules when they change their rotational–vibrational movements and cause surface heating. IR surface pasteurization has been tested for inactivation of *Listeria monocytogenes* on ready-to-eat (RTE) meats as energy efficient and clean technique.

5.3 Infrared radiation

Electromagnetic and electrotechnology-based methods include ohmic and inductive heating, radio frequency (RF) or capacitive dielectric heating, radiative or microwave (MW) dielectric heating, and infrared (IR) heating. Unlike conventional systems where heat energy is transferred from a hot medium to a cooler product resulting in large temperature gradients, electroheating involves the transfer of electromagnetic energy directly into the product, initiating volumetric heating due to frictional interaction between water molecules and charged ions (i.e., heat is generated within the product).

Along with microwave, radiofrequency, and induction, IR radiation transfers thermal energy in the form of electromagnetic (EM) waves and encompasses that portion of the electromagnetic spectrum that borders on visible light and microwaves. IR radiation can be classified into three regions, namely, near infrared (NIR), mid-infrared (MIR), and far-infrared (FIR), corresponding to the spectral ranges of 0.75–1.4, 1.4–3, and 3–1000 μm, respectively. The amount of the IR radiation that is incident on any surface has a spectral dependence because energy coming out of an emitter is composed of different wavelengths and the fraction of the radiation in each band, which is dependent on the temperature and emissivity of the emitter. The wavelength at which the maximum radiation occurs is determined by the temperature of the IR heating elements. When radiant EM energy impinges upon a food surface, it may induce changes in the electronic, vibrational, and rotational states of atoms and molecules. As food is exposed to IR radiation, it is absorbed, reflected, or scattered. The type of mechanisms for energy absorption determined by the wavelength range of the incident radiative energy are as follows: (1) changes in the electronic state corresponding to the wavelength range 0.2–0.7 μm (ultraviolet and visible rays), (2) changes in the vibrational state corresponding to the wavelength range 2.5–100 μm (FIR), and (3) changes in the rotational state corresponding to wavelengths above 100 μm (microwaves). In general, the food substances absorb FIR energy most efficiently through the mechanism of changes in the molecular vibrational state. Water and organic compounds such as proteins and starches, which are the main components of food, absorb FIR energy at wavelengths greater than 2.5 μm.

5.3.1 Mode of action when used for preservation

Fundamentals of IR radiation for food processing have been addressed in a few recent publications (Krishnamurthy *et al.*, 2008; Jun *et al.*, 2011). In general, the application of IR heating for food processing has the following advantages: (1) efficient heat transfer to the food, (2) reduced processing time and energy costs, (3) air is not heated, (4) the possibility of designing compact and easily automated equipment with high controllability and safety, (5) uniformity of heating more than in conventional heating because of IR penetration. Because of these benefits, IR radiation has been widely applied to various thermal preservation operations in the food industry mainly for dehydration and pasteurization. IR-based dehydration processes have the additional advantages of requiring simple equipment, fast transient response, and significant energy savings. Higher heat transfer and heat flux rates lead to shorter drying times and higher drying rates and suggest higher quality of final products.

Applying IR heating for surface pasteurization of solid products has the potential to become common industrial practice. Exposing a food product to an IR heating source results in an increase in the surface temperature, and the heat is conducted to the interior by conduction. If the IR exposure time is properly controlled, the surface temperature can be preferentially raised to a degree sufficient to inactivate target pathogenic microorganism without substantially increasing the interior temperature.

IR energy is known to inactivate microorganisms by conventional thermal mechanisms by causing damage in their intracellular components such as DNA, RNA, ribosomes, cell envelope, and/or cell proteins (Krishnamurthy *et al.*, 2008). Absorption of IR radiation by water molecules in the microorganism is a significant factor in the inactivation. The state, location, and amount of water inside the microorganism and bonding conditions are significant factors affecting the response of microorganism to IR heating.

5.3.2 Advantages and disadvantages of IR radiation

The application of IR radiation to food processing has gained momentum because of its higher thermal efficiency and fast heating rate/response time in comparison to conventional heating. Even though IR heating is a promising novel method because it is fast and produces heating inside the material, its penetrating powers are limited. IR radiation can be considered only as a surface treatment. Application of combined electromagnetic radiation and conventional convective heating is considered to be more efficient over radiation or convective heating alone, as it gives a synergistic effect. Being attractive primarily for surface heating applications, the combination of IR heating with microwave and other conductive and convective modes of heating holds great potential in optimizing energy usage and has efficient practical applicability in the meat processing industry (Pan and Atongulu, 2011).

5.3.3 Factors affecting interaction between infrared radiation and food

IR heating can be used to inactivate bacteria, spores, yeast, and mold in both liquid and solid foods. Efficacy of microbial inactivation by IR heating depends on the infrared power level, temperature of food sample, peak wavelength and bandwidth of infrared heating source, sample depth, types of microorganisms, moisture content, physiological phase of microorganism, and types of food materials.

5.3.4 Brief overview of available equipment

Two conventional types of IR radiators used for food process heating are electric and gas-fired heaters. These two types of IR heaters or emitters generally fit into the following temperature ranges: 343–1100 °C for gas and electric IR; and 1100–2200 °C for electric IR only. IR temperatures are typically used in the range of 650–1200 °C to prevent charring of products. The operating efficiency of an electric IR heater ranges from 40% to 70%, while that of gas-fired IR heaters range from 30% to 50%. The proper emitters and process parameters have to be selected for a particular product and process. Electrical IR heaters can be easily installed and produce a prompt heating rate. They also provide flexibility in producing the desired wavelength for a particular application. For the food sector, IR modules are manufactured in stainless steel and fitted with a wire mesh for mechanical protection. IR emitters can be switched on and off inside 1–2 s providing control from any unexpected or unwanted conveyor belt stoppages.

For example, NIR heating is generated by a halogen lamp. Halogen emits radiation near the visible light spectrum (wavelength 0.7–5 μm) with higher frequency and deeper penetration depth than typical infrared sources that emit in the mid-infrared and far-infrared regions of the electromagnetic spectrum. A typical halogen oven consists of an enclosed glass container, a power source, halogen bulbs, and a high-powered fan. The halogen light bulbs heat the air inside the container, and then the fan circulates heated air around the food being processed. Temperatures can be set to varying levels and thereby accommodate several cooking styles: broiling, roasting, steaming, grilling, air frying, dehydrating, and baking. Some halogen ovens feature built-in microwave and convection oven options, making the oven an exceedingly versatile cooking apparatus thus offering the most convenient way to cook.

Conveyorized IR broiling is a unique and innovative method based on medium-wave carbon infrared emitters that can heat meat surfaces in a targeted manner. Sandwich meat, beef patties, hamburgers, and hams can be made to look even more appetizing without additional fat. Due to high temperatures and short cooking times, the IR broiler could produce more servings per hour compared to conventional gas heating.

5.3.5 Case studies – infrared heating of meats

Since the post-process contamination of RTE fully cooked meats primarily occurs on the surface, IR heating can be used as effective post-lethality intervention step that is necessary for meat processors to ensure the final microbial safety. According to the results of the USDA study (Huang and Sites, 2008), the IR surface pasteurization was effective in inactivating *L. monocytogenes* on RTE meats such as hot dogs. The pasteurization system contained four basic elements: an infrared emitter, a hotdog roller, an infrared sensor, and a temperature controller. The infrared sensor was used to monitor the surface temperature of hotdogs. The infrared emitter, modulated by a power controller, was used as a heating source. In all experiments, the temperature of the IR emitter was below 330 °C, which can be easily achieved in the food industry. The hotdogs were surface-inoculated with a four-strain *L. monocytogenes* cocktail to an average initial inoculum of 7.32 log (CFU/g). On the average, 1.0, 2.1, 3.0, or 5.3 log reduction was observed after the surface temperature of hotdogs was increased to 70, 75, 80, or 85 °C, respectively. Holding the sample temperature led to additional bacterial inactivation. With a 3-min holding at 80 °C or 2 min at 85 °C, a total of 6.4 or 6.7 logs of *L. monocytogenes* were inactivated. The combination of time and temperature was critical in IR surface pasteurization as it is in all thermal processes. Without an additional holding period after come-up period, it was impossible to kill all bacteria if the original level of contamination was high. In addition to the ability to combat *Listeria* and extend shelf-life of cooked meat products, there is a growing interest in flame broiling and simultaneously rapid cooking methods.

5.3.6 Manufacturers of infrared equipment

Heraeus Noblelight (GA, USA) offers complete IR technology for food processors. IR heaters are completely produced by Heraeus from processing of the raw crystalline quartz to final assembly of the heater. IR heaters cover the total spectrum of technologically available wavelengths that includes nano XP, short-wave, carbon IR, medium-wave, and long-wave IR tube heaters. Heraeus built complete IR heating and control systems to match the equipment to customer specifications.

5.4 Ultraviolet radiation

5.4.1 Mode of action when used for preservation

Ultraviolet light (UV) is a portion of the electromagnetic spectrum between 100 and 400 nm. UV light can be further divided into UV-A, -B, and - C sub-diapasons that are characterized by various wavelengths and energy levels of photons and thus differ in produced effects. UV-A range from 315 to 400 nm may cause changes in human skin or tanning, whereas light in UV-B range from 280 to 315 nm affects skin and possibly lead to skin cancer. Germicidal

UV-C range from 200 to 280 nm effectively inactivates bacterial, viral, and protozoan microorganisms. The vacuum UV is from 100 to 200 nm and can be absorbed by almost all substances and thus can be transmitted only in a vacuum. When vacuum UV and UV-C photons are absorbed by oxygen atoms, the energy exchange causes the formation of ozone (Koutchma, Forney and Moraru, 2009).

Gas discharges are responsible for the light emitted from UV lamps. A gas discharge is a mixture of non-excited atoms, excited atoms, cations, and free electrons formed when a sufficiently high voltage is applied across a volume of gas. Low-pressure mercury (LPM) sources emit 90% of their UV energy at 253.7 nm. LPM sources are monochromatic, relatively cheap, and have reasonably long life time of up to 6000 h. More powerful options are high-intensity monochromatic amalgam lamps (LPA) and polychromatic medium-pressure mercury (MPM) lamps. MPM sources are mainly used in water treatment industry. The life time of amalgam lamps can be twice longer up to 12,000 h.

5.4.2 Advantages and disadvantages of UV radiation

UV light has low penetrating power because its inherent energy of photons is low in comparison with ionizing radiation and belongs to nonionizing radiation. Absorption of nonionizing radiation leads solely to electronic excitation of atoms and molecules as opposed to ionizing radiation that can lead to ionizing effects. However, the excitation energy of UV photons is much higher than the energy of thermal motions of the molecules at physiological temperatures. The latter is of the order of Boltzmann constant times the absolute temperature, which amounts to 0.026 eV/molecule (0.60 kcal/mol) in contrast to the 3.3–6.5 eV/molecule (75–150 kcal/mol) at 27°C available from UV absorption.

5.4.3 Factors affecting interaction between UV radiation and materials

UV light emitted from the gas discharge of a UV source propagates away from atoms and ions interacting with the materials through absorption, reflection, refraction, and scattering. This explains why correct positioning of UV source and distance from treated product is critical to maximize efficacy of UV treatments. Any obstruction to the path of the light such as dust, shadowing, or clumping of bacteria can reduce efficacy. So, the effectiveness of UV light is less on a rough surface than on a smooth surface.

The characteristics of UV source such as wavelength or the levels of UV light photons energy, the number of photons or UV intensity, and the exposure time will define UV process lethality. It was found that among product parameters, temperature, pH, chemical composition, and physical structure affect UV light effectiveness to control microorganisms. Additionally, efficacy of UV sources against different groups of pathogenic and spoilage organisms vary in different matrices such as air, water, surfaces, or fluid foods. Adenovirus is considered as the most UV-resistant organism in water treatment, whereas bacterial and molds spores are characterized by the highest UV resistance in air and food matrices (Koutchma, 2014).

As example, in order to calculate the dose of UV treatment in terms of inactivation of *L. monocytogenes*, the decimal reduction dose or D_{uv} value has to be known. The data in the scientific literature reported that there is a difference in UV sensitivity of bacteria depending on the surface characteristics. D_{uv} values of *Listeria* calculated based on the reported data in literature are summarized in Table 5.1.

The reported data summarized in Table 5.1 shows that UV light resistance of *L. monocytogenes* depends on the surfaces and its roughness characteristics. For a smooth surface, the decimal UV dose varies from 2.5 up to 3.5 mJ/cm^2, whereas for the product surface, the UV dose may be higher by 2 or 3 orders of magnitude and vary in the range of 200–300 mJ/cm^2. The applied UV dose on the surface to achieve specific log reduction (SLR) of *Listeria* can be determined as product of D_{uv} value and SLR.

5.4.4 Brief overview of available equipment

There are a number of commercially produced continuous and pulsed sources that emit energy in the UV range. LPM sources provide a high efficiency in inactivating cells and a number of other advantages. They emit most of their UV energy at 253.7 nm; they are monochromatic, relatively cheap, and have reasonably long service (thousands of hours) and they can run at the low surface temperature of 40 °C with no additional cooling system required. More powerful options exist for the future once validated, such as high-intensity amalgam lamps and MPM lamps. The life time of LPM amalgam lamps can be up to 12,000 h.

5.4.5 Case studies – UV in food preservation

Among known existing applications of UV light in meat processing are UV treatments of non-food and food contact surfaces and, recently, product surfaces. UV light was effective against *Escherichia coli* and *Salmonella* on pork skin, raw meats, and poultry surfaces. Precautions should be taken to avoid undesirable brown discoloration of red meat, since UV encourages metmyoglobin formation. The rate of product discoloration depends on the intensity of light and wavelength distribution. Packaged products can also be exposed to UV to avoid of risk of recontamination. The correct choice of polychromatic pulsed or monochromatic continuous source in combination with the UV filter and light permeability of the packaging material is critical when such a process is developed for red meat products.

According to USDA, contamination of foods with *L. monocytogenes* is likely to occur in all RTE meat and poultry products that are exposed to post-lethality

Table 5.1 UV resistance of *Listeria* strains on food contact and food surfaces.

Listeria strains	Medium	UV intensity and treatment time (mW/cm^2)	Reported log reduction	D_{uv} value (mJ/cm^2)	Source
Serotypes 3A, 4A, 4B, and 4C	Conveyor belts 1. Ropanyl DM 8/2 2. Volta FRMW-3.0 3. Volta FRMB-3.0 4. Ropanyl DM	5.95 mW/cm^2 for 3 s 5.53 mW/cm^2 for 3 s	7– 5.27	2.55–3.19	Morey *et al.* (2010)
L. monocytogenes strains from RTE (F4561, H7762, and 121 H7764)	Frankfurters	5 $mW/s/cm^2$ for 100 s	2.32–2.80 logs	215–178	Sommers *et al.* (2009)
L. innocua	Frankfurters with sodium diacetate and potassium lactate (SDA/PL)	1000 mJ/cm^2 followed by flash steam (0.75 s steam/121 °C) 10 mW/cm^2 for 100 s	3.19 logs	300	Sommers *et al.* (2010)
L. monocytogenes ATCC 1911	Agar surfaces	0.1–1.8 mJ/cm^2	0.9–3.9	0.48	Chun *et al.* (2009)
	Sliced ham	(unpublished data) 800 mJ/cm^2	2.57	311	
L. monocytogenes ATCC 1911	Cut pear	2 LPM, 15 W 1500–8700 mJ/cm^2	Up to 3.5	2500	Schenk *et al.* (2007)

processing environment. These post-lethality processing environments are the areas into which products are routed after complete treatments and may include slicing, peeling, dicing, re-bagging, and brining. As an example of UV technology post-lethality application, packaged RTE product can be exposed to UV so to avoid of risk of cross- or re-contamination. Knowledge of light permeability of the packaging material is critical when such a process is developed.

USDA scientists (Sommers *et al.*, 2009) have demonstrated that UV light can be used to reduce levels of *Listeria* contamination on the surfaces of RTE products. UV-C irradiation of frankfurters that were surface-inoculated with *L. monocytogenes* resulted in a 1.31, 1.49, and 1.93 log reduction at doses of 1, 2, and 4 J/cm^2, respectively. UV-C treatment had no effect on frankfurter color or texture at UV-C doses up to 4 J/cm^2. It was concluded that because the numbers of *L. monocytogenes* associated with contaminations of RTE meats are typically very low, the use of UV-C in combination with potassium lactate and sodium diacetate has the potential to reduce the number of frankfurter recalls and foodborne illness outbreaks (Sommers, Scullen and Sites, 2010).

In recent studies, UV light was reported to be effective against nonspore forming pathogenic microflora on raw meats and poultry surfaces. UV-C at 0.5 J/cm^2 reduced the initial populations of *Campylobacter jejuni*, *L. monocytogenes*, and *Salmonella Typhimurium* by 1.3–1.2 log CFU/g, respectively, on chicken breasts (Lyon, Fletcher and Berrang, 2007; Chun *et al.*, 2010). Similarly, UV treatment of raw chicken fillet at 0.2 J/cm^2 reduced *C. jejuni*, *E coli*, *Salmonella enterica* serovar *Enteritidis*, total viable counts, and *Enterobacteriaceae* by 0.76, 0.98, 1.34, 1.76, and 1.29 log CFU/g, respectively. Following UV treatment of packaging and surface materials, higher reductions of up to 3.97, 4.50, and 4.20 log CFU/cm^2 were obtained for *C. jejuni*, *E. coli*, and *S. enterica* serovar *Enteritidis*, respectively. These studies indicate that UV may be beneficial in working areas of meat and poultry processing plants to reduce the level of aerobic and nonspore forming pathogens and may be applied to decontaminate cut-up products moving on conveyors, associated packaging, and surface materials.

In order to demonstrate efficacy of UV technology to USDA-FSIS, the UV dose requirements to achieve SLR requirements need to be evaluated first. This information will allow selection of not only a correct UV source but also an effective design of a system depending on the application.

Meat processors may dedicate a full shift to cleaning disassembled equipment, conveyor belts, walls, ceilings, and floors because *Listeria* can survive for extended periods in meat processing facilities. The benefits of UV units installed in cold rooms are not limited to killing bacteria on surfaces and in the air. A substantial extension of storage life of chilled carcases is possible when UV light is used continuously and delays the onset of microbiological spoilage. Additionally, UV reduces shrinkage and retains juices, odors of cold rooms are eliminated, and mold growth on the surfaces is reduced.

Another application of UV technology is to improve the microbiological quality of brines and marinades. In meat processing, brine is a strong solution of water and salt that is used to improve texture and flavor when injected in whole muscle raw beef products. A sweetener such as sugar, molasses, honey, or corn syrup may be added to the solution for flavor and to improve browning. Also, recirculated concentrated chilled brines often used to rapidly cool processed RTE meats and poultry products. However, if not handled correctly, various types of commercial brine solutions can become contaminated with natural microflora and pose a potential microbiological hazard due to pathogens, such as *Listeria* and *E. coli,* in meat and poultry processing that can grow at low temperatures.

US Department of Agriculture regulations state that brines may be reused to chill cooked product for up to 4 weeks depending on salinity and temperature requirements. Provisions cover the following range: no temperature or sodium chloride (NaCl) controls for one production shift; reuse up to 1 week, which must maintain 9% NaCl concentration and a maximum temperature of 2.2 °C; and reuse up to 4 weeks, which requires maintaining 20% NaCl concentration and a maximum temperature of −12.2 °C (U.S. Department of Agriculture, Food Safety and Inspection Service, 1983).

Controlling microbial load of these solutions may further extend the life of brines, dramatically reducing the cost of water, salt, brine disposal, and time lost when replacing batches of brine, without fear of increased pathogen risk.

Fitting UV systems in recirculating brine chillers and meat marinade injectors are very effective at destroying natural microflora and pathogenic microorganisms. Depending on the model, some UV systems can operate very effectively through a wide range of temperatures and applications – from supercooled brines to very hot sanitation cycles. Additionally, operating costs are reduced through less frequent fluid changeover.

A number of the UV technology providers such as Aquionics, Safe Food Corporation, and SurePure offer solutions for controlling natural microbial load and pathogens by recirculating brines and marinades through UV light-based systems. According to the studies published in the *International Journal of Meat Science,* the FreshLight® 210 UV light system (Cook, Beers and Beers, 2011, Safe Food Corporation, N. Little Rock, AZ) effectively reduced the *E. coli* O157:H7 in the beef brine from 5.4 logs to 2.2 logs per mL in 30 min (12 passes) resulting in an overall bacterial reduction of greater than 99.9%.

UV water treatment unit from Aquionics Inc. (Erlanger, KY) was found to be effective against *Listeria* and lactic acid bacteria in chilling brines after 60-min circulation through the system. Additionally, the combination of UV and antimicrobial agents, such as citric acid (CA), hydrogen peroxide (HP) and dimethyl dicarbonate (DMDC), was effective in reducing *L. monocytogenes* population in fresh brine. However, only the combination of UV and HP was the most effective treatment in recirculated brines due to organic matter (Parikh *et al.*, 2011).

Marinade is a savory acidic sauce in which a food is soaked to enrich its flavor or to tenderize it. Consumers can purchase ready-to-cook poultry products that have already been marinated or basted. These products have been injected or marinated with a solution containing butter or other edible fat, broth, stock, or water, plus spices, flavor enhancers, colorings, or other approved substances. Similarly, the injection of marinade into fresh poultry products can result in a build-up of pathogens due to recirculation of the marinade through the pump and holding tank. Therefore, poultry processors are constantly searching for new ways to sanitize the recirculated marinade to eliminate the possibility of pathogens being injected into fresh poultry products.

Efficacy of the FreshLight® 210 UV light system was also demonstrated for the reduction of natural microflora and control of *S. Typhimurium* and *L. innocua* in a commercial poultry marinade at the flow rate of 20 gpm and at a turnover time of 1 min. The total bacterial count in the marinade solution was effectively reduced by approximately 1 log in less than 10 min of operational time with the FreshLight® 210 system. In terms of inactivation of *Salmonella* and *Listeria*, typical reductions were 2.5 logs at 2 passes (2 min), 3.5 logs at 16 passes (16 min), 4.7 logs at 40 passes (40 min), and 6.3 logs at 56 passes (56 min) (Beers, Rheingans and Waldrou, 2010).

Thus, the commercial UV light systems can provide the meat and poultry manufacturers with an effective way of controlling the microbial load in brines and marinade at a low cost.

5.4.6 Manufacturers of UV generation equipment

There are a large number of manufacturers of UV bulbs and UV systems that are typically supplied with automated, online wipers that keep lamps free from fouling. UV lamps require replacement every year and the quartz sleeve that houses the lamps every 2–3 years. Wiper seals typically require annual replacement. The routine downtime per system is typically 1–2 h, and most manufacturers provide factory training for customers' technical staff if required.

An example of a commercially available system to decontaminate surfaces of foods is the UV tumbling process that is offered by Reyco Systems (2014) (USA). The company incorporated either a rotating drum or screw conveyor that lifts and tumbles the product to ensure exposure to the UV source. The unit can be used to treat fresh products including meats, frozen products (meats, seafood), and cooked, refrigerated products. The C&S Equipment Company used the patented technology of Steril-Aire™ UV Emitters that are sleeved in plastic to meet the food safety requirements of food processing facilities. The patented design allows emitters to work efficiently in the cold environment of refrigerated or chiller coils, where competitive units lose their effectiveness. Examples of UV units currently used in commercial processing facilities include (1) UV tumbling drum in operation for chicken and beef fajita strips (cooked and frozen) with a capacity 6000–7000 lb/h; (2) cooked and IQF-frozen hamburger patty treatment (hooded conveyor with turnover), with a capacity 3000 lb/h; and (3) a deli meat system (custom conveyor with UVC hood) for formed deli ham logs, with a capacity of 10,000 lb/h.

5.5 High-intensity light pulses

5.5.1 Mode of action when used for preservation

Pulsed light (PL) has a broader range of the wavelengths of 170–1000 nm and often combines visible and UV photons at energies of 0.01–50 J/cm^2. PL is generated using a capacitor and delivered in several flashes of light per second, allowing fast throughput of product and low energy usage. The intensity of pulsed light is about 20,000 times the intensity of UV light. Xenon lamps are commercial sources of PL that require air or water cooling for the operation.

The following trend of susceptibility of microorganisms to PL in decreasing order was reported: Gram-negative bacteria, Gram-positive bacteria, and fungal spores. The color of the spores can play a significant role in fungal spore susceptibility. *Aspergillus niger* spores are more resistant than *Fusarium culmorum* spores because the pigment of the *A. niger* spores absorbs more in the UV-C region than that of *F. culmorum* spores, protecting the spore against UV (Gomez-Lopez *et al.*, 2007).

5.5.2 Advantages and disadvantages of high-intensity pulses

The short pulse width and high doses of the PL source may provide some practical advantages over UV sources in those situations where rapid disinfection is required. Other advantages of PL treatment are the lack of residual compounds and the absence of applying chemicals that can cause ecological problems and/or are potentially harmful to humans. Xenon flash lamps are also more environment friendly than continuous UV lamps because they do not use mercury.

Sample heating is perhaps the most important limiting factor of PL for practical applications. Heat can originate from the absorption of IR part of PL by the food or by lamp heating. Another disadvantage of PL treatments is the possibility of shadowing occurring when microorganisms readily absorb the rays.

5.5.3 Factors affecting interaction between high-intensity pulses and materials

The factors determining the interaction of PL exposure with foods are to some extent similar to UV light treatments. The critical factor affecting PL treatments is fluence incident on the sample. The energy emitted by the flash lamp is different from the energy incident on the sample. Factors such as distance from light source to target and propagation vehicle (air, water or other fluids, dust particles) affect the level of energy that ultimately reaches the target. The inactivation efficacy of PL is higher when treated products are closer to the lamp. Food composition also affects the efficacy of the decontamination by PL (Gomez-Lopez *et al.*, 2007). High protein and fat containing food products have little potential to be efficiently treated by PL. Vegetables, on the other hand, could therefore be suitable for PL treatment.

5.5.4 Case studies – High-intensity light pulses in food preservation

Pulsed light was tested for its utility to improve the microbial quality and safety of RTE vacuum-packaged ham and bologna slices inoculated with *L. monocytogenes*. PL treatment at 8.4 J/cm^2 reduced *L. monocytogenes* by 1.78 logs CFU/cm^2 in cooked ham and by 1.11 logs CFU/cm^2 in bologna. PL at 8.4 J/cm^2 did not affect the sensory quality of cooked ham, while treatments above 2.1 J/cm^2 negatively influenced the sensory properties of bologna. The combination of PL and vacuum packaging provided ham with an additional shelf-life extension of 30 days compared with only vacuum packaging. Additionally, the effectiveness of PL was demonstrated for treatments of RTE chicken frankfurters (Sommers *et al.*, 2009).

In another study (Haughton *et al.*, 2012), effectiveness of high intensity near UV/visible 395 ± 5 nm light (NUV–vis) against *Campylobacter* was assessed at three distances from the light source for up to 10 min, corresponding to doses of 0.06–18 J/cm^2 for the decontamination of raw chicken and contact surfaces. Exposure of skinless chicken fillet to NUV–vis light for 1 or 5 min at 3 cm distance reduced *C. jejuni* by 2.21 and 2.62 log CFU/g, respectively. For *Enterobacteriaceae* and total viable counts, significant reductions were achieved only on chicken fillet samples. Light treatments were significantly effective for decontaminating contact surfaces as there were no *C. jejuni* recovered from stainless steel or cutting board surfaces after NUV–vis light treatments from an initial inoculum of 2–4 log CFU/cm^2 ($P < 0.05$). The study demonstrated the potential for the use of NUV–vis light for the inactivation of *Campylobacter* spp. in liquids, on raw chicken, and contact surfaces.

5.5.5 Brief overview of available high-intensity light pulse equipment

Xenon flash lamps have an emission spectrum ranging from UV to IR light. The UV-C part of the spectrum is the most important for microbial inactivation. The pioneer company producing PL equipment for disinfection was PurePulse Technologies Inc. (San Diego, California), a subsidiary of Xenon Corp. Applications included water purification systems and virus inactivation systems for biopharmaceutical manufacturers. Nowadays, there are three commercial companies producing pilot scale and commercial systems based on PL: Claranor from

France, SteriBeam Systems from Germany, and Xenon Corporation from the United States. The detailed information regarding devices for industrial applications can be found at the websites of these companies.

5.6 Future role for UV, IR, and high-intensity light pulses in the preservation of foods

Light-based technologies are emerging as both thermal and nonthermal techniques that have much to offer for the sustainable development of society. Their application for the meat processing is energy and cost-effective and also was proven to result in products with higher safety and better quality characteristics.

Foodborne and waterborne pathogens are the largest group of microorganisms that present health hazards in the meat industry. Some of these organisms can become airborne during processing and settle on raw, semi-finished, or finished products thereby becoming amenable to control by UV and PL in the air and on surfaces. Virtually every meat processing facility can benefit from use of light technologies to control microbial hazards through treatment of air, nonfood, food contact surfaces, processing water, ingredients, raw, and finished products.

Innovative research of UV light for food applications has been grown worldwide and has presented a number of opportunities for the technology as adjunct or alternative to pasteurization, thereby enhancing the shelf-life for different groups of foods including meat products. Although there are a number of benefits for UV light in the meat and poultry industry, cost-saving opportunities for energy, processing water, and enhanced safety need to be carefully considered in each specific case for successful implementation of the technology and to assure positive benefits.

A variety of UV sources are commercially available or currently under development that can be applied for specific fruit processing purposes, whereas LPM lamps and xenon PL are currently the dominant sources of radiation for the processing and treatment of foods since they were approved by the US FDA CFR 21 179.39 (Table 5.2).

Code 21CFR179.41, issued by the Food and Drug Administration (FDA) in 1996, Department of Health and Human Services, approves the use of pulsed UV light in the production, processing, and handling of food. Pulsed light may be safely used for the treatment of foods under the following conditions. (1) The radiation sources consist of xenon flash lamps designed to emit broadband radiation consisting of wavelengths covering the range of 200–1000 nm and operated so that the pulse duration is no longer than 2 ms. (2) The treatment is used for surface microorganism control. (3) Foods treated with pulsed light shall receive the minimum treatment reasonably required to accomplish the intended technical effect. (4) The total cumulative treatment shall not exceed 12.0 (J/cm^2).

Table 5.2 US FDA CFR 21 179.39 – UV for treatment and processing foods.

Radiated food	Limitations	Use
Food and food products	Without ozone production: high fat-content food irradiated in vacuum or in an inert atmosphere; intensity of radiation, 1 W (of 2537 A. radiation) per 5–10 ft^2	Surface microorganism control

In assessing the safety of foods treated with all forms of radiation, the agency considers microbial efficacy and changes in chemical composition of the food that may be induced by the proposed treatment, including any potential changes in nutrient levels. The legal status of light treatments in the European Union has a different approach, since the legislation is not technology oriented but food and food ingredient oriented. Light technologies would fall in the scope of regulation 258/97 on novel foods and novel food ingredients, article 1, item f. The approval process is based on the risk assessment approach that includes microbiological, chemical, and toxicological safety, possible changes in composition and nutrition content, allergenicity, and dietary exposure.

Future tunable monochromatic and polychromatic sources would allow users to adjust the wavelength to specifically match the needs of the application. In recent years, UV-light emitting diodes (LEDs) have been developed with the following advantages: small size, long life time up to 10,000 h, energy efficient, easy control of emission, and no production of mercury waste and glass. The commercial UV-LED can emit in three diapasons of UVA, UVB, and UVC. The germicidal effects against several bacteria or fungi have been demonstrated and reported along with the first applications for disinfections of hospital rooms and water. Are LEDs the right new technological solutions with enormous potential for meat processing operations to fight with pathogens? The near future will show.

References

Beers, K.W., Rheingans, J.W. and Waldrou, A.L. (2010) Efficacy of the FreshLight® 210 ultraviolet light system1 for control and elimination of *Salmonella typhimurium* and *Listeria innocua* in a commercial poultry marinade (flow rate = 20 GPM and turnover time = 1 minute). *International Journal of Poultry Science*, **9** (1), 100.

Chun, H.H., Kim, J.Y. and Song, K.B. (2009) Inactivation kinetics *of Listeria monocytogenes, Salmonella enterica serovar Typhimurium*, and *Campylobacter jejuni* in ready-to-eat sliced ham using UV-C irradiation. *Meat Science*, **83** (4), 599–603.

Chun, H.H., Kim, J.Y., Lee, B.D. *et al.* (2010) Effect of UV-C irradiation on the inactivation of inoculated pathogens and quality of chicken breasts during storage. *Food Control*, **1**, 276–28.

Cook, P.E., Beers, K.L. and Beers, K.W. (2011) Efficacy of the FreshLight® 210 ultraviolet light system1 for control and elimination of *Escherichia coli* O157:H7 in a commercial beef brine solution (flow rate = 10 GPM and turnover time = 2.5 min). *International Journal of Meat Science*, **1**, p. 79.

Gomez-Lopez, V.M., Ragaerta, P., Debeverea, J. and Devlieghere, F. (2007) Pulsed light for food decontamination: A review. *Trends in Food Science & Technology*, **18**, 464–473.

Haughton, P.N., Grau, E.G., Lyng, J. *et al.* (2012) Susceptibility of *Campylobacter* to high intensity near ultraviolet/visible 395 ± 5 nm light and its effectiveness for the decontamination of raw chicken and contact surfaces. *International Journal of Food Microbiology*, **159** (3), 267–273.

Huang, L. and Sites, J. (2008) Elimination of *Listeria monocytogenes* on hotdogs by infrared surface treatment. *Journal of Food Science*, **73** (1), M27–M31.

Jun, S., Krishnamurthy, K., Irudayaraj, J. and Demirci, A. (2011) Fundamentals and theory of infrared radiation, in Edited by Zhongli Pan, Griffiths Gregory Atungulu *Infrared heating for food and agricultural processing*, CRS Press, Taylor and Francis Group, pp. 1–18. ISBN 9781420090970.

Koutchma, T. (2014) *Food Plant Safety UV Applications for Food and Non-Food Surfaces*, Elsevier. ISBN: 9780124166202.

Koutchma, T., Forney, L. and Moraru, C. (2009) *Ultraviolet Light in Food Technology: Principles and Application*, CRS Press, Taylor and Francis Group. ISBN 9781420059502.

Krishnamurthy, K., Khurana, H.K., Jun, S. *et al.* (2008) Infrared heating in food processing: An overview. *Comprehensive Reviews in Food Science and Food Safety*, **7**, 2–13.

Lyon, S., Fletcher, D. and Berrang, M. (2007) Germicidal ultraviolet light to lower numbers *of Listeria monocytogenes* on broiler breast fillets. *Poultry Science*, **86**, 964–967.

Morey, A., McKee, S.R., Dickson, J.S. and Singh, M. (2010) Efficacy of ultraviolet light exposure against survival of *Listeria monocytogenes* on conveyor belts. *Foodborne Pathogens and Disease*, **7** (6), 737–740.

Pan, Z. and Atongulu, G. (2011) *Infrared heating for food and agricultural processing*, CRS Press, Taylor and Francis Group. ISBN 9781420090970.

Parikh, P., Williams, R., Mallikarjunan, J. *et al.* (2011) Concurrent use of ultraviolet light and citric acid, dimethyl dicarbonate, or hydrogen peroxide to inactivate *Listeria monocytogenes* in chill brine. *Journal of Food Safety*, **31**, 530–537.

Reyco Systems (2014). *UV Drum*, Available at: http://www.reycosystems.com/solutions/uv-drum/ (accessed 21/04/15).

Schenk, M., Guerrero, S. and Alzamora, M. (2007) Response of some microorganisms to ultraviolet treatment on fresh-cut pear. *Food and Bioprocess Technology*, **1**, 384–392.

Sommers, C.H., Geveke, D.J., Pulsfus, S. and Lemmenes, B. (2009) Inactivation of *Listeria innocua* on Frankfurters by Ultraviolet Light and Flash Pasteurization. *Journal of Food Science*, **74** (3), 138–141.

Sommers, C.H., Scullen, O.J. and Sites, J. (2010) Inactivation of foodborne pathogens on frankfurters using ultraviolet light (254 nm) and GRAS antimicrobials. *Journal of Food Safety*, **31**, 1–12.

U.S. Department of Agriculture, Food Safety and Inspection Service (1983) *Reuse of water or brine cooling solutions on product following a heat treatment* MPI Bulletin 83-16, USDA, FSIS, Washington, DC.

CHAPTER 6

Ultrasound processing applications in the meat industry

N.N. Misra[1], Patrick J. Cullen[1] & Brijesh K. Tiwari[2]

[1] *School of Food Science & Environmental Health, Dublin Institute of Technology, Dublin 1, Ireland*

[2] *Department of Food Biosciences, Teagasc Food Research Centre, Dublin, Ireland*

6.1 Introduction

Meat is an important component of the regular diet in many countries. Meat is defined as the edible flesh of animal used as food that is composed of bundles of muscle fibres joined by strong connective tissues with fat scattered between them to provide flavour, moisture and texture. The average consumption of meat worldwide was 41.9 kg/year in 2009 and nearly 73% increase in meat consumption is expected by 2050, as per the 'World Livestock 2011: Livestock in Food Security' report (FAO, 2011; Chamorro *et al.*, 2012). The composition and high moisture content (ca. 60%) of meat create an ideal environment for the growth and proliferation of pathogenic and spoilage micro-organisms (Zhou, Xu and Liu, 2010). The initial sources of the major foodborne pathogens in meat and meat products are the asymptomatic farm animals excreting the pathogens in their faeces. *Salmonella* sp., *Campylobacter* sp. and *Escherichia coli* O157:H7 are the most commonly encountered human pathogens in meat and meat based products. The nutrient-rich environment of meat renders it an ideal food for the growth and propagation of spoilage micro-organisms and foodborne pathogens. It is therefore essential that adequate preservation technologies are applied to maintain its safety and quality (Aymerich, Picouet and Monfort, 2008; Zhou, Xu and Liu, 2010). The importance of developing novel technologies is evident from the recent large-scale foodborne illness outbreaks associated with meat consumption. The examples include the recent multistate outbreaks of *Salmonella* Enteritidis in the United States resulting from contaminated ground beef (News, 2012) and prior to that through ground Turkey in 2011 (Cargill News, 2011). Moreover, the detrimental impacts of thermal processing on colour, flavour and nutritional quality of food products cannot be overlooked (Pankaj, Misra and Cullen, 2013).

Emerging Technologies in Meat Processing: Production, Processing and Technology, First Edition.
Edited by Enda J. Cummins and James G. Lyng.

Within the past decade, two important non-thermal technologies that made it to the market were high-pressure processing (HPP) and pulsed electric field (PEF) processing. However, even a purely physical technique such as HPP, which is chemically safe, requires complex or expensive equipment and is generally incompatible with online treatments. An early known, yet under researched and emerging technology is power ultrasound processing of foods. The term 'ultrasound' is generally taken to mean that the 'frequency' of the wave is greater than the upper limit of human hearing (usually taken to be 20 kHz) (Leighton, 2007). The applications of ultrasound in food processing include such diverse areas (but not limited to): extraction, cleaning, emulsification/homogenisation, crystallisation, waste water treatment and product modification. In the context of meat processing, most of these applications are eligible with many additional potential uses. The aim of this chapter is to review the fundamentals of power ultrasound, followed by an elaborate discussion of its applications to meat and meat products processing.

6.2 Fundamentals of ultrasound processing

The history of scientific advances and discovery of ultrasound are rooted in the study of sound, with Sir Isaac Newton first proposing his theory that sound is a wave in 1687 (Newton, 1687). Sound waves are divided into three categories that encompass different frequency ranges (Figure 6.1): (1) *Audible waves* are waves that lie within the range of sensitivity of the human ear. (2) *Infrasonic waves* are waves having frequencies below the audible range. Elephants use infrasonic waves to communicate with each other, even when separated by many kilometres. (3) *Ultrasonic waves* are waves having frequencies above the audible range. From the food processing point of view, the ultrasonic region of the sound spectrum is of interest and commercial importance (Misra, Pankaj and Kadam, 2011). Ultrasound is typically divided into three regions of frequency, all above the threshold of hearing for humans. Power ultrasound is 16–100 kHz (1 Hz is 1 cycle/s). High-frequency ultrasound is 100 kHz–1 MHz, while diagnostic ultrasound is 1–10 MHz. High frequency and diagnostic ultrasound is in use for meat carcass evaluation for nearly 40 years (Pathak, Singh and Sanjay, 2011). The

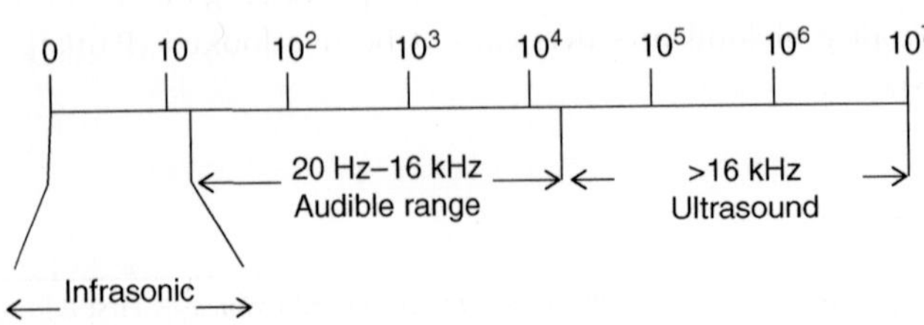

Figure 6.1 The sound spectrum.

focus of this chapter, however, is towards the application of low-frequency and high-power ultrasound.

When ultrasound propagates through any medium, it induces a series of compression and rarefaction in the molecules of the medium. Such alternative pressure changes cause formation of bubbles in a liquid medium. This phenomenon of the creation, expansion and implosive collapse of microbubbles in ultrasonically irradiated liquids is known as 'acoustic cavitation' and is schematically explained in Figure 6.2. Most liquids contain nuclei about which cavitation bubbles originate. These nuclei may consist of dispersed particles, prominences on immersed surfaces and minute gas bubbles. In fact, unless especially treated, liquids contain dissolved or entrained gas. In addition, there are two main types of cavities – transient (also called 'inertial') and stable (also called 'non-inertial'). Stable cavities are relatively long-lived gas bubbles and exist for many compression and rarefaction cycles (Ensminger, 1973b). These are produced at low

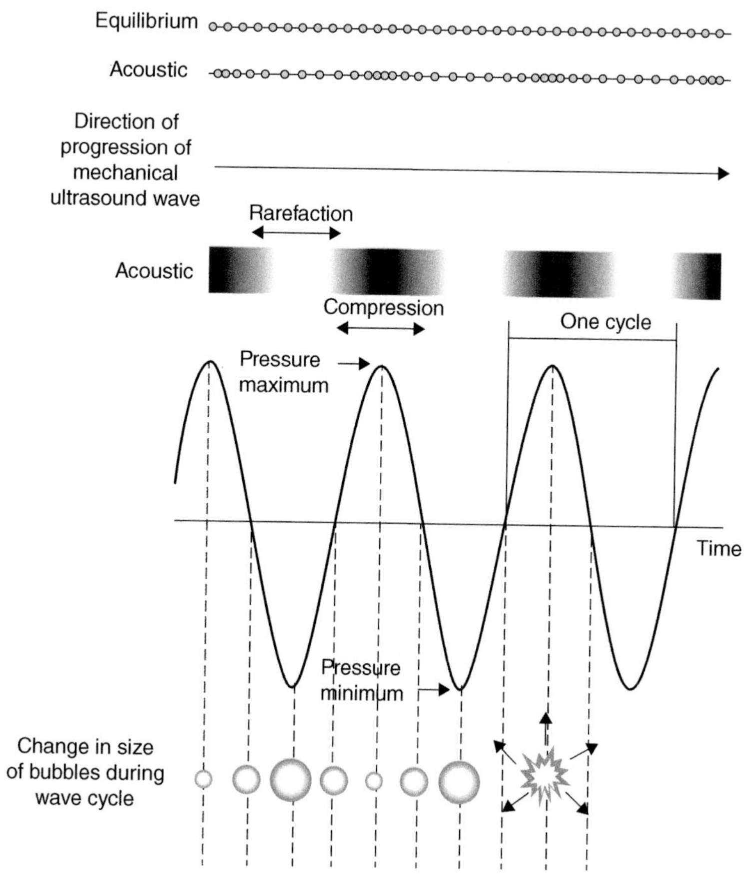

Figure 6.2 Acoustic cavitation phenomena. (Reproduced from Soria and Villamiel (2010), with permission from Elsevier.)

ultrasound intensities and usually oscillate for a number of cycles often nonlinearly about an equilibrium size without collapsing. Transient cavitation (Inertial cavitation) occurs when a cavity experiencing vibration increases in size progressively over a succession of compression and rarefaction cycles, until it collapses violently. Stable bubbles are considered to be weakly and symmetrically oscillating bubbles, whereas transient bubbles are considered to be 'active cavitation' bubbles. The collapse of cavitation bubbles is a near adiabatic process and results in the generation of high temperatures and localised high pressures within these bubbles (Suslick, Hammerton and Cline, 1986). Theoretical calculations using hydrodynamic models of cavitational collapse have given temperature and pressure estimates of 2000–10,000 K and 1000–10,000 bar (Rayleigh, 1917). However, the size of the bubbles is very small relative to the total liquid volume, so the heat that they produce is rapidly dissipated with no appreciable change in global conditions.

Most liquids contain nuclei about which cavitation bubbles originate. Once nucleation has taken place, a gas bubble will grow by rectified diffusion over many periods of the driving pressure oscillations until it breaks up. It should be noted that the bubbles could originate by virtue of dissolved gases (gaseous cavitation) or formed by vaporisation of the liquid itself (vaporous cavitation). Unlike in a single bubble system, the bubble growth in a multi-bubble system involves two processes, namely, rectified diffusion and bubble coalescence (Ashokkumar *et al.*, 2007). Occurrence of intense cavitation depends upon the relationship among the dimensions of the nuclei, the wavelength and the intensity of the field. It has be noted that gas bubbles in a liquid affect profoundly the acoustic behaviour of the liquid due to the dispersion and attenuation of the ultrasound that follows (Commander and Prosperetti, 1989).

The ability of ultrasound to cause cavitation depends on ultrasound characteristics (e.g. frequency and intensity), product properties (e.g. viscosity and surface tension), and ambient conditions (e.g. temperature and pressure). The high-shear energy created by the cavitation effect is often focused and used in several applications including particle modification in liquids, release of cellular components and molecular structures, and de-aeration of liquids and surface. When cavitation bubbles oscillate and collapse, several physical effects are generated, namely shock waves, microjets, turbulence, shear forces and so on. Acoustic streaming can be described as a steady fluid motion created under the influence of high amplitude acoustic waves, when they propagate through a dissipative fluid medium (Rayleigh, 1884). *Microstreaming* occurs due to a vigorous circulatory motion induced by the growth and shrinkage of the cavitation bubbles during cycles of rarefaction and compression (Margulis and Margulis, 2004).

The chemical effects of ultrasound have been studied extensively within the last century. Under the extreme temperature and pressure conditions, highly reactive radicals are generated. For example, if water is the medium, $H^{\bullet}$ and $OH^{\bullet}$

radicals are generated by the dissociation of water ($H_2O \rightarrow OH^- + H^+$). Both stable cavitation and an increase in the number of active bubbles can be expected to increase the amount of hydroxyl radicals generated with an increase in the ultrasound frequency. Free radicals are chemically reactive and are potentially very damaging to any cell they come in contact with (Henglein and Kormann, 1985). These primary radicals can induce a wide variety of chemical reactions in the bulk solution. The aforementioned physical and chemical effects of ultrasound make it an excellent tool for microbial decontamination. The following section describes briefly the principle of operation of an ultrasonic probe/bath, followed by discussion of important studies regarding the antimicrobial effects of ultrasound, either alone or in combination.

6.3 Ultrasound processing equipment

In 1880, the Curie brothers discovered the piezoelectric effect (Curie and Curie, 1880), which led Paul Langevin to develop one of the first uses of underwater ultrasound at a frequency of about 150 kHz (Hunt, 1982). A transducer is a device that converts one form of energy into another form. In ultrasonic applications, the transducers are tunable and designed to convert mechanical or electrical energy into high-frequency sound (vibrations). There are two main types of transducers: mechanical and electroacoustic. Mechanical transducers rely on either the flow of a liquid or gas through a siren, rotor, turbine or whistle to generate ultrasound. Electromechanical transducers are widely used in modern ultrasonic applications and are based on the inherent electrostrictive phenomenon in certain materials to produce piezoelectric or magnetostrictive transducers (Abramov, 1998). The vibrations generated by the transducer are amplified and passed on to a sonotrode or probe, often fabricated from titanium, which is in direct contact with the fluid. The sonotrode is a wear part and lasts 3–24 months, depending on the conditions and the intensity of the energy applied. Titanium is used because it has low mass and high rigidity, so less energy is lost deforming the tool and instead is transmitted to the fluid (Clark, 2008). The cost-effectiveness of ultrasound technology emerges from its energy-efficient nature, relatively simple design and low maintenance, apart from the option of easy adoption into an existing process. In an ultrasonic bath configuration, the transducer(s) is (are) bonded to the base of the tank or reactor. However, it is clear that the intensity of ultrasound waves is much greater with a probe than a bath.

Commercial standard ultrasonic equipment is being developed at a great pace, and any novel process for the application of acoustic field in industry is not possible without ultrasonic equipment manufacturers willing to build new designs according to the requirements of industries. This implies that while the technology could exhibit great promises, it will have to be carefully developed and scaled up for every single, unique application.

6.4 Ultrasound for decontamination of meat

The inactivation of micro-organisms upon exposure to an ultrasonic field is known to the scientific community since 1960s (Ensminger, 1973a). The antimicrobial action of ultrasound arises mainly from the acoustic cavitation and the physical and chemical effects propagated thereof. The inactivation of *Listeria monocytogenes, Salmonella* sp., *E. coli* and *Bacillus* sp. is well recorded in the literature (Piyasena, Mohareb and McKellar, 2003). However, a combination of ultrasound with other non-thermal approaches has often been recommended for improving efficacy. Surprisingly, studies regarding microbial inactivation in meat and meat products using ultrasound are very scarce.

In one of the early studies, Lillard (1993) reported that *Salmonella* Typhimurium attached to broiler skin was reduced by 1–1.5 log upon sonication in peptone at 20 kHz for 30 min. The author also found that sonication of chicken skin dipped in chlorine solution was more effective, allowing decrease in *Salmonella* counts by up to 2.44–3.93 log. Results of research carried out by Dolatowski and Stasiak (2002) have also shown that ultrasonication has a significant influence on microbiological decontamination of meat. Birk and Knøchel (2009) reported that ultrasound treatment combined with marination using red wine enhanced the antibacterial effect compared with ultrasound alone for *L. monocytogenes, Brochothrix thermosphacta* and *Campylobacter jejuni* in pork muscles. However, it resulted in only approximately a 1-log reduction after 10 min and was selective as well; for example, the synergistic effect was not observed for *Carnobacterium maltaromaticum*. In a recent study, Kordowska-Wiater and Stasiak (2011) reported that ultrasound treatment eliminated about 1 log CFU/cm^2 of Gram-negative bacteria (*Salmonella* sp., *E. coli* and *Pseudomonas fluorescens*) on the surface of chicken wings, while ultrasound combined with lactic acid inactivated more than 1.5 log CFU/cm^2. Haughton *et al.* (2012) evaluated the effects of sonication and thermosonication (53 ± 1 °C) on the microbial load in raw poultry using an ultrasonic bath (40 kHz, 20 W/L) and probe system (24 kHz, 20 kW/L). The group reported a maximum decrease of up to 4.72 $\log_{10}$ for *C. jejuni* following thermosonication using the probe system, which was much higher than for thermal (1.45 $\log_{10}$) or sonication treatment (3.17 $\log_{10}$) alone. The low-intensity bath unit, on the other hand, resulted in insignificant reductions for *Campylobacter*; however, up to 2.74 and 1.69 $\log_{10}$ CFU/g reductions in *Enterobacteriaceae* and total viable counts (TVCs) were recorded for thermosonication using the bath. In general, the response of vegetative pathogenic and spoilage microorganisms in meat and meat products to ultrasound can be expected to be variable and would depend on process parameters such as acoustic energy density, temperature and processing time and on product parameters such as pH, a_W, salt content and the presence of other antimicrobials.

6.5 Applications of ultrasound in meat processing

6.5.1 Tenderization of meat

Tenderness of cooked meat is generally regarded as one of the most important attributes influencing consumer appreciation of its eating quality. Any means to accelerate, ageing in meat would be highly regarded as value addition. Tenderness of meat is related to a combination of breakdown within muscle fibres, predominantly due to the activity of enzymes, and loosening of connective tissue, specifically collagen. Naturally occurring enzymes in meat, predominantly cathepsins (which are cysteine proteases) and calpains (calcium-activated proteases), tenderise meat by slowly breaking down the muscle fibres. Cathepsins are found in the lysosomes of a muscle cell and work on the bonds between actin and myosin (Barton, Bullock and Weir, 1996). The artificial tenderization methods include beating the meat; cutting it into small portions so that the strands of connective tissue are severed; marinating with vinegar, wine or salt; and enzymatic tenderization (Feiner, 2006), electrical stimulation, injection with salt solutions (more recently calcium salts), hydrodyne (explosive shock) and ultrasound treatment.

Although ultrasonics for meat tenderization was reported as early as the 1970s by Zayas and Orlova (1970), the actual mechanism is still ambiguous and the results of investigations made are also inconsistent in some cases. Got *et al.* (1999) employed high-frequency (2.4 MHz) ultrasound to treat pre-rigour meat and observed an initial increase in firmness (compression force) of raw meat compared to untreated or post-rigour sonicated meat. After 14 days of ageing, this group observed no difference between untreated and pre-rigour or post-rigour ultrasound-treated meat. Considering that water–protein interactions in meat can be modified using ultrasound (McClements, 1995), Stadnik, Dolatowski and Baranowska (2008) investigated the potential of low frequency and relatively low-intensity ultrasound (45 kHz, 2 W/cm^2) treatment at 24 h in affecting the post-mortem pH of meat and its water-holding capacity during 96 h of meat ageing. Their study revealed that ultrasound treatment did not have a significant effect on the pH of the meat. In addition, the high water-holding capacity of sonicated samples (typical of meat in an advanced post-mortem stage) suggested that acceleration of rigor mortis followed the fragmentation of protein structure of cells.

The application of ultrasound facilitates the release of myofibrillar proteins, which are responsible for binding the pieces of meat together in the formed meat products. Therefore, improved physical properties such as water-binding capacity, tenderness and cohesiveness are obtained in US-treated meat products

(Lawrie, 2006). In order to identify the extent of proteolytic activity triggered by ultrasound, if any, Xiong *et al.* (2012) applied proteolytic enzyme solutions alongside ultrasonication for culled hen muscles. They observed that the shear force values in the ultrasonicated samples (at 24 kHz frequency, 12 W/cm^2 for 15 s) were lower in comparison to control samples. Furthermore, there were insignificant differences between the cooking loss in sonicated group and enzyme-treated groups. This indicated that ultrasonication was at par with proteolytic enzyme treatments. Moreover, it opens space for new studies wherein the simultaneous use of proteolytic enzymes and ultrasonication for possible synergistic effects can be evaluated. The possible mechanisms responsible for the tenderization effect can be summarised as follows:

1 Physical disruption of the myofibrils and damage to the structure.
2 Increase in proteolysis that contributes to meat tenderization during ageing, for example, through increased release of cathepsins.
3 Fragmentation of collagen.
4 Damage to cell membranes.
5 Physical disruption of proteins through cavitation and free radical formation.

A brief summary of the important research studies carried out to investigate the effect of ultrasound on tenderization of meat have been summarised in Table 6.1. All studies to date point to the fact that power ultrasound can be considered as an alternative approach for tenderization of meat, practical industrial use of which can be realised with further studies.

6.5.2 Curing and pickling of meat

Curing, an age technology for meat preservation, involves addition of salt(s) mainly sodium chloride and nitrates or nitrites (colour-fixing agents) for chemical preservation and development of pleasant flavours through biochemical reactions. Pickling (also called brining or corning) is a process of preserving food by anaerobic fermentation in brine. Additives can also be used in brining solution to improve the nutritional value of meat apart from improving the texture of the food products and/or achieving a juicier product. Pickling and brining processes for meat overlap in their verbal use as well as function and essentially are mass transfer processes involving diffusion of salts into the meat matrix.

Traditionally, meat intended for pickling is dipped in open baths (or tank) containing the pickling brine (Tank Pickling). The pickling solution is allowed to diffuse into the meat over a time period of several days and this can exceed a month, where the meat has intact skin. Further, this method also has another disadvantage that when meat is initially placed in curing brine, the exterior muscles will be exposed to a much higher concentration of salt than that established when equilibrium between meat and brine is subsequently attained (Lawrie, 2006). In an alternative process, pickling can be established by injecting the solution into the meat (vascular pumping) through multiple syringes. However, this process

Table 6.1 Research findings in the application of ultrasound for meat tenderization, in chronology.

Muscle	Conditions	Salient results	Reference
Chicken pectoralis major	40 kHz, 2.4 kW for 15 min	After 24 h post-mortem, sonicated muscles yielded lower shear force (4.4 N) compared to untreated muscles (5.0 N)	Dickens, Lyon and Wilson (1991)
Beef semitendinosus	25.9 kHz for up to 16 min	Ultrasound-treated meat exhibited lower shear force than untreated at 2 and 4 min, but equal or higher after 8 or 16 min	Smith *et al.* (1991)
Beef muscle	2 W/cm^2 for 2 h at 40 kHz	Extensive damage to the perimysial connective tissue resulting in improved eating texture	Roberts (1992)
Cubes cut from lamb liver in buffer solution	20 kHz, 56–62 W/cm^2 intensity for 10 s–3 min	Protease (acid phosphatase) activity increased in the cell plasma fraction and then rapidly decreased. Results were in correlation with conventional ageing	Roncales *et al.* (1993)
Vacuum packaged steak	20 kHz (bath) at 1.55 W/cm^2 for 24 min	Increase in meat tenderness was not observed because of lack of focused, high-power ultrasound to induce gelatine formation	Pohlman, Dikeman and Zayas (1997a)
Beef longissimus thoracis et lumborum, semimembranosus and biceps femoris	30–47 kHz, 0.62–0.29 W/cm^2 and 30 min per side	Peak force values remained unaffected by ultrasound treatment	Lyng, Allen and McKenna (1997)
Longissimus thoracis et lumborum and semimembranosus	20 kHz (horn) at 62 W/cm^2 for 15 s intervals	Ultrasonicated steaks did not exhibit any quality improvements. Surface temperature and core temperatures rose by 12–20 and 30–40 °C, respectively	Lyng, Allen and McKenna (1998b)
Lamb longissimus thoracis et lumborum	20 kHz (horn) at 63 W/cm^2 for 15 s intervals	Bite-force tenderometry, sensory analysis and collagen solubility all showed no effect due to ultrasound treatment	Lyng, Allen and McKenna (1998a)
Semimembranosus muscle of cow	High-intensity (10 W/cm^2) and High-frequency (2.6 MHz)	No conclusive effect on meat ageing rate was observed when sonication was performed at pre-rigour stage. Post-rigour ultrasonic treatment did not induce any structural modification but slightly improved the ageing index after 6 days. No improvement in the final (day 14) ageing index was observed compared to the controls	Got *et al.* (1999)

(*continued overleaf*)

Table 6.1 (*continued*)

Muscle	Conditions	Salient results	Reference
Longissimus lumborum et thoracis and semitendinosus	24 kHz at 12 W/cm^2	Rise in pH occurred in the muscles that was distinct for each type of muscle, which was pointed to change in protein conformations. Increasing treatment time decreased meat hardness	Jayasooriya *et al.* (2007)
Beef semimembranosus	45 kHz, 2 W/cm^2 and 120 s treatment	High water-holding capacity of ultrasonicated samples suggested an acceleration of rigor mortis followed by a fragmentation of proteins structures of cell	Stadnik, Dolatowski and Baranowska (2008)
Pork loin	20 kHz, 2–4 W/cm^2 sonication and vacuum tumbling; 30–180 min	Desired microstructural and textural changes occurred, which improved water-binding capacity and diffusion of salt. Extended sonication or higher power levels caused denaturation	Siró *et al.* (2009)
Beef semitendinosus muscle	40 kHz, 1500 W ultrasonic bath; 10–60 min	Following sonication, collagenous fibres were disordered and staggered and the fibre arrangement became loose. Treatment time had a significant effect on the texture profile of muscle	Chang *et al.* (2012)

leads to leakage of the pickling solution from the openings caused by injection. Moreover, the pricks are also seen in the meat.

Ultrasound is known to enhance several mass transfer processes, including brining of meat. It is now well recognised that ultrasound allows introduction of the pickling solution uniformly into the meat tissue, under mild conditions (without drastic destruction of the muscle fibres) and in the possible shortest time. Carcel *et al.* (2007) have reported that above a critical ultrasonic intensity (51 W/cm), the uptake of brine into meat is proportional to the applied ultrasonic intensity. In their study, at the highest level of ultrasound, the total brine uptake was significantly higher than the initial water content of the meat. Siró *et al.* (2009) investigated the potential of ultrasound-assisted curing technique using low-frequency (20 kHz) and low-intensity (2–4 W/cm^2) ultrasound for improving the diffusion of sodium chloride (40 g/L) in pork loin (*Longissimus dorsi*). The group reported that sonication significantly enhanced the diffusion of salt compared to samples brined under static conditions.

From a practical industrial setup point of view, regardless of whether operated in continuous or batch mode, ultrasound transducers are recommended to be placed at periodic spaces in the brining tank and operated in pulsed (delayed) mode. This not only would lengthen the overall life time of the instrument but also accelerate the introduction of pickling solution into the meat tissue. For example, in patent WO 99/11138 (Kozari *et al.*, 1999), it is disclosed that ultrasound treatments with relaxation times proved to be very efficient in increasing the penetration time and depth of the pickling solution, increasing, at the same time, also the uniformity of pickling. Recent patent WO 2009/066154 (Nolsoe, 2009) also describes an apparatus for the pickling of additives wherein the meat is continuously travelling through the brine in tank, whose base is bonded with ultrasound transducer(s). However, it appears more suitable for brining small chunks of meat or fish.

6.5.2.1 Explaining enhanced mass transfer in an ultrasonic field

Irrespective of the type of diffusion, Fick's first law mathematically describes the flow of matter from a region of high concentration to a region of low concentration (equation 6.1):

$$j = -D\left(\hat{\imath}\frac{\partial c}{\partial x} + \hat{\jmath}\frac{\partial c}{\partial y} + \hat{k}\frac{\partial c}{\partial z}\right) \tag{6.1}$$

where $c(x, y, z)$ is the concentration field, D is the effective diffusion coefficient and j is the mass flux.

As mentioned earlier, ultrasound is known to enhance heat and mass transfer in several food processing operations, including brining of meat; that is, ultrasound increases the mass flux j when diffusion occurs. It can be concluded from equation 6.1 that the mass flux would increase if the diffusivity (D) increases or if the driving force, that is, concentration gradient increases. Ultrasound enhances the diffusivity of brine into meat, which has been demonstrated to be exponential when ultrasound intensity is raised, as shown in Figure 6.3 (Siró *et al.*, 2009).

Siró *et al.* (2009) employed an analytical solution to Fick's second law, given by equation 6.2, to estimate the diffusion coefficient of salt into pork muscle to be 3.1×10^{-9} m^2/s at 4 W/cm^2 ultrasound intensity while that for control to be 2.4×10^{-10} m^2/s at 5 °C.

$$\frac{C_{0,\mathrm{M}} - C_{t,\mathrm{M}}}{C_{0,\mathrm{M}} - C_{\mathrm{eq}}} = 1 - \sum_{n=0}^{\infty} \frac{8}{(2n+1)^2\pi^2} \times \exp\left\{\frac{-D(2n+1)^2\pi^2 t}{L^2}\right\} \tag{6.2}$$

where $C_{t,\mathrm{M}}$ is concentration of sodium chloride in the meat (g/100 g) at any given time t (s), $C_{0,\mathrm{M}}$ is the initial salt concentration in the meat, C_{eq} is the equilibrium salt concentration, D is the diffusion coefficient (m^2/s) and L is the thickness of pork loin (m).

However, equation 6.2 does not explain the effect of acoustic streaming on overall mass transfer although it enables calculation of the effective diffusivity. In order to understand the effect of this component, Fick's law can be rewritten with respect to time t (s) and to accommodate the convective effects as equation 6.3 (Gekas, 1992):

$$\frac{\partial c}{\partial t} + V\left(\frac{\partial c}{\partial x} + \frac{\partial c}{\partial y} + \frac{\partial c}{\partial z}\right) = D\left(\frac{\partial^2 c}{\partial x^2} + \frac{\partial^2 c}{\partial y^2} + \frac{\partial^2 c}{\partial z^2}\right) \tag{6.3}$$

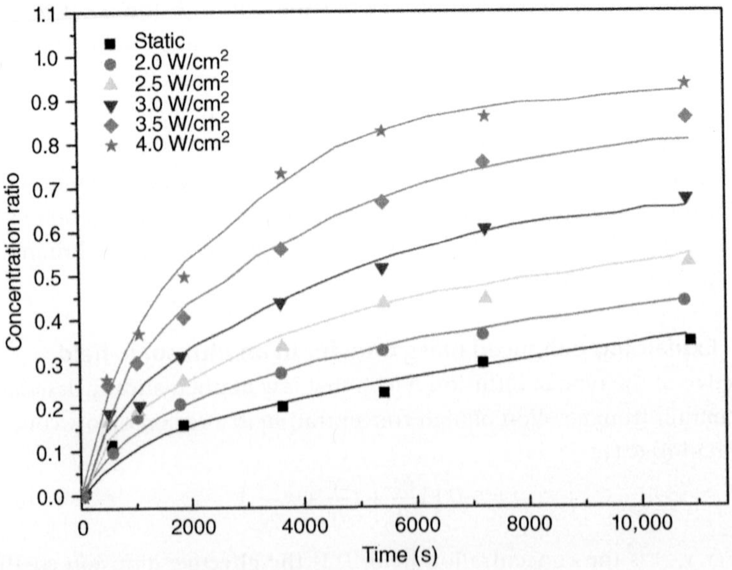

Figure 6.3 Diffusion kinetics of NaCl from a 200 g/L brine solution into pork loins as a function of different treatments. Solid lines represent the diffusion fitted to the analytical solution of Fick's second law for rectangular geometry. (Adapted from Siró *et al.* (2009) with permission.)

where $V(x, y, z)$ is the velocity vector of the particles in the medium. In an ultrasonic field, the velocity vector can be decomposed into a resultant of the convective or bulk velocity V_c, the vibrational velocity V_v and the acoustic streaming velocity V_a (Floros and Liang, 1994; Nyborg, 1965). To numerate the influence of an acoustic field on the diffusion, the contribution of bulk velocity term V_c and vibrational velocity V_v are usually ignored for obvious reasons and based on experimental observations, respectively. This essentially highlights the importance of acoustic streaming velocity V_a, which decreases the thickness of fluid–solid boundary layer by up to fivefold. The alternating pressures and microstreaming phenomenon at the interfaces disturb the diffusion boundary layer, whence decreasing the resistance to convective mass transfer (Ghafoor *et al.*, 2014). It may be noted that equation 6.3 still does not include the effects of acoustic cavitation, whose interaction with meat surface itself would be very complicated. Cavitation induced by ultrasound can cause erosion of meat fibres (rupture of myofibrils) during brining, as has been observed by Ozuna *et al.* (2013) (see Figure 6.4).

6.5.3 Sausage production

Sausages are typically made by extruding a proteinaceous meat emulsion or batter into an artificial or natural casing of a desired size and shape, and then cooking the batter within the casing. Usually sausage casings are hand or

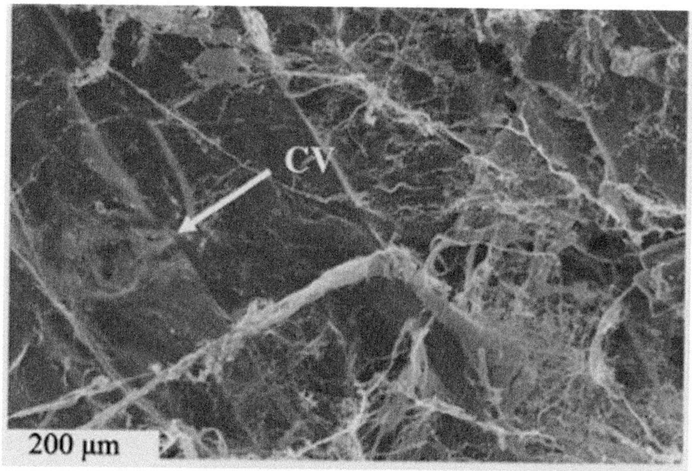

Figure 6.4 Longitudinal section observed by SEM (150×) of *Longissimus dorsi* muscle of pork meat brined for 120 min in 200 kg NaCl/m^3 solution. CV: impact of cavitation bubble implosion on myofibril. (Adapted from Ozuna *et al.* (2013), with permission from Elsevier.)

machine peeled prior to packaging, and the peels of casing so obtained are discarded. Alternatively, heating plates are employed (sometimes in conjunction with food-grade acids) to coagulate the emulsions, and/or the use of plugs at regular intervals is employed to obtain casing-less sausages. However, such processes are energy intensive and the product quality is often not very appealing.

Over the past decade, extensive work has been done in the area of polymers and materials science to develop novel extrusion processes with the aid of high-power ultrasound (Isayev, Kumar and Lewis, 2009; Guo *et al.*, 2003; Isayev, Wong and Zeng, 1990). The energy input provided by ultrasonic excitation of metal tubes by perpendicular attachment of the sonotrode can not only improve the flow behaviour of sticky or highly viscous materials due to lower drag resistance but also modify product structures (Knorr *et al.*, 2004). Likewise, if sausage manufacture is visualised as an extrusion process, it is evident from literature that application of ultrasound to the cylindrical sausage extruder unit causes stiffening of meat emulsions used for sausage stuffing (Manna *et al.*, 2000). Phenomenologically, the molecules of the raw product exposed to ultrasound shift into their normal mode of vibration, which produces frictional heat. As a result, the protein on the product surface coagulates and the desired continuous independent skin forms (Kortschack and Heinz, 2004). With this as the underpinning science, Manna *et al.* (2000) patented an ultrasonic device for the production of casing-less sausages. The fact that ultrasound below 100 kHz does not penetrate far into the tissues due to internal losses provides scope for confining the stiffening of emulsion to few millimetres below the active surface. Accordingly, the patent describes an ultrasonic tool for treatment of a cylindrical shape to create the skin. The skin resulting from ultrasonic treatment has sufficient tensile properties to hold the cylindrical shape during final processing. Thus, ultrasound-assisted processing can help elimination of casing materials and avoid re-handling (which could lead to contamination) during sausage manufacture. Kortschack and Heinz (2004) also describe in their patent a similar method to utilise ultrasound along with heat for the formation of a thin, sealed, shape-stabilising independent coagulation skin on the surface of raw sausages (see Figure 6.5).

6.5.4 Meat industry effluent treatment

Slaughterhouse operations inevitably generate waste comprised of the animal parts and wastewater as a result of washing carcasses, processing offal, and from cleaning equipment and processing area. Meat processing effluents are distinct in the sense that these are characterised by heavy loading with solids, floatable matter (fat), blood, manure and a variety of organic compounds originating from proteins.

Ultrasound has been employed for industrial waste water treatment by several workers (Mahamuni and Adewuyi, 2010) and as a pre-treatment to increase the aerobic biodegradability of distillery wastewater by Sangave and Pandit (2006). These studies indicate that ultrasonic pre-treatment enhances the biodegradability of wastewaters, including those containing resistant organic substances. It is worthwhile noting that anaerobic digestion stabilises the materials being treated while producing renewable energy. Furthermore, wastes and by-products from meat processing industry have relatively high methane production potentials (340–900 m^3 CH_4/t volatile solids added). Thus, the technology can potentially benefit the meat industries. However, the high fat content of meat processing effluent makes the biological treatment methods relatively less effective.

Figure 6.5 Production of ultrasound-assisted coating of Frankfurter-type sausages (after Heinz, unpublished results). (Adapted from Knorr *et al.* (2004) with permission from Elsevier.)

Low-frequency ultrasound as a pre-treatment step preceding anaerobic processes of meat processing industry effluent was recently investigated by Erden, Buyukkamaci and Filibeli (2010). This group observed enhanced anaerobic biodegradability of meat processing effluent when ultrasonic pre-treatment using a probe at 20 kHz, (200 W maximum power) was included in the process. Luste, Luostarinen and Sillanpaa (2009) found that ultrasound was very effective in increasing the soluble chemical oxygen demand with dissolved air flotation. In a later study, it was reported by the same group that the United States effectively hydrolysed grease and protein-containing ABPs, probably achieving lysis of grease-containing cells (Luste, Vilhunen and Luostarinen, 2011). It has also been observed that methane production potential from the co-digestion of animal by-products from meat processing industry with cattle slurry was notably higher than that of slurry alone (Luste, Heinonen-Tanski and Luostarinen, 2012).

The positive effects of sonication process as a pre-treatment to anaerobic digestion of meat processing effluent can be summarised as follows:

1. High shearing phenomena resulting from turbulence generated by cavitational collapse decreases the average particle size distribution. This ultimately increases the total available surface area for hydrolysis.
2. The particle size reduction enhances solubilisation of organic matter from solid phase to liquid phase.
3. The free radicals formed help COD (chemical oxygen demand) removal.
4. Alternatively, sonication may aid demulsification process, thereby removing the adverse effects of oil on subsequent anaerobic bioprocesses (Stack *et al.*, 2005).

6.5.5 Meat cooking

The heating effects of ultrasound in muscle resulting from the absorption of ultrasound energy were described as early as 1960s by Eggleton *et al.* (1965) and Gersten (1965). Based on this phenomenon, some attempts have been made by researchers for utilising ultrasound energy for cooking of meat. The work of Pohlman *et al.* (1997b) revealed that ultrasound cooking (20 kHz, 1000 W) results in better cooking speed (6.7 min), more moisture retention, less cooking loss (2–5 times) and better energy efficiency (0.08 W/g) for beef Longissimus and pectoralis. The authors also realised a more uniform cooking environment and less instrumental peak force work to shear than convection cooking (12.3 min, 0.22 W/g power).

At a structural level, the effects of ultrasound-assisted cooking appear as longer sarcomeres, larger diameter fibres and more myofibrillar disruption and shattering of muscle (Pohlman *et al.*, 1997b). At a macroscopic level, ultrasonication does not affect the drip, cooking and total loss, although it reduces the hardness and shear forces required for cutting by significant amount (Jayasooriya *et al.*, 2007). Based on this, it can be concluded that the damaged muscle structure resulting from ultrasonication does not affect the mechanism of moisture migration in meat.

An ultrasonic treatment apparatus involving moisture injection prior to ultrasound cooking has also been patented by Melnyczuk (2001). The inventor mentioned that the ultrasound treatment should be given before the onset of rigor mortis. Despite the several benefits identified in the literature, it is to be noted that an important disadvantage of substituting ultrasound cooking for other methods is the poorer flavour (Pohlman, Dikeman and Zayas, 1997a) compared to conventional cooking methods (e.g. roasting and grilling).

6.5.6 Cleaning

The knives used in meat processing plants are usually cleaned by hot water dipping at 82 °C. The process may appear simple, but denaturation of any proteins attached to the surface can protect the bacteria or other micro-organisms posing hazards. In addition, the efficacy of disinfection would depend on the organic matter and treatment time. In this regard, ultrasound has been evaluated to decrease the process time and increase microbial inactivation. In an early study, it has been reported that ultrasound treatment prior to water bath treatment causes lesser protein denaturation than water bath disinfection at 82 °C alone (Schütt-Abraham, Trommer and Levetzow, 1992). In a recent study, Leps *et al.* (2013) concluded that a 60 °C water bath + ultrasound for 5 s or 40 °C water bath + ultrasound + lactic acid (2%) for 5 s was effective in reducing the water temperature and treatment time for cleaning of stainless steel knives. The group also indicated that the effect of ultrasound was relatively mild compared to that of lactic acid. The inefficiency of ultrasound as a standalone treatment to remove meat biofilms from processing surfaces has also been reported by Oulahal *et al.* (2007). This group recommended the combined application of ultrasound and EDTA (ethylene diamine tetraacetic acid, chelating agent) or enzyme solutions (glycolytic and proteolytic) to remove up to 75 ± 4% and 100 ± 15% of *E. coli* and *Staphylococcus aureus* biofilms, respectively. In yet another study, the use of ultrasound (4 kW) at 65 °C for 3–6 min, with or without mechanical brushing, was considered effective for cleaning of poultry transport crates (Allen *et al.*, 2008a). It is to be noted that the plastic crates used for transporting live poultry from the farm to the processing plant are a source of contamination and cross-contamination. Allen *et al.* (2008b) also reported that ultrasonication at an immersion temperature of 60 °C reduced the counts of *Enterobacteriaceae* ($\log_{10}$ 2–3 CFU in 1–3 min) and aerobic plate count (>2 $\log_{10}$ CFU) on industrial poultry transportation crates. However, they observed that ultrasonication had insignificant effect on the microbial load. All the above-mentioned studies point to the fact that the existing cleaning procedures in meat and poultry industry can be easily modified to include ultrasound along with an antimicrobial agent for increasing the efficacy of the process.

6.6 Concluding remarks

As one of the emerging, novel food process, ultrasound can be applied to develop gentle but targeted processes to improve the quality and safety of processed foods and offer the potential for improving existing processes as well as for developing new process options. Ultrasonics has the potential to be a specialised and versatile technology with numerous applications in food processing. Extrinsic control parameters such as frequency, amplitude, ultrasonic intensity, treatment time and temperature strongly influence the observed effects of ultrasound (Deora *et al.*, 2013). Therefore, by suitably interplaying with these factors, desired changes in properties of meat and favourable effects can be achieved.

In spite of the promising published data for ultrasound processing of meat and meat products, no decisive success in the food industry has been achieved to date (Deora *et al.*, 2013). However, food industries are looking for and willing to adopt new technologies for safety reasons and continued success. For example, a beef processing plant (in Springdale, USA) recently adopted HPP technology in an effort to enhance the safety of the meat products following a *Salmonella* contamination through ground turkey in 2011 (Cargill News, 2011). Therefore, it is very likely that ultrasound-based food technologies will also make into the industries in near future, provided large-scale processing equipment are made available. This in turn can be achieved by dedicated studies keeping scale-up in focus. As a general remark, future approaches for sonication-aided decontamination should focus more on exploring hurdle technologies for synergistic effects, as well as maintaining freshness and high quality of meat (Chen *et al.*, 2012).

Although researchers have explained the phenomenon of ultrasound-induced cavitation in liquids and the hydrodynamics of bubbles, the translation of this knowledge into applications in food manufacture is yet to be manifested in several areas. In recent past, some work has been undertaken to explain the effects of ultrasound-induced acoustic cavitation when applied to food systems. However, most of the research on understanding the mechanism of ultrasound in liquid systems has concentrated on its physical effects. More specifically, the enhanced heat and mass transfer or other effects have been attributed to turbulence and physical shear. However, the capacity of free radical formation to influence food chemistry has received little attention. Therefore, further investigations in this area would be beneficial.

Commercial scale ultrasound-aided processing is still in its infancy and requires a great deal of work in order to develop the technology on an industrial scale and to more fully elucidate the effects of ultrasound on the properties of foods. Any novel process for the application of acoustic field in industry is not possible without ultrasonic equipment manufacturers willing to build new designs according to the requirements of industries. For application of power ultrasound at an industrial scale, it is essential to have energy-efficient ultrasound processors to ensure the economic feasibility of the technology. This implies that while the technology has great promises, it will have to be carefully developed and scaled up for every single, unique application. Future research should be focused on the development of non-contact ultrasound transducers or sonication bath systems with variable frequencies and the investigation of the economic feasibility of sonication as a novel food processing and preservation technique.

References

Abramov, O.V. (1998) *High-Intensity Ultrasonics: Theory and Industrial Applications*, CRC Press.

Allen, V.M., Burton, C.H., Wilkinson, D.J. *et al.* (2008a) Evaluation of the performance of different cleaning treatments in reducing microbial contamination of poultry transport crates. *British Poultry Science*, **49** (3), 233–240. doi: 10.1080/00071660802094206

Allen, V.M., Whyte, R.T., Burton, C.H. *et al.* (2008b) Effect of ultrasonic treatment during cleaning on the microbiological condition of poultry transport crates. *British Poultry Science*, **49** (4), 423–428. doi: 10.1080/00071660802262068

Ashokkumar, M., Lee, J., Kentish, S. and Grieser, F. (2007) Bubbles in an acoustic field: an overview. *Ultrasonics Sonochemistry*, **14** (4), 470–475. doi: 10.1016/j.ultsonch.2006.09.016

Aymerich, T., Picouet, P.A. and Monfort, J.M. (2008) Decontamination technologies for meat products. *Meat Science*, **78** (1–2), 114–129. doi: 10.1016/j.meatsci.2007.07.007

Barton, S., Bullock, C. and Weir, D. (1996) The effects of ultrasound on the activities of some glycosidase enzymes of industrial importance. *Enzyme and Microbial Technology*, **18** (3), 190–194. doi: 10.1016/0141-0229(95)00092-5

Birk, T. and Knøchel, S. (2009) Fate of food-associated bacteria in pork as affected by marinade, temperature, and ultrasound. *Journal of Food Protection*, **72** (3), 549–555.

Carcel, J.A., Benedito, J., Bon, J. and Mulet, A. (2007) High intensity ultrasound effects on meat brining. *Meat Science*, **76** (4), 611–619. doi: 10.1016/j.meatsci.2007.01.022

Cargill News (2011) Cargill plant resumes making ground turkey after *Salmonella*-related recall. Cargill. http://www.cargill.com/news/releases/2011/NA3053336.jsp (accessed 26 July 2013)

Cargill News (2012) Cargill ground beef recall - frequently asked questions. Cargill. http://www.cargill.com/products/ground-beef-recall/salmonella-enteriditis/index.jsp. Accessed 26 July 2013

Chamorro, A., Miranda, F.J., Rubio, S. and Valero, V. (2012) Innovations and trends in meat consumption: An application of the Delphi method in Spain. *Meat Science*, **92** (4), 816–822. doi: 10.1016/j.meatsci.2012.07.007

Chang, H.-J., Xu, X.-L., Zhou, G.-H. *et al.* (2012) Effects of characteristics changes of collagen on meat physicochemical properties of beef semitendinosus muscle during ultrasonic processing. *Food and Bioprocess Technology*, **5** (1), 285–297. doi: 10.1007/s11947-009-0269-9

Chen, J.H., Ren, Y., Seow, J. *et al.* (2012) Intervention technologies for ensuring microbiological safety of meat: Current and future trends. *Comprehensive Reviews in Food Science and Food Safety*, **11** (2), 119–132. doi: 10.1111/j.1541-4337.2011.00177.x

Clark, P.J. (2008) *An Update on Ultrasonics. Food Technology* vol July, IFT Press, USA.

Commander, K.W. and Prosperetti, A. (1989) Linear pressure waves in bubbly liquids - comparison between theory and experiments. *Journal of the Acoustical Society of America*, **85** (2), 732–746. doi: 10.1121/1.397599

Curie, J. and Curie, P. (1880) Developpment par compession de lelectricite polaire dans les cristaux hemiedres a faces inclinees *Bulletin de la Societe Francaise de Mineralogie*, **3**, 90.

Deora, N.S., Misra, N.N., Deswal, A. *et al.* (2013) Ultrasound for improved crystallisation in food processing. *Food Engineering Reviews*, **5** (1), 36–44. doi: 10.1007/s12393-012-9061-0

Dickens, J.A., Lyon, C.E. and Wilson, R.L. (1991) Effect of ultrasonic radiation on some physical characteristics of broiler breast muscle and cooked meat. *Poultry Science*, **70** (2), 389–396. doi: 10.3382/ps.0700389

Dolatowski, Z.J. and Stasiak, D.M. (2002) Bacterial contamination of meat and meat products after ultrasound treatment. *Acta Scientiarum Polonorum. Technologia Alimentaria*, **1** (1), 55–65.

Eggleton, R.C., Kelly, E., Fry, F.J. *et al.* (1965) Morphology of ultrasonically irradiated skeletal muscle, in *Ultrasonic Energy* (ed. E. Kelly), University of Illinois Press, Urbana, IL, pp. 117–136.

Ensminger, D. (1973a) *Ultrasonics-the Low- and High-Intensity Applications*, Marcel Dekker, Inc, New York, p. 570.

Ensminger, D. (1973b) *Ultrasonics, the Low- and High-Intensity Applications* vol 19. vol 4, Marcel Dekker, Inc., New York. doi: 10.1002/aic.690190442

Erden, G., Buyukkamaci, N. and Filibeli, A. (2010) Effect of low frequency ultrasound on anaerobic biodegradability of meat processing effluent. *Desalination*, **259** (1–3), 223–227. doi: 10.1016/j.desal.2010.03.055

FAO (2011) *World Livestock 2011: Livestock in Food Security*, Food and Agriculture Organization of the United Nations.

Feiner, G. (2006) *Meat Products Handbook: Practical Science and Technology, Woodhead Publishing.*

Floros, J.D. and Liang, H.H. (1994) Acoustically assisted diffusion through membranes and biomaterials. *Food Technology*, **48** (12), 79–84.

Gekas, V. (1992) *Transport Phenomena of Foods and Biological Materials*, CRC Press, Boca Raton, FL.

Gersten, J.W. (1965) Biological effects of ultrasound on neuromuscular skeletal systems, in *Ultrasonic Energy* (ed. E. Kelly), University of Illinois Press, Urbana, IL, pp. 150–159.

Ghafoor, M., Misra, N.N., Mahadevan, K. and Tiwari, B.K. (2014) Ultrasound assisted hydration of navy beans (*Phaseolus vulgaris*). *Ultrasonics Sonochemistry* **21**(1), 409–414

Got, F., Culioli, J., Berge, P. *et al.* (1999) Effects of high-intensity high-frequency ultrasound on ageing rate, ultrastructure and some physico-chemical properties of beef. *Meat Science*, **51** (1), 35–42. doi: 10.1016/S0309-1740(98)00094-1

Guo, S., Li, Y., Chen, G. and Li, H. (2003) Ultrasonic improvement of rheological and processing behaviour of LLDPE during extrusion. *Polymer International*, **52** (1), 68–73. doi: 10.1002/pi.1005

Haughton, P.N., Lyng, J.G., Morgan, D.J. *et al.* (2012) An evaluation of the potential of high-intensity ultrasound for improving the microbial safety of poultry. *Food and Bioprocess Technology*, **5** (3), 992–998. doi: 10.1007/s11947-010-0372-y

Henglein, A. and Kormann, C. (1985) Scavenging of OH radicals produced in the sonolysis of water. *International Journal of Radiation Biology and Related Studies in Physics, Chemistry, and Medicine*, **48** (2), 251–258.

Hunt, F.V. (1982) *Electroacoustics: The Analysis of Transduction, and Its Historical Background*, Acoustical Society of America, Woodbury, NY.

Isayev, A.I., Kumar, R. and Lewis, T.M. (2009) Ultrasound assisted twin screw extrusion of polymer–nanocomposites containing carbon nanotubes. *Polymer*, **50** (1), 250–260. doi: 10.1016/j.polymer.2008.10.052

Isayev, A.I., Wong, C.M. and Zeng, X. (1990) Effect of oscillations during extrusion on rheology and mechanical properties of polymers. *Advances in Polymer Technology*, **10** (1), 31–45. doi: 10.1002/adv.1990.060100104

Jayasooriya, S.D., Torley, P.J., D'Arcy, B.R. and Bhandari, B.R. (2007) Effect of high power ultrasound and ageing on the physical properties of bovine Semitendinosus and Longissimus muscles. *Meat Science*, **75** (4), 628–639.

Knorr, D., Zenker, M., Heinz, V. and Lee, D.U. (2004) Applications and potential of ultrasonics in food processing. *Trends in Food Science and Technology*, **15** (5), 261–266.

Kordowska-Wiater, M. and Stasiak, D.M. (2011) Effect of ultrasound on survival of gram-negative bacteria on chicken skin surface. *Bulletin of the Veterinary Institute in Pulawy*, **55**, 207–210.

Kortschack, F. and Heinz, V. (2004) Process for solidifying the surface of raw sausage emulsion by ultrasonic treatment. USA Patent, US 6737093 B1.

Kozari, J., Gyongyosi, J., Gyongyosi, Z. and Naszodi, Z. (1999) Procedure and instrument for the quick pickling of mainly meat. Hungary Patent WO 99/11138

Lawrie, R.A. (2006) The storage and preservation of meat: II Moisture control, in *Lawrie's Meat Science*, 7th edn, Woodhead Publishing, England, pp. 250–253.

Leighton, T. (2007) What is ultrasound? *Progress in Biophysics and Molecular Biology*, **93** (1–3), 3–83.

Leps, J., Einschutz, K., Langkabel, N. and Fries, R. (2013) Efficacy of knife disinfection techniques in meat processing. *Meat Science*, **95** (2), 185–189. doi: 10.1016/j.meatsci.2013.04.035

Lillard, H.S. (1993) Bactericidal effect of chlorine on attached salmonellae with and without sonification. *Journal of Food Protection*, **56** (8), 716–719.

Luste, S., Heinonen-Tanski, H. and Luostarinen, S. (2012) Co-digestion of dairy cattle slurry and industrial meat-processing by-products–effect of ultrasound and hygienization pre-treatments. *Bioresource Technology*, **104**, 195–201. doi: 10.1016/j.biortech.2011.11.003

Luste, S., Luostarinen, S. and Sillanpaa, M. (2009) Effect of pre-treatments on hydrolysis and methane production potentials of by-products from meat-processing industry. *Journal of Hazardous Materials*, **164** (1), 247–255. doi: 10.1016/j.jhazmat.2008.08.002

Luste, S., Vilhunen, S. and Luostarinen, S. (2011) Effect of ultrasound and addition of bacterial product on hydrolysis of by-products from the meat-processing industry. *International Biodeterioration & Biodegradation*, **65** (2), 318–325. doi: 10.1016/j.ibiod.2010.12.004

Lyng, J.G., Allen, P. and McKenna, B. (1998a) The effects of pre- and post-rigour high-intensity ultrasound treatment on aspects of lamb tenderness. *LWT - Food Science and Technology*, **31** (4), 334–338. doi: 10.1006/fstl.1997.0361

Lyng, J.G., Allen, P. and McKenna, B.M. (1997) The influence of high intensity ultrasound baths on aspects of beef tenderness. *Journal of Muscle Foods*, **8** (3), 237–249. doi: 10.1111/j.1745-4573.1997.tb00630.x

Lyng, J.G., Allen, P. and McKenna, B.M. (1998b) The effect on aspects of beef tenderness of pre- and post-rigour exposure to a high intensity ultrasound probe. *Journal of the Science of Food and Agriculture*, **78** (3), 308–314. doi: 10.1002/(sici)1097-0010(199811)78:3<308::aid-jsfa123>3.0.co;2-f

Mahamuni, N.N. and Adewuyi, Y.G. (2010) Advanced oxidation processes (AOPs) involving ultrasound for waste water treatment: A review with emphasis on cost estimation. *Ultrasonics Sonochemistry*, **17** (6), 990–1003. doi: 10.1016/j.ultsonch.2009.09.005

Manna, R.R., Russell, A.W., Voic, D., Noval, T.A.D., Ng, D. and Pantano S. (2000) Method and device for manufacturing hot dogs using high power ultrasound. USA Patent, US 6146674 A

Margulis, M.A. and Margulis, I.M. (2004) Mechanism of sonochemical reactions and sonoluminescence. *High Energy Chemistry*, **38** (5), 285–294. doi: 10.1023/b:hiec.0000041338.11770.74

McClements, D. (1995) Advances in the application of ultrasound in food analysis and processing. *Trends in Food Science & Technology*, **6** (9), 293–299.

Melnyczuk, M. (2001) Method of, and apparatus for, treating meat. Patent No: WO2001037674 A1.

Misra, N.N., Pankaj, S.K. and Kadam, S.U. (2011) *Acoustic Fields in Food Processing: Fundamentals and Applications*, 1st edn, VDM Publishing House, Germany.

Newton, I. (1687) *Philosophiae Naturalis Principia Mathematica* vol. 3, Royal Society, London, Cambridge.

Nolsoe, H. (2009) Apparatus for salting, pickling or adding additives to meat, fish or similar products, and a method for use thereof. Denmark Patent WO 2009/066154.

Nyborg, W.L. (1965) Acoustic streaming, in *Physical Acoustics* vol. 1B. vol Part B (ed. W.P. Mason), Academic Press, New York, p. 265.

Oulahal, N., Martial-Gros, A., Bonneau, M. and Blum, L.J. (2007) Removal of meat biofilms from surfaces by ultrasounds combined with enzymes and/or a chelating agent. *Innovative Food Science & Emerging Technologies*, **8** (2), 192–196.

Ozuna, C., Puig, A., García-Pérez, J.V. *et al.* (2013) Influence of high intensity ultrasound application on mass transport, microstructure and textural properties of pork meat (*L. dorsi*) brined at different NaCl concentrations. *Journal of Food Engineering*, **119** (1), 84–93. doi: 10.1016/j.jfoodeng.2013.05.016

Pankaj, S.K., Misra, N.N. and Cullen, P.J. (2013) Kinetics of tomato peroxidase inactivation by atmospheric pressure cold plasma based on dielectric barrier discharge. *Innovative Food Science & Emerging Technologies* **21**, 153–157

Pathak, V., Singh, V.P. and Sanjay, Y. (2011) Ultrasound as a modern tool for carcass evaluation and meat processing: A review. *International Journal of Meat Science*, **1** (2), 83–92. doi: 10.3923/ijmeat.2011.83.92

Piyasena, P., Mohareb, E. and McKellar, R.C. (2003) Inactivation of microbes using ultrasound: a review. *International Journal of Food Microbiology*, **87** (3), 207–216. doi: 10.1016/s0168-1605(03)00075-8

Pohlman, F.W., Dikeman, M.E. and Zayas, J.F. (1997a) The effect of low-intensity ultrasound treatment on shear properties, color stability and shelf-life of vacuum-packaged beef semitendinosus and biceps femoris muscles. *Meat Science*, **45** (3), 329–337. doi: 10.1016/S0309-1740(96)00106-4

Pohlman, F.W., Dikeman, M.E., Zayas, J.F. and Unruh, J.A. (1997b) Effects of ultrasound and convection cooking to different end point temperatures on cooking characteristics, shear force and sensory properties, composition, and microscopic morphology of beef longissimus and pectoralis muscles. *Journal of Animal Science*, **75** (2), 386–401.

Rayleigh, L. (1884) On the circulation of air observed in Kundt's tubes, and on some allied acoustical problems. *Philosophical Transactions of the Royal Society of London*, **175**, 1–21. doi: 10.1098/rstl.1884.0002

Rayleigh, L. (1917) On the pressure developed in a liquid during the collapse of a spherical cavity. *Philosophical Magazine* Series 6, **34** (200), 94–98. doi: 10.1080/14786440808635681

Roberts, R.T. (1992) High intensity ultrasonics, in *The Chemistry of Muscle-Based Foods* (eds D.E. Johnston, M.K. Knight and D.A. Ledward), Royal Society of Chemistry, Cambridge, pp. 287–297.

Roncales, P., Cena, P., Beltran, J.A. and Jaime, I. (1993) Ultrasonication of lamb skeletal muscle fibres enhances postmortem proteolysis. *Zeitschrift für Lebensmittel-Untersuchung und -Forschung*, **196** (4), 339–342.

Sangave, P.C. and Pandit, A.B. (2006) Ultrasound and enzyme assisted biodegradation of distillery wastewater. *Journal of Environmental Management*, **80** (1), 36–46. doi: 10.1016/j.jenvman.2005.08.010

Schütt-Abraham, I., Trommer, E. and Levetzow, R. (1992) Ultraschall im "Steri-Becken"? zum einsatz von ultraschall in einrichtungen zur reinigung und desinfektion von messern am arbeitsplatz in schlacht-und zerlegebetrieben. *Fleischwirtschaft*, **72**, 864–867.

Siró, I., Vén, C., Balla, C. *et al.* (2009) Application of an ultrasonic assisted curing technique for improving the diffusion of sodium chloride in porcine meat. *Journal of Food Engineering,* **91** (2), 353–362. doi: 10.1016/j.jfoodeng.2008.09.015

Smith, N.B., Cannon, J.E., Novakofski, J.E., McKeith, F.K. and O'Brien, W.D. Jr., (1991) *Tenderization of Semitendinosus Muscle Using High Intensity Ultrasound.* In: Ultrasonics Symposium, IEEE Proceedings, Orlando, FL, 8–11 Dec 1991. pp. 1371–1374. doi: 10.1109/ultsym.1991.234038

Soria, A.C. and Villamiel, M. (2010) Effect of ultrasound on the technological properties and bioactivity of food: A review. *Trends in Food Science & Technology,* **21** (7), 323–331. doi: 10.1016/j.tifs.2010.04.003

Stack, L.J., Carney, P.A., Malone, H.B. and Wessels, T.K. (2005) Factors influencing the ultrasonic separation of oil-in-water emulsions. *Ultrasonics Sonochemistry,* **12**, 153–160.

Stadnik, J., Dolatowski, Z.J. and Baranowska, H.M. (2008) Effect of ultrasound treatment on water holding properties and microstructure of beef (m. semimembranosus) during ageing. *LWT - Food Science and Technology,* **41** (10), 2151–2158.

Suslick, K.S., Hammerton, D.A. and Cline, R.E. (1986) Sonochemical hot spot. *Journal of the American Chemical Society,* **108** (18), 5641–5642. doi: 10.1021/ja00278a055

Xiong, G.Y., Zhang, L.L., Zhang, W. and Wu, J. (2012) Influence of ultrasound and proteolytic enzyme inhibitors on muscle degradation, tenderness, and cooking loss of hens during aging. *Czech Journal of Food Sciences,* **30**, 195–205.

Zayas, Y.F. and Orlova, T.N. (1970) The application of ultrasonic vibrations for tenderization of meat. *Izvestiya Vysshikh Uchebnykh Zavedenii,* **4**, 54–56.

Zhou, G.H., Xu, X.L. and Liu, Y. (2010) Preservation technologies for fresh meat - a review. *Meat Science,* **86** (1), 119–128. doi: 10.1016/j.meatsci.2010.04.033

CHAPTER 7

Application of hydrodynamic shock wave processing associated with meat and processed meat products

James R. Claus
Department of Animal Sciences, Meat Science and Muscle Biology Laboratory, University of Wisconsin-Madison, USA

7.1 Introduction

The meat and poultry industry continues to strive to provide consumers with products that deliver a consistently tender meat eating experience and in a manner that assures food safety. It is well established that meat tenderness is a highly valued consumer trait (Huff Lonergan, Zhang and Lonergan, 2010). A variety of techniques applied individually or in combination for tenderizing meat include mechanical, chemical, temperature conditioning, aging, electrical stimulation, high-pressure heat treatments, and alternative carcass positioning (Solomon *et al.*, 2006).

A widely used mechanical method is blade (needle) tenderization (Aberle *et al.*, 2012; George-Evins *et al.*, 2004; Hayward *et al.*, 1980; King *et al.*, 2009; Mandigo and Olson, 1982). Blade tenderization (BT) has been shown to be effective on connective tissues as well as on the myofibrillar structure. However, blade tenderization has come under increasing attention relative to the potential risk associated with surface bacteria, including pathogenic bacteria that are translocated to the interior (Johns *et al.*, 2011; Lemmons *et al.*, 2011). Use of blade tenderization has been linked to food safety outbreaks (Laine *et al.*, 2005) and has led to an effort on understanding the lethality associated with meat cookery (Luchansky *et al.*, 2012; Luchansky *et al.*, 2009; Shen *et al.*, 2010) as well as the potential need to combine an antimicrobial application as an intervention strategy (Echeverry *et al.*, 2010). The intervention being explored is to apply (spray) antimicrobials on the meat either immediately prior to blade tenderization and/or during tenderization. Morgan (2010) investigated the effects of antimicrobial spray application against antibiotic-resistant *Escherichia coli* O157:H7 cultures allowed surface attachment time. The two most effective spray applications associated with pathogen reduction were buffered sulfuric

Emerging Technologies in Meat Processing: Production, Processing and Technology, First Edition.
Edited by Enda J. Cummins and James G. Lyng.

acid (1.99 log, CFU/cm^2) and a combination of lauric arginate and peroxyacetic acid (1.92 log, CFU/cm^2) in samples stored 1 day after blade tenderization.

Application of exogenous proteases has been used to tenderize meat for decades. Common enzymes include bromelain (Pinto, Mazzilli and Cusin, 1999), ficin (Ramezani, Aminlari and Fallahi, 2003), and papain (Morzel *et al.*, 2006). Researchers have also explored the possibilities for improvement of meat tenderness by proteolytic enzymes from fungi and plant origin (Naveena, Mendiratta and Anjaneyulu, 2004; Shin *et al.*, 2008). Exogenous proteolytic enzymes have the potential to dissolve or break down myofibrillar proteins as well as connective tissue, resulting in tender meat. However, although these enzymes can tenderize both the myofibrillar proteins and connective tissue protein collagen, problems occur with uniformity of tenderization, overtenderization, and off-flavor notes.

Injection or marination with salt or in combination with phosphate improves tenderness (Sheard and Tali, 2004; Vote *et al.*, 2000). However, others have found that although enhanced steaks had lower instrumental shear force, enhanced steaks had more off-flavors than nonenhanced steaks, with typical descriptors of salty, metallic, or chemical, in addition to an undesirable mushy texture (Grobbel *et al.*, 2008).

Sodium chloride is the most common salt used. However, an undesirable side issue associated with its use is that sodium intake exceeds the nutritional recommendations in many industrialized countries (Ruusunen and Puolanne, 2005). In addition, most sodium phosphates contain about 30% sodium (Ruusunen, Niemistö and Puolanne, 2002). Excessive consumption has been linked to hypertension and consequently to increased risk of stroke and premature death from cardiovascular diseases (Ruusunen and Puolanne, 2005).

Certainly one of the most common methods of tenderization is associated with allowing the meat to tenderize with time postmortem based on the activity of endogenous proteases during refrigerated storage. This is done either with intact carcasses or as bone-in primals (sections) and boneless cuts often in some sort of modified atmosphere packaging (Aberle *et al.*, 2012; Smith *et al.*, 2008). Aging unpackaged meat under appropriate conditions, known as dry aging, has also been utilized (Ahnstrom *et al.*, 2006; Campbell *et al.*, 2001). Calpains, a classification of proteases, play a very significant role in meat tenderization with time postmortem (Huff-Lonergan *et al.*, 1996; Huff Lonergan, Zhang and Lonergan, 2010; Taylor *et al.*, 1995). Some have suggested that because of muscle differences, the development of specific aging times to reduce tenderness variation should be considered (Aalhus *et al.*, 2004). The endogenous enzymes help breakdown many of the structural myofibrillar proteins associated with meat toughness, but they have very limited effectiveness on the connective tissue, collagen, and are of no benefit to tenderizing elastin. Of course, tenderization does occur during the cold-chain transportation time; however this does not assure consistency in tenderness. In some cases, the presence and activity of calpastatin, an endogenous inhibitor of the calpains, prevent adequate tenderization (Huff Lonergan, Zhang and Lonergan, 2010). Therefore, even after extended storage, lack of sufficient tenderness can remain. Moreover, extending the storage time for proteolytic tenderization comes with the cost of allowing more time for bacterial spoilage.

Electrical stimulation has been used on red meats and poultry (Birkhold, Janky and Sams, 1992; Ferguson *et al.*, 2000; Hwang, Devine and Hopkins, 2003; Hwang, Lee and Thompson, 2003; Pearson and Dutson, 1985). Prerigor carcasses have been stimulated with either high or low voltage for various durations and pulses. Electrical stimulation has been used to help reduce the incidence of cold-induced toughening, which can occur when prerigor carcasses are chilled too rapidly while the pH is still high. Some evidence exists that tenderization is also associated with some physical disruption of the muscle. Electrical stimulation accelerates postmortem glycolysis so that the muscle sets up into rigor faster. Once muscles set up in rigor, this prevents sarcomere shortening, which is known to produce less tender meat. Electrical stimulation activates early proteolytic aging by stimulating the release of calcium associated with calpain activity. Electrical stimulation in some cases has been shown to result in a general reduction in calpastatin activity suggesting that it accelerated proteolysis. Calpain activity was reported to increase following stimulation although this varied with breed. Electrical stimulation has been used to accelerate rigor mortis in chickens in order to facilitate earlier deboning although inconsistency in tenderness remains an issue.

A concept of stretching muscles prerigor has been exploited commercially to improve tenderness (Sorheim and Hildrum, 2002). Prerigor stretching attempts to improve tenderness by stretching sarcomeres, thereby decreasing the overlap region between the thick and thin filaments so that when rigor is established there are fewer chemical crossbridges between actin and myosin. As a prerigor stretching process, the tendercut process (Claus, 1999; Ludwig *et al.*, 1997) involves making selected skeletal alterations prerigor to the suspended carcass in a manner that allows the muscle to stretch utilizing the weight of the remaining carcass below the cut. This concept was applied to an on-line altered suspension (OLAS) approach (Aalhus *et al.*, 1999) as well as coupled with electrical stimulation (Aalhus *et al.*, 2000). Prerigor stretching has also been exploited by hanging beef carcasses by the aitchbone in a process also known as pelvic suspension and tenderstretch (Ahnstrom, Hunt and Lundstrom, 2012; Hostetler, Carpenter and Smith, 1972; Hostetler *et al.*, 1975; Hostetler *et al.*, 1970). Sarcomere length was increased by pelvic suspension but shear force was not reduced in all muscles associated with the hind limb, and the response was influenced by gender and animal age (Ahnstrom, Hunt and Lundstrom, 2012).

Excising muscles prerigor and stretching them has also been investigated (Devine *et al.*, 2002a; Devine, Payne and Wells, 2002b; Devine, Wahlgren and Tornberg, 1999; Sorheim and Hildrum, 2002). A system that exploited this concept (Pi-Vac) operates by stretching tubes of highly elastic film to the inside walls of the packaging chamber. After inserting the muscle into the chamber, the pressure is released and the film returns to its original dimensions, which hinders the diametrical muscle expansion that would be associated with contraction of the muscle (Sorheim and Hildrum, 2002). All of the prerigor stretching methods are beneficial but are limited by the ability to stretch muscles associated with the muscle fiber orientation and/or prevent shortening. Another consideration is that even within an individual muscle, the muscle fibers do not always align in the same orientation and some areas of a muscle may almost run perpendicular to other areas. In addition, the effect of prerigor muscle stretching principally affects the myofibrillar component of tenderness and not connective tissue.

Despite the fact that a variety of tenderization techniques are available and have been used, variation in tenderness and inadequate tenderness of meat and poultry remain as issues for the industry to address (Juárez *et al.*, 2013; Lorenzen *et al.*, 2003; Solomon *et al.*, 2006). In a survey done in Canada, beef tenderness has improved over the decade starting in 2001 (Juárez *et al.*, 2013). However, they found that many of the cuts still need additional tenderization with a very significant percentage of them still being classified as tough. Along with beef, tenderness issues with pork and poultry meat still exist. The existing techniques have some limitations that have also hindered their use. Some of the tenderization techniques require additional holding periods, space, and labor (Solomon *et al.*, 2006). In addition, some of the tenderization methods require chemicals that may not be consumer label-friendly, yield undesirable flavors, and in some cases the tenderization method may put the consumer at a greater risk from a food safety standpoint.

A technology that was first conceived some 43 years ago involves the utilization of hydrodynamic shock waves (HSW; hydrodynamic pressure processing, HDP) to tenderize various fresh meat and poultry. Hydrodynamics relates to the study of the motion of fluids and the forces acting on solid bodies immersed in the fluids and the motions relative to them (Mish *et al.*, 1988). Technology associated with the application of HSW now appears to be staged for commercial introduction but is still undergoing development. A very well-written comprehensive review on HDP to improve meat quality and safety was published by Solomon *et al.* (2006). They covered the inception of HSW (Godfrey, 1970), which was based on the detonation of explosives in water through what had been last published around 2005.

The intent of this chapter is to first provide a sufficient overview of some of the early understanding of the development and applicability of the explosive-based shock wave generating approach. Building on that foundation, this chapter continues the review with the new information available that provides a greater understanding of the parameters that affect the efficacy of HSW. Finally, this chapter provides a discussion on the advances that have led to the development of electrically generated HSW systems, which are poised for commercial introduction.

7.2 Applicability of hydrodynamic shock waves on meat and meat products

Applicability falls under three major categories associated with fresh meat tenderization, utility to processed meat manufacturing, and microbiological effects. In a variety of red meat cuts and poultry, there remains the need to produce more tender meat while reducing the variability in tenderness. HSW has applicability to efficiencies in manufacturing further processed meats in terms of marinade uptake and retention. Although the main focus of HSW has been on tenderization, there is a growing body of evidence that HSW injures or kills bacteria thereby demonstrating promise for extending shelf-life and improving food safety.

In terms of tenderization, HSW has applicability to beef, pork, lamb, goat, chicken, and turkey (Claus, 2002; Claus *et al.*, 2001; Eega *et al.*, 2007a; Solomon *et al.*, 2006; Solomon, Long and Eastridge, 1997). Basically any muscle food in which tenderness is not adequate, HSW has the potential to benefit that industry.

HSW has the potential to reduce or eliminate aging time. Early studies found that HSW-treated meat to be equivalent to 21 days of aging for improving meat tenderness (Zuckerman *et al.*, 1999). HSW has applicability for tenderizing lower valued cuts associated with lower quality grades and meat from older animals as long as connective tissue is not the predominant factor influencing tenderness. In addition, HSW would be beneficial on breeds prone to producing tough meat and in meat in which the activity of calpastatin inhibits normal proteolytic aging.

Coupling HSW with other postmortem tenderization technologies also has applicability. Utilizing electrical stimulation to accelerate rigor mortis as well as muscle aging could be coupled with HSW. Once the muscles are setup into rigor and form permanent crossbridges between actin and myosin, the rigid ultrastructure of the muscles could be HSW fractured. HSW should not be applied prerigor as limited breakage in some sarcomeres would enable the remaining myofibril to contract beyond its resting length and consequently lead to less tender regions. Benefits have been demonstrated when HSW was coupled with blade tenderization (Liu *et al.*, 2006).

7.3 Approaches to the generation of hydrodynamic shock waves, and the theory and mode of action relative to muscle food applications

Shock waves are naturally produced from cosmic explosions, thunderstorms, volcanic eruptions, and meteoroid impacts (Solomon *et al.*, 2006). HSW can be produced chemically by detonating explosives or by discharging a high-voltage arc across a gap between two electrodes in water (Asay and Shahinpoor, 1993;

Cole, 1948; Long, 2003; Solomon *et al.*, 2006). A shock wave can also be produced by a nonexplosive generator system in which a piston strikes an exposed water front of a vessel containing the meat surrounded by water (Williams, Warren and Carter, 2007).

The explosive-based approach was the first employed and the most extensively studied. For reasons that will be discussed later attempts to develop a commercially viable system were abandoned. However, the research based on explosives was fundamental to developing a broad understanding of shock wave dynamics and various modes of action on meat tenderness and microbiological status.

Decades prior to any documented vision that an electrically generated shock wave could be used to tenderized meat, an electrical capacitor discharge in water had been used as a method to precisely form or shape metal (electrohydraulic forming; Fröngel, 1976). In electrohydraulic forming, the spark energy is nothing more than an electrically produced energy shock with the same behavior as that generated by explosives (Fröngel, 1976). In an electrical-based system, two electrodes with an appropriately spaced gap are submerged in a liquid (e.g., water). When the capacitor bank is discharged, an arc is established across the gap and the extremely rapid local heating in the vicinity of the arc creates a shock wave (Fröngel, 1976). An interesting modification to creating an arc simply by an appropriate gap between the electrodes is the use of a consumable wire connecting the two electrodes. Apparently for some electrohydraulic forming applications a larger spark gap is necessary and a wire is essential in order to guide the spark and insure that the pressure is directed as required onto the work piece (Fröngel, 1976).

From a meat application standpoint, the two main means of creating HSW (detonation of explosives, electrically generated) involved the formation of the shock wave within a vessel filled with water (Bowker *et al.*, 2011; Claus, 2002; Claus *et al.*, 2001; Solomon *et al.*, 2006). The literature is rather varied in the description of these systems used to generate HSW. Therefore, it is worthwhile to mention their names, common acronyms, and provide a general understanding of which ones have similar underlying operating principles. Explosive-based systems have been referred to as the HSW, Hydrodyne process, Hydrodyne-treated, HDP, hydrodynamic pressure, and supersonic HSW. Systems that generate a shock wave electrically by creating a high-energy capacitor discharge arc across an electrode gap have been referred to as a capacitor discharge system (CDS), Hydrodyne process, Hydrodyne, sparker, plasma sparking device, high-voltage arc discharge (HVADH), and TenderClass System (TCS, Claus, 2002; Cooper and Solomon, 2000; Gamble, Solomon and Long, 1998; Holzer *et al.*, 2004; Lorca *et al.*, 2003; Marriott *et al.*, 2001; Moeller *et al.*, 1999).

For this chapter, HSW will serve as a generic description and not related to any particular method of creating shock waves. For the HSW generated, the detonation of explosives will be referred to as XSW (explosive-generated shock wave). The electrically generated HSW formed from a high-energy capacitor discharge arc across an electrode gap will be referred to as ASW (arc-generated shock wave). In the case of ASW-based systems when meat is treated/exposed to HSW, it has been referred to as being pulsed. This refers to a single processing cycle in which a shock wave is generated.

Several properties associated with the formation of a shock wave and the shock wave itself are important to the understanding of the modes of action on meat tenderization and microbiological effects. These properties include the formation of an extremely high-temperature gas bubble, plasma, propagation of a high-energy compression wave, expansion waves, shear, and shock wave interactions.

Although generation of an XSW involves the formation of an extremely high-temperature gas bubble and plasma, virtually no increase in temperature of the water or meat occurs (Solomon *et al.*, 2006). Similar observations have been found using ASW systems.

Shock waves are the ubiquitous result of matter moving at velocities faster than the speed at which material in the front of the wave can move out of the way (Asay and Shahinpoor, 1993). A shock wave has a very short duration (millisecond) during which time a very high sharp pressure peak occurs (Cole, 1948). In water, a high-pressure plasma bubble expands supersonically against the surrounding fluid producing a spherically radiated shock wave (Loske *et al.*, 2002). The propagation of shock waves in a solid body is accompanied by large compressions, decompression, and shear (Asay and Shahinpoor, 1993). The energy of a shock wave dissipates (returns to ambient pressure) by the mechanism associated with expansion waves (Asay and Shahinpoor, 1993). Expansion waves are known as rarefaction waves, unloading waves, and decompression waves.

Shock waves can pass through materials that have a very similar acoustical or mechanical impedance match to that of water (Früngel, 1976; Long, 2003). If it is not an acoustical match, the shock wave will be reflected from that object. An acoustic match occurs when the bulk modulus divided by the density of the water and the object in the water are essentially identical (Long, 1993). The definition of bulk modulus is defined as $P/(\Delta V/V)$, where P is the hydrostatic pressure, ΔV is the change in volume by P, and V is the unstressed volume (Bourne, 1982). Flesh of mammals and fowl exhibits a reasonable acoustic match with water and, therefore, a hydrodynamically generated shock wave will pass through the flesh (Long, 2003). This is because meat is composed of approximately 75% water (Aberle *et al.*, 2012) and, therefore, will support the passage of the shock wave. Flesh in advance of the shock wave is rapidly compressed, which is followed by a rarefaction wave that results in a rapid stretching of the flesh (Long, 2003). The relationship between wave velocity and wave period of the ensuing shock wave likely plays a critical role in successful HSW meat tenderization (Solomon *et al.*, 2006).

Meat contains about 18.5% protein (Aberle *et al.*, 2012). The myofibrillar proteins (major contractile: myosin, actin; cytoskeletal) and stromal proteins (collagen, lesser extent elastin because of the limited amount) largely impact meat tenderness. Based on scanning electron microscopy, Zuckerman and Solomon (1998) were the first to demonstrate that the Hydrodyne process (XSW) tenderized beef at the ultrastructural level by causing myofibrillar fragmentation in the region adjacent to the Z-lines. With this evidence, the manner in which the fracture occurs has some similarity to an analogy of breaking silly putty. If the tensile stress is applied slowly, deformation (strain) can be rather large. If the applied tensile stress is rather rapid, the putty promptly breaks apart. Both the magnitude and rate of compression and tensile forces are likely involved in myofibrillar disruption.

Besides the apparent fragments, Zuckerman and Solomon (1998) found increased intramyofibrillar space with longitudinal gaps or splits in the myofibril lattice. Their work was collected using the explosive-based HSW. Confirmation of myofibrillar fragmentation was also demonstrated with an electrical-based HSW system (Figure 7.1a and b). Note unlike proteolytic produced fragmentation in which proteins would appear to degrade uniformly around site-specific ultrastructural features, relative to the transmission electron micrograph, HSW resulted in irregular, jagged fractures. Some evidence presented suggests that meat tenderization associated with using shock waves might also be the result of the connective tissue disintegration, which is responsible for the so-called background toughness in meat when not broken down (Zuckerman *et al.*, 1999). Scanning electron microscopy analysis indicated that XSW treatment disrupted the integrity of the collagen fibril network of the endomysium in both nonaged and aged samples. Data suggested that shock wave alterations to connective tissue contribute to the meat tenderization effect of XSW (personnel communication; Bowker, 2013).

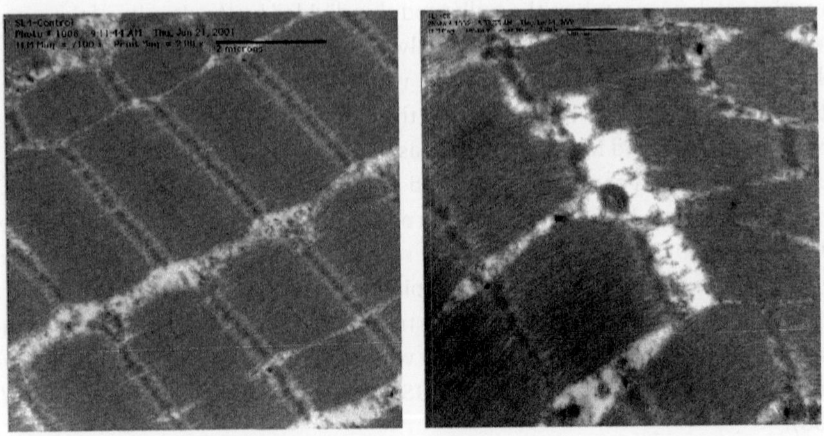

Figure 7.1 (a) Control not ASW processed (Claus, University of Wisconsin-Madison). (b) After ASW processing (Claus, University of Wisconsin-Madison).

Collagen solubility as affected by XSW, blade tenderization, and BT followed by XSW (combination) was determined on Brahman semimembranosus muscles (Bowker *et al.*, 2007c). In contrast to BT alone, XSW and the combination increased the amount and percentage of soluble collagen compared to the control.

Some of the variability in the effectiveness of tenderization using XSW was elucidated by Paroczay *et al.* (2002) to be related to time postmortem of XSW application. Also early-deboned chicken breasts HSW processed using an ASW system supports the impact of time of postmortem application. Chicken breasts were early-deboned upon exiting the chiller of a slaughter plant and stored 6 and 24 h prior to ASW (Long, Thomsen and Waits, 2007). Those stored 6 h resulted in a 60% shear force reduction (1.9 cm sample strips; 17.2 kgf control, 6.9 kgf ASW) and a 69% reduction when applied after 24 h (13.4 kgf control, 4.1 kgf ASW). In another study, early-deboned chickens ASW processed on the same day deboned (2, 4, 6 h after exiting the carcass chiller) exhibited shear force reductions of 18.3% (8.2 kgf control), 23.3% (7.3 kgf control), and 42.9% (7.7 kgf control), respectively (unpublished data, Claus, University of Wisconsin-Madison). Therefore, the improvement in ASW tenderization associated with delayed ASW application time is likely associated with contributions from proteolytic aging. The positive effect of delayed deboning time on ASW effectiveness is likely associated with a greater development of rigor (more rigid, less flexible).

With USDA Select strip loins ($n = 12$), additional postmortem aging time before ASW (12.2 kJ, one pulse) also demonstrated greater tenderization (1.5 days, 2.2% force reduction, 5.86 kgf control; 4.5 days, 20.9% force reduction, 5.22 kgf control; 7.5 days, 22.0% force reduction, 4.10 kgf, control). The percentage change declined when the strip loins were ASW processed after 10.5 days postmortem (13.5%, 3.84 kgf control). A similar study on strip loins obtained from Nelore cattle demonstrated greater ASW (12.2 kJ, two pulses) tenderization when applied after additional postmortem aging time (1.5 days postmortem, 11.2% force reduction, 8.7 kgf control; 7.5 days postmortem, 24.5% force reduction, 7.8 kgf control). An increase in effectiveness of HSW with additional time postmortem prior to HSW likely coincides with endogenous proteases having sufficient time to weaken/condition critical ultrastructural myofibrillar proteins so that they are more susceptible to mechanical fragmentation. This can be thought of similar to a glass having a minor crack being more susceptible to major failure upon subsequent stress or strain.

Solomon *et al.* (2006) pointed out in their review that XSW also seemed to have an indirect tenderizing effect on meat in that the tenderizing effect on meat sometimes did not become apparent until a few days after XSW treatment (Paroczay *et al.*, 2002; Solomon *et al.*, 2002). Steaks from XSW processed muscles continued to age beyond control steaks (control 12 days, XSW 16 days).

Focused studies associated with the XSW effect on the myofibrillar component (Bowker *et al.*, 2006; Bowker *et al.*, 2008b) as well as on the sarcoplasmic fraction (Bowker *et al.*, 2006; Bowker *et al.*, 2008c) have been published. A myofibrillar protein study using XSW processing and aging was on beef strip loins ($n = 12$) that were processed 48 h postmortem. Using sodium dodecyl sulfate-polyacrylamide gel electrophoresis and Western blotting analysis of myofibrillar proteins (Bowker *et al.*, 2008b) showed that XSW and aging decreased the intensity of the troponin-T (TnT) band and enhanced the accumulation of the 30-kDa TnT degradation product. They suggested that XSW is more effective than aging tenderization, and that XSW tenderization is caused by both protein degradation and physical disruption of the myofibril apparatus. Similarly, Bowker *et al.*, (2010c) found that a 60-kDa band intensity increased with both aging and XSW treatments in pork loins.

The sarcoplasmic fraction of meat contains water-soluble enzymes and myoglobin. Myoglobin is a pigment responsible for the color of fresh meat. Myoglobin is composed of a protein moiety (globin) and a nonprotein portion (heme ring). The heme ring contains a centrally located iron atom. Depending on the oxidative state of myoglobin and what ligand is bound to the iron in the sixth coordination site affects the pigment's color (Aberle *et al.*, 2012; Claus, 2007). Using XSW on beef strip loins had minimal effects on the sarcoplasmic proteins (Bowker *et al.*, 2012).

A complete understanding of how HSW injures or is lethal to microorganisms has not been established. Solomon *et al.* (2006) theorized that it could be related to shock wave cavitation. They suggested that following the high-pressure front produced by HSW, a tensile stress occurs resulting in the formation of microscopic vapor-filled cavities, which expand momentarily and then violently collapse creating very high-energy densities. Such cavitation would be expected to cause very high localized temperature and pressure gradients likely lethal to bacteria. Associated with the formation of a high-pressure plasma bubble causes the production of OH radicals, ions, and ultraviolet light (Loske *et al.*, 2002; Schaefer, 1998). These outcomes coupled with generated pressure pulses are useful for processes such as decontamination, disinfection, and the treatment of contaminated water (Schaefer, 1998). Therefore, numerous mechanisms may be involved in the positive microbiological effect of HSW when applied to meat.

7.4 Advantages and disadvantages of hydrodynamic shock wave

Accomplishing tenderization through the use of HSW processing has numerous advantages over more traditional methods of tenderization. Advantages discussed in this section will pertain to ASW-based HSW systems as mentioned previously an explosive-based system has too many limitations that may never be resolved to achieve a commercial viable system. ASW processing essentially produces instantaneous tenderization and also can enhance proteolysis. ASW processing does not involve use of chemicals and does not alter the color or flavor. In addition, ASW processing positively contributes to shelf-life extension and food safety.

7.4.1 Advantages of hydrodynamic shock wave processing

In contrast to natural aging (carcass aging, wet aging, and dry aging), ASW tenderization is instantaneous. As such ASW tenderization has the potential to eliminate or greatly reduce the need for postmortem aging. By eliminating or reducing the time required to achieve desirable tenderization would extend the display life at the retail market and provide consumers with a fresher product. Less cold storage space would be required and the energy cost associated with refrigeration would be reduced.

Although the interior of intact meat is essentially sterile, bacteria will be present on the exterior surface of the meat (Patel *et al.*, 2012). In contrast to blade tenderization that results in microbial translocation to the interior of the meat (Adler *et al.*, 2012; Echeverry *et al.*, 2010; Heller *et al.*, 2007; Johns *et al.*, 2011; Luchansky *et al.*, 2012; Luchansky *et al.*, 2009; Patel *et al.*, 2005), ASW does not cause any practical level of translocation from the surface and therefore does not compromise food safety. One well-documented outbreak of *E. coli* O157:H7 involved beef steaks that were blade tenderized and injected with marinade (Laine *et al.*, 2005). Translocating bacteria to the interior has the potential to put consumers at risk particularly if they choose to cook their beef to an endpoint temperature without adequate thermal lethality (time–temperature). Laine *et al.* (2005) suggested that food regulatory officials should reevaluate food safety issues presented by nonintact steak products and consider possible labeling to distinguish intact from nonintact steaks. Such food safety concerns have motivated companies to begin applying antimicrobial sprays during blade tenderization. Unlike blade tenderization, with ASW tenderization there would not be any need to consider application of antimicrobials to the meat's surface.

Undesirable changes to flavor and texture do not occur with ASW processing or impact on the nutritional attributes of the meat. Application of exogenous enzymes to tenderize meat, although very effective on both the myofibrillar proteins and collagen, can be a challenge in controlling the degree of tenderization and can result in overtenderization and mushy areas. Off-flavors associated with the use of exogenous proteases can be present. Although enhancing meat with brine solutions improves tenderness and consistency in tenderness while reducing the cost of the meat, this process adds high levels of sodium. Elevated intake levels of sodium have been associated with numerous health-related diseases including cardiovascular disease, hypertension, and development of certain cancers.

Alternative suspension methods and/or exploiting other prerigor stretching methods have limited application in terms of the breadth of muscles and the species that the methods will benefit. HSW processing can be used on all muscles.

ASW technologies offer flexibility and ease of operation to optimize tenderization on different muscles and application needs. Multiple shock wave tenderization cycles (pulses) can be easily applied as needed. ASW can be coupled with other tenderization technologies such as electrical stimulation. ASW can facilitate early deboning of chicken breasts. ASW systems have a relatively compact design that would fit into existing commercial facilities.

Further processed meats may benefit from ASW processing. The amount of marinade update and retention can be improved. The time necessary to absorb the marinade can be reduced.

There are microbiological advantages of ASW. Improving food safety and shelf-life exists with the use of ASW through a potential to reduce the number of pathogenic bacteria as well as spoilage bacteria. In addition, HSW appears not to compromise the surface integrity of intact muscles and breast fillets.

7.4.2 Disadvantages of hydrodynamic shock wave processing

HSW processing at this point cannot be used on bone-in products. Therefore, meat cuts (steaks, roasts, and bone-in chicken breast) in which the bone is to remain in the product during merchandizing cannot be HSW processed. Issues with fracturing the bones, potential damage to the HSW system, inability to maintain package integrity, and bone HSW shadowing effects prevent HSW applicability to bone-in products. Cuts in which the predominant controlling factor is associated with high collagen content or very mature collagen limit HSW applicability. Muscles having considerable fat cover on one (e.g., strip loins, pork loins) surface may require specific orientation relative to the ASW.

A potential major limitation with some designs is the requirement that the meat must be vacuum packaged. Package integrity still appears to remain as an unresolved issue. In order to have the best chance to withstand the dynamics associated with the ASW, very durable, thick packaging materials would likely be required. Such packaging could add to the cost if its only purpose was for use during ASW processing and not subsequent cold-chain distribution needs. Depending on the design of the ASW system, significant amounts of water may be required.

Practical use of high-energy HSW is on boneless meat and poultry. When applied to bone-in cuts (Figure 7.2), the shock wave deposits energy on the bone as it reflects off and can cause significant fractures or bone chips which if left within the cut would represent a physical hazard to the consumer. In addition, since the shock wave would not pass through the bone, this would likely cause a shadow effect in terms of the meat behind the bone not being equivalently tenderized.

Visible damage to the exterior of ASW processed meat can be apparent. From early efforts the exterior of the meat from inadequate packaging or in

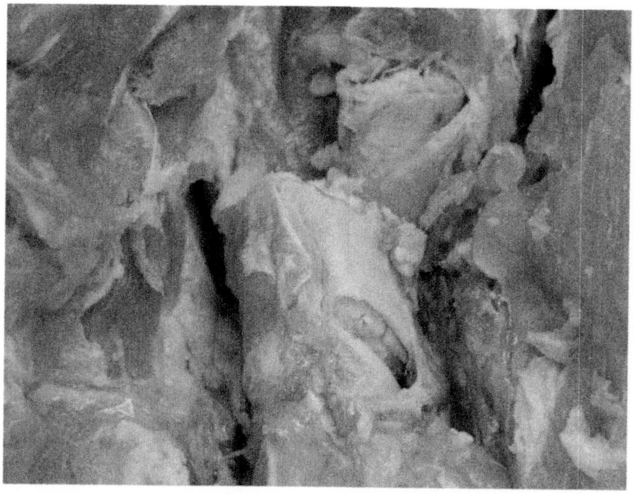

Figure 7.2 Porcine femur [drop Femur] Femur bone fracture by ASW.

Figure 7.3 Surface damage after ASW.

unpackaged meat in some cases had visible surface damage as well as adipose tissue separation (unpublished data, Claus, University of Wisconsin-Madison). The surface damage was likely associated with the shock wave unloading at the surface (Figure 7.3). Separation at the interface of the adipose and lean tissue can occur. Such separation (Figure 7.4) likely occurred because of violent meat movement as a result of the shock wave generation and/or a significant acoustic mismatch between the adipose tissue and lean. Bowker *et al.* (2011)

Figure 7.4 Muscle tissue separation at the adipose–lean interface after ASW.

Figure 7.5 Beef bottom rounds (non-ASW companion controls, left; ASW processed muscles, right).

also reported that using more than 10 pulses (ASW, 6.6 MPa, 19 l cylindrical plastic container, 64-μF capacitor bank) with the sparker set at 3.75 cm from the meat surface caused visible damage to the muscle as well as compromised the integrity of the package.

Physically restraining the meat has largely eliminated this deficiency through advanced ASW design changes. Limited visual structural differences exist between non-ASW muscles and ASW processed muscles (Figures 7.5 and 7.6).

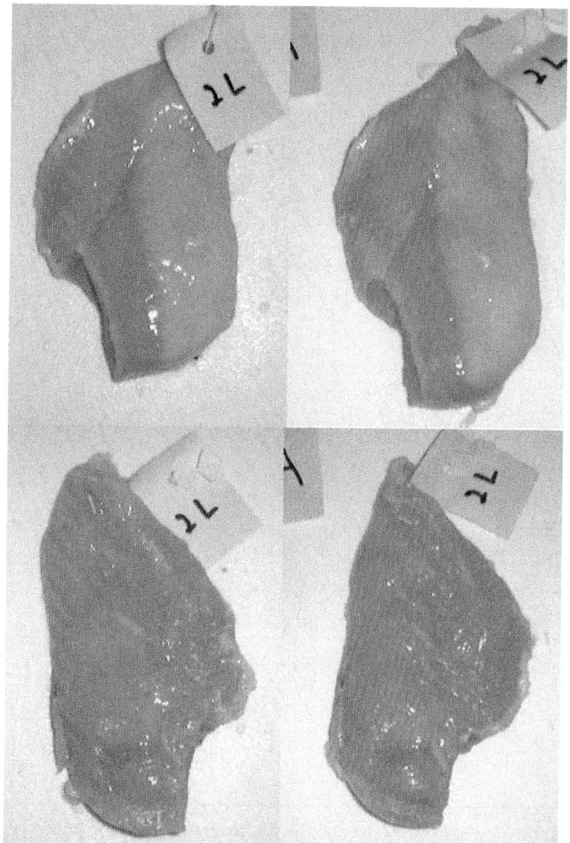

Figure 7.6 Chicken breast fillet exterior (top row), interior (bottom row); before ASW (left), after ASW (right).

7.5 Case studies: hydrodynamic shock wave treatment of meat products

This section addresses some of the research conducted on HSW with the primary focus on results collected using ASW-based systems. Case studies on the impact of HSW relative to tenderness, processed meat manufacturing and properties, and food safety are addressed.

7.5.1 Tenderness

The majority of the tenderization assessment has been performed with instrumental analysis. Therefore, mention of a few benchmarks is of value to give a perspective on the literature presented. As a benchmark based on instrumental Warner-Bratzler shear force (WBSF, kgf) if the tenderization method results

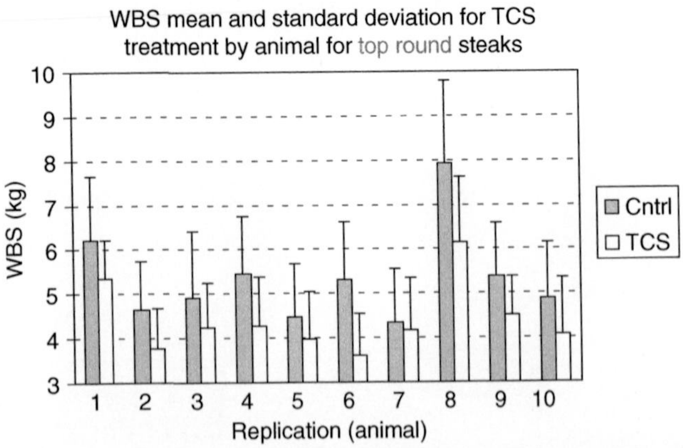

Figure 7.7 ASW response among beef animals.

Table 7.1 Warner-Bratzler shear averages (peak force, kgf) for TenderClass System (TCS) processed pork loins*.

Treatment	Individual pork loin												
	1	**2**	**3**	**4**	**5**	**6**	**7**	**8**	**9**	**10**	**11**	**12**	**Average**
Control	3.1	4.9	3.5	5.3	5.8	3.2	4.5	5	3.3	2.7	3.6	4.9	4.15
TCS (ASW)	1.8	2.5	2.2	6.2	3.6	1.7	3.1	4.2	2.5	2.4	2.7	2.9	2.98
Difference	1.3	2.4	1.3	−0.9	2.2	1.5	1.4	0.8	0.8	0.3	0.9	2.0	1.17

*TCS is an ASW system by Hydrodyne Inc. (Long, Thomsen and Waits, 2007). ASW at 72% energy and two pulses applied.
Adapted from Sagili (2003).

in at least a 10% reduction this is considered successful tenderization (Solomon *et al.*, 2006). Tenderness can also be classified into four categories (Belew *et al.*, 2003): "very tender" (<3.2 kgf), "tender" (3.2 < WBSF < 3.9 kgf), "intermediate" (3.9 < WBSF < 4.6 kgf), or "tough" (WBSF > 4.6 kgf). Some have considered a WBSF below 4.6 kgf as acceptable in tenderness. WBSF is commonly collected on cores (1/2″ diameter) obtained parallel to the muscle fibers.

The response to ASW varies among the same muscle obtained from different animals. Sagili (2003) found that top rounds (Figure 7.7; semimembranosus, USDA Select) ASW processed with one pulse at 72% energy tenderized the roasts but the response varied (control 5.35 kgf, TCS 4.41 kgf). Similarly, ASW tenderized boneless pork loins by 28% (Table 7.1), but the response was influenced by differences among pigs (Long, Thomsen and Waits, 2007; Sagili, 2003).

Changing the amount of shock wave energy and the number of times (pulses), the same meat is exposed to a shock wave influences the tenderization performance. Differences in the amount of energy (shock wave pressure) have already been extensively reported from XSW studies (Solomon *et al.*, 2006). Higher pressure fronts were not always more effective and tenderization was influenced by the type of explosive (binary vs molecular), shape, orientation, distance from the meat, reflector shape, and shock wave container support. Part of the reason for more information on such studies is based on the fact that the majority of research utilized the explosive-based systems. However, limited information is available on pulses because at that time any future development of a commercial system with more than one HSW treatment was likely not considered practical.

A number of studies conducted at the University of Wisconsin-Madison evaluated the effects of shock wave energy and the number of pulses using an ASW system (Hydrodyne Inc.: 16 kJ full-energy capacity, Long, 2003; 631 MPa potential, Lorca *et al.*, 2003). In general, after a certain minimum shock wave pressure (power, energy level), significant improvement in tenderization was influenced more by increasing the number of pulses (Sagili, 2003; Sagili and Claus, 2003). Individual reduction in shear force of 53% in a beef strip loin and 51% in a boneless pork loin have been found (Claus, Sagili and Sammel, 2002; Long, Thomsen and Waits, 2007).

Beef top rounds (semimembranosus) ASW processed with one pulse at 72% energy reduced shear force by 9.0% (control 5.4 kgf), whereas when pulsed at 76% achieved a 19.2% (control 5.6 kgf, $n=9$) shear force reduction (Long, Thomsen and Waits, 2007). These results were from two separate experiments and sets of top rounds. However, differences in animal age, application time postmortem, and other biological variability should be taken into consideration when making such a direct comparison.

USDA Choice grade beef strip loins ($n=8$) aged 21 days postmortem were ASW processed at two different energy levels. ASW improved tenderness with a similar response between energy levels (72% energy: 3.13 kgf control, 2.55 kgf ASW; 80% energy: 3.25 kgf control, 2.65 kgf ASW). Interestingly even when ASW was applied on these highly aged steaks, none of the steaks exhibited over tenderization.

A pork study (Sagili, 2003) with 112 boneless pork loins was conducted as a factorial design (3×4: pulses 0, 1, 2; energy 76%, 80%, 85%, 90%). Pork loins were ASW processed on day 2 postmortem. Results demonstrated that 76% energy was sufficient to achieve the same level of tenderization (1 day after ASW) as all other energy levels.

The effect of the number of pulses was studied on beef, pork, and chicken. Beef top rounds ($n=8$, semimembranosus) ASW processed (72% energy) produced lower shear forces than controls (5.4 kgf) when pulsed 1 and 3 times (5.0, 3.9 kgf, respectively) as reported by Long, Thomsen and Waits (2007). Beef eye of rounds (semitendinosus, $n=8$) processed at 72% energy produced lower shear forces than controls (4.5 kgf) when pulsed 1, 3, 5 times (3.6, 3.7, 3.4 kgf, respectively).

Bowker *et al.* (2011) compared 40 pulses to 80 pulses (ASW, 6.6 MPa, 19 l cylindrical plastic container, 64-μF capacitor bank) with two sparker positions (7.5 cm from meat surface) above strip loin steaks. The results indicated 80 pulses produced greater tenderization (40 compared to 80 pulses; after aging 1 day, 6.0%, 8.9%; after 7 days, 2.4%, 5.3%, respectively). Three sparkers (3.75 cm) above each strip loin roast pulsed 10 times immediately improved tenderness 25% (control, 38 N, day 0) and maintained 8% improvement (control 29.9 N) on day 7.

Two separate studies (Sagili, 2003) were undertaken to determine pulse effects on boneless pork loins. The first study pulsed aged pork (6 days postmortem, $n = 10–12$, 72% energy, 11.5 kJ; Long, 2003 ASW system). Pork pulsed once reduced shear force by 14.1% (control, 4.5 kgf, ASW 3.9 kgf) whereas when pulsed twice reduced shear force by 28.1% (control 4.2 kgf, ASW 3.0 kgf). From the energy by pulse study mentioned previously, the WBSF (1 day after ASW) was reduced after each pulse (1 pulse, 4.1 kgf; 2 pulses, 3.9 kgf) compared to the control (4.5 kgf, $n = 112$ pork loins). One pulse was sufficient to maintain a tenderness advantage 9 days after ASW (control 4.4 kgf, ASW 3.9 kgf; $n = 60$).

Strip loins (SL, 15 days postmortem, USDA Select grade, $n = 10$) were used to test uniformity of ASW (80% energy, one pulse) tenderization within a commercial demonstration unit (Sagili, 2003; Sagili and Claus, 2003). The ends of the control SL's were tested for tenderness (rib end, 3.79 kgf, more tender than sirloin end, 4.46 kgf); however, all six steaks removed in between the two controls were uniformly more tender (average 3.10 kgf) as a result of ASW.

Comparisons of HSW to blade tenderization (BT) as well as combinations of HSW and BT have been studied (Bowker *et al.*, 2005; Bowker *et al.*, 2007c; Liu *et al.*, 2006; Sagili, 2003; Sagili and Claus, 2003). A direct comparison was made between ASW and blade tenderization on USDA Select grade strip loins (Sagili, 2003; Sagili and Claus, 2003). A portion of each strip loin ($n = 9$) was commercially blade tenderized (one pass) and another portion was ASW processed (one pulse, 60% energy, 9.6 kJ). ASW provided the same level of instantaneous tenderization as blade tenderization (4.53 kgf, control; 3.52 kgf, blade tenderization; 3.38 kgf ASW).

Top rounds ($n = 12$) from Brahman cattle were also used to evaluate blade tenderization and XSW (Liu *et al.*, 2006). BT and XSW increased tenderness (18%) as well as when BT was combined with XSW (14%) compared to control steaks. Tenderness increased for XSW and BT + XSW treatment between days (day 0, day 7), whereas tenderness did not change for the BT. Tenderness increased (20%) in the controls by aging alone. Overall tenderness improvements for BT + XSW and XSW (day 7) compared to the control (day 0) were 37% and 29%, respectively. They concluded that BT provided instant tenderization but without additional improvement during aging. On the other hand XSW and BT + XSW resulted in instantaneous tenderization that also increased over time and provided the greatest amount of overall tenderization.

A comparison of the tenderization effect on XSW beef muscles was evaluated on fresh and frozen stored steaks (Solomon *et al.*, 2008). The fresh steaks had a 29% reduction in shear force. Freezing alone accounted for a 14% reduction in shear force. Compared to frozen control steaks, XSW treatment followed by freezing resulted in a 30% reduction in shear force.

Muscle fiber orientation also appears to affect HSW tenderization. The semitendinosus (ST) muscle was used to investigate the effects of fiber orientation. Muscle fibers in this particular muscle align very much parallel to the long axis of the muscle. When the ST was position horizontal (perpendicular) to the site of XSW generation, samples exhibited greater tenderization than when the sample of the ST was positioned vertically (O'Rourke *et al.*, 1999; Solomon *et al.*, 1999). Similar testing was done using an ASW (Hydrodyne Inc.). Five beef ST muscles were tested. A control was removed from the center (average 4.36 kgf). One portion of each muscle was placed perpendicular to the shock wave front while the remaining portion was positioned parallel. The height of the parallel positioned section was the same as the thickness of the perpendicularly positioned roast. The shear force of the perpendicularly positioned roast was reduced by 12.2% (3.83 kgf) while the shear force of the parallel positioned roast was reduced by 9.8% (3.94). In this case, fiber orientation did not appear to have a major effect (unpublished data, Claus, University of Wisconsin-Madison). Another study was done using 24 boneless pork loins. Pork loins (halves) were positioned horizontal with the exterior fat side facing down on the flat metal surface of the XSW processing tunnel (unpublished data, Claus, University of Wisconsin-Madison). The ventral side of the pork loin was facing up closest to the origination of the HSW. The other halves of the loins were processed separately and positioned with the anatomical medial or lateral edge of the loin facing toward the shock wave origin. Compared to the control (4.78 kgf), the XSW loins had lower average cooked shear forces (3.12 kgf, horizontal, 3.83 kgf, vertical). Although fiber orientation in this case was not controlled, the results suggest that muscle orientation relative to the HSW may influence tenderization performance.

Extent of thick and thin filament overlap was hypothesized to influence the protein degradation and tenderness improvements associated with XSW (Bowker *et al.*, 2008a). To create differences in sarcomere length, carcasses were suspended by two methods (normal and hip suspension). XSW treatment decreased shear force in longissimus steaks from control suspended carcasses (1.73 μm) but had minimal effects on hip suspended samples (2.35 μm).

In two separate experiments (Berry *et al.*, 1997) using a 10-membered trained panel, they evaluated XSW processed USDA Select strip loins. Panelists rated the XSW higher in tenderness than in controls but did not detect any difference in juiciness or flavor. Using an eight-membered trained panel, tenderization of ASW early-deboned chicken breasts was evaluated (unpublished data, Claus, University of Wisconsin-Madison). The anatomical left and right breasts were removed from the carcasses ($n = 36$) approximately 1 h postmortem and stored overnight.

One anatomical companion breast from each bird served as a control, and the other was ASW processed (one pulse, 80% energy, 12.8 kJ). The breasts were sous-vide cooked (170 F). Standard size strips were removed from each breast. Each panelist was served blind-paired control and XSW samples (paired, same bird). Panelists consistently rated the XSW as more tender than the control. A separate set of early-deboned chicken breasts ($n=72$) were processed the same and reduced the shear force by 44% (6.5 kgf, control).

Water-holding capacity is an important characteristic to consumers and processors (Bowker *et al.*, 2007b). Frozen and thawed boneless pork loins that were XSW processed resulted in a 28% decrease in shear force. However, XSW had a minimal effect on drip loss.

7.5.2 Further processed meats

Limited studies have been reported on the efficacy of exposing meat to HSW intended for processed meat manufacturing. The first work explored the effect of HSW on protein functionality (Schilling *et al.*, 1999, 2002). Beef *biceps femoris* muscles were XSW processed at three different levels (83, 104, 124 MPa). HSW reduced shear force by 19% in the muscle (3.9 kgf, control). The muscle was ground and used to manufacture frankfurters. Although shock wave processing was shown to affect the intact muscle, shock wave processing did not affect myofibrillar or sarcoplasmic protein solubility, cooking yield, or color associated with manufacturing frankfurters. In contrast, Bowker *et al.* (2007a) found that the myofibrillar and total protein solubility were higher in XSW pork samples compared to controls at day 0 but not at day 7.

XSW treatment of boneless pork loins coupled with injection and tumbling improved tenderness (Callahan *et al.*, 2008). In addition, XSW treatment improved the quality whether applied before or after the marination step. Combining XSW with injection enabled pork loins to retain more brine after overnight storage and had higher processing yields (Bowker *et al.*, 2010a).

The effect of ASW on beef and pork muscles intended for enhancement (injection) was also investigated (Claus, Sagili and Sammel, 2002). Twelve randomly selected beef semitendinosus muscles (ST, eye of rounds) were shock wave processed (72%, 454 MPa) and compared to 12 nonshock wave processed muscles. A multineedle injector was set to deliver a 12% injection (0.5% NaCl, 0.35% sodium tripolyphosphate). Each set was injected without any changes to the injector setting. The control ST took up 10.2%, whereas the HSW ST achieved a 15.1% update. Similarly, beef biceps femoris muscles ($n=20$) were processed and ASW (76%, 480 MPa) positively affected update (9.9% uptake, control; 14.0% uptake, ASW). Drip loss after storing the uncooked steaks (5 days) was also reduced, which would make sense because of the more open HSW structure and the higher amount of salt and phosphate, which are known to improve water-holding capacity. Boneless pork loins were also studied and uptake tended

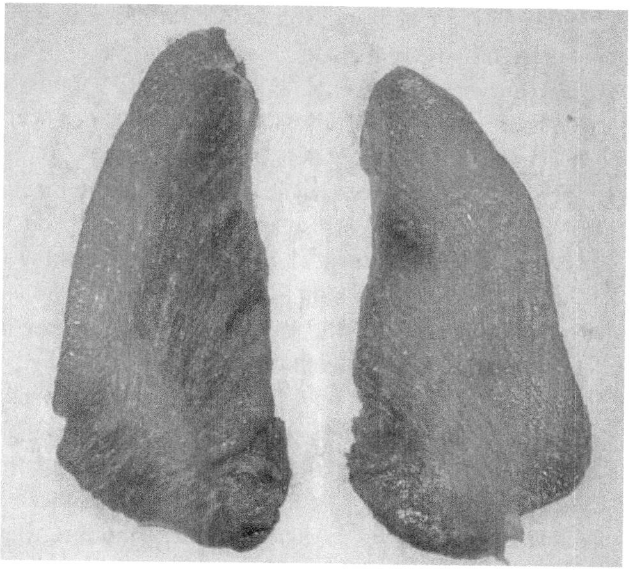

Figure 7.8 Marination update, XSW left, control, right (Courtesy of Bowker, USDA).

to be higher after HSW. The absence of a significant uptake may have been the lower shock wave energy (42%, 265 MPa) used, and the pork was relatively tender to start (2.1 kgf). Brine uptake in vacuum-tumbled turkey breasts was 1.6 times higher (11.6% vs 18.6%) for XSW processed breasts before marination (Bowker, Callahan and Solomon, 2010b). The data also clearly demonstrated that XSW processed turkey breasts had quicker brine absorption (Figure 7.8). In a 15% brine, the XSW breasts after 10 min of tumbling had an 11.0% update while the controls after 30 min had 8.8% uptake. For the same comparison but in a 30% brine, the HSW breasts picked up 15.2% and the control absorbed 14.4%. In addition, processing yield was 4.3% higher for HSW processed breasts. The more rapid uptake along with the greater retention could have the potential to contribute to reducing the manufacturing cost associated with vacuum-tumbled processed meats.

Jerky made from the chevon semimembranosus muscles XSW processed was evaluated (Eega *et al.*, 2007b). The muscle was XSW processed and then intact strips were removed and marinated in a commercial seasoning. The XSW goat jerky was rated as more tender by a trained panel than the control jerky.

7.5.3 Shelf-life and food safety

Bacterial lethality and inhibition have ranged from a high of 4.5 log to 0.2 log (Alvarez *et al.*, 2004; Holzer *et al.*, 2004; Williams-Campbell and Solomon, 2002; Zuckerman, Krasik and Felsteiner, 2002). Interestingly, the high end of the range (4.5) was found in ground beef stored (14 days) after HSW compared to the control (Williams-Campbell and Solomon, 2002). Pathogens evaluated and shown to be affected by HSW include *Listeria monocytogenes, E. coli O157:H7*, and *Salmonella Typhimurium* (Alvarez *et al.*, 2004; Patel *et al.*, 2005). In contrast, Lorca (2002) did not find any significant bacterial effect using HSW generated by explosives or electrically on ground beef.

As a nonthermal treatment to reduce bacterial populations in meat, the role of HSW remains unclear because of inconsistent reports on its bactericidal effects. Ultraviolet light is generated in the spark gap upon shock wave generation (Fröngel, 1976). Ultraviolet light exhibits bacterial lethality or bacteriostatic effects. In the HSW systems in which the meat resides in the water-filled vessel where the shock wave is generated would be exposed to UV light (Schaefer, 1998). If UV light plays a significant role, this may explain why Lorca (2002) was not able to document any microbial lethality.

Some work evaluated the ability of XSW to injure or inactive pathogens that had attached to meat (Patel *et al.*, 2012). They found XSW-induced injury (0.5 log, CFU/g) of *E. coli* O157:H7 and *S. Typhimurium*. In addition, they provided transmission electron microscopy evidence that most cell injury occurred at the cell membrane level. How the XSW causes cell damage or bactericidal effect has not been elucidated. They theorized that a cavitation phenomenon may be responsible, wherein high-pressure shock waves form microscopic vapor-filled cavities, which expand and collapse rapidly to generate high-energy density particles responsible for inactivating bacteria. Apparently Gram-negative bacteria are more susceptible to the XSW treatment than Gram-positive bacteria (Patel *et al.*, 2012). Ground beef XSW processed resulted in a reduction in *E. coli* O15:H7, although the practicability of the reduction was marginal (Podolak *et al.*, 2006). XSW was not able to kill *Salmonella* in minced chicken (Patel *et al.*, 2006).

The combination of an antimicrobial and XSW associated with frankfurters was found to inhibit *L. monocytogenes* (Patel *et al.*, 2007). After dipping in an antimicrobial solution, frankfurters were surface inoculated and then vacuum packaged followed by XSW. A synergistic effect between nisin and XSW was observed for the inhibition of *L. monocytogenes*. The reduction in aerobic cell populations during storage of antimicrobials and XSW-treated frankfurters was similar to that of *L. monocytogenes*.

If HSW was to induce the movement of pathogens into the interior of intact meat, this would represent a potential food safety risk to consumers. The potential for inducing movement of bacteria was evaluated in beef and chicken using *E. coli* (Green Fluorescent Protein labeled, Figure 7.9) and rifampicin-resistant *E. coli* (Lorca *et al.*, 2003; Lorca *et al.*, 2002). HSW treatment induced the movement of both marker bacteria into the steak (XSW; maximum depth of 250 μm or 1/4 mm) and chicken breast (ASW; 0–100 μm beyond untreated surface). For steaks they indicated that because popular steak-cooking techniques involve the application of heat from the exterior surface of the steak to achieve internal

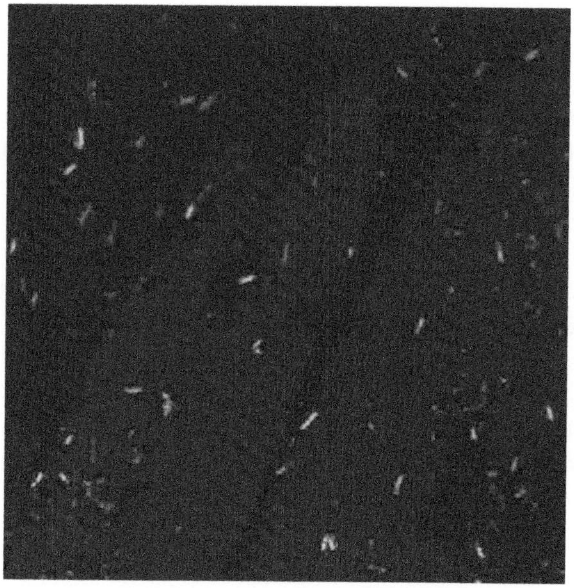

Figure 7.9 Visualization of surface bacteria – *E. coli* (Green Fluorescent Protein labeled) (Courtesy of Lorca, PhD dissertation presentation; Lorca, 2002).

temperatures ranging from 55 to 82 °C, the extent of bacterial penetration observed in HSW treated steaks does not appear to pose a safety hazard to consumers. Similarly, the extent of bacterial penetration in chicken was not considered to pose any safety hazard to consumers.

7.6 Developmental advances in hydrodynamic shock wave equipment

The generation and understanding of a high-pressure shock wave pertinent to the development of a system for improving meat tenderness and food safety has been most extensively studied on explosive-based systems. Extensive studies were conducted to evaluate the effect of different levels of explosives, type of explosives, shape of explosives, and impact of changes to the water-filled shock wave processing vessel, and this information can be found in the review by Solomon *et al.* (2006).

An attempt to develop a commercially viable XSW system was met with numerous unresolved engineering problems as well as with a reduction in tenderization efficacy associated with performance. As Solomon *et al.* (2006) pointed out, although the effectiveness of an explosive-based XSW process was well documented, commercialization was hampered by four technical problems: safety, throughput, packaging, and performance. They mentioned that apart from the obvious safety concerns when using explosives, the use of explosives necessitated a batch-type processing system. A batch system would not be compatible with the high throughput of most modern meat processing plants. In addition, the meat had to be vacuum packaged and all the air had to be removed from each package to prevent rupturing or tearing during the process. Performance deteriorated as designed changes were made to improve the integrity of the supporting structures holding the water-filled processing vessel. Other challenges include the inability to easily expose the meat to more than one shock wave treatment and/or make changes to the kinetic energy based on tenderization objectives.

Although some issues with packaging were mentioned, in normal manufacturing practices a certain level of package failures would be anticipated. The impact of meat as such coming in contact with potential explosive-related residues, incomplete consumption of the explosive constituents, or packaging film residues would need to be studied. Under the very rapid and high pressure associated with shock wave generation, air pockets in the package or trapped under the package would likely lead to holes associated with the heat generated due to rapid compression of the air.

Systems that have the most commercial viability are designed to produce an ASW. Such systems have similarities (e.g., capacitor discharge sparker based, ability to change the discharge energy, repeated pulses) and also have subtle differences in terms of controlling the properties of the shock wave. In the ASW systems, differences exist in terms of the extent the shock waves are focused. Some of the subtle differences are likely proprietary and deal with designs to maximize the effects of a compression wave, shock wave interactions, and the value of the unloading wave. Interestingly, from the review provided by Solomon *et al.* (2006), the research based on XSW supported the value of designing systems to put more emphasis on an unloading wave.

When a shock wave is formed, it expands spherically interacting with the meat as well as reflecting off of the vessel and other structures within the water. Similar to dropping two objects in water separated by some distance when the waves meet the wave fronts interact resulting in a greater peak wave. The collapse of the gas bubble also causes another shock wave. Thus, a multitude of shock wave interactions occur in an ASW system. Capitalizing on shock wave interactions that occur in the meat likely is of value.

A unique distinction among the HSW systems pertains to the location of the meat relative to the generation of the shock wave. In all of the XSW studies, the meat was located in the water where the shock wave was produced. For the two ASW systems that will be described, one positions the sparker electrodes in the water-filled vessel above the meat, while the other system generates the shock wave in water in a manner in which the water is not in contact with the meat.

A system developed by the German Institute of Food Technologies (DIL, Figures 7.10 and 7.11) is designed with a flat-bottomed, cylinder-shaped steel

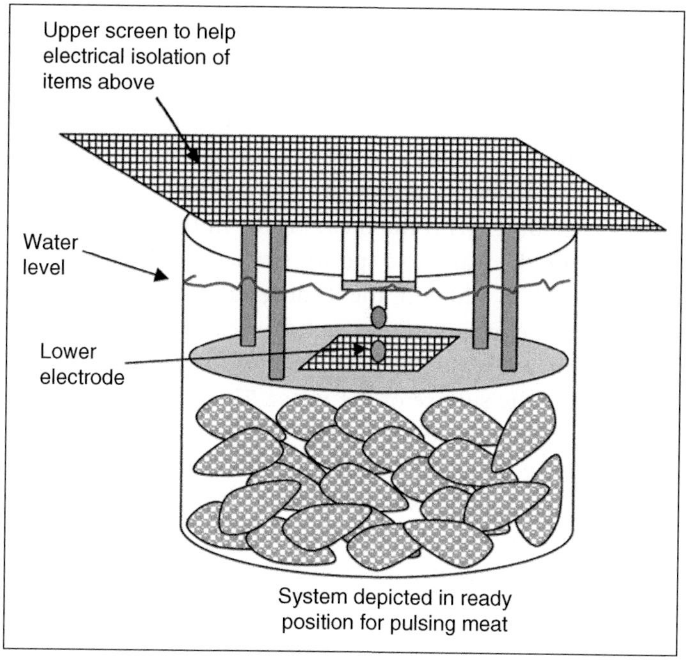

Figure 7.10 Schematic of DIL's semi-industrial prototype shock wave system.

Figure 7.11 Semi-industrial prototype shock wave system (Courtesy of DIL).

vessel with the shock wave being generated with a sparker system. Similar to the explosive-based system, the site of shock wave origination is above the meat such that the shock wave passes through the meat, reflects off of the bottom of the vessel, and then passes back through the meat. In addition, this system has an innovative mechanical way of placing the sparker into the processing bowl. The sparker electrodes are mounted onto and just above a flat screen as part of the sparker module. With the meat loaded in the processing vessel, the module is lowered into the water to a specific strategic distance from the bottom of the processing vessel. In addition, the leading edge of the module establishes a separation between the meat and the sparker electrodes. Although lean meats are more dense than water and should sink, the screen provides assurance that the meat is below the surface of the water. The sparker module also has a flat upper screen that helps isolate the electrical system and limits water from being expelled from the vessel associated with the shock wave gas bubble. By this movable module design, this should facilitate the further development of a more continuous ASW processing system where, for instance, a series of vessels loaded with meat could be advanced one at a time under the sparker module.

As indicated previously, one potential existing limitation is the meat would be in direct contact with water unless it was vacuum packaged. With the assumption that appropriate package integrity can be maintained, use of such packaging would be an additional cost to the user. Nevertheless, a potential advantage of this system is that the packaged meat would have some exposure to ultraviolet light generated from the sparker system. Such exposure has the potential to inactivate or injury bacteria thereby improving food safety and/or extending product shelf-life. A semi-industrial version of this system is now available. However, no published data was found in the scientific literature on this system in terms of tenderization efficacy.

Hydrodyne Inc. first conceived development of a toroid-based ASW system (Figure 7.12). This system was to capitalize on an acoustic converging wave design (Long, 2001; Long and Ayers, 2001). The toroid concept was designed with the goal of being able to have a continuous processing system. However, rather than pursuing this exact design, they worked on the development of a food tube that was coupled with a shock wave generating chamber. The food tube had an opening in the side of the metal tube that was sealed with an acoustically transparent diaphragm that served to separate the shock wave chamber water from contents of the food tube. This food tube system concept demonstrated ASW tenderization, but excessive visual damage was often apparent and as such this design was abandoned.

However, Hydrodyne Inc. continued down the path of exploring the utility of the shock wave generating chamber as referred to as a shock wave generating head (Figure 7.13). By this design, meat is placed on the rubber diaphragm (drum head) that is acoustically matched with water. The inside of the chamber was milled out to form basically a parabola (Figure 7.14). The placement of

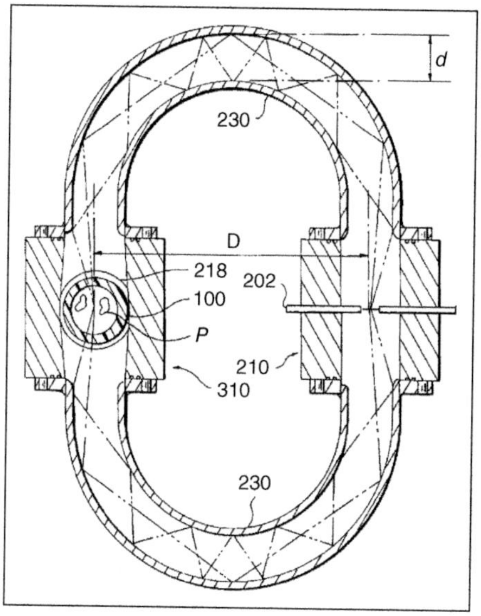

Figure 7.12 Toroid design (Long and Ayers, 2001).

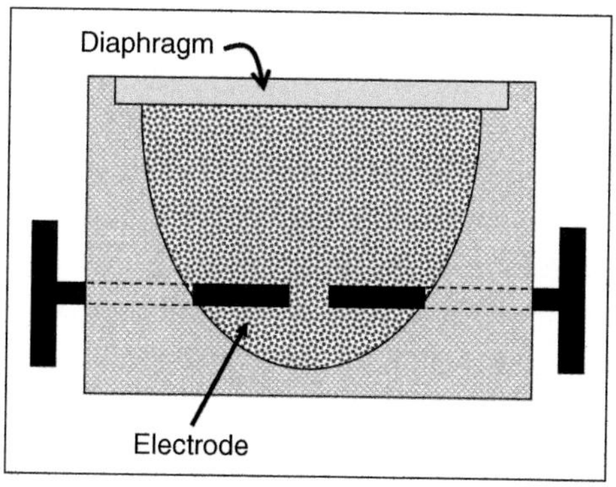

Figure 7.13 Water-filled shock wave generating head.

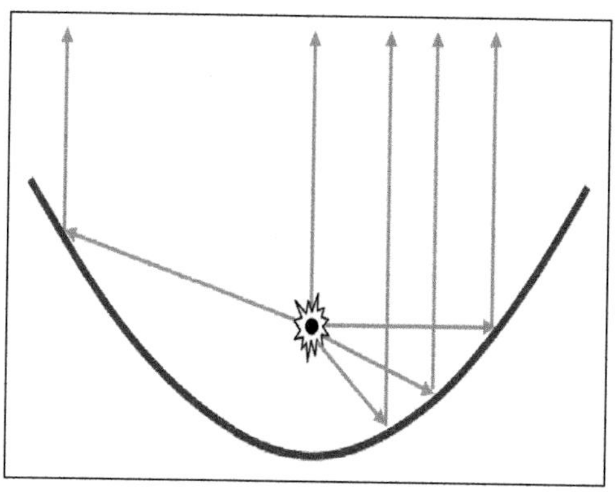

Figure 7.14 A parabola shaped bowl with shock wave origin occurring at the focus of the parabola.

the gap between the electrodes is at the focus of the parabola. By this design when the initial shock wave is generated and expands spherically, for that part of the shock wave that proceeds in the direction of the chamber upon contacting the chamber wall, it is reflected back more or less in parallel to the parabola's vertical axis. Although an initial shock wave is created, depending on interpretation of what happens to the wave front, additional waves are created albeit all of this happens in a fraction of a millisecond. Part of the pressure front would directly head toward the diaphragm, and some of the pressure front would first move in the opposite direction before reflecting off of the bottom of the chamber. Regardless of the interaction with the chamber wall, the shock waves would be redirected essentially in parallel toward the diaphragm.

Interestingly it was not anticipated that the generation of a shock wave with a HVADH system (16 kJ) would result in the significant movement of the pulsed meat. As noted in the patents (Long, 2002, 2003), unrestrained meat was propelled upwardly through a ceiling tile in a room with a 10 ft ceiling. At this stage, considerable attention was given to the value of both the compression wave effect and the negative or tension wave effect. Long (2002) realized that if a system could be designed to maximize compression followed immediately by very rapid unloading (negative, tension wave), this would be very beneficial to tenderization. Long indicated that the compression wave will transverse water and meat. If the wave passes through the meat to an air boundary, the wave pressure unloads. As the wave passes through the meat, if the unloading occurs at the surface of the meat, unlike water, the meat can support the tension and the tension wave progresses back through the meat breaking muscle fibers.

So coupled with the interest in the negative wave and the recognition that a meat restraint system was needed to be incorporated, development continued in that direction. The first generation of a meat restraint system (Long, 2002) used an inflated tire inner tube (Figure 7.15). This was followed by a plastic retaining ring (Figure 7.16). Tenderization performance was promising but neither restraint system was able to consistently retain all of the meat. Development continued using somewhat of a processing tunnel-type design to study side walls described as metallic picture frames over a material with a reasonable mechanical impedance match with water (Figures 7.17 and 7.18). Later, positioning and retaining pistons were added to the ends of the tunnel.

These developments led to what is referred to as the (Hydrodyne Inc.). One aspect not previously mentioned was the need to establish the active tenderization area of the ASW system. Sagili and Claus (2003) determined that equivalent tenderization occurred up to 1 in. beyond the lateral edge of the opening of the shock wave generating chamber. The TCS described by Claus (2002) is composed of a pulser (capacitors and electronic control module), shock wave generating head, and meat processing tunnel (Figures 7.19 and 7.20).

As noted in the schematic of the TCS, the shock wave generating chamber was repositioned above the meat processing tunnel. This system served as a

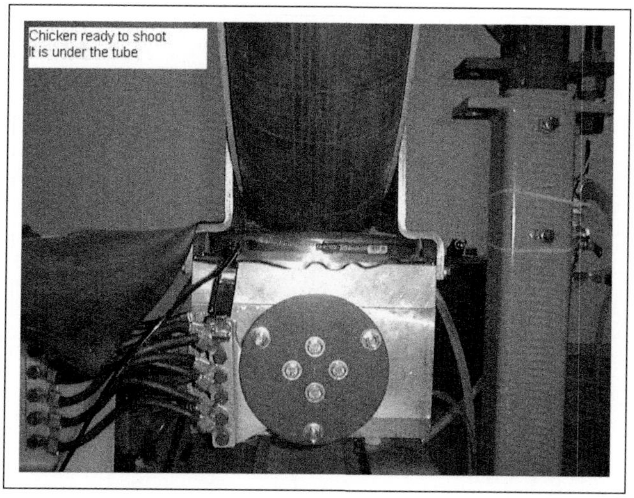

Figure 7.15 Inner tube meat restraint.

Figure 7.16 Plastic meat retaining ring.

demonstration unit only. To use this system, unpackaged meat was hand placed into one of the processing feed tunnels (tunnel in the foreground). An indexer moved the meat into position so that another indexer in the tunnel aligned with the chamber could move the meat under the diaphragm. When positioned there, indexers on the opposite side moved in to compress and retain the meat in the active tenderization area. Proprietary developments led to the exclusion of the

Figure 7.17 Shock wave tunnel development.

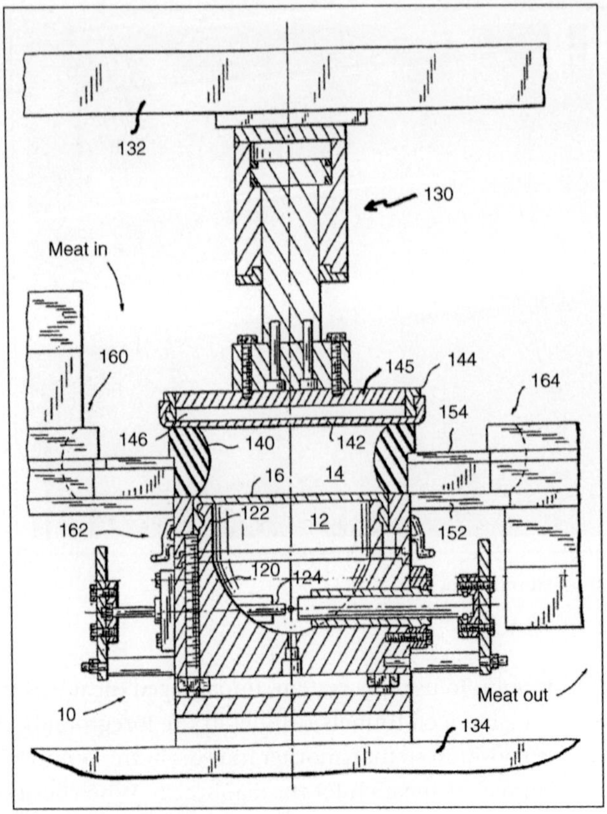

Figure 7.18 Principal design of a shock wave processing system (Long, 2003).

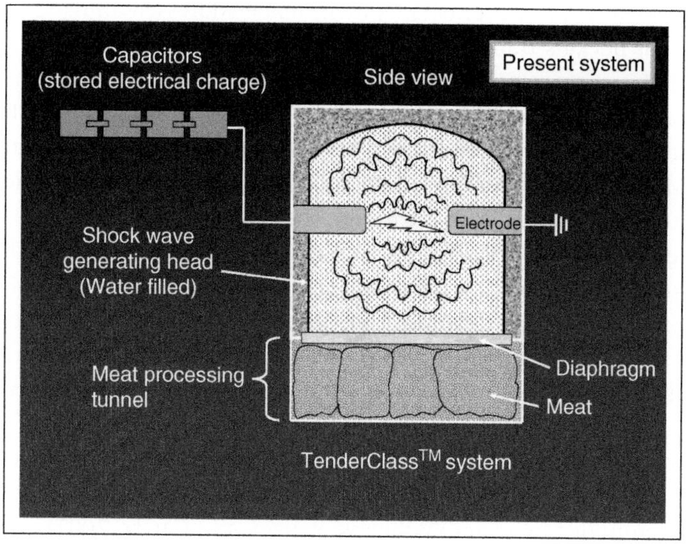

Figure 7.19 Schematic of the TenderClass System (Hydrodyne Inc.)

Figure 7.20 Commercial demonstration unit (TenderClass System, Hydrodyne Inc.)

negative-side panels. Another development made addressed an issue associated with small air pockets trapped at the surface of the meat. With the very rapid rise in high pressure, trapped air would occasionally cause localized burning of the lean or fat (Long and Thomsen, 2006). This problem was solved by flushing the chamber with a nonoxidizing gas. From a commercialization standpoint, the remaining feature under development is a meat-moving system such that a continuous flow of meat can pass through the active area.

A sparker system based off of a few patents (Schaefer, 1998, 2004a,2004b) has also been tested (Bowker *et al.*, 2011). One of the most unique aspects of this system is the future potential to strategically place multiple shock wave generating electrodes over the surface of the meat. However, the current design requires the meat to be vacuum packaged and package durability will likely remain a limitation.

Another approach apparently only patented involves a nonexplosive generator system in which a piston is mounted onto a vessel (Figure 7.21) containing the meat surrounded by water (Williams, Warren and Carter, 2007). The piston is mounted in a manner in which it has direct access to the surface of a column of water within the vessel. When the piston strikes the water column, it creates a shock wave. In the piston system, the piston travels at 50–150 m/s resulting in an

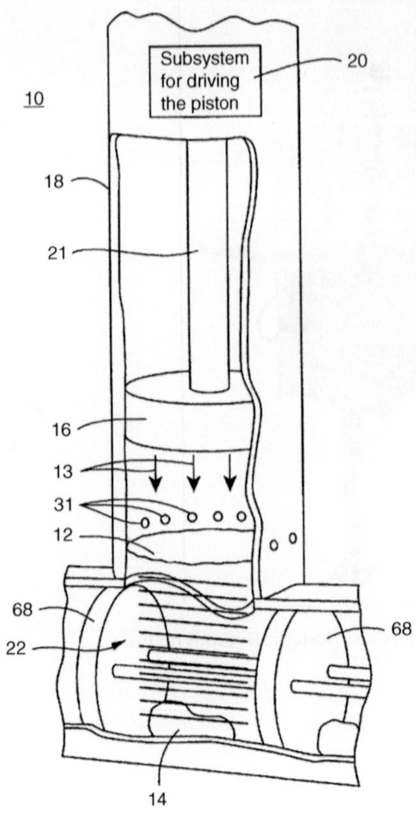

Figure 7.21 Piston pressure generator (Williams, Warren and Carter, 2007).

impact pressure of 70–200 MPa. The inventors' claimed a 17–25% improvement in tenderness compared to XSW tenderization. However, no published data was found in the scientific literature on this system in terms of tenderization efficacy. This system appears to only be in an experimental stage of development.

7.7 Brief overview of available hydrodynamic shock wave equipment and manufacturers of hydrodynamic shock wave equipment

The two systems that appear to be at the forefront of commercial entry are the semi-industrial shock wave system designed by the German Institute of Food Technologies (DIL) and the TCS by Hydrodyne Inc. (Table 7.2). According to DIL, the semi-industrial shock wave system is ready for use by the meat-industry. Their processing specification sheet indicates that this system is capable of processing 10,000 lb/h. A single processing batch contains about 88 lb of meat. They are also pursuing the development of a continuous ASW system to facilitate improvement in processing efficiency.

The only remaining component to be developed for the TCS is a meat-moving system that will enable continuous meat processing. Basically the meat-moving system will enable a continuous flow of meat through the active area of the processing the tunnel associated with shock wave generating chamber. Since numerous systems already exist for moving meat in fabrication plants through the use of conveyors, auger systems, and pumps, identifying and refining a meat moving system to couple with the TCS is expected to proceed with limited development time. The continuous TCS is designed to process 15,000 lb/h.

A system by Phoenix Science and Technology Inc. is very early on in the developmental stage (Table 7.2). Current testing has only demonstrated potential applicability of their system.

Table 7.2 Contact information for companies involved with the development and manufacturing of hydrodynamic shock wave processing equipment.

German Institute of Food Technologies (DIL)	Hydrodyne Inc.	Phoenix Science and Technology Inc.
Contact person: Prof. Dr Stefan Toepfl/Dr Tomas Bolumar Prof.-von-Klitzing-Str. 7 49610 Quakenbrueck (Germany). E-mail: s.toepfl@dil-ev.de/t.bolumar@dil-ev.de http://www.dil-ev.de/en.html	Contact person: William J. McKenna (President) 7221 Preservation Court Fulton, MD 20759 E-mail: wmckenna@mckenna-associates.com	Contact person: Raymond Schaefer, PhD (President, Founder) 27 Industrial Avenue, Chelmsford, MA 01824 Phone: 978-367-0232; Fax: 978-367-0236 E-mail: info@phoenixsandt.com; RSchaefer@PhoenixSandT.com http://www.phoenixsandt.com/

References

Aalhus, J.L., Best, D.R., Costello, F. and Jeremiah, L.E. (1999) A simple, on-line processing method for improving beef tenderness. *Canadian Journal of Animal Science*, **79** (1), 27–34.

Aalhus, J.L., Dugan, M.E.R., Robertson, W.M. *et al.* (2004) A within-animal examination of postmortem ageing for up to 21 d on tenderness in the bovine longissimus thoracis and semimembranosus muscles. *Canadian Journal of Animal Science*, **84** (2), 301–304.

Aalhus, J.L., Larsen, I.L., Dubeski, P.L. and Jeremiah, L.E. (2000) Improved beef tenderness using a modified on-line carcass suspension method with, or without low voltage electrical stimulation. *Canadian Journal of Animal Science*, **80**, 51–58.

Aberle, E.D., Forrest, J.C., Gerrard, D.E. and Mills, E.W. (2012) *Principles of Meat Science*, 5 edn, Kendall/Hunt Publishing Co., Dubuque, Iowa.

Adler, J.M., Geornaras, I., Belk, K.E. *et al.* (2012) Thermal inactivation of *Escherichia coli* O157:H7 inoculated at different depths of non-intact blade-tenderized beef steaks. *Journal of Food Science*, **77** (2), M108–M114.

Ahnstrom, M.L., Hunt, M.C. and Lundstrom, K. (2012) Effects of pelvic suspension of beef carcasses on quality and physical traits of five muscles from four gender-age groups. *Meat Science*, **90** (3), 528–535.

Ahnstrom, M.L., Seyfert, M., Hunt, M.C. and Johnson, D.E. (2006) Dry aging of beef in a bag highly permeable to water vapour. *Meat Science*, **73** (4), 674–679.

Alvarez, U.M., Loske, A.M., Castano-Tostado, E. and Prieto, P.E. (2004) Inactivation of *Escherichia coli* O157:H7, *Salmonella Typhimurium* and *Listeria monocytogenes* by underwater shock waves. *Innovative Food Science and Emerging Technologies*, **5**, 459–463.

Asay, J.R. and Shahinpoor, M. (1993) *High-pressure shock compression of solids*, Springer-Verlag Inc., New York.

Belew, J.B., Brooks, J.C., McKenna, D.R. and Savell, J.W. (2003) Warner-Bratzler shear evaluations of 40 bovine muscles. *Meat Science*, **64**, 507–512.

Berry, B.W., Solomon, M.B., Johnson, W.L. *et al.* (1997) Application of the Hydrodyne process to strip loins from U.S. Select grade beef. *Journal of Animal Science*, **75** (Suppl. 1), 128.

Birkhold, S.G., Janky, D.M. and Sams, A.R. (1992) Tenderization of early-harvested broiler breast fillets by high-voltage postmortem electrical-stimulation and muscle tensioning. *Poultry Science*, **71** (12), 2106–2112.

Bourne, M.C. (1982) Food Texture and Viscosity: Concept and Measurement, Food Science and Technology- A Series of Monographs (eds G.F. Stewart, B.S. Schweigert and J. Hawthorn), Academic Press, Inc, San Diego, California, p. 327.

Bowker, B., Liu, M., Callahan, J. and Solomon, M. (2010a) Effect of hydrodynamic pressure processing on the processing and quality characteristics of moisture-enhanced pork loins. *Journal of Food Science*, **75** (4), S237–S244.

Bowker, B.C., Callahan, J.A. and Solomon, M.B. (2010b) Effects of hydrodynamic pressure processing on the marination and meat quality of turkey breasts. *Poultry Science*, **89** (8), 1744–1749.

Bowker, B.C., Eastridge, J.S., Paroczay, E.W. and Solomon, M.B. (2008a) Influence of sarcomere length on aging and hydrodynamic pressure processing of beef muscle. *Journal of Animal Science*, **86** (E-Suppl. 2, abstract), 43.

Bowker, B.C., Fahrenholz, T., Paroczay, E.W. and Solomon, M.B. (2006) Effect of Hydrodynamic Pressure Processing and Aging on Sarcoplasmic Proteins Isolated from Beef Strip Loins. Institute of Food Technologists Annual Meeting Book of Abstracts, Paper No. 039I-22. June 24–28, 2006, Orlando, Florida.

Bowker, B.C., Fahrenholz, T.M., Paroczay, E.W. *et al.* (2008b) Effect of hydrodynamic pressure processing and aging on the tenderness and myofibrillar proteins of beef strip loins. *Journal of Muscle Foods*, **19** (1), 74–97.

Bowker, B.C., Fahrenholz, T.M., Paroczay, E.W. and Solomon, M.B. (2008c) Effect of hydrodynamic pressure processing and aging on sarcoplasmic proteins of beef strip loins. *Journal of Muscle Foods*, **19** (2), 175–193.

Bowker, B.C., Fahrenholz, T.M., Sarnoski, P.J. and Solomon, M.B. (2012) Alterations in the sarcoplasmic protein fraction of beef muscle with postmortem aging and hydrodynamic pressure processing. *Journal of Food Science*, **77** (6), C594–C602.

Bowker, B.C., Liu, M., Eastridge, J.S., Paroczay, E.W., Callahan, J.A. and Solomon, M.B. (2007a) *Effect of Hydrodynamic Pressure Processing and Aging on the Tenderness and Protein Characteristics of Pork Loins*. American Meat Science Association Reciprocal Meat Conference Proceedings, Paper No. 92.

Bowker, B.C., Liu, M., Eastridge, J.S., Paroczay, E.W., Callahan, J.A. and Solomon, M.B. (2007b) Effect of hydrodynamic pressure processing on the tenderness and drip loss of pork loins. Institute of Food Technologists Annual Meeting Book of Abstracts, Paper No. 094-005.

Bowker, B.C., Liu, M., Fahrenholz, T.M., Callahan, J.A., Vinyard, B.T. and Solomon, M.B. (2005) *Blade Tenderization and Hydrodynamic Pressure Processing Effects on Protein Characteristics in Top Rounds from Brahman Cattle*. Proceedings of the 51st International Congress of Meat Science and Technology, August 7–12, 2005, Baltimore, Maryland, Paper No. Th52.

Bowker, B.C., Liu, M.N., Eastridge, J.S. *et al.* (2010c) Effect of postmortem aging and hydrodynamic pressure processing on pork loin quality. *Journal of Muscle Foods*, **21** (2), 379–398.

Bowker, B.C., Liu, M.N., Solomon, M.B. *et al.* (2007c) Effects of hydrodynamic pressure processing and blade tenderization on intramuscular collagen and tenderness-related protein characteristics of top rounds from Brahman cattle. *Journal of Muscle Foods*, **18** (1), 35–55.

Bowker, B.C., Schaefer, R.B., Grapperhaus, M.J. and Solomon, M.B. (2011) Tenderization of beef loins using a high efficiency sparker. *Innovative Food Science and Emerging Technologies*, **12** (2), 135–141.

Callahan, J.A., Liu, M., Bowker, B.C., Paroczay, E.W., Eastridge, J.S. and Solomon, M.B. (2008) *Application of HDP before or after Marination to Improve the Quality of Pork Loins*. American Meat Science Association Reciprocal Meat Conference Proceedings, Paper No. 63.

Campbell, R.E., Hunt, M.C., Levis, P. and Chambers, E. (2001) Dry-aging effects on palatability of beef longissimus muscle. *Journal of Food Science*, **66** (2), 196–199.

Claus, J.R. (1999) The Tendercut Process-A simple technique to increase meat tenderness. *Meat&Poultry*, **September**, 68–73.

Claus, J.R. (2002) Shock treatment- shock waves are an effective tool for tenderizing meat. *Meat&Poultry*, **48** (12), 61–63.

Claus, J.R. (2007) *Color Changes in Cooked Beef*. Fact sheet, National Cattlemen's Beef Association.

Claus, J.R., Sagili, J.V.V. and Sammel, L.M. (2002) *Tenderization of Beef and Pork with Shock Waves Produced with a Capacitor Discharge System*. Proceedings of the 55th Reciprocal Meat Conference. American Meat Science Association, Champaign, IL. http://www.meatscience.org (accessed March 20, 2012).

Claus, J.R., Schilling, J.K., Marriott, N.G. *et al.* (2001) Tenderization of chicken and turkey breasts with electrically produced hydrodynamic shockwaves. *Meat Science*, **58**, 283–286.

Cole, R.H. (1948) *Underwater Explosions*, Princeton University Press, Princeton, NJ.

Cooper, C. and Solomon, M.B. (2000) Meat tenderization by means of plasma sparking device. Canadian Patent #5,432,756.

Devine, C.E., Payne, S.R., Peachey, B.M. *et al.* (2002a) High and low rigor temperature effects on sheep meat tenderness and ageing. *Meat Science*, **60** (2), 141–146.

Devine, C.E., Payne, S.R. and Wells, R.W. (2002b) Effect of muscle restraint on sheep meat tenderness with rigor mortis at 18°C. *Meat Science*, **60** (2), 155–159.

Devine, C.E., Wahlgren, N.M. and Tornberg, E. (1999) Effect of rigor temperature on muscle shortening and tenderization of restrained and unrestrained beef m-longissimus thoracicus et lumborum. *Meat Science*, **51** (1), 61–72.

Echeverry, A., Brooks, J.C., Miller, M.F. *et al.* (2010) Validation of lactic acid bacteria, lactic acid, and acidified sodium chlorite as decontaminating interventions to control *Escherichia coli* O157:H7 and *Salmonella Typhimurium* DT 104 in mechanically tenderized and brine-enhanced (nonintact) beef at the purveyor. *Journal of Food Protection*, **73** (12), 2169–2179.

Eega, K.R., Lee, J.H., Solomon, M.B. *et al.* (2007a) Effect of hydrodynamic pressure processing on chevon quality characteristics. *Journal of Animal Science*, **85** (Suppl. 1), W129.

Eega, K.R., Lee, J.H., Solomon, M.B. *et al.* (2007b) Quality characteristics of jerky made from hydrodynamic pressure processed (HDP) chevon and beef. *Journal of Animal Science*, **85** (Suppl. 1), W130.

Ferguson, D.M., Jiang, S.T., Hearnshaw, H. *et al.* (2000) Effect of electrical stimulation on protease activity and tenderness of M-longissimus from cattle with different proportions of Bos indicus content. *Meat Science*, **55** (3), 265–272.

Fröngel, F.B.A. (1976) *High Speed Pulse Technology* Volume III. Capacitor discharge engineering. (Vol. III), Academic Press, New York.

Gamble, H.R., Solomon, M.B. and Long, J.B. (1998) Effects of hydrodynamic pressure on the viability of Trichinella spiralis in pork. *Journal of Food Protection*, **61** (5), 637–639.

George-Evins, C.D., Unruh, J.A., Waylan, A.T. and Marsden, J.L. (2004) Influence of quality classification, aging period, blade tenderization, and endpoint cooking temperature on cooking characteristics and tenderness of beef gluteus medius steaks. *Journal of Animal Science*, **82** (6), 1863–1867.

Godfrey, C.S. (1970) Apparatus for tenderizing food. U.S. Patent US3492688.

Grobbel, J.P., Dikeman, M.E., Hunt, M.C. and Milliken, G.A. (2008) Effects of different packaging atmospheres and injection-enhancement on beef tenderness, sensory attributes, desmin degradation, and display color. *Journal of Animal Science*, **86** (10), 2697–2710.

Hayward, L.H., Hunt, M.C., Kastner, C.L. and Kropf, D.H. (1980) Blade tenderization effects on beef longissimus sensory and instron textural measurements. *Journal of Food Science*, **45** (4), 925–935.

Heller, C.E., Scanga, J.A., Sofos, J.N. *et al.* (2007) Decontamination of beef subprimal cuts intended for blade tenderization or moisture enhancement. *Journal of Food Protection*, **70** (5), 1174–1180.

Holzer, Z., Berry, B., Campbell, A. *et al.* (2004) Effect of koshering and hydrodynamic pressure on beef color, odor, and microbial loads. *Journal of Muscle Foods*, **15** (1), 69–82.

Hostetler, R.L., Carpenter, Z.L. and Smith, G.C. (1972) Improving beef tenderness - elevated-temperature aging and suspension by obturator foramen. *Journal of Animal Science*, **35** (1), 202-Abstract.

Hostetler, R.L., Carpenter, Z.L., Smith, G.C. and Dutson, T.R. (1975) Comparison of post mortem treatments for improving tenderness of beef. *Journal of Food Science*, **40** (2), 223–226.

Hostetler, R.L., Landmann, W.A., Link, B.A. and Fitzhugh, H.A. Jr., (1970) Influence of carcass position during rigor mortis on tenderness of beef muscles: comparison of two treatments. *Journal of Animal Science*, **31**, 47–50.

Huff-Lonergan, E., Mitsuhashi, T., Beekman, D.D. *et al.* (1996) Proteolysis of specific muscle structural proteins by μ-calpain at low pH and temperature is similar to degradation in post mortem bovine muscle. *Journal of Animal Science*, **74** (993–1008).

Huff Lonergan, E., Zhang, W. and Lonergan, S.M. (2010) Biochemistry of postmortem muscle - lessons on mechanisms of meat tenderization. *Meat Science*, **86** (1), 184–195.

Hwang, I.H., Devine, C.E. and Hopkins, D.L. (2003) The biochemical and physical effects of electrical stimulation on beef and sheep meat tenderness. *Meat Science*, **65** (2), 677–691.

Hwang, I.H., Lee, J.M. and Thompson, J.M. (2003) The role of electrical stimulation in meat tenderness. *Food Science and Biotechnology*, **12** (2), 191–197.

Johns, D.F., Bratcher, C.L., Kerth, C.R. and McCaskey, T. (2011) Translocation of surface-inoculated *Escherichia coli* into whole muscle nonintact beef striploins following blade tenderization. *Journal of Food Protection*, **74** (8), 1334–1337.

Juárez, M., Larsen, I.L., Klassen, M. and Aalhus, J.L. (2013) Canadian beef tenderness survey: 2001–2011. *Canadian Journal of Animal Science*, **93** (1), 89–97.

King, D.A., Wheeler, T.L., Shackelford, S.D. *et al.* (2009) Effect of blade tenderization, aging time, and aging temperature on tenderness of beef longissimus lumborum and gluteus medius. *Journal of Animal Science*, **87** (9), 2952–2960.

Laine, E.S., Scheftel, J.M., Boxrud, D.J. *et al.* (2005) Outbreak of *Escherichia coli* O157:H7 infections associated with nonintact blade-tenderized frozen steaks sold by door-to-door vendors. *Journal of Food Protection*, **68** (6), 1198–1202.

Lemmons, J.L., Lucia, L.M., Hardin, M.D. *et al.* (2011) Evaluation of *Escherichia coli* O157:H7 translocation and decontamination for beef vacuum-packaged subprimals destined for nonintact use. *Journal of Food Protection*, **74** (7), 1048–1053.

Liu, M.N., Solomon, M.B., Vinyard, B. *et al.* (2006) Use of hydrodynamic pressure processing and blade tenderization to tenderize top rounds from Brahman cattle. *Journal of Muscle Foods*, **17** (1), 79–91.

Long, J., Thomsen, P. and Waits, D. (2007) Shock wave treatment of meat. U.S. Patent 7,244,459 B2.

Long, J. B. (1993) Tenderizing meat. U.S patent 5,273,766.

Long, J. B. (2001) Continuous shock wave food processing with shock wave reflection. U.S. Patent US6168814 B1.

Long, J.B. (2002) Shock-wave treatment. U.S. Patent US2002 0072318 A1.

Long, J.B. (2003) Shock wave treatment of meat. U.S. Patent US6669546 B2.

Long, J.B. and Ayers, R. A. (2001) Shock-wave food processing with acoustic converging wave guide. U.S. Patent US6224476 B1.

Long, J.B. and Thomsen, P. (2006) Treatment of meat. U.S. Patent US2006 0115560.

Lorca, T.A. (2002) The affects of explosively and electrically generated hydrodynamic shock waves on the bacterial flora of beef and poultry. Virginia Tech Dissertation http://scholar.lib.vt.edu/theses/available/etd-07292002-153950/ (accessed March 16, 2012).

Lorca, T.A., Claus, J.R., Eifert, J.D. *et al.* (2003) Penetration of surface-inoculated bacteria as a result of electrically generated hydrodynamic shock wave treatment of boneless skinless chicken breasts. *Poultry Science*, **82**, 1205–1210.

Lorca, T.A., Pierson, M.D., Claus, J.R. *et al.* (2002) Penetration of surface-inoculated bacteria as a result of hydrodynamic shock wave treatment of beef steaks. *Journal of Food Protection*, **65** (4), 616–620.

Lorenzen, C.L., Miller, R.K., Taylor, R.F. *et al.* (2003) Beef Customer Satisfaction: Trained sensory panel ratings and Warner-Bratzler shear force values. *Journal of Animal Science*, **81**, 143–149.

Loske, A.M., Alvarez, U.M., Hernandez-Galicia, C. *et al.* (2002) Bactericidal effect of underwater shock waves on *Escherichia coli* ATCC 10536 suspensions. *Innovative Food Science and Emerging Technologies*, **3**, 321–327.

Luchansky, J.B., Porto-Fett, A.C.S., Shoyer, B.A. *et al.* (2012) Fate of shiga toxin producing O157:H7 and non-O157:H7 *Escherichia coli* cells within blade-tenderized beef steaks after cooking on a commercial open-flame gas grill. *Journal of Food Protection*, **75** (1), 62–70.

Luchansky, J.B., Porto-Fett, A.C.S., Shoyer, B.A. *et al.* (2009) Thermal Inactivation of *Escherichia coli* O157:H7 in blade-tenderized beef steaks cooked on a commercial open-flame gas grill. *Journal of Food Protection*, **72** (7), 1404–1411.

Ludwig, C.J., Claus, J.R., Marriott, N.G. *et al.* (1997) Skeletal alteration to improve beef longissimus muscle tenderness. *Journal of Animal Science*, **75**, 2404–2410.

Mandigo, R.W. and Olson, D.G. (1982) Effect of blade size for mechanically tenderizing beef rounds. *Journal of Food Science*, **47** (6), 2095–2096.

Marriott, N.G., Wang, H., Solomon, M.B. and Moody, W.G. (2001) Studies of cow beef tenderness enhancement through supersonic-hydrodynamic shock wave treatment. *Journal of Muscle Foods*, **12** (3), 207–218.

Mish, F.C., Gilman, E.W., Lowe, J.G. *et al.* (1988) *Webster's Ninth New Collegiate Dictionary*, Merriam-Webster Inc., USA.

Moeller, S., Wulf, D., Meeker, D. *et al.* (1999) Impact of the Hydrodyne process on tenderness, microbial load, and sensory characteristics of pork longissimus muscle. *Journal of Animal Science*, **77**, 2119–2123.

Morgan, J.B. (2010) Integral Antimicrobial Solution Application Systems on Ross Blade Tenderizers. In Proceedings of the 63rd Reciprocal Meat Conference: American Meat Science Association, Champaign, IL.

Morzel, M., Gatellier, P., Sayd, T. *et al.* (2006) Chemical oxidation decreases proteolytic susceptibility of skeletal muscle myofibrillar proteins. *Meat Science*, **73**, 536–543.

Naveena, B.M., Mendiratta, S.K. and Anjaneyulu, A.S.R. (2004) Tenderization of buffalo meat using plant protease from Cucumis trigonus Roxb (Kachri) and Zingiber officinale roscoe (Ginger rhizome). *Meat Science*, **68**, 363–369.

O'Rourke, B.M., Calkins, C.R., Rosario, R.T. *et al.* (1999) Relationship of muscle fiber orientation during Hydrodyne treatment to tenderness, aging, and proteolysis of beef. *Journal of Animal Science*, **77** (Suppl. 1), 45.

Paroczay, E.W., Solomon, M.B., Berry, B.W. *et al.* (2002) *Evaluating tenderness improvement using hydrodynamic pressure processing (HDP)* IFT Annual Meeting Book of Abstracts, 185, Institute of Food Technologists, Chicago.

Patel, J., Macarisin, D., Sanglay, G. and Murphy, C. (2012) Inactivation and injury of pathogens on intact beef treated with hydrodynamic pressure. *Innovative Food Science and Emerging Technologies*, **14**, 38–45.

Patel, J.R., Bhagwat, A.A., Sanglay, G.C. and Solomon, M.B. (2006) Rapid detection of Salmonella from hydrodynamic pressure-treated poultry using molecular beacon real-time PCR. *Food Microbiology*, **23** (1), 39–46.

Patel, J.R., Sanglay, G.C., Sharma, M. and Solomon, M.B. (2007) Combining antimicrobials and hydrodynamic pressure processing for control of *Listeria monocytogenes* in frankfurters. *Journal of Muscle Foods*, **18** (1), 1–18.

Patel, J.R., Williams-Campbell, A.C., Liu, M.N. and Solomon, M.B. (2005) Effect of hydrodynamic pressure treatment and cooking on inactivation of *Escherichia coli* O157:H7 in blade-tenderized beef steaks. *Journal of Muscle Foods*, **16** (4), 342–353.

Pearson, A.M. and Dutson, T.R. (1985) *Advances in Meat Research Volume 1 - Electrical Stimulation*, The AVI Publishing Co., Inc, Westfort, Connecticut.

Pinto, E.S.M.E.M., Mazzilli, R.N. and Cusin, F. (1999) Composition of hydrolysates from meat. *Journal of Food Composition and Analysis*, **12**, 219–225.

Podolak, R., Solomon, M.B., Patel, J.R. and Liu, M.N. (2006) Effect of hydrodynamic pressure processing on the survival of *Escherichia coli* O157:H7 in ground beef. *Innovative Food Science & Emerging Technologies*, **7** (1–2), 28–31.

Ramezani, R., Aminlari, M. and Fallahi, H. (2003) Effect of chemically modified soy proteins and ficin-tenderized meat on the quality attributes of sausage. *Journal of Food Science*, **68** (1), 85–88.

Ruusunen, M., Niemistö, M. and Puolanne, E. (2002) Sodium reduction in cooked meat products by using commercial potassium phosphate mixtures. *Agricultural and Food Science in Finland*, **11**, 199–207.

Ruusunen, M. and Puolanne, E. (2005) Reducing sodium intake from meat products. *Meat Science*, **70** (3), 531–541.

Sagili, J.V.V. (2003) Electrically-generated hydrodynamic shock wave effects on tenderness and other quality attributes of beef and pork. MS thesis. UW. 154 pp. Unpublished Research, University of Wisconsin, Madison, Wisconsin.

Sagili, J.V.V. and Claus, J.R. (2003) *Tenderization Effects of Electrically Produced Hydrodynamic Shock Waves on Top Round and Strip Loin Muscles of Beef.* Poster Session: Post-Harvest Effects on Meat Quality. In 56th Annual Reciprocal Meat Conference. University of Missouri, Columbia, MO. June 15–19: American Meat Science Association, Champaign, IL. http://www.meatscience.org/ (accessed 21 January 2016).

Schaefer, R.B. (1998) Sparker source systems. US Patent 5,841,737,1998.

Schaefer, R.B. (2004a) High efficiency and projection reflectors for light and sound. US Patent 6, 672, 729, 6.

Schaefer, R.B. (2004b) High efficiency long lifetime sparker sources. US Patent 6, 687, 189, 3.

Schilling, M.W., Claus, J.R., Marriott, N.G., Solomon, M.B., Eigel, W.N. and Wang, H. (1999) *Hydrodynamic Shock Wave Effects on Protein Functionality.* Proceedings of the 52nd Reciprocal Meat Conference. American Meat Science Association. Champaign, IL Graduate Student Research Poster Competition. M.S. Division, 123.

Schilling, M.W., Claus, J.R., Marriott, N.G. *et al.* (2002) No effect of hydrodynamic shock wave on protein functionality of beef muscle. *Journal of Food Science*, **67** (1), 335–340.

Sheard, P. and Tali, A. (2004) Injection of salt, tripolyphosphate and bicarbonate marinade solutions to improve the yield and tenderness of cooked pork loin. *Meat Science*, **68** (2), 305–311.

Shen, C., Adler, J.M., Geornaras, I. *et al.* (2010) Inactivation of *Escherichia coli* O157:H7 in nonintact beefsteaks of different thicknesses cooked by pan broiling, double pan broiling, or roasting by using five types of cooking appliances. *Journal of Food Protection*, **73** (3), 461–469.

Shin, H.G., Choi, Y.M., Kim, H.K. *et al.* (2008) Tenderization and fragmentation of myofibrillar proteins in bovine longissimus dorsi muscle using proteolytic extract from Sarcodon aspratus. *LWT - Food Science and Technology*, **41**, 1389–1395.

Smith, R.D., Nicholson, K.L., Nicholson, J.D.W. *et al.* (2008) Dry versus wet aging of beef: Retail cutting yields and consumer palatability evaluations of steaks from US Choice and US Select short loins. *Meat Science*, **79** (4), 631–639.

Solomon, M.B., Berry, B.W., Eastridge, J.S. and Paroczay, E.W. (1999) Muscle fiber orientation during Hydrodyne treatment on tenderness improvement in semitendinosus muscles. *Journal of Animal Science*, **77** (Suppl. 1), 91.

Solomon, M.B., Berry, B.W., Paroczay, E.W., Callahan, J.A. and Eastridge, J. S. (2002) *Effects of Hydrodynamic Pressure Processing (HDP) and Aging on Beef Tenderness.* IFT Annual Meeting Book of Abstracts, Anaheim, CA, June 15–19, 2002, 111. Chicago: Institute of Food Technologists.

Solomon, M.B., Liu, M.N., Patel, J. *et al.* (2008) Tenderness improvement in fresh or frozen/thawed beef steaks treated with hydrodynamic pressure processing. *Journal of Muscle Foods*, **19** (1), 98–109.

Solomon, M.B., Liu, M.N., Patel, J.R. *et al.* (2006) Hydrodynamic pressure processing to improve meat quality and safety, in *Advanced Technologies For Meat Processing* (eds L.M.L. Nollet and F. Toldrá), CRC Press, Taylor & Francis Group Chapter 9, pp. 222–239.

Solomon, M.B., Long, J.B. and Eastridge, J.S. (1997) The Hydrodyne: A new process to improve beef tenderness. *Journal of Animal Science*, **75** (6), 1534–1537.

Sorheim, O. and Hildrum, K.I. (2002) Muscle stretching techniques for improving meat tenderness. *Trends in Food Science & Technology*, **13** (4), 127–135.

Taylor, R.G., Geesink, G.H., Thompson, V.F. *et al.* (1995) Is Z-disk degradation responsible for postmortem tenderization? *Journal of Animal Science*, **73**, 1351–1367.

Vote, D.J., Platter, W.J., Tatum, J.D. *et al.* (2000) Injection of beef strip loins with solutions containing sodium tripolyphosphate, sodium lactate, and sodium chloride to enhance palatability. *Journal of Animal Science*, **78** (4), 952–957.

Williams-Campbell, A.M. and Solomon, M.B. (2002) Reduction of spoilage microorganisms in fresh beef using hydrodynamic pressure processing. *Journal of Food Protection*, **65** (3), 571–574.

Williams, J., Warren, P.A. and Carter, J. (2007) Non-explosive shockwave generator system and method for hydrodynamic pressure processing of food products. U.S. Patent US2007 0157729 A1.

Zuckerman, H., Berry, B.W., Eastridge, J.S. and Solomon, M.B. (1999) Microstructure changes in connective tissue of bovine semimembranosus muscle caused by hydrodynamic shock-waves. *Journal of Scanning Microscopy*, **21**, 128–129.

Zuckerman, H., Krasik, Y.A. and Felsteiner, J. (2002) Inactivation of microorganisms using pulsed high-current underwater discharges. *Innovations in Food Science and Emerging Technology*, **3**, 329–336.

Zuckerman, H. and Solomon, M.B. (1998) Ultrastructural changes in bovine longissimus muscle caused by the Hydrodyne process. *Journal of Muscle Foods*, **9**, 419–426.

CHAPTER 8
Robotics in meat processing

Kompal Joshi[1], Tomas Norton[2], Jesús M. Frías[1] & Brijesh K. Tiwari[3]
[1] *School of Food Science and Environmental Health, Dublin Institute of Technology, Dublin, Ireland*
[2] *Engineering Department, Harper Adams University, Newport, Shropshire, UK, TF108NB*
[3] *Department of Food Biosciences, Teagasc Food Research Centre, Dublin, Ireland*

8.1 Introduction

A robot is a programmable machine that is capable of carrying out complex actions automatically. Robotics is the branch of science which deals with the design, construction, operation and application of robots. In today's modern life, robotic technology has contributed in many ways to improving society. Robotic systems exploit artificial intelligence to replace many labour-intensive operations in several industries. The automobile industry uses robots for the whole production process. In the electronics industry, the use of robots to assemble circuit boards and computers is very common. These days in industrial fields such as construction or ship building, automatic climbing robots are increasingly being considered in situations where maintenance and inspection involves dangerous operations. Scientists are currently investigating micro-robotics in the medical field with a potential to provide better medical technology in the near future. Other specialized field robots are being developed to work in many conditions such as in mines, underwater or farm fields (Prarthana, Bhavana and Mani Shankar, 2013).

When it comes to the food industry, the use of robots has been traditionally limited to picking and palletization. One of the earliest applications of robots in the 1980s involved packaging of assorted chocolates into trays (Wallin, 1997). Today, however, robots are used in material handling and secondary or tertiary packing. Recent developments with faster computers and sophisticated sensors have made it possible to use robotics in the meat processing sectors, where their application has reduced processing costs, occupational injuries, improved efficiency and hygiene associated with meat products. Compared to other industries, the working environment in the meat industry is not very conducive to robotics due to the noisy, damp and cold conditions. Slaughtering animals and cutting meat into pieces and disposing waste is an intensive physically demanding task. Due to these hostile conditions, labourers are not readily available and

Emerging Technologies in Meat Processing: Production, Processing and Technology, First Edition.
Edited by Enda J. Cummins and James G. Lyng.

Table 8.1 Advantages and disadvantages of automation in meat industry.

Advantages	Disadvantages
Reduced floor space due to overhead mounting	High initial investment
Better quality end product and hygienic processing	Not flexible like human beings
Highly efficient and quick production	Not completely reliable
Suitable for working in all kinds of harsh environments	Complex in handling biological material
Reduced wastage	Lacks quick decision-making capability
Produces more uniform end product	
Easy to clean and can withstand corrosion and acid wash	
Lightweight yet rigid construction	

the demand for automation has thus increased. Automation requires high financial investment and the payback can be expected only after a long time and when the volume of meat processed is very high. However, the advantages of automation are higher with few disadvantages, thus the need of automation cannot be ignored in the meat processing industry. The objective of this chapter is to review the application of robotics in the meat industry and the advancements that have been made until now. The various advantages and disadvantages that come to mind when you think of automation of the meat industry listed in Table 8.1. Case studies on automating a number of key processes are discussed.

8.2 Application of robotics in meat processing

The development of robotic technology for the meat industry has gained interest because of the decline in the labour force, unpleasant working environment and ever-increasing hygiene requirements. The important challenge faced by scientists doing research around meat processing is to develop a technology for safe, wholesome and convenient products that offer great value. In beef production, the first reported use of robotic equipment was in splitting a complete carcass into carcass sides. A number of manufacturers produce such equipment which replaces this particular arduous manual process. Several manual operations have been mechanized, improving both process efficiency and hygiene in the meat processing industry.

For many years, the meat industry has been industrialized in terms of organization and use of production lines. The meat industry is taking inspiration from other industries where robot-assisted and computer-controlled manufacturing systems are more common. Now, to maintain the increasing demand with globalization, the meat industry is also ensuring the use of robots where it is justifiable (Figure 8.1). Traditionally, plant modernization in the meat industry

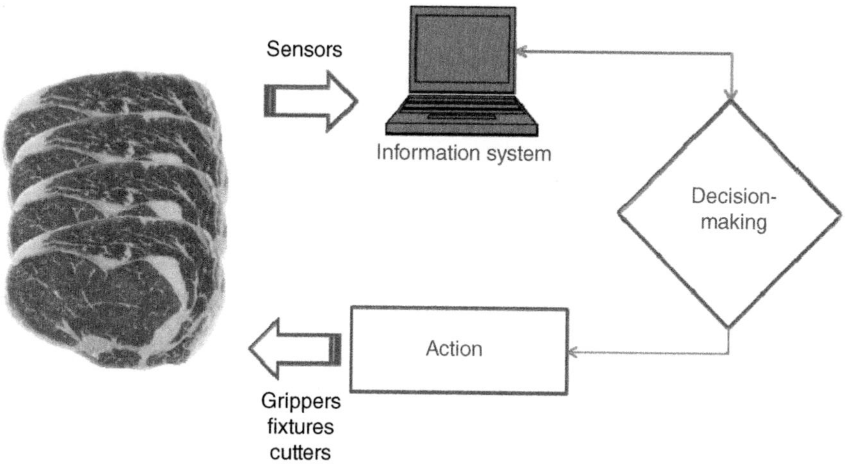

Figure 8.1 Schematics of robotic system in the meat industry (Adapted from Purnell, 1998).

started by replacing a few arduous processes with machines rather than changing the whole production line. However, new plants dedicated to meat processing have fully robotized processing lines from animal handling to packaging. This helps in improving production, hygienic processing, consistency and accuracy.

8.2.1 Advantages and disadvantages of robotics for meat processing

Tasks in food manufacturing plants often require the handling and manipulation of individual products, which can be challenging given the variations in meat properties. The mechanization or robotization of the meat sector is heterogeneous, depending on the animal species or breeds. The meat industry is one of the most hazardous and unpleasant industries in the food processing sector. The main reason for robotization is to reduce the difficulty and danger in working environments (Matthieu *et al.*, 2014). Using robots leads to lower production costs (Purnell, 1998). Automation can result in improved worker safety; it will also reduce repetitive strain injury. Performing the same motion work for prolonged periods in chilled environment causes fatigue and damage to joints and muscles of people. Automation can also result in improved food hygiene, whereas human operation can lead to product contamination. Robots can be designed to work in harsh environmental conditions maintaining hygienic conditions, which will increase the overall product quality (Tarrant, 1998). They are also made to withstand the effect of cleaning agents such as acids, cleaning enzymes and bleach.

There can be wide variations in size and shape of single product types; hence, robots can be used in the meat industry to improve precision and hygiene requirements. Especially as the food safety regulations are becoming more stringent. Robots have limited decision-making ability and cannot act accordingly. There is also a problem of limited motion with robots as they are built to suit a specific task. Humans can make a decision extremely rapidly, whereas robots have to be given a specific standard as criteria to accept/reject conditions. The components which are in direct contact with food are limited to be made out of specific material such as stainless steel or select plastics. Product surface can exhibit a very low coefficient of friction; therefore, handling tools design can be very challenging.

8.3 Mechatronic and robotic systems in the food industry

When automation started in the meat industry, there were many companies which developed carcass splitting equipment. The performance of this equipment was not satisfactory in terms of accuracy of splitting and hygiene problems associated with deposition of dust on the carcass surface (Madsen and Nielsen, 2002). Beef and lamb carcass grading systems used in the past were based on visual assessment of fat cover; however, it suffered from lack of objectivity (Tarrant, 1998). Semi-automated equipment for grading beef carcases were made using Vision Image Analysis (VIA) in Denmark, Germany, Australia and Canada. Since then, many new inventions have been made in terms of automation in the meat industry, but only a few have been used successfully in commercial environments. The application of robotics in various meat processing stages is outlined as follows.

8.3.1 Animal reception

Live animals are brought to the abattoir in trucks and unloaded. Here animals are rested for a day or two before slaughter. Animals should be handled considerately because stressed animals can affect the quality of final meat. Animals are cleaned properly before slaughtering to remove dirt from the skin. Abattoirs are specially designed to reduce the stress caused to animals as this maintains the meat quality. Proper bedding, water to drink and special lighting are a few of the special arrangements provided. In the case of poultry processing, the birds were picked up by hand and placed on moving conveyors, but now mechanical harvesters in which moving rubber fingers collect and encourage the birds to move to the conveyor (Warriss, 2000).

8.3.2 Stunning and bleeding

Stunning is the process through which animals are rendered unconscious before decapitation to effect bleeding. The stunning process must render the animal

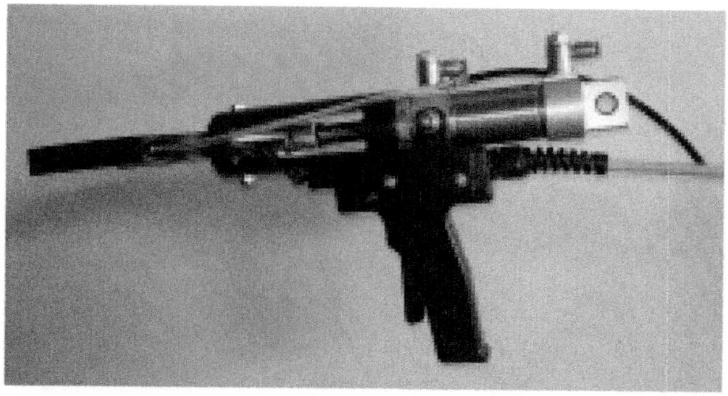

Figure 8.2 Manual stunning suitable for halal.

unconscious immediately as it can affect the quality of meat. Stunning can be done by various methods, which can be categorized as mechanical stunning using bolt pistol, percussion stunner (free bullet), electrical stunning by giving an electric shock (Figure 8.2) and chemical stunning by inducing anaesthetic gas such as carbon dioxide. As soon as the animal is rendered unconscious, the animal must be killed by exsanguination; this is done by cutting the blood vessels in the neck or the blood vessels entering the heart (sticking). In poultry, electrical stunning is used to render birds unconscious long enough to allow automated neck cutting and reduce carcass damage.

There are many systems to convey animals to the stun station, of which most commonly used is a V-shaped conveyor for pigs. Electrical stunning is carried out by human operators for animal welfare reasons due to complexity of accurate placing of the stun electrodes on a moving animal. However, alternate stunning methods in which carbon dioxide is used to render unconsciousness have been fully automated. A company called Automated Control Concepts, Inc. made automated CO_2 stunning equipment consisting of a chamber below the ground level. An elevator lowers the group of pigs into carbon dioxide atmosphere where stunning occurs. After appropriate dwell time, the pigs are brought back up. The elevator tilts a bit allowing pigs to slide onto a conveyor belt where they proceed for exsanguination (Figure 8.3).

8.3.3 Dressing, evisceration, cutting and boning

Dressing involves the separation of the head, feet, hide (in the case of sheep, goats and cattle), excess fat, viscera and offal (edible and inedible) from the bones and edible muscular tissue. It has been shown that image processing and 3-D profiling can used for sensing these parameters (Singh, Potgieter and Wei Liang, 2012).

Industrial Research Ltd, New Zealand, has marketed a robot called Envirobot used for critical Y-cuts in connection with dehiding and cutting of lamb carcasses

Figure 8.3 Automated pig stunning.

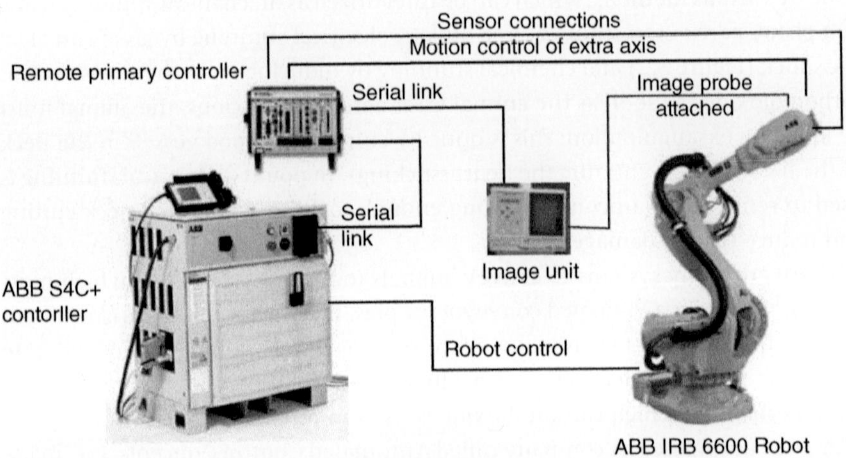

Figure 8.4 Robotic carcass splitting system (Adapted from Li *et al.*, 2003).

and for sawing through the breastbone on the cattle carcasses. Envirobot is made to withstand the harsh corrosive and chilled environment in the slaughterhouse.

Robotic carcass splitting system (Figure 8.4) consists of a robotic system and controller. The major work is done by ultrasound image processing unit along with remote primary control unit. Initially, the position of backbone is detected *via* analysis of images using ultrasound unit (Figure 8.5). The primary controller after processing data generates a traverse coordinate used to control horizontal positioning and rotary angle of the carcass splitting saw. While the robot arm is moving along the spine, the saw position and rotation about its axis are automatically corrected with ultrasound feedback data. The image acquisition, communication and other control processes need to be fast enough to match splitting operation requirements.

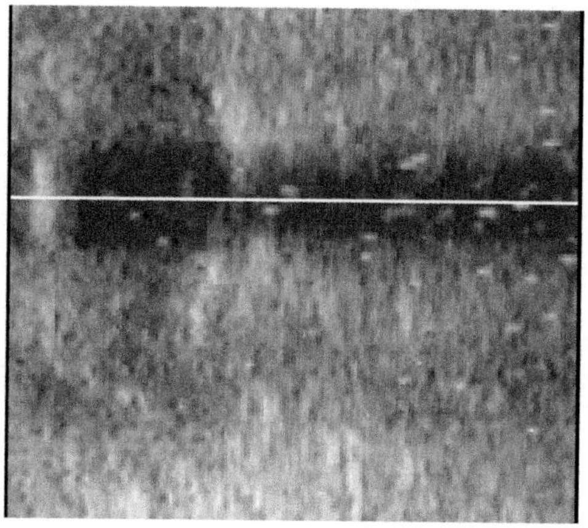

Figure 8.5 Spine position detected by ultrasound image (Adapted from Li *et al.*, 2003).

Robotics in online quality analysis and control in the food industry is a major step which should be taken care of at every processing stage. Non-destructive analysis of meat products can be done to identify the physical, mechanical, chemical properties as well as abnormalities or deformities in the product. It is a beneficial tool for online quality analysis of meat products. Thermal imaging uses the principle that every body emits infrared radiations. A thermal imaging system consists of a camera, an optical system, detector, signal processing and image processing system. The TI detectors are heated by incoming infrared radiation and temperature changes, which is measured by any temperature-dependent mechanism such as thermoelectric voltage, resistance or pyroelectric voltage. The microbolometer is an IR-sensitive detector which functions as a thin film transistor (TFT). The TI camera system converts radiation emitted from a sample in a specified waveband (e.g. 8–12 μm) into an electrical signal, which is then processed into an image. The thermal image may be coupled with computer software for enhanced interpretation. The criteria of band selection are based on various factors including the operating distance, indoor–outdoor operation, temperature and emissivity of the bodies of interest. The thermograph obtained can be further analyzed using a wide range of image processing techniques.

Beef is aged at low temperatures for a long period of time; during this period, sometimes microbial growth can occur. To overcome this problem, the aging process and microbial growth should be monitored. Various chemical sensors can be used for this purpose. Electronic nose is being used as a quick quality sensing tool in online processing. It has great potential for use in other applications such as sensory quality, shelf life, spoilage, off favours and taints. With a few more advancements in this field, the electronic nose could turn out to be the most beneficial tool in automation (Ghasemi-Varnamkhasti *et al.*, 2009).

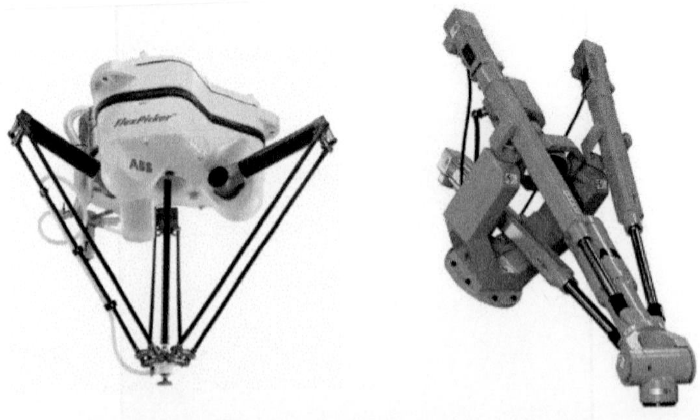

Figure 8.6 Robot used for picking in the meat industry (Adapted from Anderson, 2004).

In meat packing, online composition evaluation is an important step. Online fat/lean ratio is an important parameter in pricing, procurement and also helps in improving the product overall quality. Using electronic probes for fat measurement and muscle depth is an easy and widely used method worldwide. However, this method is very strenuous and not widely accepted; therefore, cost-effective and suitable automation in the grading systems are done. This system has a sophisticated sensing system to find probing location. It consists of various sensors to measure the physical parameters like force required to probe into the flesh, to perform insertion and retraction along a straight line. The overall processing consists of several operations that are synchronized to perform the operation with required specifications (Goldenberg and Lu, 1997).

8.3.4 Packaging

Beyond handling unwrapped products, robotic packaging systems have been successfully implemented in primary packaging, casing, stacking of cases into pallets and arranging the pallets into warehouse. Generally, packaging in the meat industry involves packing sausages and processed meats. After primary processing, placing processed products into individual containers is the first step in packaging which uses picking robotic hands (Totosaus et al., 2007) (Figure 8.6).

8.4 Case studies

8.4.1 Automation in lamb processing

There has been significant automation in processing of lamb, mostly because the lamb carcass is relatively flexible compared to others. Full automation has not been achieved yet, but combining human strength and automation at different levels has proved to be beneficial.

It was found that the first automated stunner for sheep and lamb evolved from pig stunning equipment, which was developed in Europe (Longdell, 1996). Today, the most prevalent method followed in EU for stunning is electrical stunning (Food Chain Evaluation Consortium (FCEC), 2014).

The most notable automation processes developed are pelt removal and cuts assisting evisceration. This pelt removal technique was developed in Ireland and New Zealand in the 1990s; four main parts in the process are automated as follows.

8.4.1.1 Y-cutting

It is a primary pelt removing task. The cut as shown in Figure 8.7(a), marked by bold line, splits the pelt and separates it from flesh underneath. This is done by human operators where they perform the delicate task of sensing and delicate positioning operations. The cutting tool developed uses an air-powered reciprocating knife to both cut and clear the pelt.

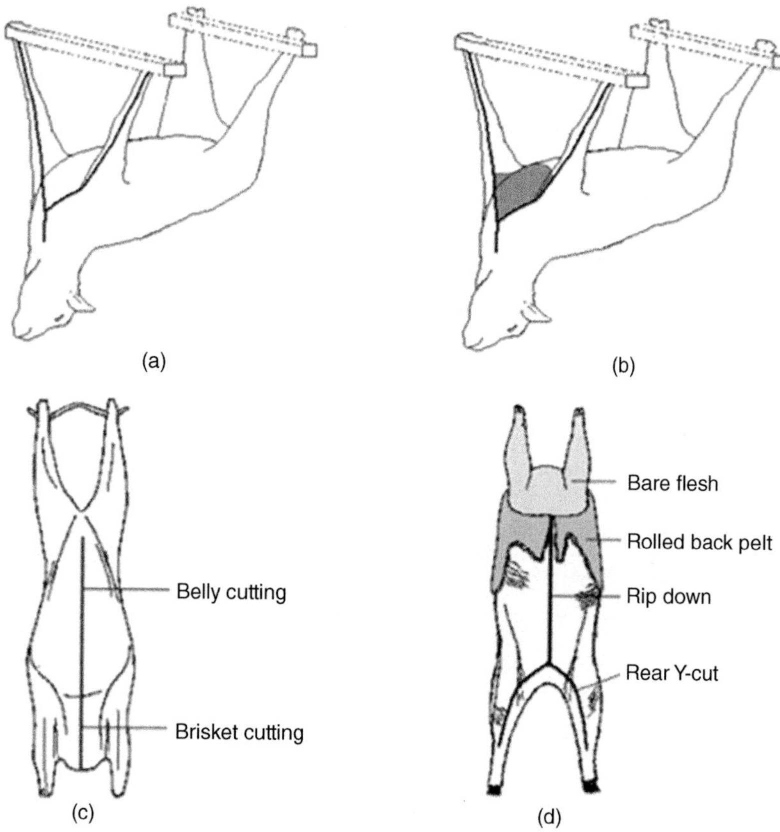

Figure 8.7 (a) Y-cut – marked in bold black line (b) Brisket cut – flap lifted and shifted back. (c) Belly cut (d) Carcass opening cut – rip down and rear Y-cut followed by complete pelt removal.

8.4.1.2 Brisket cleaning

It is pelt removal task which occurs after Y-cutting. The flap is lifted and pushed back as shown in Figure 8.7(b). Full automation of this process would require a very complex sensing system. There should be precision in cutting. Semi-automated with a cooperative robot or 'cobot' is used for providing the brutal force and human operator doing complex sensing task. It is pneumatically activated where a human operator controls it with a joystick and performs the task of guiding.

8.4.1.3 Rip down and rear Y-cut

The rip down and rear Y-cut is for further pelt removal (Figure 8.7(d)). A deep cut is made on the belly to help in removing the pelt.

8.4.1.4 Carcass opening cut

Carcass opening is the first step in the evisceration process where the complete pelt is removed and the carcass hangs by its rear legs. The cut involves slitting the belly wall (Figure 8.7(c)) and then shearing through the brisket (Templer, Nicholls and Nicolle, 1999).

8.4.1.5 Shoulder fleecing machine

The process of removal of shoulder meat is called shoulder fleecing. It also includes removing the scapula and foreleg bones, from the neck and trunk vertebrae and the rib bones of a lamb carcass. There is a machine (Figure 8.8) designed by MIRINZ to do this, which does this task in four steps. Initially,

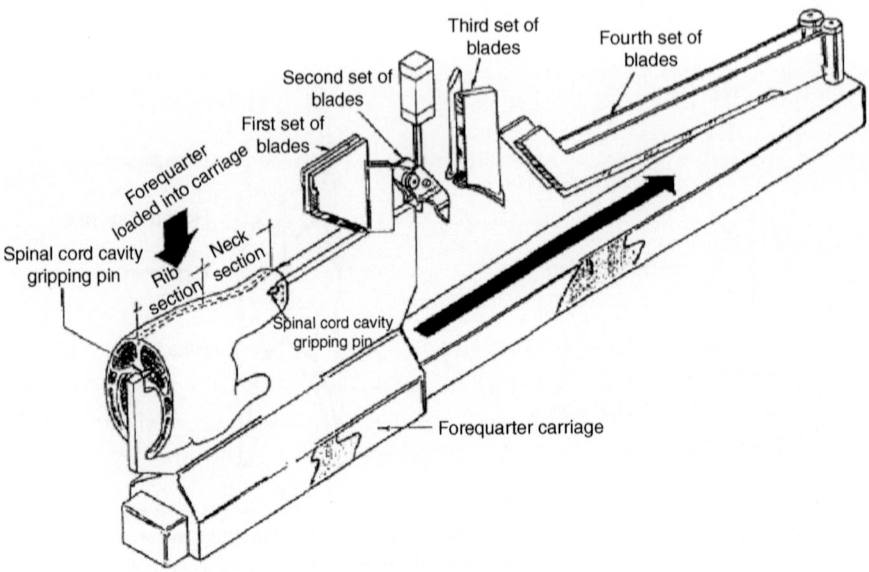

Figure 8.8 Layout of MIRINZ shoulder fleecing machine (Adapted from Longdell, 1996).

a knife cuts the entire shoulder barrel, one on each side of the spine. In the next step, the meat from the neck bone is removed with the help of a pair of specially designed contour knives. Following this, a third pair of knives removes the muscle from thoracic vertebrae and portion of ribs close to vertebrae. In the final step, a pair of knives, specially designed and curved, is used to separate the meat from vertical portion of the rib bones (Longdell, 1996).

Robotic Technologies Ltd (RTL) is a joint venture company, 50% owned by both Silver Fern Farms and Scott Technology Ltd. They developed a fully automated lamb boning room and commercialized it in 2011. Initially, X-ray scans, analyzes and provides accurate data for the cut point in the primal cutting system. The X-ray data is fed to Saddle and hindquarter machines of the boning room vision to deliver cut position accuracy. The components of this system include primal cutting, forequarter processing, middle processing and hindquarter processing (Figures 8.9, 8.10). Since 2003, MLA (Meat and Livestock Australia) has been supporting RTL for this project (Starling, 2011).

8.4.2 Automation in pig processing

Pig carcasses are the most uniform large carcass and pork has a high demand worldwide, which has rendered this industry being more automated than any other industry. Also pig slaughtering, cutting and boning are hard and labour-intensive work. Figure 8.11 shows the various processes involved in the processing of pigs. The pre-slaughter handling of pigs is important for ensuring good meat quality.

For pigs, stunning can be carried out by electric shock or anaesthesia with carbon dioxide (60–70%). Stunning with a bolt pistol is not done because of the problem with skull penetration. Sticking is done by cutting the cervical vein and one of the arteries (Barbut, 2001).

Scalding is an important process to obtain good results in dehairing. It is done by raising the temperature of the skin to loosen the root hair so that it comes out easily. Scalding can be done by three main methods:

- Immersion in hot water
- Sprinkling hot water
- Steam scalding.

Temperature (60–65 °C) is an important aspect in this method, as a too high temperature can lead to bad meat quality, whereas a low temperature will result in problems in dehairing. Scalding using steam is the most preferred method as it is more hygienic. The problems associated with contamination with water are also avoided Figure 8.12.

8.4.2.1 Dehairing

This automated dehairing machine by Meat Processing Systems, the Netherlands, is ideal for dehairing around 250 pigs per hour. It consists of a stable, special

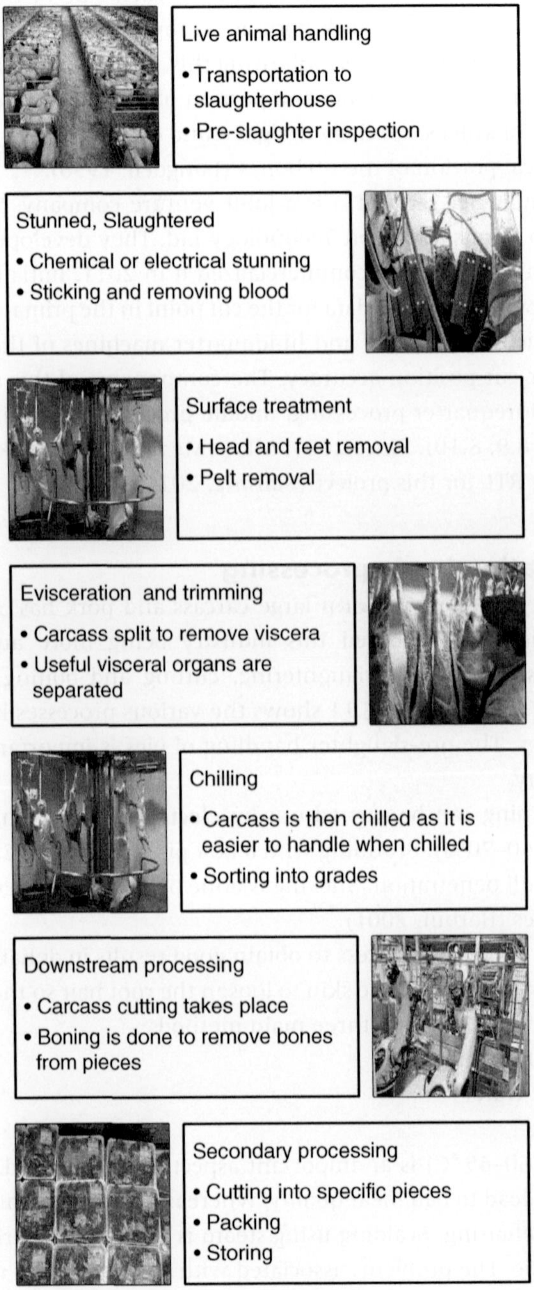

Figure 8.9 Process and sub-process in lamb slaughtering.

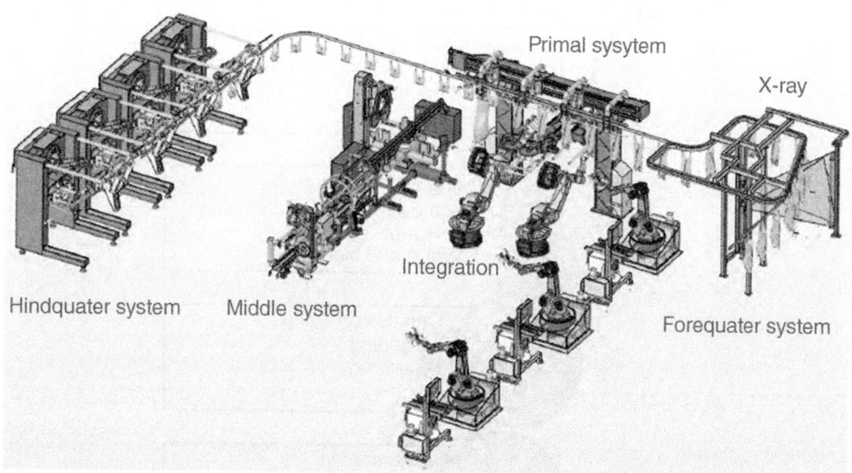

Figure 8.10 Fully automated lamb boning room – RTL Vision (Adapted from Starling, 2011).

cast iron frame and U bars and heavy scraper rollers. Continuous flow of water ensures automatic removal of hair and nails. A perforated mesh is provided at the bottom of the machine to eliminate hair and nails (Figures 8.13 and 8.14).

8.4.2.2 Cutting

An automated line was developed by the Danish Meat Research Institute (DMRI) and consisted of eight automatic machines. The Danish pig slaughter programme started in 1998 and the complete slaughter line was made available by 2006. In the beginning of the automated line, a series of machines will cut the throat, loosen the fat end, divide the hind legs, open the carcass and perform evisceration. The evisceration equipment developed by DMRI cuts the diaphragm, tears off flare fat and loosens the intestinal tract such that the entire intestines and organs are removed from the carcass prior to separation. As the separation is conducted outside the carcass, the chances of contamination are less and product is safer (Nielsen, Fertin and Christensen, 2005). A measuring station is present so that robots can adapt themselves to the variations in carcass size. An automated carcass cutting machine was also developed by DMRI. This machine helps in cutting the half carcass into fore-end, middle and hind leg (Hinrichsen, 2009).

A company (ATTEC) based in Denmark has made automatic and semi-automatic lines for primal pork cutting. In the semi-automatic version, all cuts are performed by circular knives, resulting in clean cuts and no bone fragments from sawing, optimum yield without product loss. In fully automatic cutting lines for pork, the carcass is cut into four parts: hind feet, leg, middle and shoulder. The automatic measuring system measures each half carcass ensuring

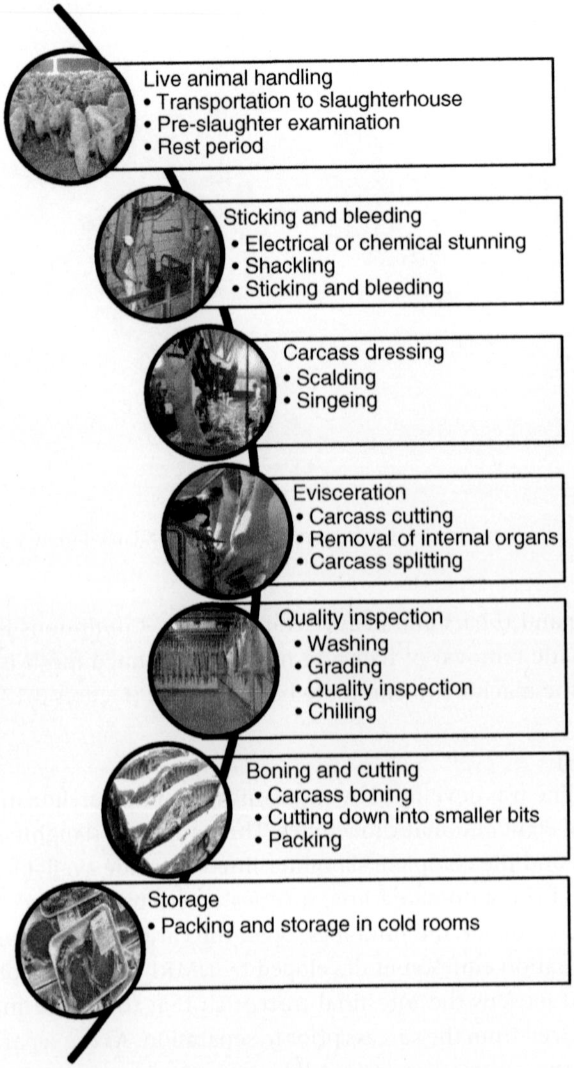

Figure 8.11 Process and sub-processes involved in pig slaughtering.

optimum registration, maximum yield and labour saving. This processing line is based on servo electric and mechanical principle (http://www.attec.dk/).

ATTEC together with DMRI has developed equipment for cutting the rib tops and dividing the loin and belly automatically. The conveyor is equipped with a grasping device, which pulls the rib tops out to their full length. This straightens out the backbone, simplifying the subsequent sawing off of the rib tops. After having been fixed in the conveyor, the middle passes through measuring stations, which measures the length and any inaccuracy in the earlier carcass splitting (Figures 8.15 and 8.16). Based on these measurements, the saws are adjusted to

Figure 8.12 Pig scalding using steam (Adapted from http://www.mps-group.nl/en).

Figure 8.13 Dehairing machine.

cut off the belly from the loin and subsequently cut off the loin from the rib tops (Madsen and Nielsen, 2002).

8.4.2.3 Boning

The machine for removing surface bones – neck bone and riblet – was developed in cooperation with the Dutch division of the American company of Townsend Engineering. After this operation, two to three operators are required to carry out final trimming of the boned fore-ends. For removing the interior bones – shank bones, humerus and shoulder bone – another machine is used (Figure 8.17).

Figure 8.14 Interior of dehairing machine (Adapted from http://www.mps-group.nl/en).

Figure 8.15 Fully automated cutting line.

8.5 Future role for robotics in the processing of meat and meat products

When thinking about ways that industrial robot arms are used, the meat industry may not spring to mind. Most likely, people first think of the automobile or electronics manufacturing industry. Automation can be found right along the production line in the meat processing factory – from slaughter through to the final packaging and despatch. The meat industry is currently using industrial

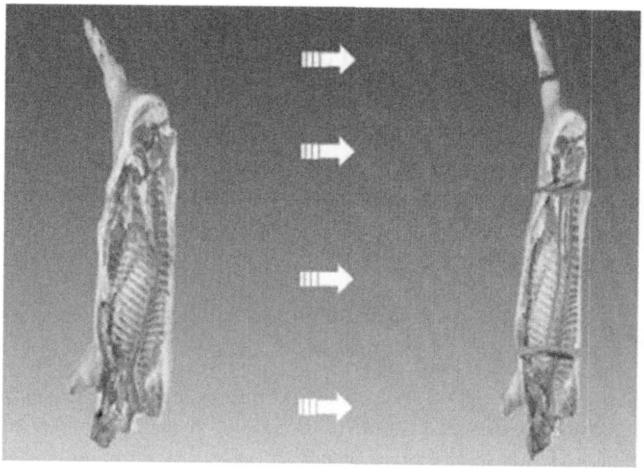

Figure 8.16 Cuts made on half pig carcass.

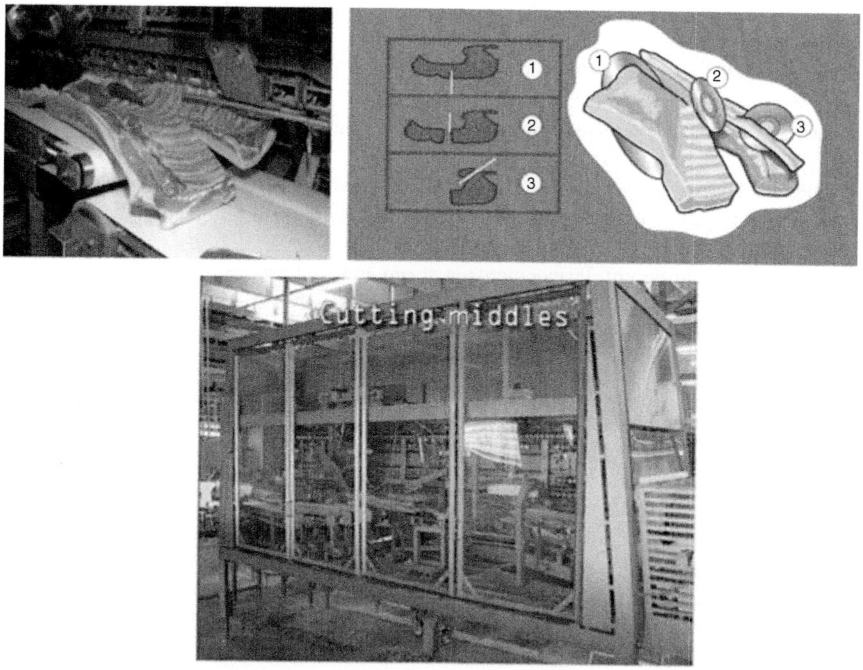

Figure 8.17 Middle cutting machine cuts off rib tops and separates the loin and belly automatically.

robotic arms, and is planning to integrate even more automation in the near future. The plan for the industry is to make these kinds of robotic industrial arms more available to smaller markets and make them more universal for meat processing. While this process is highly automated already, the meat industry would like to further automate their system, and is currently devising leaner, more efficient ways to break down and debone various animal carcasses.

Robotics holds the promise of reducing the processing cost by helping speed up processing lines, making production more efficient and reducing labour requirement. The time is not far away when all the processes in meat processing, starting from the primary processing, secondary processing, packing and despatch, will be fully automated.

References

Anderson, E.W. (2004) Design of a low cost, high speed robot for poultry processing. School of Mechanical Engineering, Georgia Institute of Technology 2005, Thesis dissertation.

Barbut, S. (2001) *Poultry Products Processing: An Industry Guide*, CRC Press. ISBN 9781587160608.

Food Chain Evaluation Consortium (FCEC) (2014) *Civic Consulting*, Agra CEAS Consulting. http://ec.europa.eu/food/animal/welfare/index_en.htm (accessed 02 February 2016).

Ghasemi-Varnamkhasti, M., Mohtasebi, S.S., Siadat, M. and Balasubramanian, S. (2009) Meat quality assessment by electronic nose (Machine Olfaction Technology). *Sensors*, **9** (8), 6058–6083.

Goldenberg, A.A. and Lu, Z. (1997) Automation of meat pork grading process. *Computers and Electronics in Agriculture*, **16** (2), 125–135.

Hinrichsen, L. (2009) Manufacturing technology in the Danish pig slaughter industry. *Meat Science*, **84** (2), 271–275.

Li Z., Ring P., MacRae K. and Hinsch A (2003) *Control of Industrial Robots for Meat Processing Applications*. Proceedings of the Australasian Conference on Robotics and Automation. ABB Flexible Automation.

Longdell, G.R. (1996) Recent developments in sheep and beef processing in Australasia. *Meat Science*, **43**, 165–174.

Madsen, K.B. and Nielsen, J.U. (2002) Automated meat processing, in *Meat Processing: Improving Quality* (eds. J.P. Kerry, J.F. Kerry and D.A. Ledward), Elsevier. ISBN 9781855735835.

Matthieu, A., Franck, S., Laurent, S., Kévin, S., Grigore, G. and Youcef, M. (2014) *Robotic Solutions for Meat Cutting and Handling*. European Workshop on Deformable Object Manipulation, March 20, 2014, Lyon, France.

Nielsen, J., Fertin, C. and Christensen, H. (2005) Up-to-date equipment for pig slaughtering, cutting and boning and their influence on product safety. *Tehnologija Mesa*, **46** (1/2), 62.

Purnell, G. (1998) Robotic equipment in the meat industry. *Meat Science*, **49**, S297–S307.

Prarthana, B.K., Bhavana, Y. and Mani Shankar, N. (2013) A venture to the latest robotic technological research: A review. *Advances in Robotics & Automation*, **2** (103), 2.

Singh, J., Potgieter, J. and Wei Liang, X. (2012) Ovine automation: Robotic brisket cutting. *Industrial Robot: An International Journal*, **39** (2), 191–196.

Starling S (2011) Robotic Technologies Limited. Meat and Livestock Australia Limited ABN 39 081 678 364 (MLA).

Tarrant, P.V. (1998) Some recent advances and future priorities in research for the meat industry. *Meat Science*, **49**, S1–S16.

Templer, R.G., Nicholls, H.R. and Nicolle, T. (1999) Robotics for meat processing – from research to commercialisation. *Industrial Robot: An International Journal*, **26** (4), 290–296.

Totosaus, Alfonso, V. Kuri, L. Nollet, T. Boylston, F. Chen, P. C. Coggins, M. B. and Gloria *et al.* (2007). Packaging of fresh and frozen poultry. Handbook of Meat, Poultry and Seafood Quality, 475–486.

Wallin, P.J. (1997) Robotics in the food industry: An update. *Trends in Food Science & Technology*, **8** (6), 193–198.

Warriss, P.D. (2000) *Meat Science: An Introductory Text*, CABI Publishing, Wallingford, UK.

PART II
Novel Packaging and meat functionality

CHAPTER 9

Packaging systems and materials used for meat products with particular emphasis on the use of oxygen scavenging systems

Malco Cruz-Romero & Joseph P. Kerry
Food Packaging Group, School of Food and Nutritional Sciences, University College Cork, Cork, Ireland

9.1 Introduction

Modern consumers have expectations and demands concerning retailed meat products that far exceed those which were required by shoppers some 20–30 years ago. Issues such as product freshness, product presentation, quality attributes, sensory properties, convenience, nutritional status, safety issues and shelf-life stability must be addressed in presenting meat products to the consumer as not only is each factor important in its own right but also due to the interplaying relationship that exists between these factors.

The extremely demanding and competitive meat market in developed countries is continually looking for novel, sophisticated and stable products which retain to the greatest extent possible the eating quality of the fresh product, as well as safer and better tasting processed and convenience-style muscle-based foods. The role that packaging plays in addressing all of this, and much more besides, cannot be underestimated.

The fundamental aspects of all packaging materials and systems is that in an economic manner, they must contain, protect, preserve, inform (throughout the entire distribution process from point of manufacture to points of consumer usage) and provide convenience (at many different levels) while acknowledging the constraints placed upon their usage from both legal and environmental perspectives. As these fundamental principles apply to all forms of packaging materials and systems, it follows that irrespective of the specific level at which the packaging is industrially applied (primary-sales packaging, secondary-collation and handling packaging or tertiary-transport packaging), all must conform to

Emerging Technologies in Meat Processing: Production, Processing and Technology, First Edition.
Edited by Enda J. Cummins and James G. Lyng.

these same principles (Cruz-Romero and Kerry, 2008). Packaging materials used for packaging raw and cooked meat products must conform to the fundamental principles outlined above and satisfy all that is required of the product from both technical (containment, protection and preservation) and sales (cost, convenience, sales information, labelling information, legal requirements and environmental requirements) perspectives (Cruz-Romero and Kerry, 2011).

The food packaging area is one in which availability and demand is continuously subject to changes such as modifications in consumer habits, distribution developments, new technologies, new materials, increasing international distribution and, last but not least, increasing environmental constraints (Sonneveld, 2000). Packaging is a global exercise and the application of optimal packaging is the most important approach towards maintaining the well-being and safety of consumers, and in the promotion of successful commerce (WPO, 2011).

In today's global market, it has become imperative for companies to explore ways to improve their productivity in terms of maintaining safety, using advanced packaging materials and solutions, implementing flexible and standardized technology, adopting proven management principles around good packaging practice and through the reduction of product and resource wastage. FAO (2011, 2013) estimated that global food loss and waste is approximately 1.3 billion tonnes annually. This wastage has not only an enormous negative impact on the global economy and food availability but also major negative environmental impacts. The direct economic cost of food wastage derived from agricultural products (excluding fish and seafood), based on producer prices only, is about US$750 billion, equivalent to the GDP of Switzerland (FAO, 2013). In Europe, households are also becoming smaller. Packages of food for five or six people are no longer appropriate and excess product offered is likely to result in organic waste. Portioned plastic packaging can eliminate this waste and can be emptied more easily. Plastic closures or zipper systems also keep food fresher for longer. Modified atmosphere packaging (MAP) can increase the shelf life of meat products, allowing distribution of fresher products even in areas far removed from the site of product manufacture. The extension in product shelf life and the improved containment of the meat product reduces food waste, a parameter that contributes significantly to sustainability issues pertaining to the packaged meat product.

It was estimated that the fresh and processed meat market in Europe is worth 271 billion Euros (FFT (Food For Thought), 2013). It had been reported that the US demand for meat, poultry and seafood packaging would increase 3.0% per annum to US$9.2 billion in 2015 (Freedonia, 2011). One of the reasons for increased packaging demand is the ongoing shift to case-ready packaging by many retailers as a means of reducing in-store labour costs. Case-ready packaging generally uses more expensive trays and high barrier films, thus bolstering packaging sales and value gains. Packaging demand will also be aided by the increasing use of value-added features (e.g. in-pack valve systems, zipper closures and consumer-friendly packaging which possess ancillary items such as spoons, forks, knives, condiment sachets and active systems); the expanding number of meat and poultry items presented in smaller, more convenient sizes and the growing range of products that are further processed. Many such products, which are preferred by consumers for their easy preparation and storage capabilities, tend to use more packaging relative to their volume than larger unprocessed items (Freedonia, 2011).

9.2 Case-ready packaging

Case-ready meat packs is defined as any meat that is packaged in a meat processing plant and ready to be placed directly into display cases upon arrival at a retail store (AMI (American Meat Institute), 2009). Case-ready packaging is an efficient, economical and safer centralized packaging (Brody, 2007). Unlike meat cut at the retail store – which carries only a retail brand – case-ready fresh meats carry a wide array of brands, which encourages companies to produce meat products to the highest possible safety and quality standards. Now, fresh products are readily identifiable and consumers can purchase the brands they like repeatedly (AMI (American Meat Institute), 2009). Case-ready meat products can be centrally processed within HACCP-certified facilities, packaged and shipped long distances to the respective retailers with anticipated shelf lives of 16–21 days, thereby providing sufficient time for product to move through the supply chain and into consumers' homes well before the expiration date (Young, 2009). Therefore, case-ready meat products have the potential of transforming a meat product line into a valued-added product line that can result in increased profit margins (Young, 2009).

Case-ready meat programs utilize new and improved production techniques, processing and packaging technologies and innovative retail marketing strategies in ways that have transformed how retailers order, stock and merchandize meat products. Consequently, the availability of case-ready meat has dramatically altered how consumers now purchase and prepare meat products in their homes. New marketing strategies involve segmentation of the market, thus providing consumers with meat product choices that better reflect their preferences and needs. Moreover, case-ready meat products are simpler to prepare at home. Examples of commercially available case-ready raw and cooked meat products are shown in Figure 9.1.

9.2.1 Benefits of case-ready packaging system

Case-ready meats offer numerous benefits to retailers and consumers (AMI (American Meat Institute), 2009). Consumers benefit from case-ready meat products in two important ways: they no longer have to prepare meat products

Figure 9.1 Examples of commercially available case-ready raw and cooked meat products. (a) Modified atmosphere packaged and (b) skin or vacuum packaged meat products.

prior to cooking and consumers can also choose the types of meat they most prefer, and these packages can either be prepared or placed in the freezer for later use, with no apparent decline in quality. Retailers also benefit from case-ready meat products as prices received for these value-added products increases profit. In addition, food safety is enhanced as a result of eliminating cross-contamination of meat from various sources The centralization of meat packing brings with it benefits in terms of economies of scale, less wastage, less weight loss, better and more consistent standards in quality control, reduced loss at the point of sale and enhanced product presentation at the point of sale. Shelf-life extension of case-ready meat products can be achieved through the use of various forms of vacuum packaging or MAP formats.

The retail benefits of case-ready packaging systems (AMI (American Meat Institute), 2009) are as follows:

- Retail stores can order the meat cuts that meet market demands, thereby eliminating the requirement for carcass handling and butchering operations in-store.
- Case-ready products arrive at store already prepared and do not require additional handling, cutting or processing.
- Because these packages are produced and packaged at the processing plant, they have longer shelf-lives than meat products packaged within the retail store, which generally lasts about 1–5 days after it is opened, cut and repackaged.
- Case-ready products allow retail stores to manage inventory more effectively and reduce 'out of stocks', which is frustrating to consumers when they have a demand for that product.
- Case-ready products allow retail stores to control unit costs, provide consistent products every day and ensure that key meat prices are competitive with other foods.

For consumers, case-ready meats also offer distinct benefits (AMI (American Meat Institute), 2009) which are as follows:

- Many case-ready meats are packaged in tamper-proof packages, which can offer consumers an additional measure of confidence.
- Case-ready packaging should be leak-proof and easily handled when sourced from either chilling or freezing compartments in-store.
- Less meat manipulation is required for case-ready meats as full trimming and meat preparation has taken place, thereby providing convenience and eliminating waste.
- Because these products are often branded, consumers can easily identify products for which they have developed a loyalty based upon individual tastes and features, which make purchasing decisions easier.
- Case-ready processing and packaging allows for precise control of product aging, which can enhance the consumer eating experience.
- Low-oxygen (O_2) packaging of case-ready allows for the elimination of O_2, which causes off-flavours, off-colours and off-odours.

9.2.2 Packaging requirements for case-ready meat products

Product quality and shelf life of meat products are influenced by the product's physical, chemical and biological characteristics; the processing conditions used, packaging characteristics chosen and their effectiveness on application and the environment to which the meat product is exposed during distribution and storage (Brown and Williams, 2003).

Packaging companies have responded to the growing demand for fresh, pre-prepared/-packed and ready-to-eat foods through the development of lightweight, high-barrier films that are manufactured for gas-flush processing, thereby helping to reduce the growth of bacteria and mould, while protecting physicochemical characteristics such as product composition, nutritive quality and colour.

The key to successful packaging of meat products is to select the package material and design that best satisfy competing needs with regard to product characteristics, marketing considerations (including distribution and consumer needs), environmental and waste management issues and cost (Kirwan and Strawbridge, 2003). Balancing so many factors is difficult and also requires product-specific analysis.

Factors to be considered when choosing a packaging system for case-ready meat products include properties of the packaging material, physical form and physicochemical characteristics of the meat products intended for packaging, possible meat product/package interactions, the intended market for the product and the desired product shelf life. Other factors to take into consideration include environmental conditions during storage and distribution, product end-use, eventual package disposal and costs related to the package throughout the production and distribution process (Marsh and Bugusu, 2007). Therefore, when selecting a packaging material required for use in packaging meat products, various requirements need to be met (Table 9.1 and Table 9.2).

Packaging materials used for packaging meat products are usually plastic films, more specifically, laminates, in which film polymer layers with good O_2-barrier properties like polyamide or polyethylene terephthalate are adhered to a polymer

Table 9.1 Properties of packaging materials required for use in packaging meat products.

- Fully contains meat products, thereby preventing all possible contamination (microbial, chemical and particulate)
- Fit for use and fit for purpose, thereby satisfying legal contact requirements
- Protects products from physical damage
- Appropriate barrier properties to gases, water vapour, volatiles and light
- Possesses antifogging properties to enhance in-pack product appearance
- Tamper evident, yet easy open
- Offers convenience throughout the lifetime of the product; from the point of product manufacture through to final product usage and package disposal
- Conforms to all legislative requirements
- Enhances sales/consumer appeal through the provision of appropriate information
- Robust to deal with all usage and storage conditions
- Address environmental concerns in packaging design and usage
- Be an efficient process, thereby making the packaging process more cost-effective

Table 9.2 Primary function of various basic materials used in packaging meat products.

Basic materials	Abbreviation	Primary function
Aluminium	Al	High barrier (moisture and gas)
Amorphous polyester	APET	Rigidity, high barrier
Crystallized polyethylene terephthalate	CPET	Rigidity, high temperature resistance, gas barrier
Ethylene-vinyl acetate	EVA	Sealing layers
Ethylene-vinyl alcohol	EVOH	Gas barrier
High-density polyethylene	HDPE	Moisture barrier, rigidity, microwave capability, sealing layers
Low-density polyethylene	LDPE	Sealing layers
Oriented polyamide	OPA	Gas barrier
Oriented polyethylene terephthalate	OPET	High temperature resistance, flexibility, puncture resistance
Oriented polypropylene	OPP	Moisture barrier, flexibility, puncture resistance
Polyamide (nylon)	PA	High temperature resistance, flexibility , toughness, some gas barrier
Acrylonitrile	PAN	Rigidity, some gas barrier
Polyethylene terephthalate (polyester)	PET	Moisture barrier, rigidity, microwave capability
Polypropylene	PP	Rigidity
Polystyrene	PS	Rigidity, gas barrier
Polyvinyl chloride	PVC	Moisture barrier, gas barrier
Polyvinylidene chloride	PVdC	

Adapted from Linde (2013).

layer possessing good humidity barrier performance and sealing properties such as polyethylene or polypropylene. Oxidative rancidity is one of the main factors limiting the quality and acceptability of meat products, affecting the sensory quality relating to flavour and odour (Argyri, Panagou and Nychas, 2012). Packaging materials with good O_2-barrier properties are used for low O_2 VP and MAP applications (Argyri, Panagou and Nychas, 2012). Typically, the higher the barrier properties of the packaging material used by the meat processor, the greater the cost of the packaging material. Gas flushing, another way to reduce headspace O_2, is, similarly, an added expense to packaging system used. While satisfactory shelf life can be obtained by these methods, it is desirable in many cases to increase the shelf life and/or avoid the costs of using the highest available barrier film by using a commonly available film with moderately high barrier properties, and this can be obtained using O_2 scavengers (see Section 9.).

Packaging formats for meat food products vary significantly and are linked with the method of production, storage and consumer preference (Tucker, 2006). Packaging products need to offer not only sufficient protection for food products but also flexibility and ease of handling in order to reduce unnecessary waste (Eliasson, 2008).

Much of the quality of meat and processed meat products is defined by the packaging of the end product, including the atmosphere surrounding the meat. Ideally, packaging should delay spoilage by inhibiting bacterial activity, allow maturation to continue by not inhibiting enzymatic activity, thereby improving tenderness (Anonymous, 2012).

9.2.3 Product characteristics associated with the quality of case-ready MAP meat products

The microbiology of case-ready meat is a significant issue, as it has a bearing on both the safety and eating quality of the meat in question; however, it can also impact negatively meat colour (Thoden van Velzen and Linnemann, 2008). Colour is usually independent of quality, texture, eating characteristics, spoilage condition, or temperature/time history, but it looms large as one of the most important factors to be considered in red meat packaging, especially beef, since consumers have been using colour to judge beef quality for many decades (Brody, 2007).

Consumers' perceptions and evaluations of meat quality attributes such as colour and shelf life influence purchasing decisions, and these product attributes can be affected by the packaging system used (Grebitus *et al.*, 2013). MAP extends the shelf life of meat products provided that the starting quality of the meat is of premium quality and that the cold chain is maintained during the lifetime of the product. The appearance of the fresh retail meat is a major determinant of its appeal to consumers and, consequently, sales of the meat product. Oxidative processes in meat lead to the degradation of lipids and proteins, which, in turn, contribute to the deterioration of flavour, texture and colour of displayed fresh retail meat (Decker *et al.*, 1995). These biochemical changes in meat limit the shelf life of fresh meat products.

The muscle pigment myoglobin, responsible for fresh beef colour, can exist in one of three chemical forms, namely, deoxymyoglobin, oxymyoglobin and metmyoglobin (Kropf, 1993). Oxymyoglobin, the cherry-red form of the pigment is associated with freshness. Over time, oxymyoglobin is oxidized to metmyoglobin, which is brown and undesirable from a consumer perspective. When exposed to O_2 in atmospheric air, the purple muscle pigment myoglobin becomes loosely oxygenated and converts into oxymyoglobin. Elevated O_2 levels in the meat package (typically in the commercial range of 70–80% O_2) converts myoglobin into oxymyoglobin and as long as the high O_2 content is maintained under the created in-pack partial pressure and refrigerated temperatures, the oxymyoglobin remains relatively stable and the desirable bright red colour is maintained for a limited but extended period.

Following MAP and during subsequent storage, several quality aspects must be considered simultaneously in order to obtain an optimal product quality and shelf life. Colour stability is also one of the important characteristics associated with the quality of MAP cured, cooked ham, since it is the primary quality cue sought by the consumer, who may use it as a freshness indicator and a signal to product wholesomeness (Nannerup *et al.*, 2004). Larsen *et al.* (2006) stated that the colour of meat products such as cooked ham is of major importance for consumer acceptance. Therefore, the colour stability of packaged cured meat products during storage and retail display is commercially important (Møller *et al.*, 2003; Solomon, 2004). Discolouration problems can be encountered when the packaged products contain small amounts of O_2 and are then exposed to inappropriate ultraviolet (UV) light levels associated with display cabinet illumination, as well as normal retail lighting, leading to photo-oxidation reactions. Changing from vacuum to MAP has introduced headspace volume and residual O_2 in the package as new factors influence the colour of cooked meat products. The oxygen permeability of the packaging material has a significant influence on the colour stability of processed meat products (Lee, 2010).

Nannerup *et al.* (2004) studied the relative importance of the packaging conditions (% residual-O_2, product headspace (P-H) volume ratio, temperature and light intensity) on the colour stability of cured, cooked ham. These authors showed that the parameters studied interacted in a complex manner to influence product colour with factors such as % residual-O_2 and P-H volume ratio – that is shown to be the most critical parameters. The use of packaging films with O_2 permeabilities of less than $100\,cm^3/m^2/24\,h$ at standard temperature and pressure (STP) is generally used for vacuum packaging or MAP applications by the meat industry.

While low O_2 MAP and vacuum packaging technologies are used for O_2-sensitive foods such as processed meat products, residual O_2 is typically present in most meat packs after sealing. Oxygen scavengers are secondary packaging devices that can help absorb this residual O_2, thereby minimizing the negative impacts of O_2 on product quality (Vermeiren *et al.*, 2003).

9.2.4 Gases used in MAP of meat products

The gas atmosphere surrounding a particular food product can be altered to minimize biotic and abiotic spoilage processes that are detrimental to product quality. The primary gases predominantly used in MAP are nitrogen (N_2), O_2 and carbon dioxide (CO_2), but other secondary gases can also be usually included in MAP systems. In Europe, for instance, secondary gases such as argon, nitrous oxide and helium are permitted for MAP packaging, while debate is still ongoing regarding the use of carbon monoxide and hydrogen in small quantities (Lee, Yam and Piergiovanni, 2008). Oxygen and CO_2 are of most significant in MAP meat products and the relative proportions of each of them affect changes in meat colour and microbiological quality. Primary gases can be used either alone or in mixtures. Table 9.3 shows the properties of the main gases used in MAP of meat products. The combination of gases used for MAP application depends

Table 9.3 Properties of the main gases used in MAP packaging of meat products.

Gas	Properties	Use in MAP meat products
Oxygen	Moderately soluble in water and fat, moderately permeable through plastic film, supports growth of aerobic spoilage microorganisms; oxidizes fat and other sensitive components; leads to oxidized forms of myoglobin (oxymyoglobin of bright red colour and metmyoglobin of brown colour; inhibits strict anaerobic bacteria	Usually excluded from package atmosphere to improve the quality preservation of cooked meat products; avoids anaerobic condition, blooms the raw meat colour
Carbon dioxide	Highly soluble in water and fat, most permeable through plastic film; suppresses aerobic spoilage microorganism, may cause acid taste at higher concentration	Delays microbial deterioration mostly caused by Gram-negative bacteria and moulds; produces tight contact contour package due to dissolution of CO_2 on the product
Nitrogen	Inert, low solubility in water and fat, least permeable through plastic film	Displaces oxygen so as to prevent oxidation and aerobic microbe growth; inert filler gas to prevent package collapse and to dilute oxygen or CO_2 properly
Argon	Inert, more soluble in water and fat than nitrogen, higher density than nitrogen; it competes with oxygen due to similar size and properties	Displaces oxygen so as to prevent oxidation and aerobic microbial growth

Adapted from Lee, Yam and Piergiovanni (2008).

on many factors such as the type of product being packed, packaging materials used, storage temperature, fat and moisture content, bacterial population and colour requirements (red meat). The packaging system selected must have sufficient headspace to provide enough gas to interact with the entire product. The headspace must contain a reservoir of CO_2 to compensate for the gas absorbed by the product and for that lost across the packaging material (Parry, 1993). The longer the required shelf life, the larger the headspace required. The gas mixtures and gas volume recommended for typical raw and cooked meat products are listed in Tables 9.4 and 9.5.

Carbon dioxide is highly soluble in water and fat, and the solubility increases greatly with decreasing temperature (Sivertsvik, Rosnes and Bergslien, 2002). The effectiveness of the gas is always conditioned by the storage temperature, resulting in increased inhibition of bacterial growth, as storage temperature is decreased (Gill and Tan, 1980; Molin, 2000). The selective bactericidal effect of CO_2 on bacterial growth is complex, and four antimicrobial mechanisms associated with CO_2 have been identified (Parking and Brown, 1982; Daniels,

Table 9.4 Typical gas mixtures used in MAP of some raw, cooked, cured and processed meat products.

Product category	Recommended gas mix						Raw, cooked, cured and processed food items	Typical types of packaging	
	Retail			Bulk				Retail	Bulk
	O_2 (%)	CO_2 (%)	N_2 (%)	O_2 (%)	CO_2 (%)	N_2 (%)			
Raw red meat	70	30	—	65	35	—	Beef, goat, hare, lamb, pork*, rabbit, veal (*Recommended bulk gas mix for pork 80% CO_2: 20% N_2)	Tray sealed and thermoformed packages	Bag-in-box, master pack
-Venison and wild boar	80	20	—	80	20	—	Venison and wild boar		
- Primal	NA	NA	NA	50	50	—	Beef, goat, hare, lamb, pork*, rabbit, veal (*Recommended bulk gas mix for pork 80% CO_2: 20% N_2)		
Raw poultry and game	—	30	70	—	100	—	Capon, chicken, duck, goose, grouse, guinea hen, partridge, pheasant, pigeon, poussin/Cornish hen, quail and turkey	Tray sealed and thermoformed packages	Bag-in-box, master pack
Cooked, cured and processed meat products	—	30	70	—	50	50	Bacons, beef burgers, black pudding, charcuterie, chopped pork and ham, cooked sausages, corned beef, frankfurters, luncheon meats, meat jerky, meat slices, ox tongue, pastrami, pâtés, pepperoni, potted meats, rillettes, roast meats, salami, smoked reindeer, smoked venison, terrines, wurst sausages	Tray sealed and thermoformed packages, flow packages	Bag-in-box, master pack
Cooked, cured and processed fish and seafood products	—	30	70	—	70	30	Bloaters, Bombay duck, buckling, cod's roe, cold smoked fish, fish galantine, fish rillettes, fish terrines, hot smoked fish, kippers, potted fish and shellfish, salted cod, salted anchovies, salted caviar, salted fish roes, salted jellyfish, seafood pâtés, smoked haddock, smoked halibut, smoked mackerel, smoked salmon, smoked trout, taramasalata	Tray sealed and thermoformed packages, flow packages	Bag-in-box, master pack

(continued overleaf)

Table 9.4 (*continued*)

Product category	Recommended gas mix						Raw, cooked, cured and processed food items	Typical types of packaging	
	Retail			Bulk				Retail	Bulk
	O_2 (%)	CO_2 (%)	N_2 (%)	O_2 (%)	CO_2 (%)	N_2 (%)			
Cooked, cured and processed poultry and game bird products	—	30	70	—	30	70	Capon galantine, chicken ballotine, chicken roll, cured game birds, cured poultry, duck ballotine, duck pâté, duck galantine, smoked chicken, smoked duck, smoked poussin, smoked turkey, turkey bacon, turkey ballotine, turkey galantine, turkey roll	Tray sealed and thermoformed packages, flow packages	Bag-in-box, master pack
Convenience food products	—	30	70	—	50	50	Battered: fish, seafood, meats and poultry	Tray sealed and thermoformed packages, flow packages	Bag-in-box, master pack
							Bouchée/breaded: fish, seafood, meats and poultry		
							Burritos, enchiladas, falafels, filled crêpes, pancakes and rolls, kebabs, omelettes, pasties, pizzas, pasta and pies containing meat, poultry, fish and seafood, quiche, roule au fromage, sandwiches, satays, sausage rolls, soufflés, spring rolls, stuffed pita bread, tacos, tostadas, vol au vents		

Adapted from Pbi-Dansensor (2013).

Table 9.5 Recommended gas mixtures and gas volume for meat and meat products.

Product	Gas mixture	Gas volume per 100 g product	Typical shelf life		Storage temperature (°C)
			Air	MAP	
Raw red meat	60–80% O_2 +20–40% CO_2	100–200 ml	2–4 days	5–8 days	2–3
Raw light poultry	40–100% CO_2 +0–60% N_2	100–200 ml	4–7 days	16–21 days	2–3
Raw dark poultry	70% O_2 +30% CO_2	100–200 ml	4–7 days	7–14 days	2–3
Sliced cooked meat	30%CO_2 +70% N_2	50–100 ml	4–7 days	2–5 weeks	4–6
Sausages	20–30% CO_2 +70–80% N_2	50–100 ml	4–7 days	2–5 weeks	4–6

Adapted from Linde (2013).

Krishnamurthi and Rizvi, 1985; Dixon and Kell, 1989; Farber, 1991) and are as follows:

1. Alteration of cell membrane function, including effects on nutrient uptake and absorption
2. Direct inhibition of enzymes or decreases in the rate of enzyme reactions
3. Penetration of bacterial membranes, leading to intracellular pH changes
4. Direct changes in the physicochemical properties of proteins.

Obviously, a combination of all of these activities accounts for the overall bacteriostatic and fungistatic effects observed. In short, a certain amount of CO_2 must dissolve into the product to inhibit bacterial growth (Gill and Penney, 1988).

Nitrogen is typically used as an inert gas filler to dilute the concentration of O_2 and CO_2 in the package. The use of N_2 has no direct effect on meat colour and it provides no benefit to the meat product in terms of bactericidal or bacteriostatic effects but helps to maintain the package stability when CO_2 is absorbed into the meat.

9.2.5 Brief overview of packaging materials used in MAP for case-ready meat products

MAP requires an appropriate container designed or selected to complement the characteristics of the food or operate effectively with the distribution and storage conditions applied to the product in question. Typical materials used for packaging case-ready meat products are shown in Table 9.6. Depending on the container manufacturer and the automatic dispensing equipment (which is described in more detail in Section 9.2.8) used for the product, these can be classified as follows (Catalá and Gavara, 2000):

- Vertical packaging machinery for three- or four-seal pouches

Table 9.6 Typical materials used for packaging chilled meat products.

Pack type	Bottom web materials	Top web materials (where applicable)
Flexible vacuum pack	PA-LDPE, coextruded as 5-layer film	
Flexible MAP pack	PA-LDPE	OPA-LDPE
	PA-EVOH-LDPE	
	PA-EVOH-PA-LDPE	PET-PVdC-LDPE
	PP-EVOH-LDPE	
	LDPE-EVOH-LDPE	
Rigid vacuum pack	APET	OPA-LDPE
	PVC or PVC-LDPE	PET-PVdC-LDPE
	PS-EVOH-LDPE	OPA-LDPE-EVOH-LDPE
		PET-LDPE-EVOH-LDPE
	PVC	OPA-LDPE
Rigid MAP pack	PVC-LDPE or PVC-EVOH-LDPE	PET-PVdC-LDPE
	APET	OPA-LDPE-EVOH-LDPE
	APET-LDPE or APET-EVOH-PE	PET-LDPE-EVOH-LDPE
	PS-EVOH-LDPE	
Skin packs	PVC-LDPE	Laminates of several combinations of up to seven or more layers but incorporating EVOH as gas barrier
	PS-EVOH-LDPE	
	APET	
	APET-LDPE	

Adapted from Robertson (2013).

- Horizontal packaging machinery for wrapping or preformed trays in pouches (flow pack)
- Horizontal tray thermoforming machinery with thermoformed rigid lids or, flexible film seals.

The packages for these products are generally high-barrier plastic laminations, either bags or lidding material sealed onto rigid trays. Most packaging is performed on thermoform/fill/seal machines using PVC/PVdC/LPDE or PET/PVdC/LPDE for the lidding material (Zagory, 1997). A tray lidding film selector guide is presented in Figure 9.2.

Two common tray packages used for cooked meat products are as follows (Dawson and Stephens, 2004):

1. A non-barrier expanded polystyrene (EPS) tray, overwrapped with a barrier film
2. An expanded polystyrene tray with a built-in barrier and barrier lidding seal to the tray.

The package layer structure is as follows:

- Tray: HIPS (PS)/PS foam/HIPS/adhesive/barrier film/sealer
- Barrier film: 50.8–76.2-µm-thick LLDPE-/adhesive-/PVdC-coated nylon/heat-seal coating

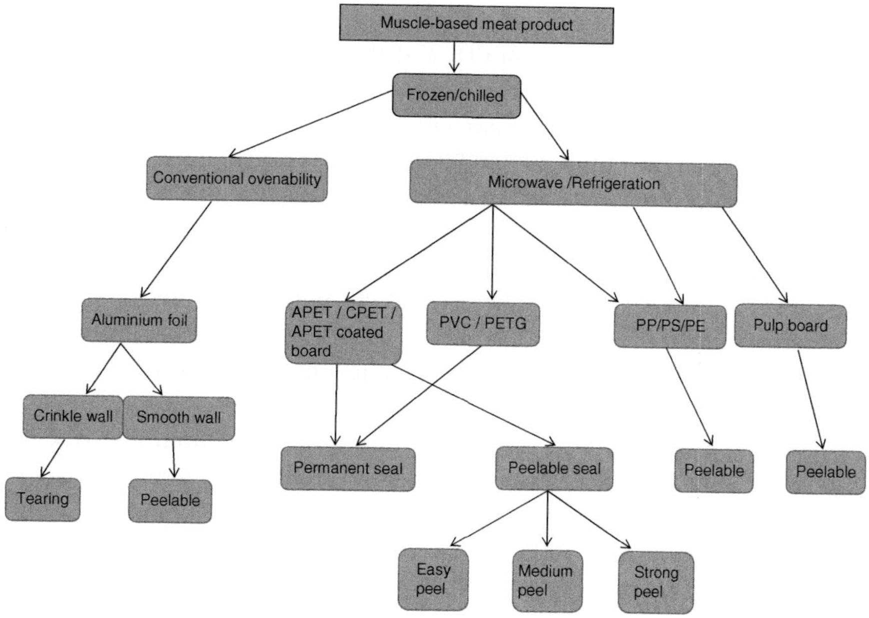

Figure 9.2 Tray film sealing selector guide (Adapted from DuPont Teijin Films, 2005).

- Lidding: PVdC-coated PET/LLDPE/63.5-μm-thick EVA.

The barrier materials used are either PVdC or EVOH. Nylon is used in packages of hot wings and roasted chicken lines because of its gas barrier properties, toughness, heat resistance and forming properties. PET is used for printability, clarity and relatively low cost compared to nylon. LLDPE adds bulk, toughness and is low in cost, while EVA provides a heat-sealing layer and seal strength on cooling (Dawson and Stephens, 2004).

Visiongain estimated that the global MAP & CAP packaging market, consisting of the packaging materials outlined above, would reach US$10.80 billion in 2013 (Visiongain, 2013).

9.2.6 Packaging systems used in case-ready MAP meat products

A variety of packaging systems exist for MAP case-ready products such as MAP high O_2, MAP low O_2, MAP very low O_2, peelable-lid/rebloom, vacuum-skin format and bulk gas flushing.

A combination of active packaging and MAP is now being used, not only for single sales unit packages but also to store fresh meat in sacks or pouches containing multiple sales units (masterpack or motherpacks) by distributors. The packs, once sealed, are placed in a corrugated paperboard box. The masterpack or motherpack system contains four to six conventional retail packs, overwrapped in O_2-permeable films. The masterpack is evacuated to remove O_2 and then back-flushed with a gas mixture of CO_2 and N_2. Generally, gas mixtures range from 80% to 100% CO_2 with N_2 used as the remaining gas (Ščetar, Kurek and Galic, 2010). The anoxic environment created by the use of O_2 scavengers allows for the shelf life of the meat to be extended and guarantees an adequate re-oxygenation of the myoglobin pigment on the surface of the meat when finally exposed to O_2. The meat products contained in the motherpack blooms in approximately 30 min after the motherpack is opened and the individual meat packs are exposed to atmospheric O_2 (Buffo and Holley, 2005). Headspace composition within a package changes dynamically during the distribution period and must be controlled continuously in order to achieve the desired meat quality and safety. Therefore, for successful master packaging, the use of O_2 scavengers is critical in order to remove residual O_2 from the package atmosphere. Even a small amount of residual O_2 is conducive to the formation of metmyoglobin, which is stable and undesirable in fresh meat products.

To achieve the maximum storage shelf life of master packaged meat products, the following are required (Buffo and Holley, 2005):

1 Superior initial quality of the meat
2 A consistent temperature of −1.5 °C to be maintained throughout the cold chain
3 Headspace gas composition consistently maintained at 100% CO_2 during the distribution period without the presence of any residual O_2.

Meat processors use a variety of gas (CO_2, N_2 or O_2) combination strategies for packaging raw or cooked meat products, including high-O_2, low-O_2 or vacuum packaging. The packaging system chosen depends on a variety of factors such as tray or package configuration, meat cut type and size, distribution life required, retail case shelf life required, consumer expectations and cost. No one particular system fits all of the needs in the marketplace (AMI (American Meat Institute), 2009). Case-ready meats are typically sold in one of two ways: vacuum packaged, in which the plastic packaging conforms to irregularly shaped meat and O_2 is removed or in modified MAP. The main packaging methods in MAP of meat products are gas flushing and compensated vacuum.

Gas flushing is a process by which air inside a pack is replaced by a continuous gas flow of specified composition before closing the packaging. The major advantage of gas flushing is the speed achieved by the flushing equipment; however, this method replaces air by dilution, so this method has an efficiency limit in terms of removing O_2 (the typical residual O_2 remaining in packs after flushing being typically around 2–5%). This level of residual O_2 is too high for muscle foods that are sensitive to O_2.

Compensated vacuum removes air by applying a vacuum and then the gas mixture is blown or released into the pack. This method is slower compared to gas flushing. However, air is completely removed by vacuum and there is no dilution of the final gas mix. The residual O_2 obtained by this method is generally lower than 0.5%.

There are four types of vacuum packaging technologies used for meat products:

- Heat shrinking of a flexible material around primal cuts
- Use of a preformed plastic pouch in an evacuation chamber
- Application of a preformed plastic tray to which, after filling, a film is placed and the atmosphere is extracted prior to sealing.
- Vacuum skin packaging (VSP).

VSP is an alternative to conventional vacuum packaging for retail portions. A further enhancement of this approach is in the use of VSP materials, which are peelable in nature. In a number of these VSP systems, air is evacuated and packs are sealed using an outer peelable O_2-impermeable film to maintain the O_2-free environment of the pack. Prior to retail display, the outer impermeable film is removed and the meat blooms on exposure to air (Šcetar, Kurek and Galic, 2010).

9.2.7 Advantages and disadvantages of MAP for case-ready meat products

MAP of fresh case-ready meat products is the most attractive technology to both the retailer (extended shelf life in the store) and the consumer (quality and convenience). However, the quality of MAP products is very much dependent upon carefully controlled temperatures and handling throughout the production, transportation and storage of the raw material and product. Therefore, major changes can occur in the storage and distribution of raw materials and in the marketing of food products. New developments in packaging materials, machinery and sensor technology have provided further improvements in food shelf life, organoleptic quality, product range and safety. Some advantages and disadvantages of MAP are listed in Table 9.7.

9.2.8 Brief overview of packaging equipment used for MAP of case-ready meat products

The machinery used on packaging and filling lines for MAP case-ready meat products are as follows:

Horizontal-form-fill-seal (HFFS) – These machines are also called flow packaging or flow-wrapping. This packaging system uses packaging materials from single films to laminate constructions for the production of flexible pouches or containers. Air is generally removed through continuous gas flushing.

Vertical-form-fill-seal (VFFS) – Similar process to HFFS; however, in this system the product is weighed and fed vertically into a partially formed pack prior to sealing.

Thermoform-fill-seal (TFFS) and Pre-formed tray and lidding film (PTLF) – TFFS systems produce packaging consisting of a semi-rigid tray hermetically sealed

Table 9.7 Advantages and disadvantages of modified atmosphere packaging for case-ready meat products.

Advantages	Disadvantages
• Shelf life increase by possibly 50–400% • Reduced economic losses due to longer shelf life • Decreased distribution costs (fewer deliveries over longer distribution distances are required) • Provides a high-quality product • Easier separation of sliced products • Centralized packaging and portion control • Improved presentation – clear view of product and all-round visibility • Little or no need for chemical preservatives • Sealed packages, barrier against product re-contamination and drip from package • Odourless and convenient packages • Reduction in production and storage costs due to better utilization of labour, space and equipment	• Capital cost of gas packaging machinery • Added costs (costs of gases and packaging materials) • Temperature control necessary • Different gas formulation for each product type • Special equipment and training required • Product safety to be established • Increased pack volume – adversely affects transport costs and retail display space • Loss of benefits once the pack is opened or leaks • Dissolved CO_2 into the food could lead to pack collapse and increased drip • Potential growth of foodborne pathogens due to temperature abuse by retailers and consumers

Adapted from Parry (1993); Sivertsvik, Rosnes and Bergslien (2002); Floros and Matsos (2005).

by a flexible lid material. Packaging is then divided by cutting units. In this system, air is removed by vacuum. PTLF systems are similar to TFFS, except for the use of preformed trays instead of semi-rigid trays.

Vacuum chamber packaging – This packaging application uses preformed bags and pouches typically and operates on the compensated vacuum principle. It is used only for small-scale product production as packaging must be brought into the chamber manually prior to evacuating and heat sealing.

Snorkel-type vacuum packaging – This technology also employs the vacuum compensated system for the production of bag-in-box packaging for catering and pouched product format. The preformed empty bags are placed on the sealing bar and air is removed by snorkels that subsequently fill bags with the gas mixture chosen before heat sealing (Baroni, Baroni and Torri, 2009).

Some of the leading companies in the MAP & CAP market are Air Products PLC, CVP Systems, Curwood, EMCO Packaging Systems, GEA Convenience-Food Technologies (formerly GEA Convenience Food Technologies), Harpak-Ulma Packaging LLC, HayssenSandiacre, ILAPAK, Landec Corporation, LINPAC, Maxwell Chase Technologies, Multisorb Technologies, Orics Industries, OYSTAR, Paper Pak Industries, Paragon Print and Packaging Group, RAP, Reiser, PolyOne Corporation (formerly Spartech), StePac, Cryovac Sealed Air, Amcor, Eastman, Dupont, Færch Plast and other leading MAP & CAP Companies in the Market (Visiongain, 2013).

9.2.9 Novel packaging systems used for case-ready MAP meat products

The Cryovac® Grip & Tear™ bags combines shrink and toughness without compromising oxygen-barrier properties. It can be used with a variety of non-abrasive applications, including fresh red meat, whole chickens and deli meats. The easy open end-seal bag is available in widths from 5½ to 18 in. The enhanced toughness gives the bags better performance against material punctures under normal distribution handling. The technology offers possible down-gauging opportunity for processors using heavier gauge bags for slightly more abusive products and runs on existing Cryovac® rotary chamber vacuum systems.

Mylar® cook from DuPont Teijin Films reduce cooking time in both conventional oven and microwave oven. It generates value-added 'trayless' ready-meal solutions, no-touch cooking experience removes the hassle of preparing fish, chicken and meat and eliminates the risks of handling raw food, self-basting flavour and texture enhancement, food hygiene control from factory to fork, portion control with no cross-contamination between products, packing methods range from pre-made bags through to form fill-seal and even vacuum thermoforming. As the consumer never touches the raw meat product, the meat packer can deliver food safely from factory to fork.

The Mylar® COOK range has huge benefits for the food service industry. Products can be cooked in a microwave, conventional oven or even a contact grill. Reduced cooking time is a big advantage to food service, but there is a further benefit as mixed products can be cooked at the same time, on the same equipment, without the risk of cross-contamination. Concerns over food hygiene are addressed as the product is safely encapsulated in its packaging direct from factory until delivery to the customer.

The Mylar® COOK range are thermoformable films (there is a choice of capping web providing peelability, venting or weld seals to the pack), films for HFFS or VFFS (self-venting through to permanent heat-seal technology which will remain intact throughout the cooking process) and films for pre-made bags (pre-made bags for small- and medium-size application; heat-seal technology can be tailored from permanent-seal through to self-venting during the cooking process).

Cryovac® has developed a packaging system termed Marinade on Demand™, where the packaging system has two separate compartments: one for the meat and one for the marinade; consequently, the product is marinated in a sealed environment. Marinade on Demand™ is a two-part thermoformed roll-stock package that creates an easy, sanitary and controlled marinating process for consumers who only remove the product from the package for cooking after the desired marination time (McMillin and Belcher, 2012).

A novel packaging film has recently been developed (FreshCase®, Curwood Inc., Division of Bemis Company Inc., Neenah, Wisconsin) that has the potential

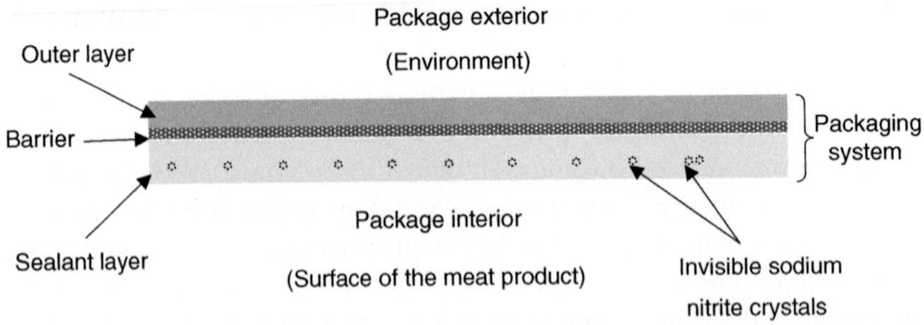

Figure 9.3 Cross-sectional representation of a nitrite-containing film (Adapted from Siegel, 2013).

to overcome consumer concerns and provide a bright cherry-red colour under anaerobic conditions (Claus and Du, 2013). The film contains embedded crystals of sodium nitrite. Sodium nitrite on the film surface comes in contact with naturally occurring meat enzymes and the nitrite is reduced to nitric oxide (NO) gas. The bright red colour in meat is formed from the reaction of NO with the myoglobin present in the meat (Goerich, 2013). The nitric oxide binds to the sixth coordinate position on the iron in the heme ring. If the iron is in the ferrous state (reduced, Fe^{2+}), the raw meat pigment that is formed is nitric oxide myoglobin (NOMb, bright red). If nitric oxide binds to iron in the oxidized state (Fe^{3+}), the pigment formed is nitric oxide metmyoglobin (NOMMb, brown). This latter pigment typically occurs in the presence of O_2. In anaerobically packaged fresh meat, the meat continues to respire (consume O_2) and has good reducing ability. As such, the NOMMb will convert into NOMb during refrigerated storage time.

Yang *et al.* (2012) reported that redness (CIE a-values) was greater in beef steaks that were vacuum packed in FreshCase® packaging systems compared to samples vacuum packed in films not containing nitrite.

A diagram of the structure of nitrite-containing film (e.g. FreshCase®) is shown in Figure 9.3. Typical films are coextruded or laminated structures that possess oriented polyester or nylon as an outer packaging layer and a nitrite-containing polyolefin (e.g. polyethylene or ethylene vinyl acetate) as the sealant layer.

Five conditions must be met in order to use nitrite-containing film for case-ready packaging of raw meat (Siegel, 2010):

1 Fresh raw meat with an adequate ability to reduce metmyoglobin and consume O_2.
2 Vacuum packaging to remove excess air from inside the package. This eliminates O_2 and collapses the package so the nitrite-containing film makes intimate contact with the viewing surface of the meat.
3 The film itself must deliver a sufficient amount of nitrite to the film-to-meat interface.

4 The film must provide a good barrier to the ingress of O_2 after packaging.
5 Refrigerated storage time is needed to allow for colour development.

The amount of nitrite incorporated into the film and the amount that migrates from the film surface into a packaged product surface is precise and repeatable. The amount of nitrite needed for the desired effect to be produced is dependent on the species of meat. Pork and poultry require lower amounts of nitrite than beef does because they naturally contain a lower concentration of myoglobin. One mole of nitric oxide (NO) from sodium nitrite ($NaNO_2$) is required to nitrosylate one mole of myoglobin. Darker coloured beef products contain 3–5 mg/g of myoglobin compared to lighter coloured pork muscles that have 1–3 mg/g myoglobin. A typical film specification for beef delivers sodium nitrite at a level of 1000 μmoles/m^2 (Siegel, 2013). Pork and poultry films deliver much less. Typical FreshCase® films deliver 70–90 mg/m^2 for beef, 15–20 mg/m^2 for pork and 5–10 mg/m^2 for poultry. The ingressing level of sodium nitrite for a typical beef application, where the cut of meat is about 1 in. in thickness, is less than 3 ppm. Since the nitrite converts into nitric oxide after dissolution into the meat juices, nitrite is not detectable in the finished product (Siegel, 2010; Siegel, 2013).

Saddle pack-style packaging system uses special rollstock material to package individual meat or poultry portions on a rollstock machine. Perforations in the film separate product portions, thereby allowing consumers to remove only the quantity of product required for use. The product format can be flat or folded to create a smaller-size retail package. The vacuum packaging eliminates a need for repacking before freezing while preventing leaks and freezer burn and allows for options of easy opening and printing the top web in up to eight colours (McMillin and Belcher, 2012).

9.3 Theory of MAP/oxygen scavenging technology for meat products

High-performance packaging has been developed with strong barrier characteristics, and also in certain instances active and intelligent components, respectively, adapt to changes in pack contents and environment and extend shelf life or provide information on the pack and its contents to promote supply chain efficiencies (WPO, 2008). Consumers are highly influenced by the convenience attributes of meat food products, whether it is simplicity or multiplicity of use, ease of storage, portability or even features that help to save time when making purchasing decisions, such as colour coding which can range from helping to segregate products based on identity through to being used to aid in the reading and interpretation of product information (e.g. the traffic light system used for product labelling).

Food packaging has dramatically shifted from traditional forms and uses to more advanced packaging applications. Advanced packaging systems employ smart packaging systems which may be active, intelligent or other in operation. The advanced packaging technologies market is witnessing major growth at present due to the increase in demand for convenience food such as ready-to-cook meal, stricter food safety regulations and demands for more sustainable packaging. Increasing health awareness among consumers, food wastage and manufacturers' concern for longer shelf life of food products and supply chain inefficiencies are other factors fuelling the growth of this market. Furthermore, consumer confidence with regard to packaged food, organized food safety regulations, threat from bioterrorism and increased applications in ready-to-eat meals and frozen foods have created huge opportunities for the advanced packaging technology market. In 2010, the smart packaging technology market had the highest growth rate within the packaging sector, estimated at a compounded annual growth rate (CAGR) of 10.5–12.1% from 2010 to 2015, respectively (Market and Markets, 2011). Modified atmosphere technology accounted for largest share (~54%) of the total advanced packaging market. Out of the total global market for advanced packaging, the contribution to the food and beverage sectors is 51% and 19%, respectively. Some of the major companies involved in the development of advanced packaging technologies include Multisorb Technologies (U.S.), Sealed Air Corporation (U.S.), Ball Corporation (U.S.), Amcor limited (Australia), Timestrip PLC (U.K.), Sysco Corporation (U.S.), PakSense Incorporated (U.S.), Eastman Chemical Company and M&G USA Corporation. These, and other companies not listed here, have launched new active and intelligent packaging technologies as a part of their major packaging development strategy (Market and Markets, 2011).

Cooked meat products or cook-chill convenience-style meat products are O_2-sensitive chemical entities, and for this reason, these products are usually stored under some form of vacuum or under MAP using a gas mixture of 60–80% N_2:20–40% CO_2 (Mullan and McDowell, 2003; Hogan and Kerry, 2008); however, residual O_2 content in MAP packs is often 0.5–2% (Smiddy, Papkovsky and Kerry, 2002; Anonymous, 2012). These meat products provide an excellent substrate for microbial growth, allowing contaminant bacteria to proliferate rapidly while conditions, particularly temperature and the gaseous environment, remain favourable. Under these conditions, such microbial growth will eventually cause spoilage but may also pose a health hazard (Nychas and Eleftherios, 2000).

As highlighted above, the application of vacuum packaging or MAP techniques do not always facilitate complete removal of O_2 (Vermeiren *et al.*, 2003). Oxygen that permeates through the packaging film or is trapped within the meat product or between meat slices (if processed) cannot be removed by these techniques. Therefore, by absorbing the residual O_2 using O_2 scavengers in cooked meat products that are vacuum packed or MAP packed, negative quality changes can be minimized (Hogan and Kerry, 2008). Oxygen scavengers protect ingredients from oxidative changes, protecting the colour, flavour, aroma and nutritional properties of the meat product while reducing or eliminating microbial growth, thereby, increasing the shelf life of the meat product.

Deterioration of meat products occurs in the following order: discolouration, oxidative rancidity and microbial degradation. Colour is synonymous with freshness and is the most important product attribute at the point of sale (Anonymous, 2012). Meat discolouration can be prevented by excluding all O_2 from the pack. Lipid oxidation results in rancidity, off-flavours and off-odours and deterioration in the colour and texture properties of meat. Cooked meat products are more susceptible to oxidation than fresh meat products are. The rate of oxidation of lipids is influenced by light, O_2 levels, high temperature, presence of catalysts and water activity. By controlling these parameters, lipid oxidation can be significantly reduced (Anonymous, 2012).

The main advantage of using O_2 scavengers is that they are capable of reducing O_2 levels to less than 0.01%, which is much lower than the typical 0.3–3.0% residual O_2 levels achievable by MAP and, consequently, delay oxidative reactions. Oxygen scavengers can be used alone or in combination with MAP (Day, 2003), but commercially in the cooked meat industry, it is more common to remove most of the O_2 content using MAP technology and then use the O_2 scavenger to eliminate the residual O_2 present in the packaged cooked meat product.

Oxygen scavengers have a current global market value of around US$1150 million, with meat packaging being the main end-use market, followed by snack foods and ready meals (Plastemart, 2013).

There are many types of O_2 scavengers available to cover a wide range of applications. Oxygen scavengers are the most commercially important sub-category of active packaging for food products and the most well-known O_2 scavenger takes the form of small sachets containing various iron-based powders containing an assortment of catalysts. These chemical systems often react with water supplied by the food to produce a reactive hydrated metallic reducing agent that scavenges O_2 within the food package and irreversibly converts it into a stable oxide. The iron powder is separated from the food by keeping it in a small, highly O_2 permeable sachet that is labelled 'Do not eat' and includes a diagram illustrating this warning (Day, 2008). While O_2 scavengers are not edible, they are not toxic, they create no harmful gases and the O_2 scavenger does not remove the fresh smell or taste of the packaged meat product.

Non-metallic O_2 scavengers have also been developed to alleviate the potential for metallic taints being imparted to food products. Non-metallic scavengers include those that use organic reducing agents such as ascorbic acid, ascorbate salts or catechol. They also include enzymatic O_2 scavenger systems using either glucose oxidase or ethanol oxidase, which could be incorporated into sachets, adhesive labels or into packaging film surfaces, card and closure liners (Day, 2003; Shuppakul *et al.*, 2003).

9.3.1 Commercial scavenging systems used for case-ready meat packs

Ageless® (Mitsubishi Gas Chemical Co., Japan) is the most common O_2 scavenging system based on iron oxidation. These scavengers are designed to reduce O_2 levels to less than 0.1% in packs. Other examples of O_2-absorbing sachets include ATCO® (used commercially as O_2 scavengers in pre-packed, cooked and cooked sliced meat products – Emco Packaging Systems, UK); Standa Industrie (France), FreshPax®, FreshMax® and FreshCard™ (Multisorb Technologies, Inc., USA) and Oxysorb (Pillsbury Co., USA).

Cryovac® O_2 scavenging systems (Cryovac® OS systems) are polymer-based O_2 scavenging film developed by Cryovac, Sealed Air Corporation, USA. This UV light-activated O_2 scavenging film, unlike other scavenging systems such as sachets and labels, the transparent oxygen scavenger is extruded as a layer on the packaging film. The Cryovac® OS systems can reduce headspace O_2 levels achieving less than 0.1% O_2 levels in just 3–10 days and is comparable in effectiveness with O_2 scavenging sachets. The Cryovac® OS films can be used in both wet and dry product applications ranging from fresh, dried, cured and cooked meat products. The Cryovac® OS O_2 scavenging films have applications in a variety of food products, including dried or smoked meat products and processed meats (Hogan and Kerry, 2008). Another benefit of the Cryovac® OS system is that it offers scavenging on demand; Cryovac® has designed its triggering systems (like the model 4104A) to activate OS films in conjunction with the existing packaging equipment and verified rapidly. The Cryovac® Model 4212 Scavenging Initiation System (SIS) is a patented process specifically designed to trigger the O_2 scavenging capabilities of Cryovac® OS films. OS films are passed through the Cryovac® Model 4212 SIS and exposes the OS film to 254 nm wavelength, activating the unique O_2 scavenging performance of the film. Once the optimum triggering dose is set for the customer's product and conditions, the 4212 will automatically adjust the amount of UV light exposure that the OS film receives. Residual O_2 is effectively removed from the package atmosphere, allowing food manufacturers to deliver fresher products to customers. The system is designed to be used with existing packaging lines. The Cryovac® Model 4212 SIS is an add-on accessory. The system is typically added above or behind the existing packaging lines. In most instances, OS film may be used for up to 30 min after being triggered, minimizing film waste.

For example, Multisorb Technologies provides a wide range of O_2 scavengers for use in the meat industry: FreshPax® (a sachet containing an O_2 scavenger blend), FreshMax® a patch or self-adhesive O_2 scavenger blend. FreshCard™ a backer card with an O_2 scavenger blend incorporated into the structure. These O_2 scavenger packets, composed of food-grade materials, remove O_2 from a package and protect packaged foods from spoilage, rancidity and mould growth while retaining colour, nutrients, texture and flavour of the cooked meat product, thereby increasing product shelf life. FreshPax® comes in a grease-proof breathable film and is formulated differently depending on the food product and how the food is stored. FreshCard™ uses a Solid Lay-Down process that gives the cards an extremely flat profile and prevents the active material from spilling out even if the card is cut accidentally. In addition, both sides of the FreshCard™ are available in up to four colours to act as an added marketing tool within the package and can be used to give nutritional or recipe information. The packets can be dispensed from Multisorb's Active-Pak automation equipment, which features a packet spool supply system offering automated N_2 purge with N_2 supply and O_2 level monitoring with alarm.

MAPLOX™ (Modified Atmosphere Packaging Low-O_2 System for Case-Ready Meats) combines FreshPax® CR Packets with the APA-series of dispensers. It is an easy-to-use, low-O_2 packaging program designed to extend the saleable life of fresh meat at retail and help you to realize significant savings from central location packaging. It provides consumers with fresh and appealing whole muscle beef, ground beef, pork, lamb, veal and other value-added meat products.

In a typical case-ready package, a number of trays are over-wrapped and hermetically sealed in a master bag. The atmosphere of the master bag is typically modified down to $<1\%$ residual O_2. Additional O_2 remains in the primary package, in the foam tray, dissolved in the meat and bound to haemoglobin. The FreshPax® CR oxygen scavenger used in the MAPLOX™ program contains an absorbent formulation that literally scavenges residual O_2 out of the master bag while simultaneously taking the O_2 out of the primary tray packs. The FreshPax® CR absorber removes the residual O_2 within 24–48 h, reducing O_2 content in the master bag to below 0.1% and maintaining this level despite O_2 ingress due to permeation. When combined with low temperature storage and distribution, FreshPax® CR effectively inhibits oxidative degradation and spoilage of the meat. When the master bag is opened, the meat blooms to the appropriate colour. FreshPax® CR typically extends the shelf life of case-ready meat for 15–21 days.

MAPLOX™ can be applied to case-ready meats such as beef (whole muscle and ground), pork (chops, sausages, etc.), lamb, veal and value-added products (Kabobs, Roasts, Pinwheels, etc.). Clariant (Clariant International Ltd, Switzerland) provides to the meat industry Oxy-Guard™ O_2 scavenger packets in a variety of sizes based on the size of the package and the residual air inside. Clariant's Oxy-Guard O_2 scavengers are available globally in a variety of sizes, from 20 to 2000 cubic centimetres (cc) based on the size of the packaging and the residual air inside. Oxy-Guard™ is available as either individual packets or a continuous strip for automatic insertion. Clariant's Oxy-Guard packaging inserts contain a combination of salt and iron powder, which combine to maintain oxygen levels at less than 0.01% inside sealed packages (Charlotte, 2013). This level is reached within 24 h at normal storage temperature. As the salt absorbs moisture from the packaging, the iron begins to oxidize, initiating the O_2 absorption process.

Oxy-Guard™ O_2 scavenging packets absorb available O_2 inside meat packaging to prolong shelf life, maintain product quality and reduce loss; maintain taste, texture and freshness while preventing mould, microbial growth and rancidity in packaged foods (Clariant, 2013).

EMCO Packaging Systems supplies the only European-produced O_2 absorber and is available in two forms: self-adhesive form (label/sticker) (50, 100 and 200 cc capacities) and sachet form (30–1000 cc). The ability to minimize the O_2 concentration within the food pack, and to maintain it at this low level, offers far-reaching benefits of substantially extended shelf life and the optimization of product appearance, taste and nutritional value. The labels are much easier and more reliable to apply into food packs and labels become integral to the packaging virtually eliminating consumer interaction, potential misuse or accidental ingestion. This makes them far superior to sachets and very much the retailer's preferred option. The smaller capacity labels are supplied as a complement to gas flushing where cooked meat manufacturers are looking for methods to eliminate the remnants of O_2 (typically 1% or less) where the residual O_2 would react with UV light in retail chiller cabinets and cause discolouration of the cooked meat. The large capacity labels allow packers to use them as a complete alternative to gas flushing instead of merely being a complement to it. The new larger labels have the capacity to reduce O_2 headspace from 21% down to 0%.

Cooperation between ALBIS Plastic GmbH (Hamburg, Germany) and RPC Bebo Plastik GmbH (Bremervörde, Germany) have created a rigid O_2 absorber pack by incorporating the absorber technology from ALBIS PLASTIC (SHELFPLUS® O_2) with multilayer packaging supplied by RPC Bebo Plastik for a chili con carne container. The additive batch SHELFPLUS® O_2 absorbs both the O_2 that penetrates through the outside of the packaging and also the residual O_2 in the headspace. SHELFPLUS® O_2 is available on various base materials such as PP, PE or also EVA. A PA-based type is also under development. All types can be used in both flexible and rigid plastic packaging (AIPIA (Active and Intelligent Packaging Association), 2013).

Innovative packaging systems, such as those described for O_2 scavenging systems, have the potential to increase packaging costs, and so restrict options for commercialization, especially for small- and medium-sized businesses. However, these cost increases are counterbalanced by reductions in wastage due to the improved quality and shelf life of meat products (Lee, 2010). The benefits of using O_2 scavengers in packaged meat products are shown in Table 9.8.

9.4 Future role for novel packaging systems in the preservation of meat

Consumers expect safe food and the food industry have a responsibility to meet their expectations. In today's world of global markets and stiff product

Table 9.8 Benefits of using technologies such as O_2 scavengers in packaged meat products.

- Extends shelf life of packaged meat products
- Prevents growth of aerobic pathogenic and spoilage microorganisms (including moulds)
- Eliminates or minimizes the use of antioxidants and antimicrobial additives such as BHA, BHT, sulphur dioxide, sorbates and benzoates in meat products
- Significantly improves preservation quality of polyunsaturated fatty acids due to oxidation
- Helps retain fresh flavour of packaged meat products
- Prevents oxidation of spice oleoresins present in spices used as seasoning of meat products
- Prevents oxidation of vitamins such as vitamins A, C and E
- O_2 scavengers used in conjunction with MAP/vacuum packaging remove residual oxygen and also absorb O_2 that permeates the package system used for packaging meat products

competition, along with increasing consumer demands for higher food quality, ever-stringent food safety expectations and a greater offering of convenience, it has become imperative for companies to explore ways to improve their productivity in terms of maintaining safety, using advanced packaging materials and solutions, implementing flexible and standardized technology, adopting proven management principles around good packaging practice and through the reduction of product and resource wastage. Furthermore, environmental impact of meat production is placing pressure on food processors to reduce manufacturing waste and to adopt a more sustainable approach to manufacturing. Huge interest in product wastage is now becoming a 'hot political button' due to issues pertaining to resource security and environmental concerns, especially with respect to food. Resources are becoming more scarce, making material wastage taboo (Kerry, 2014). Case-ready meat programs utilize new and improved production techniques, processing and packaging technologies and innovative retail marketing strategies in ways that have transformed how retailers order, stock and merchandize meat products. Consequently, the availability of case-ready meat has dramatically altered how consumers now purchase and prepare meat products in their homes. Consumers are demanding more sophisticated muscle-based products and packaging from meat processors. Convenience and food safety are the paramount issues of the current and future marketplace of meat products. Convenience parameters may be related to productivity, processability, warehousing, traceability, display qualities, tamper resistance, ease of opening and cooking preparation (Han, 2014). A variety of new packaging technologies are being developed to meet or exceed consumer expectation. These novel packaging systems will play a significant role in the preservation of meat as huge interest in safer and high-quality meat products will be demanded as the global economy shrinks and meat prices increase.

Retailers will require ever longer shelf lives for retail meat products. These pressure points will eventually create more niches for MAP/oxygen scavenger packaging systems. The use of preservation technologies such as MAP combined with active packaging O_2 scavengers could result in synergetic actions that would increase the safety and shelf life of the packaged meat products. Oxidation is a major contributor to meat quality deterioration, thus limiting shelf life. Its rate and extent can be reduced or stopped by MAP/O_2 scavenger packaging systems. Currently, consumer desire for minimally processed, yet *safe*, meat products, the opportunities for the use of MAP and O_2 scavengers' active packaging may increase. Active packaging will continue to increase its market share in the packaging meat industry, as it can deliver cost-effective spoilage and safety control from minimal environmental impact (Cruz-Romero and Kerry, 2011).

References

AIPIA (Active and Intelligent Packaging Association) (2013) *Co-operation Creates Rigid Oxygen Absorber Pack*. Available at: http://www.aipia.info/news_220_cooperation-creates-rigid-oxygen-absorber-pack.php (accessed 04 February 2016)

AMI (American Meat Institute) (2009) *Case-Ready Meats Modified Atmosphere Packaging*. Available at: http://www.meatami.com/ht/a/GetDocumentAction/i/50958 (accessed 04 February 2016)

Anonymous (2012) A practical look at some of the issues in meat packaging. *International Meat Topics*, **3** (4), 7–9.

Argyri, A.A., Panagou, E.Z. and Nychas, G.-J.E. (2012) Advances in vacuum and modified atmosphere packaging of poultry products, in *Advances in Meat, Poultry and Seafood Packaging* (ed. J.P. Kerry), Woodhead Publishing Limited, Cambridge, pp. 205–247.

Baroni, B., Baroni, M.R. and Torri, L. (2009) *Modified atmosphere*, Artek S.n.c, Porto Ceresion.

Brody, A.L. (2007) Case-ready packaging for fresh meat. *Food Technology*, **61** (3), 70–72.

Brown, H. and Williams, J. (2003) Packaged food product quality and shelf life, in *Food Packaging Technology* (eds R. Coles, D. McDowell and M.J. Kirwan), Blackwell Publishing Ltd., Oxford, pp. 65–94.

Buffo, R.A. and Holley, R.A. (2005) Centralized packaging systems for meats, in *Innovations in Food Packaging* (ed. J.H. Han), Elsevier Academic Press, London, pp. 227–236.

Catalá, R. and Gavara, R. (2000) Plastic materials for modified atmosphere packaging, in *Trends in Food Engineering* (eds J.E. Lozano, C. Añon, E. Parada-Arias and G.V. Barbosa-Canovas), Technomic Publishing Co., Inc., Lancaster, pp. 311–325.

Charlotte, N.C. (2013) New Technology Preserves Food, Beverages Without Chemicals. Food Product Design, July. Available at: http://www.foodproductdesign.com/news/2013/07/new-technology-preserves-food-beverages-without-c.aspx (accessed 04 February 2016)

Clariant (2013) *Oxy-Guard™ Oxygen Scavengers*. Product sheet. Available at: http://www.desiccants.clariant.com/fileadmin/pages/clariant/user/cdp-pdf/Oxy-Guard_Eng_ProdSheet.pdf (accessed 04 February 2016).

Claus, J.R. and Du, D. (2013) Nitrite-embedded packaging film effects on fresh and frozen beef color development and stability as influenced by meat age and muscle type. *Meat Science*, **95**, 526–535.

Cruz-Romero, M. and Kerry, J.P. (2008) Crop-based biodegradable packaging and its environmental implications. *CAB Reviews: Perspectives in Agriculture, Veterinary Science, Nutrition and Natural Resources*, **3** (082), 25.

Cruz-Romero, M. and Kerry, J. (2011) Packaging of cooked meats and muscle-based convenience-style food products, in *Processed Meats: Improving Safety Nutrition and Quality* (eds J. Kerry and J. Kerry), Woodhead Publishing Limited, Cambridge, pp. 666–705.

Daniels, J.A., Krishnamurthi, R. and Rizvi, S.S.H. (1985) A review of effects of carbonic dioxide on microbial growth and food quality. *Journal of Food Protection*, **48**, 532–537.

Dawson, P.L. and Stephens, C. (2004) Poultry Packaging, in *Poultry Meat Processing and Quality* (ed. G.C. Mead), Woodhead Publishing Limited, Cambridge, pp. 135–163.

Day, B.P.F. (2003) Active packaging, in *Food Packaging Technology* (eds R. Coles, D. McDowell and M.J. Kirwan), Blackwell Publishing Ltd., Oxford, pp. 282–302.

Day, B.P.F. (2008) Active packaging of food, in *Smart Packaging Technologies for Fast Moving Consumer Goods* (eds J. Kerry and P. Butler), John Wiley & Sons, Ltd., Chichester, pp. 233–245.

Decker, E.A., Chan, W.K.M., Livisay, S.A. *et al.* (1995) Interactions between carnosine and the different redox states of myoglobin. *Journal of Food Science*, **60**, 1201–1204.

Dixon, N.M. and Kell, D.B. (1989) The inhibition by CO_2 of the growth and metabolism of microorganisms. *Journal of Applied Bacteriology*, **67**, 109–136.

Dupont Teijin Films (2005) *Tray Sealing*. Which Mylar® polyester do I need? Available at: http://usa.dupontteijinfilms.com/marketspaces/downloads/packaging/DTF_Selector_Guide_English.pdf (accessed 04 February 2016).

Eliasson, A. (2008) Packaging for a Better Planet. Swedish innovations at the forefront of sustainable packaging. *Food Quality & Safety: Farm to Fork safety*, **October/November**. http://www.foodqualityandsafety.com/article/packaging-for-a-better-planet/?singlepage=1 (accessed 30 May 2016).

FAO (2011) *Global Food Losses and Waste*. Extent, Causes and Prevention. Available at: http://www.fao.org/docrep/014/mb060e/mb060e00.pdf (accessed 04 February 2016).

FAO (2013) *Food Wastage Food Print*. Impact on natural resources. Summary report. Available at http://www.fao.org/docrep/018/i3347e/i3347e.pdf (accessed 04 February 2016).

Farber, J.M. (1991) Microbiological aspects of modified atmosphere packaging – a review. *Journal of Food Protection*, **54**, 58–70.

FFT (Food For Thought) (2013) *The Fresh and Processed Meat Market in Central Europe (Edition 2013)*. Available at: https://www.fft.com/The_Fresh_And_Processed_Meat_Market_in_Central_Europe_FFTReport_num_R-CEE-1019.html (accessed 04 February 2016)

Floros, J.D. and Matsos, K.I. (2005) Introduction to modified atmosphere packaging, in *Innovations in Food Packaging* (ed. J.H. Han), Elsevier Academic Press, London, pp. 159–172.

Freedonia (2011) *Meat, poultry and seafood packaging. US industry study with forecast for 2015&2020*, The Freedonia Group, Cleverland, p. 302. Available at: http://www.freedoniagroup.com (accessed 04 February 2016).

Gill, C.O. and Penney, N. (1988) The effect of the initial gas volume to meat weight ratio on the storage life of chilled beef packaged under carbon dioxide. *Meat Science*, **22**, 53–63.

Gill, C.O. and Tan, K.H. (1980) Effect of carbon dioxide on growth of meat spoilage bacteria. *Applied and Environmental Microbiology*, **39**, 317–319.

Goerich, W.K. (2013) *Vacuum Packaging Fresh Beef with Nitrite-Containing FreshCase® Film*. Available at: http://curwoodinnovates.com/Files/KnowledgeBase/ItemDownload/freshcase-whitepaper.pdf (accessed 04 February 2016)

Grebitus, C., Jensen, H.H., Roosen, J. and Sebranek, J.G. (2013) Fresh meat packaging: Consumer acceptance of modified atmosphere packaging including carbon monoxide. *Journal of Food Protection*, **76**, 99–107.

Han, J.H. (2014) A review of food packaging technologies and innovations, in *Innovations in Food Packaging*, 2nd edn (ed. J.H. Han), Academic Press, London, pp. 3–12.

Hogan, S.A. and Kerry, J.P. (2008) Smart packaging of meat and poultry products, in *Smart Packaging Technologies for Fast Moving Consumer Goods* (eds J. Kerry and P. Butler), John Wiley & Sons, Ltd., Chichester, pp. 233–245.

Kerry, J.P. (2014) New packaging technologies, materials and formats for fast-moving consumer products, in *Innovations in Food Packaging*, 2nd edn (ed. J.H. Han), Academic Press, London, pp. 549–584.

Kirwan, M.J. and Strawbridge, J.W. (2003) Plastics in food packaging, in *Food Packaging Technology* (eds R. Coles, D. McDowell and M.J. Kirwan), Blackwell Publishing Ltd., Oxford, pp. 174–240.

Kropf, D.H. (1993) Colour stability: Factors affecting the colour of fresh meat. *Meat Focus International*, **1**, 269–275.

Larsen, H., Westad, F., Sørheim, O. and Nilsen, L.H. (2006) Determination of critical oxygen level in packages for cooked sliced ham to prevent color fading during illuminated retail display. *Journal of Food Science*, **71**, S407–S413.

Lee, K.T. (2010) Quality and safety aspects of meat products as affected by various physical manipulations of packaging materials. *Meat Science*, **86**, 138–150.

Lee, D.S., Yam, K.L. and Piergiovanni, L. (2008) *Food packaging science and technology*, CRC Press, Boca Raton.

Linde (2013) *The Ultimate Combination for Freshness*. MAPAX® modified atmosphere packaging. Available at: http://www.vakuumverpacken.de/GB/pdf/MAPAX%20Booklet%20e%20Sept%202006.pdf (accessed 04 February 2016)

Market and Markets (2011) *Global Active, Smart and Intelligent Packaging Market by Products, Applications ,Trends and Forecasts (2010–2015)*. Available at: http://www.marketsandmarkets.com/Market-Reports/smart-packaging-324.html (accessed 04 February 2016)

Marsh, K. and Bugusu, B. (2007) Food packaging and its environmental impact. *Food Technology*, **61** (4), 46–50.

McMillin, K.W. and Belcher, J.N. (2012) Advances in the packaging of fresh and processed meat products, in *Advances in meat, poultry and seafood packaging* (ed. J.P. Kerry), Woodhead Publishing Limited, Cambridge, pp. 173–204.

Molin, G. (2000) Modified atmospheres, in *The microbiological safety and quality of food* Vol. 1 (eds B.M. Lund, T.C. Baird-Parker and G.W. Gould), Aspen Publishers, Inc., Gaithersburg, Maryland, pp. 214–234.

Møller, J.K.S., Jakobsen, M., Weber, C.J. *et al.* (2003) Optimisation of colour stability of cured ham during packaging and retail display by a multifactorial design. *Meat Science*, **63**, 169–175.

Mullan, M. and McDowell, D. (2003) Modified atmosphere packaging, in *Food Packaging Technology* (eds R. Coles, D. McDowell and M.J. Kirwan), Blackwell Publishing Ltd., Oxford, pp. 303–339.

Nannerup, L.D., Jakobsen, M., van den Berg, F. *et al.* (2004) Optimizing colour quality of modified atmosphere packed sliced meat products by control of critical packaging parameters. *Meat Science*, **68**, 577–585.

Nychas, G.-J.E. and Eleftherios, H. (2000) Spoilage of meat, in *Encyclopedia of Food Microbiology* (eds R.K. Robinson, C.A. Batt and P.D. Patel), Academic Press, London, pp. 1253–1260.

Parking, K.L. and Brown, W.D. (1982) Preservation of seafood with modified atmospheres, in *Chemistry and Biochemistry of Marine Food Products* (eds R.E. Martin, G.J. Flick, C.E. Hebard and D.R. Ward), AVI Publishing, Westport, CT, pp. 453–465.

Parry, R.T. (1993) Introduction, in *Principles and applications of modified atmosphere packaging of food* (ed. R.T. Parry), Blackie Academic & Professional , Glasgow, pp. 1–18.

Pbi-Dansensor (2013) *A Guide to MAP Gas Mixtures*. Available at: http://www.modifiedatmospherepackaging.com/~/media/Modifiedatmospherepackaging/Pictures/MAP_Poster.ashx (accessed 04 February 2016)

Plastemart (2013) *Over 2.5 mln Tons of Polymer Used in Thin-Walled Plastics Packaging in Europe*. Available at: http://www.plastemart.com/Plastic-Technical-Article.asp?LiteratureID=1552&Paper=polymer-used-in-thin-walled-plastics-packaging-in-Europe-leading-injection-moulder-of-plastics-packaging (accessed 04 February 2016)

Robertson, G.L. (2013) Packaging of flesh foods, in *Food Packaging: Principles and Practice*, 3rd edn, (ed. G.L. Robertson), CRC Press Taylor & Francis Group, Boca Raton, pp. 445–476.

Ščetar, M., Kurek, M. and Galic, K. (2010) Trends in meat and meat products packaging – a review. *Croatian Journal of Food Science and Technology*, **2**, 32–48.

Shuppakul, P., Miltz, J., Sonneveld, K. and Bigger, S.W. (2003) Active packaging with emphasis on antimicrobial packaging and its implications. *Journal of Food Science*, **68**, 408–420.

Siegel, D.G. (2010) *Retail Packaging for Fresh Meats. Meat Processing Technology Series*, American Meat Science Association, Illinois.

Siegel, D.G. (2013) *Vacuum Packaging Fresh Meat with Nitrite Containing Films*. Available at: http://curwoodinnovates.com/Files/KnowledgeBase/ItemDownload/freshcase-description-by-siegel.pdf (accessed 04 February 2016)

Sivertsvik, M., Rosnes, T. and Bergslien, H. (2002) Modified atmosphere packaging, in *Minimal Processing Technologies in The Food Industry* (eds T. Ohlson and N. Bengtsson), Woodhead Publishing Limited, Cambridge, pp. 61–86.

Smiddy, M.A., Papkovsky, D.B. and Kerry, J.P. (2002) Evaluation of oxygen content in commercial modified atmosphere packs (MAP) of processed cooked meats. *Food Research International*, **35**, 571–575.

Solomon, J. (2004) Eliminating oxygen. *Meat Poultry*, **9**, 38–41.

Sonneveld, K. (2000) What Drives (Food) Packaging Innovation? *Packaging Technology and Science*, **13**, 29–35.

Thoden van Velzen, E.U. and Linnemann, A.R. (2008) Modified atmosphere packaging of fresh meats-sudden partial adaptation caused an increase in sustainability of Dutch supply chains of fresh meats. *Packaging Technology and Science*, **21**, 37–46.

Tucker, G. 2006. Thermal processing of ready meals. In: D-W Sun (Ed.), *Thermal Food Processing New Technologies and Quality Issues* (pp.363-385). Boca Raton: CRC Press.

Vermeiren, L., Heirlings, L., Devlieghere, F. and Debevere, J. (2003) Oxygen, ethylene and other scavengers, in *Novel Food Packaging Techniques* (ed. R. Ahvenainen), Woodhead Publishing Limited, Cambridge, pp. 22–49.

Visiongain (2013) *World Modified & Controlled Atmosphere Packaging (MAP & CAP) Market 2013–2023*. Available at: http://www.reportlinker.com/p01594429-summary/World-Modified-Controlled-Atmosphere-Packaging-MAP-CAP-market.html (accessed 04 February 2016)

WPO (Word Packaging Organisation) (2011) *Packaging – An Important Tool for a Sustainable Society (Position Paper)*. Available at: http://www.worldpackaging.org/i4a/doclibrary/index.cfm?category_id=1 (accessed 04 February 2016).

WPO (World Packaging Organisation) (2008) *Market statistics and future trends in global packaging*, WPO/PIRA International Ltd. Downloaded from www.worldpackaging.org (accessed 04 February 2016).

Yang, X. (2012) An evaluation of the effectiveness of Freshcase® technology to extend the shelf life of beef and pork. Thesis: Colorado State University.

Young, G.A. (2009) Case-ready meats: An outsourcing decision for food retailers. Thesis: Michigan State University.

Zagory, D. (1997) Modified atmosphere packaging, in *The Wiley Encyclopedia of Packaging Technology*, 2nd edn (eds A.L. Brody and K.S. Marsh), John Wiley & Sons, Inc., New York, pp. 650–656.

CHAPTER 10
Smart packaging solutions encompassing nanotechnology

Maeve Cushen & Enda J. Cummins
UCD School of Biosystems and Food Engineering, Belfield, Dublin, Ireland

10.1 Introduction

The continued increase in the global population is placing sustained pressure on the world's food resources, forcing food commodity prices to increase (Rosegrant, Tokgoz and Bhandary, 2013). This upward pressure on food commodity prices is also being exacerbated by food wastage by consumers. The European Commission (2010) reported a yearly food waste disposal of 292,000 tonnes in Ireland. Improvements in food packaging have the potential to alleviate this pressure by allowing food to be stored and remain fresh for longer. Packaging provides food with a physical barrier from its surrounding environment. This can protect it from physical damage and also the influences of its surroundings – such as moisture, light, oxygen, microbes, mechanical stresses, foreign bodies and dust. In future, food packaging is likely to be a lot more than a physical container providing protection to the contents from the immediate environment. Different types of packaging are being investigated including those where migration of particles into the food matrix is warranted and necessary and those where migration is not required.

For a manufacturer to invest in a new packaging material, the benefits of the packaging must be sufficient to warrant this investment. Prolonging the shelf life of food products has the added incentive of extending the geographical export market boundaries. High-value products are more likely to warrant investment in packaging to prolong the product shelf life. For consumers to accept the product in a new packaging (that uses unfamiliar technologies), the benefits of this product over a traditionally packaged counterpart must be communicated to them (Grebitus *et al.*, 2013). Recent developments have seen new smart packaging concepts (including nanotechnology) being developed. These may play a role in the preservation and protection of perishable food products including high-value meat products.

Emerging Technologies in Meat Processing: Production, Processing and Technology, First Edition.
Edited by Enda J. Cummins and James G. Lyng.

10.2 Smart packaging

Smart packaging is defined as a material which benefits an associated food, while maintaining traditional packaging functions and is a collective term to describe a range of novel packaging techniques and their associated materials, which can be subdivided into improved packaging, active packaging and intelligent packaging. This has garnered much interest in the meat industry, where intact meat cuts can be susceptible to microbial contamination (from exposure to airborne entities or post-processing handling).

At the most basic level of smart packaging, improved packaging imparts an effect on the food in a passive manner. As the concepts and technologies have advanced over the years, intelligent packaging has evolved; materials have been developed to interact with the food itself, retrieving specific information and either communicating with the relevant personnel and the consumer or initiating an active role in response. The language involved in describing smart packaging and its subgroups is often interpreted differently. As with any emerging technologies, absolute definitions are not yet agreed upon. For instance, the titles of *Smart* and *Intelligent* may be interchangeable to some and this can cause confusion. In this chapter, the encompassing title smart packaging is used to describe three subgroups of improved, active and intelligent.

Unsurprisingly, the meat industry is expressing interest in smart packaging. Meat, as well as fish, is a perishable, high-value food. It is also a food that is often associated with foodborne illnesses as a result of pathogenic contamination (Painter *et al.*, 2013) as pathogens, as well as spoilage microorganisms, thrive on the high water availability and nutritious meat.

10.2.1 Theory of smart packaging

The same preservation challenges exist for smart packaging that have existed for the last number of decades for the traditional packaging of meat. Moisture, light, oxygen, microbes, mechanical stresses, foreign bodies and dust can all lead to, or create, the conditions for the deterioration of meat. These unfavourable conditions accelerate changes in some attributes of the meat: odour, colour change, microbial contamination, often detectable by the consumer. The aim of smart packaging is to protect foods from negatively impacting entities more effectively than traditional packaging means. Traditional methods of preserving foods from microbial growth have seen food being pretreated before it is packaged. Such traditional methods aim to reduce the level of microbes in the packaged food and include thermal processing, drying, irradiation and the addition of antimicrobial agents or salt. Other methods relied on delaying microbial growth by storing the product in cold conditions (refrigeration) or by packing the food in a modified atmosphere. The severity of treatment is generally regarded as an indication of effectiveness, irradiation being the most severe and most effective

(Quested *et al.*, 2010). The majority of these techniques cannot be applied to a number of high-value perishable foods such as fresh fish, poultry and meats because consumers demand that these foods are minimally processed (Bartosz, 2013; Olsen, Grunert and Sonne, 2010). This has seen the widespread adoption of refrigeration and modified atmosphere packaging (MAP) for such products. On their own, the basic protective function of conventional food packaging is to prolong the shelf life. However, shelf life can be further increased by taking additional steps. Continued improvements in food packaging technologies and the growth of 'smart packaging' are providing further options to counter antimicrobial spoilage, while avoiding pretreatment of these high-value perishable foods.

10.2.2 Modes of action

The methods in which these smart packaging systems work are as diverse as the systems themselves. Some similarities exist over the three subgroups of improved, active and intelligent such as methods of material incorporation into the food packaging material matrix. Modes of action depend on the packaging matrix that the novel entity is associated with in the system and also on the function of the entity. The range of complexities in smart packaging is huge, with some acting as a simple physical barrier and others having the potential to react to certain conditions in the food.

10.2.3 Improved packaging

Food packaging materials provide essential barriers to environmental factors. The integrity of these barriers is an indication of the material's ability to contain and preserve a food. Consumers' acceptance of this improved technology is due to its passive nature, which does not require (nor sees any benefit from) contact with the food. As such, consumers do not associate this form of packaging with a risk of contamination from the packaging material. Improved packaging systems involve the incorporation of a filler material into a food packaging material matrix. Depending on the filler and the packaging matrix, these have the potential to improve gas barrier properties (Majeed *et al.*, 2013), flexibility (Abdollahi *et al.*, 2013), mechanical strength (Ozcalik and Tihminlioglu, 2013), light barrier properties (Mallakpour and Madani, 2012) and moisture barrier properties (Bilbao-Sainz *et al.*, 2011). In many cases, the filler will be multifunctional, improving two or more of the aforementioned properties (Armentano *et al.*, 2013).

Of the three subgroups of smart packaging, improved packaging materials are not only the most commercially available but also the most acceptable to consumers. This is due to the design of the fillers and their interaction with the relevant packaging material, limiting their migration into the food. There is no benefit in having the active entity close to the food to impart the effect; in some cases, an inner layer of material not containing filler can be placed on the food contact surface of the film and so that the likelihood of migration is minimised.

Mechanical properties and elastic modulus of the polymer matrix at a molecular level influence the strength and flexibility of the polymer, particularly the crystalline or amorphous portions. The addition of a filler can interrupt the regular structure and change the polymer chain arrangement, and when the interspatial arrangement of the polymer chains is changed, their interaction with one another changes. This can lead to less crystalline regions and more strength due to the resistance that flexibility offers.

Traditional packaging materials are being replaced by bionanocomposites which rely upon biodegradable materials and incorporate nanomaterial fillers (Sorrentino, Gorrasi and Vittoria, 2007). The nanofillers assist in protecting the food produce and extending the shelf life of the product. Various barrier properties of food packaging materials can be improved with the addition of nanofillers. During processing, nanoparticles may partake in cross-linking with the polymer chains within food packaging matrices, which increases mechanical strength and thermal resistance (Chisholm *et al.*, 2005). The reduced use of plastic packaging means less processing and fossil fuel consumption during polymer production. In addition, biodegradation of the bionanocomposites results in a more environmentally friendly product (Sozer and Kokini, 2009). The use of biodegradable packaging has been touted as a green environmentally friendly initiative; however, the use of bionanocomposites has raised concerns regarding the incidental release of nanoparticles into the environment following polymer degradation. It is clear uncertainties remain regarding the use of nanomaterials, particularly with regard to ecotoxicity.

The inclusion of nanoscale fillers reinforces the polymer matrix and has resulted in the development of Food Contact Materials (FCMs) with improved flexibility, gas barrier properties, temperature control and moisture stability. The use of nanoscale fillers in composite films represents a radical, promising alternative to conventional polymer composites (Alexandre and Dubois, 2000; Giannelis, 1996; Sinha Ray *et al.*, 2002). A low fill rate (i.e. 5%) can have a significant effect with significant enhancements in modulus, dimensional stability and solvent or gas resistance compared to conventional polymers. In addition, density and transparency can remain largely unchanged while improving surface properties and recyclability (Sorrentino, Gorrasi and Vittoria, 2007). The inclusion of particulate nanomaterials can result in improved gas barrier properties of composites. For a larger polymer particle matrix with nanoscale fillers, the route through which the gas must travel is significantly longer due to the winding pathway the gas must travel compared to conventional polymer matrix.

Clay montmorillonite is commonly used as a nanoscale filler in the production of nanocomposites. Clay montmorillonite has a plate-like morphology consisting of stacked silicate sheets with a high aspect ratio (ratio of length to thickness).

This high aspect ratio has resulted in the formation of reinforced films, which are widely studied as part of composite materials (Rhim and Ng, 2007). Laponite is another filler used in polymers, which has been researched recently and which has a higher aspect ratio than that of montmorillonite (Chung *et al.*, 2010). The effectiveness of nanoparticulate filler can be greatly influenced by the aspect ratio, due to the extended pathway for the diffusion of a gas around filler particles and therefore acting as a barrier to gas exchange in the polymer (Adame and Beall, 2009; de Azeredo, 2009, Sinha Ray and Okamoto, 2003).

TiO_2 nanoparticles are effective UV-blocking additives due to their ability to absorb short-wavelength light and to scatter and reflect other longer wavelengths; in addition to this, it is highly photostable (Duncan, 2011; Yang, Zhu and Pan, 2004). Their widespread use in the cosmetics industry as sunscreens is testament to this. Although they appear white in colour at particle sizes in the micro range, they become transparent at the nanolevel (Newman, Stotland and Ellis, 2009). In principle, food packaging films incorporating TiO_2 nanoparticles may have the ability to protect foods from the oxidising effects of UV light while maintaining good optical clarity. Certain fatty acids in foods are susceptible to lipid oxidation, which can give rise to volatiles associated with spoilage. The presence of susceptible fatty acids in meat could justify the improvement of the light barrier properties of meat packaging materials.

Nanocomposites that retard gas exchange are commercially available. They are being used to contain CO_2 in carbonated beverages (Cushen *et al.*, 2012). This technology also has the potential to be used to retard the influx of unwanted gases into sealed meat packaging. MAP and vacuum packaging are both widely used packaging systems for meats that limit gas resources to microorganisms. The maintenance of specific ratios of gases is fundamental in MAP. Provided the package is sealed, the point of gas transfer is through the interspatial areas of the polymer packaging itself; the gases inside the enclosed package seek to equilibrate with the surrounding atmosphere gases. The gas barrier properties of the packaging materials can be improved with the incorporation of a high-aspect-ratio filler during the extrusion stage of plastic processing. When evenly distributed and correctly aligned, perpendicular to the direction of diffusion, the high-aspect-ratio filler acts as a hurdle course for moving gases (Thompson *et al.*, 2012).

Fresh meat and processed meat products (such as brined goods or patties) may require different characteristics from a food packaging material in terms of moisture permeability. Low water vapour permeability is a desirable trait in a meat packaging material. As with gas exchange, moisture movement through a barrier material can negatively impact the quality of the meat product by changing the optimised environment brought about by MAP or allowing for excessive moisture loss from the meat. Water vapour permeability of a packaging material can be reduced and optimised with the addition of fillers. In processed meats, where the available water has been minimised, the influx of moisture to the relatively *dry* package environment is undesirable, hence a filler being used for the purpose of improving moisture barriers, often in conjunction with barriers to other entities such as gas exchange.

Renowned for having high water vapour permeability and being incompatible with high-moisture foods, biodegradable packaging materials and their use with muscle foods is reviewed by Cutter (2006). The author discusses how edible starch films can retard the migration of moisture into meat or poultry products during storage and also reduce drip loss in the meat by binding some of the available water (Wong, Camirand and Pavlath, 1994) and the sheet-like arrangement of chitosan added to lauric acid films provided an efficient barrier to moisture exchange (Wong *et al.,* 1992). This is a significant development considering that the usefulness of biodegradable polymers depends largely on the oxygen transmission rate, which tends to be influenced by humidity (Sanchez-Garcia, Gimenez and Lagaron, 2007; Duncan, 2011).

Modified humidity packaging can be used to control dehydration and condensation of free water within food packages. This type of packaging would be better suited to the high water availability of fresh meat and poultry. Reduction in visible packaging condensation is done by reducing the surface tension of the water at the interface of the headspace and packaging. Although this type of packaging does not alter the humidity within the package, the lack of water adherence leads to a consumer perception of a fresher product; this has been found to be the case in the use of this packaging with freshly prepared salads (Muriel-Galet *et al.,* 2013). Packaging materials with high water vapour permeability can be used to minimise the quality losses due to water levels equilibrating between the meat product and the headspace.

Consumers expect certain attributes from particular packaging materials; for a highly variable product such as fresh meat where discerning consumers assess the quality of the cut based on visual cues such as marbling and/or colour, transparency in packaging is essential. The addition of fillers to improve packaging materials is usually limited so that the transparency of the film is maintained. High fill rates are more likely to cause opacity or a haze in the resulting packaging material. In general, the improvement of materials at very low fill levels are more likely to be obtained when the filler is in nanoform due to the high reactivity and superior dispersion that can be achieved with nanofillers.

10.2.4 Active packaging

Some packaging materials are designed to release antimicrobial compounds, antioxidants and/or flavours in an attempt to improve quality and product shelf life. Such products are termed as 'Active'. Active and intelligent packaging seem to be the next generation of packaging technology after nanofiller technology. Recent research has seen a focus on the active packaging and its benefits to the food industry.

Active packaging (as the name suggests) is not a passive technology and interacts directly with the food imparting an active effect such as oxygen absorption or antimicrobial activity. Antimicrobial packaging either kills microbes present or inhibits the growth of existing colonies on the surfaces of the packaged foods. Antimicrobial packaging is made by incorporating an antimicrobial entity or a suite of complementary antimicrobials into a food packaging matrix during plastic film casting. Although there have been new developments in other methods of manufacture which involve coating the inner surface of the packaging material with an antimicrobial entity.

Unlike improved packaging, there is a benefit in physical contact between the active food packaging surface and the food. In fact, for an active packaging material to be functional, this contact is required (Vermeiren, Devlieghere and Debevere, 2002). This intimate contact is likely to raise concerns over migration potential.

A clay product, kaolinite, has been incorporated into food packaging matrices (HDPE and LDPE) by Busolo *et al.* (2010). Its potential to extend the shelf life of the foods, including meat, is twofold, in a passive manner, improving the gas barrier performance of the polymer and the iron in the Kaolinite plays an active role by trapping and reacting with passing oxygen. TiO_2/NiO_2 is reported to rupture microbial cell membranes of some foodborne bacteria and to disrupt the enzymatic functions of the bacteria (Amna *et al.*, 2013); furthermore, a synergistic effect was seen between TiO_2 and NiO_2.

Surface microbial contamination of foods can be controlled by combining packaging materials with active ingredients, with some exhibiting antimicrobial effects. Contact between the active packaging and the food surface is critical for this to be effective (Vermeiren, Devlieghere and Debevere, 2002). Such developments can result in a significant extension of product shelf life enhancing food safety and quality with reduced wastage.

Silver is a metal which has been used in active packaging. The antimicrobial behaviour of silver increases as the particle size reduces (Baker *et al.*, 2005). Silver exhibits antimicrobial effects in its three forms: as the metal (including in nanoparticulate form), as a compound and as the free dissolved ion (Knetsch and Koole, 2011). Although it is a relatively recent discovery, it is largely accepted that the behaviour mechanism responsible for silver's antimicrobial effect involves the release of silver ions from the surfaces of nanoparticles (Klasen, 2000; Knetsch and Koole, 2011). Nanoparticles present a far greater surface area than larger particle sizes and provide a more concentrated release of ions – theoretically increasing its antimicrobial effectiveness over traditional silver applications. As an antimicrobial, it is more effective against aerobic bacteria than anaerobic strains (Lu *et al.*, 2013). The release of ions from their surfaces happens as part of the natural ageing of the particles through natural oxidative reactions. The rate of ion release is highly dependent on the surrounding environment that the nanoparticles are in (pH, water availability, light, etc.). Ionic silver's potential for toxicity compared to that of silver nanoparticles is different because the latter is an almost infinite source of the former that, environment permitting, readily releases ions (Quadros *et al.*, 2013). The nanoparticles may continue to release ions for a long period while the silver ions' antimicrobial potential is short lived. Silver compounded in a salt can significantly reduce the release of silver ions (and hence its antimicrobial potential) due to the hydrophobic, insoluble nature of many of them. Antimicrobial active packaging can have antimicrobial nanoparticles either anchored to the packaging surface (Ruparelia *et al.*, 2008) or incorporated within the polymer matrix (Sheikh *et al.*, 2011). It is highly likely that exposure to particulate nanomaterials will occur as a result of loose nanoparticles being on the surface of the packaging material and subsequently being transferred to the associated food item as is reported in Quadros *et al.* (2013). It is also likely that particulate nanomaterials within the polymer matrix will migrate to the packaged food.

A lot of the active packaging technologies are attributed to improvements in the understanding and applications of nanotechnologies; however, not all active packaging is associated with nanotechnology or uses nanomaterials. It is important to note that the current definition of a nanomaterial is still under review and subject to change. Due to the knowledge gaps around nanomaterial toxicity, the particle size range (50% or more of the particles are in the size range of 1–50 nm) is arbitrary (European Commission (EU), 2011). In general, the nanotechnologically associated active packaging technologies are more effective, novel and advanced, with improved, active and intelligent loosely representing the advancements of nanotechnologies in recent years.

The composition of the nanomaterial itself will ultimately determine any effect on the food product. The action of the nanomaterial may also be influenced by the polymer matrix, including particle release rate. While research is ongoing, many commercially successful products have been developed. Potential benefits include antimicrobial activity, O_2 barrier and improved gloss, improved modulus and strength, decreased water permeability and O_2 scavenging effects.

Due to cytotoxicity, nanomaterials have been shown to kill eukaryotic cells (Long *et al.*, 2006; Nel *et al.*, 2006) and inhibit the growth of prokaryotic cells (Brayner *et al.*, 2006; Thill *et al.*, 2006). This has particular importance for the meat industry in the control of different spoilage microorganisms and pathogens (prokaryotic cells). It is likely that cell nutrients are held at the surface of the nanomaterial which deprives the cell of valuable nutrients resulting in ultimate death (Geys, Nemery and Hoet, 2010). Nano-metals (e.g. silver, copper, gold) in particular have been identified as having unique antimicrobial properties with some products already on the market (Bouwmeester *et al.*, 2007). Commercialised products have also employed metal oxides where there is potential for light-activated microbial inactivation (Bouwmeester *et al.*, 2007).

Ag nanoparticles incorporated into packaging materials can extend the shelf life of food materials, with much reported regarding extension of the shelf life of fruit and vegetables (Hu and Fu, 2003; Li *et al.*, 2009). TiO_2-coated films have been shown to have an effect against *Escherichia coli* and therefore can benefit the shelf life of fresh produce, including meat (Chawengkijwanich and Hayata, 2008). Fernández *et al.* (2009) studied the antimicrobial effects of Ag nanoparticles in absorbent pads. Significant antimicrobial effects were noted against *E. coli* and *Staphylococcus aureus* using MAP for poultry meat. In addition, the freshness of fish samples were determined by using a xanthine amperometric sensor based on calcium carbonate nanoparticles (Shan *et al.*, 2009). Nanopackaging is particularly suited to assist in preserving high-value perishable products such as meat products.

There are many examples of packaging, or the broader category of food contact materials, that contain nanomaterials on the global market. However, in the EU, few examples exist that have passed migration tests. Their use in the EU is subject to risk assessment including migration testing on a case-by-case basis (EU 10/2011). This demonstrates the importance that migration results have on deciding the fate of a food contact materials in the EU, whether it is acceptable or not and therefore any chance of achieving commercial success. Regulations in the food industry are understandably rigorous as research on this new packaging type is very limited and little scientific data is available as to nanoparticle biocidal properties or their propensity to migrate into food that they are in contact with. It is important that these research gaps are filled prior to any adoption in industry. Although the research into the novel exciting properties has been done, the range of complexities within the nanoparticle realm determines the expansive range of behavioural properties and hence, any application to use nanoparticles in conjunction with food packaging materials must undergo a case-by-case risk assessment.

Not all active packaging is related to novel compounds; active packaging can also enlist commonly used compounds in novel ways. Contini *et al.* (2013, 2014) showed that the use of citrus extract could be used to delay oxidative deterioration in cooked meats, and a related study found that orientation of the adhered molecules, and specifically the availability of the active entities of the compounds, determined the antioxidant potential (Contini *et al.*, 2013, 2014).

10.2.4.1 Polymers used in conjunction with the particulate nanomaterials

Polymers commonly used in conjunction with nanoparticles have included polyamides, nylons, polyolefins, ethylene-vinylacetate copolymer, polystyrene, epoxy resins, polyurethane and polyethylene terephthalate (PET) (Šimon, Chaudhry and Bakos, 2008). The polymer choice can have a significant bearing on the efficacy of the components of the active packaging matrix. In particular, the rate of release of bioactives will be influenced by the polymer density (Cruz *et al.*, 2008). The polymer may be designed for a timed release of bioactive or they may exist as bound components in the polymer. Research has centred on harmonising nanoparticle size and concentration with required release rate. The use of multiple nanoparticles in one film and the surface contact area between the nanopackaging material and the food are also important considerations which need to be optimised to improve safety and shelf life. Multiple nanoparticles can have synergistic effects and complement one another in a nanocomposite.

Two main approaches have been used to produce nanomaterial polymers (Yang, Wang and Wang, 2007); the first involves the production of polymer materials at the nanoscale, and the alternative and most commonly used is to form a nanocomposite by introducing particulate nanomaterials into a polymer matrix.

10.2.4.2 Risks and chemical migration

When the active entities themselves exhibit novel biocidal properties to which no humans or any ecosystem has been exposed, concerns exist regarding long-term effects on human health, particularly in relation to gut microflora or bioaccumulation issues leading to organ failure as well as environmental issues. Aquatic filter feeding organisms are believed to be the most vulnerable to adverse effects of this technology as well as the food chains and ecosystems to which they contribute. Because of these concerns, European Legislation has put a blanket ban on the use of nanoparticles in food packaging materials.

The potential migration of particulate nanomaterials from FCMs into the food matrix has given rise to public health concerns. Migration studies are currently limited, therefore, resulting in significant data gaps in relation to the potential human health risk from using nanopackaging materials for foodstuffs. Given their role as packaging materials, it is usually not intentional that nanopackaging would give rise to significant exposure to migrating particles. Evaluation of potential human exposure is currently determined by migration tests to determine likely migrating levels of substances, followed by exposure assessment. The potential hazard characterisation (i.e. dose–response) is conducted based on results of animal studies. Further investigation is required to determine, more accurately, human exposure to migrating particles in food through contact with nanopackaging materials. The significance of the risk to human health is currently not known and will depend on several factors, including toxicity of the nanomaterial used, rate of migration of harmful substances and the level consumption of the food material.

Šimon, Chaudhry and Bakos (2008) has attempted to characterise migration potential of nanomaterials in a food packaging matrix using basic information pertaining to the particles themselves (e.g. density, diameter) and the polymer matrix. Following known models for diffusion, it is concluded that polymer dynamic viscosity and nanoparticle size can greatly influence migration potential and hence influence potential human exposure levels.

A biopolymer incorporating montmorillonite (nanoclay) was found to result in a significant increase in minerals in vegetables (Avella *et al.*, 2005). In addition, increases in silicon were detected in food samples when nanoclay was used as part of the packaging matrix; however, it was noted that levels did not exceed the EC norm for contact materials. Due to continued lack of knowledge regarding a dose–response relationship for nanomaterials, the precautionary principle is sure to dictate limited uptake until such time as material safety is demonstrated. A significant milestone in reaching this was the opinion of the European Food Safety Authority (EFSA) Scientific Panel on FCMs, Enzymes, Flavourings and Processing Aids (CEF) regarding the use of titanium nitride nanoparticles in plastic containers at 20 mg/kg, where they conclude migration did not represent any toxicological risk (National Nanotechnology Initiative, 2009). Huang *et al.* (2011) reported the use of food storage bags containing Ag nanoparticles resulted in some migration to four food simulants, with varying results depending on storage time and temperature. This highlights that data gaps currently exist in relation to our understanding of migration potential and this is sure to impact on the uptake of this technology.

Where migration is likely to occur, greater restrictions and regulation are sure to follow to ensure greater safety and negate potential negative consumer sentiment. This is likely to stifle uptake and progress in this technology. Developments in nanotechnologies will be costly and slow, as the methods are still been perfected while material consistency and potential risks also need to be considered.

Recent developments have seen greater emphasis on active compounds that do not migrate compared to those that do migrate from the polymer into the food material. Under legislation, those that migrate are considered as food additives. All necessary safety data is required before such materials can be used and included in the food additives positive list. Nano-Ag-embedded baby bottles have used bound-active particulate nanomaterials in FCMs which have an active effect without migrating into the digested food (Alfadul and Elneshwy, 2010). While some FCMs are designed to release active particles, other FCMs may not be designed to do so, but in any event may have some active effect.

As with any emerging technology, new risks and questions around safety arise. Both the health impact of ingesting nanoparticles and their effect on the environment is not yet fully understood. In addition, the use of these materials will increase human exposure to nanoparticles in the long term. Therefore, the aim of nanopackaging manufacturers should be to minimise the nanomaterial use, where possible, to a point where the antimicrobial effect is upheld to an acceptable level. In addition, manufacturers should seek to perfect the nanomaterial, through modification, size selection and so on, so that migration from the composite is minimised.

In theory, the ideal nanopackaging would incorporate nanoparticles which are small enough to be effective but large enough to be retained by the packaging. However, size is not the only influence on migration. It would appear migration is largely subject to physical attributes of the food packaging matrix, physical attributes of the nanoparticles and storage conditions (Šimon, Chaudhry and Bakos, 2008), although many uncertainties remain. If the relationships between the various parameters can be defined, through scientific literature or laboratory measurements, a mathematical model to simulate migration could also prove useful in the development of nanopackaging materials and for industry use. Uncertainties and variations in the model inputs can be accounted for by the use of inputs in the form of probability density distributions, instead of single figures. Model inputs can therefore be processed through a series of functions to arrive at an output and parameter uncertainty can be accounted for using Monte Carlo simulation (Cushen *et al.*, 2014). The complexity, cost and time-consuming nature of conventional migration testing are acknowledged in Article 32 of Commission Regulation (EU) No 10/2011 (EU 10/2011), and the validity of using calculations and modelling to assess migration is affirmed and recognised. Risk assessment-driven predictive modelling can identify the most influential factors influencing migration, ultimately determine the mass of migrating compound, and where modelling exposure, it can determine if the levels consumed by the population are above or below a specific safe limit (if available in the literature). Given the importance of agriculture and food in the EU and the opportunities presented by emerging nanotechnologies in food packaging, there is a need to investigate the safety concerns of food nanopackaging.

10.2.5 Intelligent packaging

Intelligent packaging is generally considered the most advanced and sees materials interacting with the food itself, retrieving specific information and either communicating with the relevant personnel and the consumer or initiating an active role in response. It is this communication element that distinguishes it from improved and active packaging; however, it may also employ the more basic functional properties of improved and active technologies (Armentano *et al.*, 2013).

The pre-packaging of meat somewhat limits consumers' ability to fully assess the meat, particularly the freshness of the meat. The detection of texture, odour and the presence of a biological slime layer are all hindered by the pre-packaging of meat. When using MAP, headspace analysis is carried out systematically at various stages of the food chain to ensure packaging integrity, but probe insertion is destructive. To minimise product destruction, sampling is minimised; however, this means that the samples taken represent a large number of units and so a product failure could mean that a large number of corresponding packages are destroyed or repackaged (Puligundla, Jung and Ko, 2012). The only indication of freshness that the consumer relies on is often the best before date. There is a potential for meat producers to mislead the consumer into thinking the product is fresher for longer than it is. Even if producers estimate a genuine shelf life, this estimation is made in the assumption that certain storage conditions are upheld. In the event that the storage conditions are not maintained to the expected standard, microbial growth and quality deterioration could be accelerated and the given shelf life becomes redundant. This concept has led to research in the area of intelligent packaging systems. These communicate information to the consumer in real time. The detection of marker chemical compounds is usually at the foundation of these technologies. These systems are often highly product specific because microbial contamination and hence microbial metabolites depend on the food. Ripesense (Sharrock and Henzell, 2010) and OnVu (Brizio and Prentice, 2014) are established quality indicators being used on the fruit and vegetable market. They communicate product ripeness to the consumer based on a logo colour change that responds to a ripening hormone exuded by the fruit. This highlights the food-specific nature of this technology; however, intelligent packaging can all benefit from systems that are irreversible, easy to read to the untrained eye, non-invasive, reliable/accurate, of appropriate sensitivity, tamper-proof, non-toxic and not added significantly to the overall price of packaging (Duncan, 2011).

The level of volatile basic nitrogen (VBN) is a very good indicator of beef quality because it directly relates to off-flavours that are produced over time (Byeon et al., 2009; Han *et al.,* 2012). The presence of biogenic amines in poultry is a spoilage indicator (Kuswandi *et al.,* 2014). Biogenic amines and volatile amines, which include VBN, are produced from the enzymatic decarboxylation of specific amino acids due to microbial enzyme activity (Kuswandi *et al.,* 2014). Han *et al.* (2012) developed a VBN prediction mechanism based on beef colour measurement; however, the VBN measurement was done using a laboratory titration and not on a packaging bound sensor.

Having noted the product-specific nature of quality indicators, there are also broad spectrum quality indicators in development. Oxygen sensors and carbon dioxide sensors have been developed. They are used independently of one another according to the food product. The increase of O_2 in an enclosed container (where O_2 levels had originally been low or eliminated) can indicate the loss of packaging integrity. If this elevation in oxygen is detected by an in situ oxygen sensor and converted into a detectable response, the consumption of a spoiled or potentially contaminated product can be avoided. Photoactivated indicator inks for use in package oxygen detection have been developed by Lawrie, Mills and Hazafy (2013). They describe the development of an ultraviolet-light-activated O_2-sensitive dye and suggest its suitability for use in MAP where there is an absence of O_2 or need for O_2 concentration control such as in ready-to-eat cooked meat packages.

Puligundla, Jung and Ko (2012) review the potential food packaging applications for existing CO_2 monitoring/measuring/detecting technology. They find that dry optical sensors containing pH-sensitive dye indicators are suited to CO_2 monitoring and/or spoilage indicators, whereas photonic crystal sensors also show promise due to their low cost, long lifetime and effective indication with quick response times. CO_2 is a microbial metabolite and so the presence of elevated levels in an enclosed meat package could indicate high microbial counts. Lactic acid bacteria are common spoilage microorganisms in meat and meat products (Pothakos *et al.*, 2013; Macè *et al.*, 2013). A number of pH indicating dyes were assessed by Kim and Lee (2012) for their suitability for use as a time temperature indicator. It is noted that the colour transition range of the substance must be correlated to the microbiological activity transition from fresh to spoiled.

If the intelligent packaging system is calibrated and engineered correctly, the system has the potential to visually change in a way that is detectable and recognisable by consumers at a point where the food is no longer safe to eat. Work by Duncan (2011) suggests that similar technology could be developed for the detection of food allergens, contaminants and adulterants.

The use of Information Communication Technologies (ICT) can further enhance the benefits of using smart packaging and in particular, intelligent packaging materials. ICT and intelligent packaging can integrate different stages of the supply chain, giving more control to the producer, processor and retailer. Real-time information can be logged in cloud infrastructure and queried by industry stakeholders or even consumers. Communication with consumers in this way can be a powerful marketing tool where consumers are given access to the storage history of a specific product unit or carbon footprint of the farm. Food packaging is not just important for the end users of the product; the interpretation of the intelligent packaging materials can be automated. In the high-throughput environment of some meat processing plants, food packaging can hold information important to the logistics of the processing plant. Automated readers of the intelligent food packaging materials can aid the decision-making process of waste stream management. For example, if there is a packaging-borne sensor designed to assure vacuum packaging integrity of pre-packed bacon, visual recognition software in conjunction with ICT could be employed to detect defective units. Decision-making tools incorporated into the plant ICT system could reroute defective units for repackaging, thus minimising waste.

Time–temperature logs have the ability to help with biosecurity and traceability. Expiry dates are set based on underlying assumptions regarding time and temperature. However, these conditions are seldom maintained potentially resulting in reduced shelf life and in extreme circumstances the potential for the formation of pathogens and toxins in the food. Intelligent FCMs are the next generation of active FCMs as research into this field is rapidly extending around the world. Research into nanosensors as part of the packaging solution has gained traction with nanosensors build to detect environmental abuse and specific harmful food pathogens and chemicals, and thereby alerting producers, retailers and consumers (Neethirajan and Jayas, 2011). Nanosensors integrated into food packaging can detect specific environmental changes, including indicators of microbial metabolism. In addition, different stimuli can have the impact of changing the food product environment using this new intelligent FCMs technology. This is achieved by the release of substances into the environment of the food product to limit contamination (Pereira de Abreu et al., 2012). These stimuli may include environmental changes such as in pH, pressure, temperature and light (Otles and Yalcin, 2008). In addition, the products of microbial metabolism may act as a stimulus to alter storage conditions. Changes in a product's temperature, light or O_2 exposure can also be alerted to consumers. Given the continuous monitoring facility, these sensors may provide a more realistic evaluation of the condition of the food and may even eliminate the need for expiry dates on certain products. It also has the added advantage for manufacturers in ensuring the product has reached the consumer in the condition required while also identifying potential weaknesses in the distribution chain by identifying incidences of time/temperature abuse and mishandling. It has been reported that nanosensors have already been commercialised in certain countries (Nachay, 2007). The current high cost of these sensors has limited their role out in Europe. In addition, retailers and food companies have been cautious to use the technology until security and safety has been demonstrated. The use of sensors of this kind is not yet widespread throughout Europe due to high costs and cautious uptake. In addition, restrictions may be in place where chemicals are used as part of the intelligent packaging.

Many of these developments in intelligent packaging systems are still in their infancy but are sure to become more prominent in their usage in future years. Current EU legislation is likely to be stringent and limit commercial availability of this technology until such time as all potential human and environmental risks are investigated and a better understanding of the technology is achieved.

10.3 Conclusion

The development of smart packaging has presented new opportunities to the meat industry. Smart packaging can present great opportunities to control product quality and improve the supply chain efficiency. Among the recent advances include the potential to enhance traditional packaging materials with antimicrobial fillers. Novel antimicrobial meat packaging has the potential to offset food spoilage which can facilitate the extension of export boundaries allowing access to new markets. Delaying food spoilage also benefits consumers in reducing food waste.

Advances in nanotechnologies present many opportunities for the meat industry. However, uncertainties around their safe use and disposal have led to some negative consumer sentiment. This issue is particularly sensitive for the food sector, with many people drawing comparisons with the ongoing

GM controversy/debate. It is important that current knowledge gaps regarding safety are investigated to address consumer concerns regarding the use of nanotechnologies. There is a need to determine potential migration from food packaging materials and subsequent human exposure levels.

The manner in which smart packaging systems work are as diverse as the systems themselves. The packaging matrix as well as the function and reactivity of nanoparticles determine the functionality of the smart packaging. The range of smart packaging materials is large and expanding, with some acting as a simple physical barrier and others having the potential to react to certain conditions in the meat. One of the major advantages of smart packaging is that it acts on the contents of the package without the need to interfere with the contents. Advances in new packaging concepts continue at a fast pace and are sure to play a continued role in the monitoring and preservation of meat quality.

References

Abdollahi, M., Alboofetileh, M., Rezaei, M. and Behrooz, R. (2013) Comparing physico-mechanical and thermal properties of alginate nanocomposite films reinforced with organic and/or inorganic nanofillers. *Food Hydrocolloids*, **32** (2), 416–424.

Adame, D. and Beall, G.W. (2009) Direct measurement of the constrained polymer region in polyamide/clay nanocomposites and the implications for gas diffusion. *Applied Clay Science*, **42** (3–4), 545–552.

Alexandre, M. and Dubois, P. (2000) Polymer-layered silicate nanocomposites: Preparation, properties and uses of a new class of materials. *Materials Science and Engineering, R: Reports*, **28** (1-2), 1–63.

Alfadul, S.M. and Elneshwy, A.A. (2010) Use of nanotechnology in food processing, packaging and safety – review. *African Journal of Food, Agriculture, Nutrition and Development*, **10** (6), 2719–2739.

Amna, T., Hassan, M.S., Yousef, A. *et al.* (2013) Inactivation of foodborne pathogens by NiO/TiO_2 composite nanofibers: A novel biomaterial system. *Food and Bioprocess Technology*, **6** (4), 988–996.

Armentano, I., Bitinis, N., Fortunati, E. *et al.* (2013) Multifunctional nanostructured PLA materials for packaging and tissue engineering. *Progress in Polymer Science*, **38** (10–11), 1720–1747.

Avella, M., De Vlieger, J.J., Errico, M.E. *et al.* (2005) Biodegradable starch/clay nanocomposite films for food packaging applications. *Food Chemistry*, **93** (3), 467–474.

Baker, C., Pradhan, A., Pakstis, L. *et al.* (2005) Synthesis and antibacterial properties of silver nanoparticles. *Journal of Nanoscience and Nanotechnology*, **5** (2), 244–249.

Bartosz, G. (2013) *Food Oxidants and Antioxidants: Chemical, Biological, and Functional Properties*, CRC Press.

Bilbao-Sainz, C., Bras, J., Williams, T. *et al.* (2011) HPMC reinforced with different cellulose nano-particles. *Carbohydrate Polymers*, **86** (4), 1549–1557.

Bouwmeester, H., Dekkers, S., Noordam, M. *et al.* (2007) *Health Risks of Application of Nanotechnologies and Nanoparticles within the Food Production Chain. Health Impact of Nanotechnologies in Food Production*, RIKILT Institute of Food Safety and RIVM National Institute for Public Health and the Environment.

Brayner, R., Ferrari-Iliou, R., Brivois, N. *et al.* (2006) Toxicological impact studies based on *Escherichia coli* bacteria in ultrafine ZnO nanoparticles colloidal medium. *Nano Letters*, **6** (4), 866–870.

Brizio, A.P. and Prentice, C. (2014) Use of smart photochromic indicator for dynamic monitoring of the shelf life of chilled chicken based products. *Meat Science*, **96** (3), 1219–1226.

Busolo, M.A., Fernandez, P., Ocio, M.J. and Lagaron, J.M. (2010) Novel silver-based nanoclay as an antimicrobial in polylactic acid food packaging coatings. *Food Additives & Contaminants: Part A*, **27** (11), 1617–1626.

Byeon, K. E., An, S. R., Shim, S. D., Lee, J. Y., Hong, K. W., Min, S. G., and Lee, S. J. (2009) Investigation on beef quality indicator of off-flavor development during storage. *Korean journal for food science of animal resources*, **3**, 325–333.

Chawengkijwanich, C., and Hayata, Y. (2008). Development of TiO2 powder-coated food packaging film and its ability to inactivate Escherichia coli *in vitro* and in actual tests. *International Journal of Food Microbiology*, **123** (3), 288–292.

Chisholm, N., Mahfuz, H., Rangari, V.K. *et al.* (2005) Fabrication and mechanical characterization of carbon/SiC-epoxy nanocomposites. *Composite Structures*, **67** (1), 115–124.

Chung, Y.-L., Ansari, S., Estevez, L. *et al.* (2010) Preparation and properties of biodegradable starch-clay nanocomposites. *Carbohydrate Polymers*, **79** (2), 391–396.

Contini, C., Àlvarez, R., O'Sullivan, M. *et al.* (2014) Effect of an active packaging with citrus extract on lipid oxidation and sensory quality of cooked turkey meat. *Meat Science*, **96** (3), 1171–1176.

Contini, C., Katsikogianni, M.G., O'Neill, F.T. *et al.* (2013) Storage stability of an antioxidant active packaging coated with citrus extract following a plasma jet pretreatment. *Food and Bioprocess Technology*, **7** (8), 2228–2240.

Cruz, J.M., Sanches Silva, A., Sendón García, R. *et al.* (2008) Studies of mass transport of model chemicals from packaging into and within cheeses. *Journal of Food Engineering*, **87** (1), 107–115.

Cushen, M., Kerry, J., Morris, M. *et al.* (2012) Nanotechnologies in the food industry – Recent developments, risks and regulation. *Trends in Food Science & Technology*, **24** (1), 30–46.

Cushen, M., Kerry, J., Morris, M. *et al.* (2014) Evaluation and simulation of silver and copper nanoparticle migration from polyethylene nanocomposites to food and an associated exposure assessment. *Journal of Agricultural and Food Chemistry*, **62** (2), 1403–1411.

Cutter, C.N. (2006) Opportunities for bio-based packaging technologies to improve the quality and safety of fresh and further processed muscle foods. *Meat Science*, **74** (1), 131–142.

de Azeredo, H.M.C. (2009) Nanocomposites for food packaging applications. *Food Research International*, **42**, 14.

Duncan, T.V. (2011) Applications of nanotechnology in food packaging and food safety: Barrier materials, antimicrobials and sensors. *Journal of Colloid and Interface Science*, **363** (1), 1–24.

European Commission (2010) *Bio Intelligence Service Preparatory Study on Food Waste Across EU 27*: http://ec.europa.eu/environment/eussd/pdf/bio_foodwaste_report.pdf (accessed September 2013)

European Commission (EU) (2011) Commission recommendation of 18th October 2011 on the definition of nanomaterials (2011/696/EU). *Official Journal of the European Union*, **54**, 1e38, No. L 275/38.

Fernández, A., Soriano, E., López-Carballo, G. *et al.* (2009) Preservation of aseptic conditions in absorbent pads by using silver nanotechnology. *Food Research International*, **42** (8), 1105–1112.

Geys, J., Nemery, B. and Hoet, P.H.M. (2010) Assay conditions can influence the outcome of cytotoxicity tests of nanomaterials: Better assay characterization is needed to compare studies. *Toxicology in Vitro*, **24** (2), 620–629.

Giannelis, E.P. (1996) Polymer layered silicate nanocomposites. *Advanced Materials*, **8** (1), 29–35.

Grebitus, C., Jensen, H. H., Roosen, J. and Sebranek, J. G. (2013) *Consumer Acceptance of Fresh Meat Packaging with Carbon Monoxide*. Animal Industry Report: AS 659, ASL R2756.

Han, J.Y., Jung, M.K., Dong, S.S. and S.L. Ju, (2012) Application of fuzzy reasoning to prediction of beef sirloin quality using time temperature integrators (TTI's). *Food Control*, **24** (1), 148–153.

Hu, A.W. and Fu, Z.H. (2003) Nanotechnology and its applications in packaging and packaging machinery. *Packaging Engineering*, **24**, 22–24.

Huang, Y., Chen, S., Bing, X. *et al.* (2011) Nanosilver migrated into food-simulating solutions from commercially available food fresh containers. *Packaging Technology and Science*, **24** (5), 291–297.

Kim, K. and Lee, S.J. (2012) New enzymatic time-temperature integrator (TTI) that uses laccase. *Journal of Food Engineering*, **113** (1), 118–123.

Klasen, H.J. (2000) Historical review of the use of silver in the treatment of burns. I. Early uses. *Burns*, **26** (2), 117–130.

Knetsch, M.L.W. and Koole, L.H. (2011) New strategies in the development of antimicrobial coatings: The example of increasing usage of silver and silver nanoparticles. *Polymers*, **3** (1), 340–366.

Kuswandi, B., Jayus, A. and Oktaviana, R. *et al.* (2014) A novel on -package sticker sensor based on methyl red for real time monitoring of broiler chicken cut freshness. *Packaging Technology and Science*, **27** (1), 69–81.

Lawrie, K., Mills, A. and Hazafy, D. (2013) Simple inkjet-printed, UV-activated oxygen indicator. *Sensors and Actuators B: Chemical*, **176**, 1154–1159.

Li, H., Li, F., Wang, L. *et al.* (2009) Effect of nano-packing on preservation quality of Chinese jujube (*Ziziphus jujuba* Mill. var. inermis (Bunge) Rehd). *Food Chemistry*, **114** (2), 547–552.

Long, T.C., Saleh, N., Tilton, R.D. *et al.* (2006) Titanium dioxide (P25) produces reactive oxygen species in immortalized brain microglia (BV2): Implications for nanoparticle neurotoxicity. *Environmental Science & Technology*, **40** (14), 4346–4352.

Lu, Z., Rong, K., Li, J. *et al.* (2013) Size-dependent antibacterial activities of silver nanoparticles against oral anaerobic pathogenic bacteria. *Journal of Materials Science: Materials in Medicine*, **24** (6), 1465–1471.

Macè, S., Joffraud, J.-J., Cardinal, M. *et al.* (2013) Evaluation of the spoilage potential of bacteria isolated from spoiled raw salmon (*Salmo salar*) fillets stored modified atmosphere packaging. *International Journal of Food Microbiology*, **160** (3), 227–238.

Majeed, K., Jawaid, M., Hassan, A. *et al.* (2013) Potential materials for food packaging from nanoclay/natural fibres filled hybrid composites. *Materials & Design*, **46**, 391–410.

Mallakpour, S. and Madani, M. (2012) Transparent and thermally stable improved poly(vinyl alcohol)/Cloisite Na^+/ZnO hybrid nanocomposite films: Fabrication, morphology and surface properties. *Progress in Organic Coatings*, **74** (3), 520–525.

Muriel-Galet, V., Cerisuelo, J.P., Lopez-Carballo, G. *et al.* (2013) Evaluation of EVOH-coated PP films with oregano essential oil and citral to improve the shelf life of packaged salad. *Food Control*, **30** (1), 137–143.

Nachay, K. (2007) Analyzing nanotechnology. *Food Technology*, **61**, 34–36.

National Nanotechnology Initiative (2009) *National Nanotechnology Initiative: Research and Development Leading to a Revolution in Technology and Industry, Supplement to President's FY 2010 Budget*, vol. **40**, Office of Science and Technology Policy, Washington, DC. Available at: http://www.nano.gov/NNI_2010_budget_supplement.pdf (accessed August 2010).

Neethirajan, S. and Jayas, D. (2011) Nanotechnology for the Food and Bioprocessing Industries. *Food and Bioprocess Technology*, **4** (1), 39–47.

Nel, A., Xia, T., Madler, L. and Li, N. (2006) Toxic potential of materials at the nanolevel. *Science*, **311** (5761), 622–627.

Newman, M.D., Stotland, M. and Ellis, J.I. (2009) The safety of nanosized particles in titanium dioxide – and zinc oxide – based sunscreens. *Journal of the American Academy of Dermatology*, **61** (4), 685–692.

Olsen, N.V., Grunert, K.G. and Sonne, A.-M. (2010) Consumer acceptance of high-pressure processing and pulsed-electric field: A review. *Trends in Food Science & Technology*, **21** (9), 464–472.

Otles, S., and Yalcin, B. (2008) Intelligent food packaging. *LogForum*, **4** (4), 1–9.

Ozcalik, O. and Tihminlioglu, F. (2013) Barrier properties of corn zein nanocomposite coated polypropylene films for food packaging applications. *Journal of Food Engineering*, **114** (4), 505–513.

Painter, J., Hoekstra, R., Ayers, T. *et al.* (2013) Attribution of foodborne illnesses, hospitalizations, and deaths to food commodities by using outbreak data, United States, 1998–2008. *Emerging Infectious Diseases*, **19** (3), 407–415.

Pereira de Abreu, D. A., Cruz, J. M., and Paseiro Losada, P. (2012) Active and Intelligent Packaging for the Food Industry. *Food Reviews International*. **28** (2), 146–187.

Pothakos, V., Nyambi, C., Zhang, B., Papastergidis, A., De Meulenaer, B. and Devlieghere, F. (2013) *Spoilage potential of psychrotropic lactic acid bacteria (LABnn Leuconostoc gasicomitatum and Lactococcus piscium under different packaging conditions*. Paper presented at the Microbial Spoilers in Food, Quimper, France, 1–3 July 2013.

Puligundla, P., Jung, J. and Ko, S. (2012) Carbon dioxide sensors for intelligent food packaging applications. *Food Control*, **25** (1), 328–333.

Quadros, M.E., Pierson, R., Tulve, N.S. *et al.* (2013) Release of silver from nanotechnology-based consumer products for children. *Environmental Science & Technology*, **47** (15), 8894–8901.

Quested, T.E., Cook, P.E., Gorris, L.G.M. and Cole, M.B. (2010) Trends in technology, trade and consumption likely to impact on microbial food safety. *International Journal of Food Microbiology*, **139** (Supplement), 29–42.

Rhim, J.-W. and Ng, P.K.W. (2007) Natural biopolymer-based nanocomposite films for packaging applications. *Critical Reviews in Food Science and Nutrition*, **47** (4), 411–433.

Rosegrant, M.W., Tokgoz, S. and Bhandary, P. (2013) The new normal? A tighter global agricultural supply and demand relation and its implications for food security. *American Journal of Agricultural Economics*, **95** (2), 303–309.

Ruparelia, J.P., Chatterjee, A.K., Duttagupta, S.P. and Mukherji, S. (2008) Strain specificity in antimicrobial activity of silver and copper nanoparticles. *Acta Biomaterialia*, **4** (3), 707–771.

Sanchez-Garcia, M.D., Gimenez, E. and Lagaron, J.M. (2007) Novel PET nanocomposites of interest in food packaging applications and comparative barrier performance with biopolyester nanocomposites. *Journal of Plastic Film and Sheeting*, **23** (2), 133–148.

Shan, D., Wang, Y., Xue, H. and Cosnier, S. (2009) Sensitive and selective xanthine amperometric sensors based on calcium carbonate nanoparticles. *Sensors and Actuators B: Chemical*, **136** (2), 510–515.

Sharrock, K.R. and Henzell, R.F. (2010) Ethylene ripening of pears by unconventional means: Use of experimental thimble-sized ethylene capsules inside cartons and clamshells. International Symposium Postharvest Pacifica 2009 - pathways to quality: V International Symposium on Managing Quality in Chains. *Acta Horticulturae*, **880**, 339–346.

Sheikh, F.A., Kanjwal, M.A., Saran, S. *et al.* (2011) Polyurethane nanofibers containing copper nanoparticles as future materials. *Applied Surface Science*, **257** (7), 3020–3026.

Šimon, P., Chaudhry, Q. and Bakos, D. (2008) Migration of engineered nanoparticles from polymer packaging to food – a physicochemical view. *Journal of Food and Nutritional Research*, **47** (3), 105–113.

Sinha Ray, S., Yamada, K., Okamoto, M. and Ueda, K. (2002) Polylactide-layered silicate nanocomposite: A novel biodegradable material. *Nano Letters*, **2** (10), 1093–1096.

Sinha Ray, S., and Okamoto, M. (2003). Polymer/layered silicate nanocomposites: a review from preparation to processing. *Progress in Polymer Science*, **28** (11), 1539–1641.

Sorrentino, A., Gorrasi, G. and Vittoria, V. (2007) Potential perspectives of bio-nanocomposites for food packaging applications. *Trends in Food Science & Technology*, **18** (2), 84–95.

Sozer, N. and Kokini, J.L. (2009) Nanotechnology and its applications in the food sector. *Trends in Biotechnology*, **27** (2), 82–89.

Thill, A., Zeyons, O., Spalla, O. *et al.* (2006) Cytotoxicity of CeO_2 nanoparticles for *Escherichia coli*. Physico-chemical insight of the cytotoxicity mechanism. *Environmental Science & Technology*, **40** (19), 6151–6156.

Thompson, M.S., Agarwal, S., Mukhopadhyay, P. and Gupta, R.K. (2012) Diffusion through polymers containing platelike nanomaterials, in *Graphite, Graphene, and Their Polymer Nanocomposites* (eds P. Mukhopadhyay and R.K. Gupta), CRC Press, p. 467.

Vermeiren, L., Devlieghere, F. and Debevere, J. (2002) Effectiveness of some recent antimicrobial packaging concepts. *Food Additives & Contaminants, Part A: Chemistry, Analysis, Control, Exposure & Risk Assessment*, **19** (4 supp 1), 163–171.

Wong, D.W.S., Camirand, W.M. and Pavlath, A.E. (1994) Development of edible coatings for minimally processed fruits and vegetables, in *Edible Coatings and Films to Improve Food Quality* (eds J.M. Krochta and M.O. Nisperos-Carriedo), CRC Press Llc.

Wong, D.W.S., Gastineau, F.A., Gregorski, K.S. *et al.* (1992) Chitosan-lipid films: Microstructure and surface energy. *Journal of Agricultural and Food Chemistry*, **40** (4), 540–544.

Yang, K.K., Wang, X.L. and Wang, Y.Z. (2007) Progress in Nanocomposite of Biodegradable Polymer. *Journal of Industrial and Engineering Chemistry*, **13** (4), 485–500.

Yang, H., Zhu, S. and Pan, N. (2004) Studying the mechanisms of titanium dioxide as ultraviolet-blocking additive for films and fabrics by an improved scheme. *Journal of Applied Polymer Science*, **92** (5), 3201–3210.

CHAPTER 11
Probiotic functionality in meat

Muhammad Issa Khan[1,2], Cheorun Jo[2] & Ubaid-ur-Rahman[1]
[1] *National Institute of Food Science and Technology, University of Agriculture, Faisalabad, Pakistan*
[2] *Department of Animal Biotechnology, Seoul National University, Seoul, Republic of Korea*

11.1 Introduction

The term "let food be thy medicine and medicine be thy food" was used nearly 2500 BC by Hippocrates and, possibly, he could not be more accurate in forecasting the trends and preferences of the present-world consumers. Consumer's food preferences have demographic changes based on their education as a result of factors including modern health worries, negative effect of medicines, and desire for a healthier life style (Chen, 2011). Nowadays, consumers demand healthy and natural food products meeting the required nutrition and safety levels. Functional foods are foods or dietary components providing health benefits beyond the provision of basic nutrition (Kołozyn-Krajewska and Dolatowski 2012). They are classified into four distinct groups (Khan *et al.*, 2013): (1) foods that are naturally rich in biologically active compounds (e.g., oat bran has dietary fibers); (2) foods having scientific support but lack any approved health claims (e.g., fish rich in omega-3); (3) foods that have been fortified or enriched with active components (e.g., calcium-fortified juice); (4) whole foods associated with the reduced risk of disease (e.g., probiotics that have been shown to improve gastrointestinal health (Goldin, 1999)). Meat is a highly nutritious food rich in proteins, essential amino acids, vitamin, and minerals and current consumer trends is toward meat products with incorporated functional ingredients (Zhang *et al.*, 2010). Lactic acid bacterial (LAB) are known for more than a century to have beneficial health impacts on human health and incorporation of LAB into meat during fermentation and hydrolysis results in the production of bioactive peptides that are beneficial for human health (Arihara, 2006). The original observation of the positive role played by certain bacteria was first introduced by Russian scientist and Nobel laureate Élie Metchnikoff, who in the beginning of the 20th century suggested that it would be possible to modify the gut flora and to replace harmful microbes with useful microbes. The term probiotics was first used by Lilly and Stillwell (1965) to describe "substances secreted by one microorganism which stimulate the growth of another." Sperti (1971)

Emerging Technologies in Meat Processing: Production, Processing and Technology, First Edition.
Edited by Enda J. Cummins and James G. Lyng.

also used this term to describe "the tissue extracts which stimulated microbial growth." Parker in 1974 described probiotics as "organisms and substances which contribute toward intestinal microbial balance." Fuller in 1989 redefined probiotics as "A live microbial feed supplement which beneficially affects the host animal by improving its intestinal microbial balance." He emphasized the requirement for the probiotics to be viable. The most appropriate definition, according to FAO/WHO in October 2001, is "probiotics are live microorganisms which when administered in adequate amounts confer a health benefit on the host." According to this definition, probiotics have the following characteristics: It must be alive, deliver a measured physiological benefit, not restricted to food applications or oral delivery and shouldn't limit the mechanism of action. Probiotics protect us in two ways by maintaining the balance of good and bad bacteria in digestive tract and by protecting our immune system.

(a) *Balance between good and bad bacteria in the intestinal tract*
Our digestive system requires a balance between beneficial and harmful bacteria for proper functioning. A healthy digestive tract eliminates the harmful substances such as toxic chemicals, harmful bacteria, and other waste generated during the digestion process, and absorbs the required nutrients to deliver them in the body to the specific place of their utilization. All the bad bacteria should not be killed as the body also needs bad bacterial in addition to good ones. Problem occurs if the number of bad exceeds the good. This imbalance leads to several digestive ailments including infection of urinary tract, fatigue, diarrhea, and muscle pain. Probiotics are thought to play a vital role in maintaining the balance of both types of bacterial populations in our digestive tract.

(b) *Protection of the immune system*
Probiotics are also claimed to protect the immune system from attack by different pathogens. This function of probiotics is very important for a healthy lifestyle. A poor immune system causes several allergic reactions, infections (infectious diarrhea, skin and vaginal infections, and *Helicobacter pylori*), and autoimmune disorders (ulcerative colitis, rheumatoid arthritis, and Crohn's disease).

Probiotics are associated with the maintenance of a normal intestinal milieu, detoxification of colon, stimulation of the immune system, prevention of lactose intolerance, cholesterol lowering capacity, and production of several essential metabolites responsible for good intestinal health (Resta-Lenert and Barrett, 2003). The proposed mechanism for probiotics to deliver possible health benefits have been summarized in Figure 11.1.

11.2 Ecology of gastrointestinal tract (GIT)

The human gastrointestinal tract has a complex and diverse ecosystem. The surface area of an average human GI system is approximately 150–200 m^2, which is

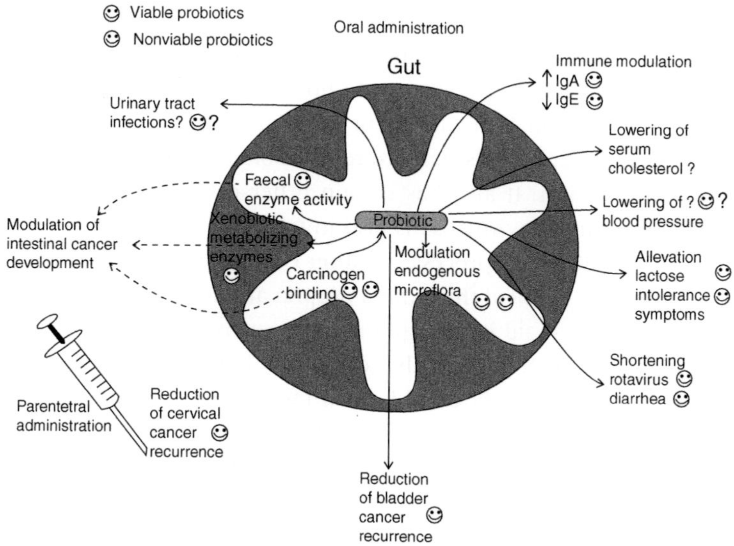

Figure 11.1 Proposed mechanism of viable and nonviable probiotic health effects. (From Salminen *et al.* (1999). Reproduced with permission of Elsevier.)

huge when compared to 2 m^2 skin surface of our whole body (Waldeck, 1990). The circular folding triples the surface area, the folding of epithelium gives rise to a 7- to 10-fold increase, and microvilli formation results in a 15- to 40-fold increase in surface area. Gut ecosystem facilitates the adhesion and interaction of microbes. A large surface for adhesion of microbes is provided by the mucosal membrane in the small intestine providing sufficient space for adhesion, interaction, and subsequent colonization during the digestive process. Barriers formed by the mucosal epithelium, if healthy and stable, ensure good health to the host and also ensure healthy intestinal functions and immunological resistance. Lymphoid tissues, present in the gut, are the largest immune organ of the human body and provide internal defense against infective and pathogenic agents. According to an estimate, 80% ($\sim 10^{10}$) of all immunoglobulin producing cells are present in the small intestine (Shanahan, Thomson, and Amyes, 1994). It is another fact that the gut flora is itself a prerequisite for stimulation of the immune system associated with mucosal membrane and augmentation of the immune cells. The digestion process involves physical and chemical digestion. In physical digestion, the size of the food particles is reduced and ultimately the surface area of the ingested food is increased for maximum contact with digestive enzymes. However, in chemical digestion, complex food molecules are broken down into their subunits by enzymatic hydrolysis. Proper physical digestion helps in efficient chemical digestion. Digestion is the combination of several processes including ingestion, mastication, and deglutition. The digestive tract consists of alimentary canal, a tube ranging from mouth to anus, along with several organs and glands that are responsible for the secretion of different hormones. The GI tract is normally composed of the stomach, small and large intestines, but the process of designation includes the whole alimentary canal. The alimentary canal has four layers made up of certain tissues. The first layer is called the mucosa and is made up of columnar epithelium. This layer is further composed of goblet cells that are responsible for the secretion of mucus that provides lubrication. It also has certain other cells that secrete a variety of enzymes and hormones. Mucosa is also involved in the absorption of some end products of the digestion process. Another component, present beneath the surface of epithelium, is called the lamina propria and is composed of blood and lymph capillaries, which aid in absorption. The mucosal boundary is called the muscularis mucosa, which provides better contact between the mucosal lining and contents of the alimentary canal by contraction. The second main layer is called the submucosa, which is present beneath the mucosa and mainly contains areolar connective tissues having major blood vessels, nerves, and lymph nodes. The digestive functions are generally controlled by the submucosal nerve plexus. The muscularis consists of layers of longitudinal and circular smooth muscles. Circular layers of smooth muscles squeeze to provide segmentation in the intestine, whereas longitudinal muscle layer cause peristalsis. The fourth layer is called serosa or fibroserous layer that is a covering membrane of the alimentary canal in the peritoneal cavity. After the completion of the digestion and absorption process, defecation and exfoliation take place.

Approximately 10^{14} microbial cells are found in average human adult (Luckey and Floch, 1972), which is approximately 10 times greater when compared to the cells of our body altogether. Such a large population of microbes suggests the potential of regulatory effects in the human body, specifically in the colon where this number is the maximum, that is, 5–10^{11} per gram. Consisting of more than 400 species, these naturally occurring microbes belong to diversified genera. The large intestine is dominated by Gram-positive anaerobic genera specifically *Bifidobacterium, Bacteroides,* and *Eubacterium.* Miscellaneous groups such as *lactobacilli, peptostreptococci,* and *Clostridia* perform vital roles such as maintenance of a healthy gut mucosa and short-chain fatty acid. The important role of the gut flora in human physiology was previously overlooked and not considered in the past, and the main area of interest was centralized on studying enteric pathogens and factors that could cause gastrointestinal malfunction (also referred to as dysbiosis). But now we have realized the importance of a healthy intestinal ecosystem, in addition to a well-balanced microflora, which functions as an important barrier against the uptake and subsequent proliferation of pathogens, antigens, and injurious entities from our lumen. In addition, the intestinal epithelium imbibes antigens, and immune system is triggered by a special immune response mechanism in Peyer's patches and intestinal villi (Heyman *et al.,* 1986). The main physiological functions of gut flora include barrier function, stimulation of immune system, maintenance and circulation of mucosa nutrition, improvement in the bioavailability of nutrients, and betterment of bowel motility. However, despite the advancements in science and technology, the actual mechanism involved behind these complex reactions and the way of microbial interaction is still unknown (Holzapfel *et al.*, 1998).

11.3 Identification of potential microorganisms

Taxonomic classification and identification is the first step to be followed during the selection of probiotics, giving suggestions about the origin, habitation, and physiology of the specie. On the whole, these characters have unique and vital outcomes on the selection of the probiotic microbes (Morelli, 2007).

Taxonomic classification and the association of probiotics are anchored in the comparative study of genes that have information encoded in the form of ribosomal RNA (rRNA). Large databases for sequencing of 16S and 23S ribosomal RNA make identification and classification of potential probiotics a simple, fast, and reliable practice. Closely related strains are differentiated by employment of DNA-based methods such as restriction enzyme analysis (REA), plasmid profiling, randomly amplified polymorphic DNA (RAPD), pulse-field electrophoresis (PFGE), and ribo-typing. Probiotic products consist of bacteria or yeasts that help in the restoration of balance of good and bad gut microflora. The main detected gut microflora includes *Clostridia, Enterococci, Lactobacilli, Escherichia coli*, and many more. Probiotics are of many types and have specific health benefits that can be determined by their specific functional role in the gut. Health benefits of some commonly used probiotics are discussed in the following lists.

(a) *Lactobacillus*

Lactobacilli have more than 50 species and are mostly present in our digestive, excretory, and genital systems. Fermented foods such as meat sausages and yogurt also contain plentiful amounts of these bacterial strains. *Lactobacilli* are known to play an important role in the treatment and prevention of a number of diseases. They take part in curing several yeast infections, protecting the urinary tract, and preventing several other diseases including irritable bowel syndrome, diarrhea, traveler's diarrhea, *Clostridium difficile* infection, lactose intolerance, respiratory disorders, and skin infections. *Lactobacillus GG* effectively reduces irritable bowel syndrome and abdominal pain in children of 5–14 years of age. It also prevents diarrhea. *Lactobacillus bulgaricus, Lactobacillus casei*, and *Streptococcus thermophilus* reduce the chances of diarrhea among adults. *Lactobacillus GG* also prevents lung infection in children between 1 and 6 years of age. *Lactobacillus gasseri* and *Lactobacillus rhamnosus* containing foods are also helpful in the prevention of bacterial vaginosis.

(b) *Bifidobacteria*

Bifidobacteria have more than 30 species that make up 90% of the total healthy bacteria found in the colon. They are present in the intestine immediately after birth, particularly in the breastfed infants. Several studies confirm the role of *Bifidobacteria* in preventing dental cavities and glucose intolerance. *Bifidobacterium infantis* 35624 is very helpful for the people with irritable bowel to reduce abdominal pain, bowel dysfunctioning, straining, bloating, incomplete evacuation, and passage of gas. *Bifidobacterium lactis* Bb12 has beneficial effects on body metabolism. It is associated with lowering blood LDL-cholesterol level, diabetes, and glucose intolerance in women during pregnancy.

(c) *Saccharomyces boulardii*
This is the only yeast probiotic that is involved in controlling diarrhea, *C. difficile*, and acne. It is also used to decrease the side effects of *H. pylori* infection treatment.

(d) *S. thermophilus*
S. thermophilus is normally reported to produce lactase enzyme in the body and improves lactose tolerance in lactose intolerants.

(e) *Enterococcus faecium*
E. faecium is a habitant of human and animal intestinal tract. Two strains of this bacterium, *E. faecium* M-74 and *E. faecium* SF68, have probiotic functionality.

(f) *Leuconostoc*
This bacterium has been widely used in the production of various probiotic foods and can be ingested as live or dead bacteria. Moreover, metabolites of *Leuconostoc* are also used in food processing or normal functioning of the digestive system.

11.4 Selection of probiotics

Before selection as probiotics, it is necessary that the considered microbe should possess certain selection criteria and contain some specific physicochemical properties proposed by FAO and WHO in 2002. This report contains sets of standards for a particular microbe to be safely used as a probiotic in food products. These guidelines also comprise of strain-specific health effects of microorganisms selected as probiotics. The following criteria are generally used for the selection of microbial strains as probiotics.

1 *Origin:* Origin of the selected strain must be the human GI-tract.
2 *Safety:* Microbial strain must be nonpathogenic and subtle to antibiotics and not dock resistance against antibiotics. In addition, the strain should not produce biogenic amines.
3 *Withstand host's natural barriers:* The strain should be able to withstand the environmental conditions within the intestinal tract. It should be resilient to pH change, bile salts, and protease enzyme while *in vitro* screening.

4 *Adherence to intestinal epithelium:* It must be capable of adhering to the intestinal cells and block sites occupied by pathogens.
5 *Commercial propagation:* They must have an ability to grow under certain commercial conditions without losing their viability during the storage.
6 *Functional properties*: Strains should possess the specific functional properties that meet the definition of probiotics. The four major categories include the following:
   i. *Appropriateness:* The strain should cover proper taxonomic identification, must be of human origin, nontoxic, and of GRAS (generally recognized as safe) status.
   ii. *Technological suitability:* Probiotic microorganisms should have high production, growth rates and recovery, adequate storage stability, and desired viability (10^6–10^8 CFU/g).
   iii. *Competitiveness:* Probiotic strains must compete with natural microflora, resistant to antimicrobial agents, acids, and bacteriocins.
   iv. *Performance and functionality:* Probiotic strains must have certain potential health benefits and should be antagonistic to pathogens, antimicrobial substances, and other toxic compounds. They should also be able to produce beneficial bioactive compounds.

The proposed characters for theoretical selection of probiotics microorganism are summarized in Figure 11.2.

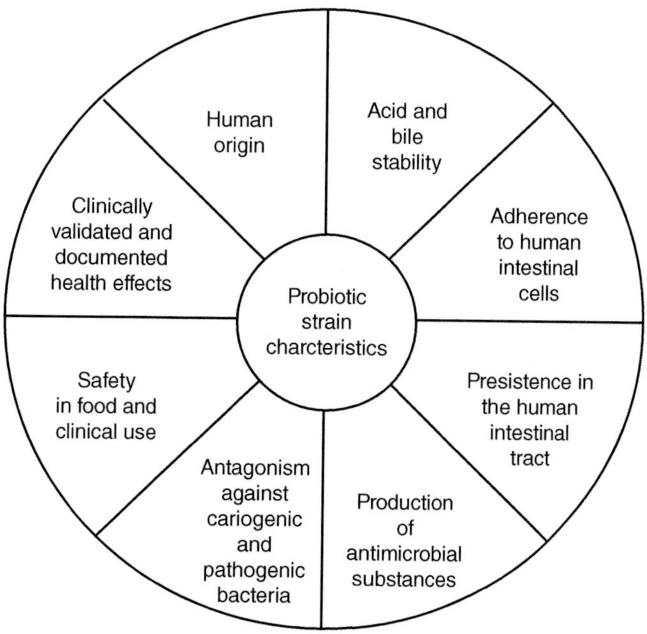

Figure 11.2 Theoretical basis for selection of probiotic microorganism. (From Saarela *et al.* 2000. Reproduced with permission of Elsevier.)

11.4.1 Safety of human origin probiotics

The probiotic microbes of human origin have been found to be the most successful ones with regard to both safety and commercial value, possibly because probiotic microbiota performs at their best in conditions similar to their habitat (Human GIT). A number of human origin probiotics are also applied in commercial products (Ouwehand, 2007). Lately the specific course of action taken by a probiotic strain has been pointed out to be more important than the origin of the strain by the FAO/WHO expert panel, mainly because of the ambiguity about the origin of GIT microflora, since infants are born with an intestinal tract almost bacteria-free. Probiotic strains in commercial dairy products have been used for a long time without any safety issue. Probiotic consumption is rarely reported to cause any injurious effects, despite the fact that particular LAB have been found and isolated from local infections and blood (Salminen, 2006). Human intestinal flora used as potential probiotics include *Lactobacillus* Spp. (*Lactobacillus acidophilus, Lactobacillus fermentum, Lactobacillus buchneri,* and *Lactobacillus crispatus*), *Bifidobacterium* Spp. (*Bifidobacterium bifidum, B. adolescentis, B. Pseudocatenulatum,* and *B. dentium*), and *Enterococcus* Spp. (*E. faecium* and *E. faecalis*) (Hammes and Hertel, 1998).

11.4.2 Antibiotic resistance aspects

Transfer of antibiotic resistance genes via plasmids is yet another important aspect to be considered before using probiotic microbes. Evidence from different studies on this view is fairly conflicting; some results propose that some strains of *Lactobacillus* and *Bifidobacterium* (Matteuzzi *et al.*, 1983) expressed resistance to applied antibiotics. Contradictory to these studies, findings by Moubareck *et al.* (2005) showed no resistance in the *Bifidobacterium* strains. Marteau (2002) reported that the chances of transfer of antibiotic-resistant genes depend upon the following conditions: genetic material type (e.g., plasmids), nature and amounts of donor and recipient strains, their interactions, and environmental conditions, that is, the presence of an antibiotic may facilitate the growth of antibiotic-resistant mutants. Consequently, the antibiotic resistance of the probiotic organisms needs to be checked and verified to limit the chances of such gene transfers with the endogenous bacteria, before its application as a probiotic.

11.4.3 Physiological aspects

The functionality of probiotics is studied and confirmed initially by means of *in vitro* studies and using trials and finally verification by controlled human trials. In order to establish the preference for probiotics, several functionality aspects must be brought under consideration, which may include survival under acidic and gastric juice conditions, tolerance to bile (to be able to survive small bowel environment), adherence to simulated mucosal surfaces and perseverance in the human GIT, immunostimulation, nonproinflammatory, antagonistic against pathogenic microbes such as *Salmonella, H. pylori, C. difficile,* and *Listeria monocytogenes* and antimutagenic and anticarcinogenic characteristics.

11.4.4 Technological/industrial aspects

Feeding trials have reported the vanquishing of the probiotic strains from the GIT in a couple of weeks once the ingestion of the strain is stopped (Alander *et al.,* 1999). Thus, the perseverance of probiotic strains inside the human GIT has provoked some serious questioning. Yet a temporal presence increases the probability of probiotic strains to confer health benefits to their host. Conformity to the essential safety and functional criterion is only a milestone and different aspects such as production and processing of probiotic products have to be dealt with afterward. Some of the vital technological characters, for commercial applications, whose presence in the potential probiotic is considered mandatory, are survival under applied processing conditions, permanence after production and storage, acceptable sensorial attributes, and resistance against phage viruses. Superior viability and physiological performance of probiotic microbes are the established prerequisites for them to function at their best. Nonetheless, non-viable probiotic microorganisms have also been known to confer some health benefits to the host, probably because they have the abilities to bind carcinogens and induce immunomodulative responses (Salminen *et al.,* 1998). So it is quite adequate if some probiotic microbes get a chance to grow in sufficiently high numbers during production of the product that they will provide their host with some health benefits. In the long run, this phenomenon no longer helps the situation as the viability of microbes keeps on decreasing as the time of storage increases. Meanwhile the production of off-flavor compounds by a microbe is considered a negative trait as it destroys the sensorial attributes of the product.

11.4.5 Food technology's perspective

Food technology has to perform its role to make sure that the food milieu meets the necessary survival needs of the probiotic microorganism, that the probiotic constituents are present in an adequately high amount, and that they hold optimal physiological activity. Moreover, if the changes in the consistency of the food milieu or the culture are inexorable or inadequate, then it is critical to make sure that such actions do not adversely affect the probiotic characteristics and efficiency. Nevertheless, it is not economically and morally practicable to continuously confirm the safety of the usage of these probiotics by human-based clinical studies. This problem has been resolved by designing such *in vitro* replicas that can depict the physiological behavior of various probiotic cultures. It has been observed that some characteristics of the probiotic cultures rely on the formulation technique and are comparatively easier to perform by *in vitro* resistance against imitated gastric juice and bile acid treatment, growth features, and properties of the revived culture and more importantly the span of lag-phase and immune markers expression by human peripheral blood cells such as gamma interferon and interleukin-12 (Haller *et al.*, 1997). The above methods were shown to affect the probiotic activity as they altered the growth conditions in the culture.

11.4.6 Problems during selection of probiotics

The problems being faced by researchers in this regard are technological, microbiological, and legislative. It has been a long time since the establishment of the significance of particular technological and physiological properties of potential probiotic strains was theorized and identified. Although many criterions have been acknowledged and proposed (Ouwehand, 2007), no universal agreement exists among researchers with regard to the most important criteria for selection.

(a) *In vitro* tests

In vitro trials are recommended before any subsequent *in vivo* animal or human studies. These tests are carried out to confirm the authentication of *in vitro* tests with *in vivo* trials. Common *in vitro* tests include acidity and bile salts resistance, adherence to mucus and human epithelial cells, hydrolase activity, and effectiveness to control pathogenic microorganisms such as *C. difficile, H. pylori, Listeria monocytogenes,* and *Salmonella.*

(b) *In vivo* survival

After different *in vitro* trials, probiotics are tested for their *in vivo* survival. Probiotics should be stable to the hostile conditions in the stomach and gastrointestinal tract. Probiotics should colonize, flourish, and employ some beneficial effects after ingestion. They should also be stable to different stresses during formulation, processing, packaging, and storage. They must have excellent organoleptic attributes.

11.5 Probiotic meat products

The increasing anxieties among consumers about food safety issues have provoked the development of foods that promote health and well-being in addition to their dietary impacts (Tyopponen *et al.*, 2003). Functional foods are those that have a valuable impact on human health beyond nourishment. The effectiveness of functional foods is due to additional active ingredients and elimination or substitution of undesirable mixes in their composition (Erkkila *et al.*, 2001b). The marketing of functional food products started in the 1960s and the trend continued further in the 1970s to reduce harmful ingredients from food products. In the 1980s, the pattern proceeded with the reduction or amputation of food additives, which prompted the addition and induction of valuable constituents such as vitamins, mineral compounds, and probiotic microorganisms in the 1990s (Khan *et al.*, 2011). Probiotics contribute an enormous share of the functional food market, being extensively used in meat, cereal, dairy, and beverage industries (Santos *et al.*, 2003). Utilization of probiotics is most promising in raw fermented meat products as these kinds of products are consumed without prior heat treatment, which would lead to the death of probiotic microorganisms (Andersen, 1998; Sameshima *et al.*, 1998). However, the addition of probiotic bacteria is also a big challenge for processors due to the sensitivity of probiotic bacteria to different spices, salts, flavors, and other ingredients used in the sausage formulations (Erkkila *et al.*, 2001a). The addition of these beneficial microorganisms requires that the strains that are stable to fermentation remain in a reasonable number to survive the pH of the stomach and play a beneficial role while passing through the intestine (Lucke, 2000). Moreover, the preparation of probiotic meat items infers, considering the suitability of the probiotic culture to the target consumer, functionality of intestine, survival rate of probiotics during processing, and maintenance of sensory properties in the probiotic product (Ferreira, 2003).

11.5.1 Probiotics for meat products

Probiotics are mostly the strains from *Bifidobacterium* and *Lactobacillus* species. Others include some species of *Saccharomyces, Lactococcus, Propionibacterium,* and *Enterococcus* (Zhang *et al.*, 2010). Several studies highlighted the potential utilization of probiotic bacteria in meat products. Arihara *et al.* (1998) used *L. gasseri* to increase the microbiological safety of fermented meat. Furthermore, Sameshima *et al.* (1998) used *L. rhamnosus* and *L. paracasei* subsp. *paracasei* in fermented meat products for the same purpose. Pennacchia *et al.* (2004) also reported the potential of *L. plantarum* and *L. paracasei* as meat probiotics. It has also been investigated by Erkkila and Petaja (2000) that probiotic strains of *L. rhamnosus* GG. *L. rhamnosus* LC-705, *L. rhamnosus* VTT, and *L. rhamnosus* VTT-97800 can be used in the formulation of frankfurters. They also used *L. gasseri, L. casei, L. paracasei* subsp. *paracasei, L. rhamnosus,* and *B. lactis* for the production of salami. In addition, the mixture of a traditional starter culture, *Bactoferm* TSPX and probiotic strains of *L. casei* LC-01 and *B. lactis* Bb-12 can also be used to ferment meat products (Andersen, 1998). *Lactobacillus sakei* Lb3 and *Pediococcus acidilactici* PA-2 showed significant survival in fermented sausages during processing and storage, when used as probiotics (Vuyst *et al.*, 2008). *In vitro* studies reveal that *L. casei* and *L. paracasei* consumed with fermented sausages showed functional properties in the body (Rebucci *et al.*, 2007). Macedo *et al.* (2008) found that *L. casei, L. rhamnosus,* and *L. paracasei* can be used in fermented dry sausages without disturbing the sensory and technological attributes of the final product. Khan *et al.* (2011) reported that *Lactobacillus* species particularly, *L. plantarum* and *L. casei* also have great potential to use as probiotics in meat sausages. The potential probiotic microorganism and their strains are summarized in Table 11.1.

Table 11.1 Microorganism species most commonly used as starter cultures in fermented meat products.

Microorganisms	Genus and species
Lactic acid bacteria	*Lactobacillus acidophilus**, *L. alimentarius*†, *L. brevis*, *L. casei**, *L. curvatus*, *L. fermentum*, *L. plantarum*, *L. pentosus*, *L. sakei*, *Lactococcus lactis*, *Pediococcus acidilactici*, *P. pentosaceus*
Actinobacteria	*Kocuria varians*‡ *Streptomyces griseus*, *Bifidobacterium* sp.*
Staphylococcus	*S. xylosus*, *S. carnosus* subsp. *carnosus*, *S. carnosus* subsp. *utilis*, *S. equorum*†
Halomanadaceae	*Halomonas elongata*† (tested in dry cured ham)
Enterobacter	*Aeromonas* sp.
Mold	*Penicillium nalgiovense*, *P. chrysogenum*, *P. camemberti*
Yeast	*Debaryomyces hansenii*, *Candida famata*

*Used as probiotic cultures.
†Used in commercial tests in industrial scale (Laboratorium Wiesby, Niebull and Rudolf Müller and Co).
‡Formerly known as *Micrococcus varians*.
Source: From de-Macedo, Pflanzer and Gomes (2012), http://cdn.intechopen.com/pdfs-wm/39591.pdf. Used under CC BY 3.0 license. http://creativecommons.org/licenses/by/3.0/.

11.5.2 Technological characteristics of probiotic cultures for meat products

For use in fermented meat products, probiotic microbes need to maintain their feasibility toward the unfavorable conditions produced during fermentation: pH reduction (<5.0), increase in salt concentration (2–3%), elevated nitrite contents (approximately 120 ppm), and reduced water activity (<0.85). The probiotic cultures should possess certain other technological characteristics: fast growth rate during fermentation process, resistance against lypholization and freezing processes, and provision of longer shelf life to the product and maintenance of the organoleptic properties of the final product (Erkkila *et al.*, 2001a; Vuyst *et al.*, 2008). Probiotic cultures can be included in fermented sausages as a component of the starter culture or as an extra culture during the mass blending (Figure 11.3).

11.5.2.1 Lactic acid production

Production of lactic acid in probiotic fermented meat products is the main characteristic of probiotic cultures, particularly *Lactobacillus*. Lactic acid production causes acidification of the product that is responsible for safety and incorporation of desired sensory attributes of the fermented meat products. Lower pH coagulates the myofibrillar proteins, improves color (bright red) and flavor of the meat, and provides firmness to the product by enhancing the cohesiveness. It also prevents meat spoilage by inhibiting the growth of pathogenic bacteria.

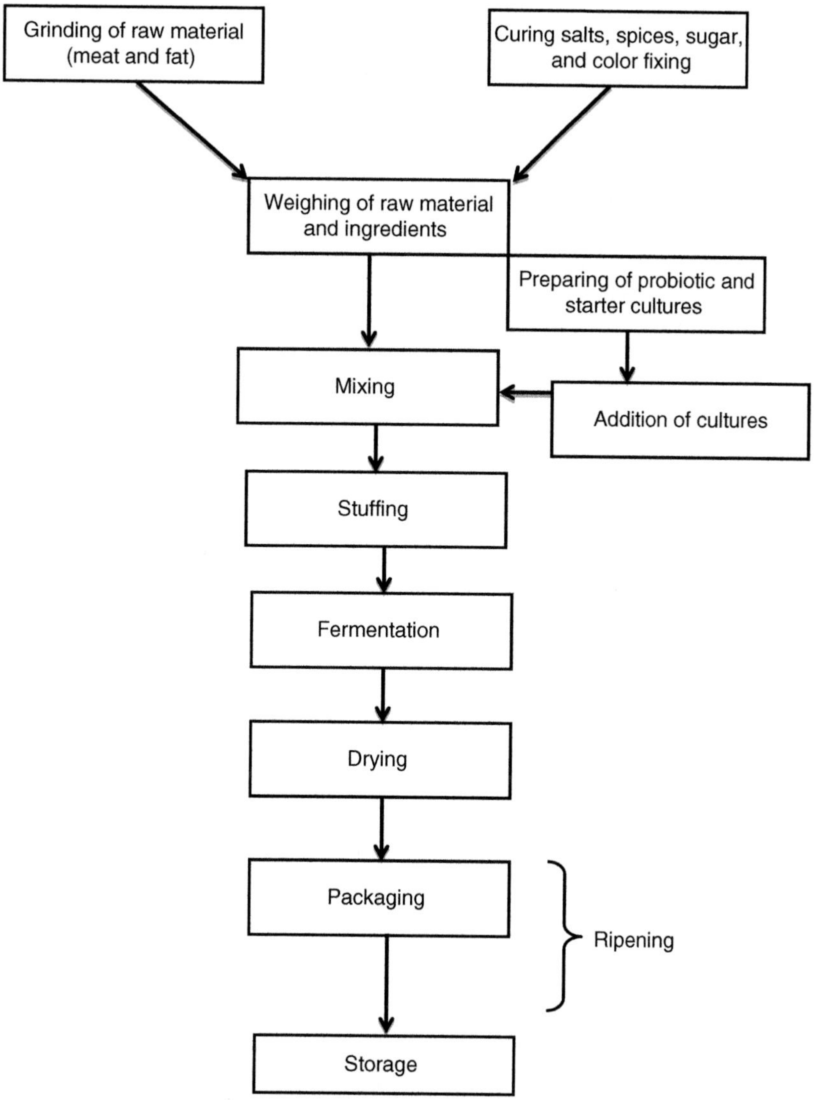

Figure 11.3 Basic flowchart of the processing of fermented dry sausage with the addition of probiotic cultures. (Reprinted from de-Macedo, Pflanzer and Gomes (2012).)

Moreover, a sudden decline in pH avoids the accumulation of harmful biogenic amines in the product during the storage. However, it is of utmost importance that these probiotics must produce the L (+) isomer lactic acid instead of D (−) isomer lactic acid, due to higher efficiency of L (+) lactic acid to control pathogens. In addition, D (−) lactic acid isomer can cause several health concerns because it cannot be metabolized in the human body (Ammor and Mayo, 2007; Vuyst *et al.*, 2008).

11.5.2.2 Resistance to salt and nitrites

In Japan, 3% sodium chloride and 200 ppm sodium nitrite is added to fermented sausages to enhance the keeping quality and to inhibit the microbial growth in the final product (Arihara and Itoh, 2000). Both of these salts can also reduce the efficiency of the added probiotics. To overcome this problem, resistant strains should be used for the preparation of meat products with certain probiotic properties (Papamanoli *et al.*, 2003). Sameshima *et al.* (1998) studied the stability of 202 *Lactobacillus* strains against sodium chloride and sodium nitrite and found that strains of *L. rhamnosus*, *L. paracasei* ssp. *paracasei*, and *L. acidophilus* showed resistance against these salts. Similar findings were obtained by Macedo *et al.* (2005) who reported the stability of *L. rhamnosus*, *L. casei*, and *L. paracasei* against sodium chloride (3%) and sodium nitrite (200 ppm).

11.5.2.3 Production of bacteriocins

Bacteriocins are microbiologically derived proteins that hinder the growth and development of certain pathogenic Gram-positive bacterial strains, particularly *Listeria monocytogenes*. Bacteriocin-producing probiotic cultures used in meat products do not produce toxins in the body after consumption and act as natural preservatives without disturbing the quality characteristics of the product. Several strains of lactic-acid-producing bacteria, for example, *Lactobacillus curvatus*, *Lactobacillus brevis*, *L. sakei*, *L. plantarum*, and *L. casei* are responsible for bacteriocin production (Leroy *et al.*, 2006).

11.5.3 Physiological characteristics of meat probiotics

11.5.3.1 pH resistance

Resistance to acidity is the main characteristic of probiotics that demonstrates their ability to survive under the acidic conditions in the stomach (Erkkila and Petaja, 2000; Annuk *et al.*, 2003). Low pH in the stomach is quite harmful for lactic acid bacteria, since its viability is incredibly hindered at pH below 4.5. Such hindrance is identified with a decrease in the intracellular pH of microorganisms due to the production of nondissociated lactic acid, which penetrates the bacterial membrane and enhances bacteriostatic and bactericidal effects by collapsing the electrochemical gradient (Ammor and Mayo, 2007). Selection of suitable probiotic strains depends on its ability to resist the pH of stomach. Meat has buffering properties so it helps the added probiotics to survive in the stomach (Tyopponen *et al.*, 2003).

Studies indicate that *L. sakei*, *L. pentosus*, *Pediococcus pentosaceus*, and *P. acidilactici* have resistance against low pH up to 4. In addition, efficiency of these probiotics is not adversely affected at this pH (Erkkila and Petaja, 2000). Other researchers reported that some species of probiotics can survive at pH 3 for a time of 2–4 h, which ensures their survival in gastrointestinal tract. Macedo *et al.* (2005) used *Lactobacillus paracasei* in salami and found 20% expansion in the initial count at pH 3 during the 4 h. Pennacchia *et al.* (2004) also studied the ability of different *Lactobacillus* species extracted from salami to survive at low pH and found that *Lactobacillus casei* was more effective to tolerate the pH 2.5 for 3 h. They also exposed the potential of *L. paracasei* and *L. rhamnosus* to resist the acidic conditions of stomach.

11.5.3.2 Resistance to bile salts

Bile salts are mainly responsible in the intestinal defense system to inhibit microbial growth (Charteris *et al.*, 2000). Bile salts destroy the fatty acid layer of the bacterial cell membrane. Some species of *Lactobacillus* contain bile salt hydrolase that hydrolyzes bile salts (Papamanoli *et al.*, 2003). However, this ability of *Lactobacillus* depends on several other factors including stress reaction system of the microorganism, presence of components that strengthen cellular membrane, metabolic system, and transportation of amino acids and biosynthesis of fatty acids (Taranto *et al.*, 2006). The normal concentration of bile salts in the intestine is 0.3%, so it is considered as the critical concentration for the selection of probiotics (Erkkila and Petaja, 2000). Macedo *et al.* (2005) also reported the ability of *L. paracasei* to bile salts at 0.3% level. Meat itself plays a dynamic role to protect microorganisms against bile (Ganzle *et al.*, 1999). Added probiotics are encapsulated by the meat during the processing and do not make any contact with the bile salts.

11.5.3.3 Detoxification of biogenic amines

Biogenic amines are generated in the meat products due to the decarboxylation of amino acids by naturally occurring or added microorganisms. These biogenic amines can produce several harmful toxins, particularly in persons having amino oxidase deficiency. Meat provides the ideal conditions for biogenic amine-producing bacteria because of its high protein content. Moreover, fermented meat products take time to ripen so it is also beneficial for these pathogenic microorganisms to produce biogenic amines in meat. Some studies demonstrate the ability of *Lactobacillus* to produce amino acid decarboxylase that averts the accrual of biogenic amines in the product during the storage. *Lactobacillus plantarum* and *L. casei* also possesses detoxification capacity of biogenic amines in fermented meat products (Leroy *et al.*, 2006; Ammor and Mayo, 2007). Ergonul and Kundakç (2011) used three different types of probiotic starter cultures (*Lactobacillus acidophilus, L. casei,* and their blend) in Turkish sausages and reported lower level of biogenic amines compared to the nonprobiotic product. Putrescine contents ranged about 1.98 ppm during manufacturing and increased up to 35.48 ppm during refrigerated storage of 8 months, while cadaverine concentration ranged between 0.96 and 18.50 ppm, histamine contents from 1.41 to 10.84 ppm, and tyramine ranged from 1.75 and 9.36 ppm during the same period of storage.

11.5.4 Addition of probiotic cultures in fermented meat products

The addition of probiotic cultures in sausages can be accomplished in three ways: as liquid inoculum, lyophilized cultures, or in high concentration. However, using lyophilized cultures delays the fermentation time and viability of the culture is lost in the product. This loss can be prevented by microencapsulation of the culture prior to lyophilization. This method can also be used when probiotic strains are hindered by different sausage ingredients (Leroy *et al.*, 2006).

11.5.4.1 Culture microencapsulation

Microencapsulation protects the bacterial cells by producing a polymeric layer around the cells and thus increases the viability of the microbes. LAB cultures are usually microencapsulated by emulsification and extrusion processes. Microcapsules produced by extrusion processes have 60 times greater size (2–3 mm in diameter) than of those produced by emulsification. The widely used materials for microencapsulation include alginate, k-carrageenan, guar gum, starch, whey proteins, xanthan gum, and gelatin. Muthukumarasamy and Holley (2006) used alginate for the microencapsulation of *Lactobacillus reuteri* ATCC 55730 in fermented sausages and concluded that alginate did not impose any adverse effects on the sensory characteristics of the final product. They also proposed that despite invisibility with the naked eye, they were detected by panelists as fat particles because of color and size similarity. Alginate-based microcapsules can also be used for *Lactobacillus reuteri* and *Bifidobacterium longum* to enhance their survival in fermented meat (Rivera-Espinoza and Gallardo-Navarro, 2008). Recently, succinylated β-lactoglobulin tablets have also been introduced to protect *B. longum* after several *in vivo* and *in vitro* trials (Poulin *et al.*, 2011). Heidebach *et al.* (2009) also reported the higher viability of *Lactobacillus* F19 when compared to *Bifidobacterium* Bb12 during freezing after encapsulation with casein. They also concluded that the survival rate of these strains can be further enhanced by microencapsulation with rennet-prompted gelation (Heidebach *et al.*, 2010).

11.5.5 Fermented meat as a probiotic product

The selection of probiotics for utilization in fermented meat products depends on the type of product and the processing technique. Generally, the species selected as probiotic starter culture are chosen from strains characteristically transcendent in meat products and, consequently, adapted to this environment. Therefore, these species show prominent metabolic activity that plays a vital role in providing desirable sensory properties to the product (Leroy *et al.*, 2006). Given the adversative conditions of the meat for various microbes, including probiotics, a few studies recommend the use of lactic acid bacteria as commercial starter cultures for fermented meat products as they provide the same sensory and technological properties to the product as the conventional starter cultures, and also give significant health effects (Lucke, 2000). *L. fermentum*, *L. plantarum*,

L. brevis, and *Pediococcus pentosaceus* are the most commonly used meat probiotics (Pennacchia *et al.*, 2004). In addition, *L. sakei* and *P. acidilactici* also have great potential to be used as meat probiotics, because of high resistance to bile and acidic conditions (Erkkila and Petaja, 2000).

11.5.5.1 Probiotic fermented sausages

Fermented sausages are quite suitable for the addition of probiotic microbes since mild heat treatment is needed for their production, which provides the ideal conditions for probiotics to survive (Ammor and Mayo, 2007; Khan *et al.*, 2011). Care should be taken while designing the sausages that the optimum number of probiotics should be maintained. This can be accomplished by controlling the pH (e.g., <5.0), ripening time (e.g., >1 month), and temperature (Vuyst *et al.*, 2008; Khan *et al.*, 2011). In the meat sector, the probiotic cultures are added to achieve several health benefits, besides nutritional benefits (Hammes and Hertel, 1998). They can also be used as protective cultures, since they can survive under unfavorable situations and deliver different organic acids and bacteriocins (Maragkoudakis *et al.*, 2009). Probiotics are also called functional starter cultures as they are associated with product safety and improve the nutritional and sensory characteristics of the product (Leroy *et al.*, 2006). The feat of adding probiotic microorganisms has been confirmed after the consumption of these probiotic products. It is not necessary that these beneficial bacteria show the same effects in all products because their efficiency is totally dependent on environmental conditions. A number of studies tell us about some of the suitable strains of probiotics that could be incorporated in meat products (Table 11.2) to acquire desired properties (Hammes and Hertel, 1998; Tyopponen *et al.*, 2003). The estimated count of cells that have to be added in the product for intestinal colonization is about 10^9–10^{10} CFU/g of product. Studies also conclude that the estimated viable count of these cells ranges from 10^6 to 10^8 CFU/g in 1 g of feces. In this manner, in a fermented meat product having 10^8 CFU/g of probiotic bacteria, the minimum daily intake may be 10–100 g of product (Penna, 2002; Tyopponen *et al.*, 2003). Rivera-Espinoza and Gallardo-Navarro (2008) also suggested that about 10^8–10^9 CFU/g of viable cells must be ingested in the product to obtain desirable physiological impacts. The key benefits of using probiotic meat products include improvement in the digestion and absorption of nutrients in the gastrointestinal transit, preventing lactose intolerance, betterment in immune system, reduced cholesterol level, and control of diarrhea and colon cancer (Heenan *et al.*, 2002; Tharmaraj and Shah, 2003). Most of the studies report the effects of using probiotic bacteria in meat products on product safety, but very few studies demonstrate the beneficial health impacts of consuming probiotic meat products (Vuyst *et al.*, 2008). The probiotic strains used as a starter culture in fermented meat products require certain functional and sensory properties (Hammes and Hertel, 1998). In addition to this, product design is very important to ensure the viability of added probiotics in the optimum range. Process controls such as pH reduction,

Table 11.2 Examples of potential utilization of probiotic bacteria in meat products.

Microbial strain	Meat product	Findings
Lactobacillus sake, *Pediococcus acidilactici*	Meat starter cultures	*L. sake* and *P. acidilactici* in simulated gastrointestinal conditions had the strongest capacity for surviving acidic conditions and 0.30 bile salts; the strains may be regarded as potentially probiotic
Lactobacillus rhamnosus GG, LC-705 and E-97800	Dry sausage	Strains seem suitable for the production of dry sausages; *L. rhamnosus* E-97800 had the faster growth rate and acidification
NTMS cultures, including *Lactobacillus acidophilus* LAFTIkL10, *L. paracasei* LAFTIkL26, *L. paracasei* 5119, *Lactobacillus* sp. L24 and *Bifidobacterium lactis* LAFTIkB94	Hungarian salami	It was concluded that NTMS cultures may be used to increase the safety of Hungarian salami because these cultures gave strong inhibition of both *E. coli* O111 and *L. monocytogenes*
Lactobacillus curvatus, *Lactobacillus plantarum*	Two types of Greek naturally fermented dry sausages	58% of *L. curvatus* strains and all *L. plantarum* strains isolated from sausages were resistant to 0.3% bile salts and had antimicrobial activity
Potential probiotic 20 strains *Lactobacillus* (*brevis, plantarum, casei, paracasei, pentosus, reuteri, curvatus, sakei, zeae*)	Fermented sausages	*Lactobacillus* strains capable of surviving the pH of the stomach and the environment of the intestine, which make them potential probiotics and may be eventually used as probiotic starter cultures in novel dry-fermented sausages manufacture
Potential probiotic *L. plantarum/pentosus* (MF1291, MF1298, MF1300)	Scandinavian-type fermented sausage	Strains were able to produce a fermented sausage with similar quality as sausage fermented with the commercial starter culture and were able to survive and establish in an environment similar to the human GIT, inhibit potential pathogenic bacteria *in vitro*, and were considered safe to be used with regard to their antibiotic resistance pattern
1000 strains isolated from Iberian dry-fermented sausages (363), and human (337) and pig feces (300)	Iberian dry-fermented sausages	Most of these strains are promising probiotic meat culture candidates suitable for Iberian dry-fermented sausages
Potential *L. fermentum* HL57, *L. reuteri* PL519, *L. reuteri* PL542	Dry-fermented Iberian sausages	*L. fermentum* HL57, *L. reuteri* PL519, and *L. reuteri* PL542 were identified as a potential probiotic meat starter culture suitable for manufacture of dry-fermented Iberian sausages

Table 11.2 (*continued*)

Microbial strain	Meat product	Findings
Probiotic *Lactobacillus fermentum* HL57 *Pediococcus acidilactici* SP979	Iberian dry-fermented sausages	Iberian dry-fermented sausages inoculated with *L. fermentum* HL 57 and *P. acidilactici* SP979 may be considered as functional products given the counts of these strains found at the end of processing
Lactobacillus casei ŁOCK 0900 and *Lactobacillus casei* ŁOCK 0908	Dry-fermented pork loins	Probiotic strains of *Lactobacillus casei* ŁOCK 0900 and ŁOCK 0908, added to the cured pork loins, grew and achieved a count of 107 log cfu/g, and in the samples with the additive of 0.2% glucose, their count was 108 log cfu/g
Lactobacillus acidophilus Bauer and *Lactobacillus casei* Bif3/IV	Dry-fermented pork loins	The results of the experiment showed that probiotic strains had a good ability to survive in dry-fermented pork loins and this can be used for dry-fermented loin production
Probiotic *L. casei* ŁOCK 0900	Dry-fermented pork loins	*L. casei* ŁOCK 0900 strain had a good ability to survive in dry-cured loins and it can be used for dry-fermented loins production

Source: From Kolozyn-Krajewska and Dolatowski (2012). Reproduced with permission of Elsevier.

prolonged ripening, and excessive heating are also very important in maintaining the viability and obtaining desirable beneficial results (Vuyst *et al.*, 2008). *L. gasseri* JCM1131 is usually responsible for providing safety to the probiotic fermented meat products (Arihara *et al.*, 1998). The convenience of *L. paracasei* subsp. *paracasei* FERM P-15121 and *L. rhamnosus* FERM P-15120 has also been widely reported as potential meat probiotics (Sameshima *et al.*, 1998). Probiotic fermented sausages can also be successfully formulated by using a mixture of a traditional starter culture (*Bactoferm* T-SPX) and a probiotic culture of *L. casei* LC-01 or *B. lactis* Bb-12 (Hammes and Hertel, 1998). Lactic acid bacteria are considered the most successful probiotics used in meat sausages. *L. casei* and *L. plantarum* are mostly used as starter cultures in sausages, but they also have potential to be used for the manufacturing of probiotic sausages (Vuyst *et al.*, 2008). Furthermore, using functional starter cultures would provide more health benefits and improve taste and product safety, compared to traditional starter cultures (Ammor and Mayo, 2007).

11.6 Functionality of probiotics

The positive effects of probiotics on immunomodulation and regulation, reinforcement of the immune system, alterations in GIT flora, prevention of diarrhea, and cutbacks in fecal enzyme activities have been confirmed by a number of clinical studies. Some of the most pronounced stages and effects of probiotics are discussed as follows.

11.6.1 Adhesion and subsequent colonization

Adhesion of probiotics is a prerequisite to the colonization of probiotics in the GIT. Adhesion determinants of *Bifidobacteria* could be glycoproteins or carbohydrate chains that attach to the wall of the GIT. Hydrophobicity is also an important factor for the adhesion properties of *Bifidobacteria*. Multiple adherence factors from the bacterial surface are responsible for the linkage of probiotics to the intestinal mucosa. Cytokines are known as diminutive proteins secreted by immune cells, and many other types of cells. Each cytokine may have multiple activities in unusual circumstances and cell types. Their action depends on binding with some suitable receptors present on the cell surface and by stir up changes in the growth and development patterns, and activity of the target cells. Different production patterns of cytokines are applied to segregate T helper cell (Th) responses to dissimilar classes. These cells are of three types. Th1-type cells generate IFN-γ, IFN-α, IL-2, and IL-12, and also endorse cell-mediated immunity (Kidd, 2003). The second type of T helper cells (Th2-type) produce different interleukins including IL-4, IL-5, IL-6, and IL-13, and trigger antibody-mediated immune response and eosinophils cells (Wu *et al.*, 2007). Th3-type cells function against Th1 and Th2, and they downregulate the inflammatory retort through TGF-β production and Tr1 cells are responsible for maintaining the balance of immunity through the production of IL-10 (Izcue and Powrie, 2008). Cytokines are categorized according to their production, including proinflammatory, inflammatory, and anti-inflammatory interleukins. Proinflammatory cytokine, TNF-α, is responsible for activating local inflammation concerned with the immune system (Bertazza and Mocellin, 2008). TNF-α is involved in the production of professional immune cells such as T cells, macrophages, NK cells, and epithelial cells. These cells are capable of inactivating macrophages and T cells (Barksby *et al.*, 2007). These effects are totally dependent on the strains to be used as probiotics, and a specific contour of cytokines is established in a systematic way for each new strain. The second one is inflammatory cytokines that include IL-1α/β (Barksby *et al.*, 2007), TNF-α (Bertazza and Mocellin, 2008), IFN-γ, and IL-6 (Gabay, 2006). While IFN-γ, produced by N cells and T cells, also have antiviral effects and are associated with the activation of macrophages, suppression of Th2 responses and involved in the systemic responses of inflammation including fever, synthesis of hepatic proteins, and weight loss. IL-1 and IL-6 are also liable to mediate the same response in systematic inflammation. IL-2, produced by T cells, proliferation is promoted, but concentration may fluctuate in different immune supporting organs. Mast and T cells also produce IL-4 and IL-5, respectively. IL-4 is known for the activation of B cells and is involved in the control of IgE and suppressing of Th1 cells, while IL-5 enhances the growth of eosinophils. IL-6 is produced by macrophages, endothelial cells, and T cells, and is concerned with the growth and division of T cells and B cells (Gabay, 2006). IL-12 is responsible for IFN-γ production and also differentiates Th1- and Th2-type cells. It also plays a significant role in providing linkages between innate and adaptive immunity. Phagocytes and dendritic cells produce IL-12 during infection and are linked with NK-cell activation (Trinchieri, 2003). IL-10 is considered as an anti-inflammatory or regulatory cytokine and can be produced by many types of cells. All through infection, it inhibits the activity of Th1 cells, NK cells, and macrophages, required for optimal pathogen clearance, but in addition contributes to tissue damage (Couper *et al.*, 2008).

11.6.2 Modulation of immune system

Probiotics adhere to the intestinal tract in order to buy the time needed for their activity and thus are able to show pronounced immunomodulation effects than those of nonadhesive microorganisms. The enteric flora, including greater than 500 bacterial species, comprise approximately 95% of the total number of cells in the human body and contribute significantly to the host's resistance to infectious disease. Furthermore, adhesion facilitates the contact with lymphoid tissues, which controls and ultimately affects the local and systemic immune system. Only the adherent probiotics are believed to produce effective immune responses (Salminen *et al.*, 1996). Adhesion also reduces the competition of probiotics with other pathogenic microbes: Such diminishing competition induced by LAB and *bifidobacteria* has been observed by Coconnier *et al.* (1998), who used Caco-2 and HT-29-MTX cell lines (*in vitro* model systems resembling human colon carcinoma cell lines) to study this inhibitory action on pathogenic flora. Many studies have indicated that different probiotics differ in their adhesive potentials. Interaction of adhesive probiotic strains with the gut-associated lymphoid tissue stimulates the immune system. Many human trials have concluded that probiotic bacteria positively affect their host's immune system. The concept of probiotic therapy is based on the maintenance of a normal and healthy composition of natural gut inhabitants in the body. Food supplementation with probiotics have been proven to have certain beneficial effects on health including normalization of intestinal permeability, improvement of barrier functions of the intestine, prevention of intestinal inflammation by maintaining the composition of intestinal microbiota, and strengthening of the immune system (Salminen *et al.*, 1998). An imbalance of intestinal microbes is associated with a number of inflammatory conditions such as rotavirus diarrhea (Isolauri *et al.*, 1994), rheumatoid arthritis (Malin *et al.*, 1996a), and atopic disease. These bacteria are responsible for providing strong intestinal defense and regulation of homeostasis mechanisms. Presence of these bacteria is reported throughout the GI tract, but they are found in higher concentrations in the large intestine (Salminen *et al.*, 1998). The GI tract is usually sterile at the time of birth, but microbial colonization starts immediately after birth. These microbes are established in the gut in a systematic way during earlier years of life (Benno and Mitsuoka, 1986). It is very necessary to avoid certain inflammation and infections such that the immune system has the potential to differentiate between the pathogenic and the nonpathogenic indigenous microorganisms. Furthermore, it is also reported through scientific studies that resident microflora generate endogenous stimuli essential for maturation and normal functioning of the immune system (Toms and Powrie, 2001). At the same time, disparities have been observed for probiotic microbes with respect to the effects that they command on the immune system modulation (Isolauri *et al.*, 1994). *L. johnsonii* LJ-1 and *L. salivarius* UCC 118 have been shown to initiate production of immunoglobulin A and an amplified phagocytic activity. These strains did not produce a negative impact on host health such as inflammation, but on the contrary they were associated with positive transitional modification that proved to be favorable for their host. Extracts derived from *B. lactis* Bb-12 and *L. rhamnosus* GG were observed to have a reduction on lymphocyte proliferation. In another study involving these two strains conducted on children suffering from severe food allergy, children who ingested *L. rhamnosus* GG and *B. lactis* Bb-12 demonstrated a noteworthy mitigation in allergy symptoms compared to the children who did not receive such probiotics. *L. rhamnosus* GG has also been shown to alleviate inflammation induced by milk in milk-allergic subjects; in addition, stimulation of the immune system was observed in nonallergic subjects (Pelto *et al.*, 1998). Furthermore, for this strain, promotion of the immune response for the production of immunoglobulin A in patients suffering from Crohn's disease (Malin *et al.*, 1996) and augmentation in the antibody secreting cell response in children with rotavirus diarrhea has also been observed.

11.6.3 Antagonism and competitive inhibition

A hierarchy exists in microorganisms and this applies to probiotics also, so it is of great importance for probiotic microbes to be antagonistic toward other microbes, especially pathogenic microbes. For this to occur, either production of antimicrobial compounds is done by probiotic microbes or the pathogenic microbes are kept at hand by virtue of their competitive nature. An *in vitro* analysis was performed to equate the antioxidant and antagonistic activity of *Bifidobacterium* and *Lactobacillus* strains against various pathogens. *L. paracasei* 8700:2, *L. rhamnosus* GG, *L. fermentum* ME-3, *L. plantarum* 299v, *L. acidophilus* La5, *B. longum* 46, and *B. lactis* Bb12 were used to study their antagonistic characteristics against *E. coli*, *Salmonella enterica*, *H. pylori*, *Shigella sonnei*, and *C. difficile*. Results depicted the highest efficiency of GG and bifidobacteria against *E. coli* while *Salmonella* growth significantly suppressed *Lactobacilli* strains 299v, ME-3, and 8700:2. *Shigella sonnei* was controlled by ME-3 and *bifidobacteria* in anaerobic milieu. *L. paracasei*, *L. plantarum*, and *L. rhamnosus* showed moderate antagonistic potential against

H. pylori. However, *C. difficile* was not significantly controlled by these probiotics. A positive correlation was also found among lactic acid production, pH, and antimicrobial activity of all probiotics. These strains were further recommended for clinical trials for the treatment and prevention of several urinary and gastrointestinal tract infections (Hutt *et al.*, 2006). Bacteriocin production has also been rigorously studied by researchers. Lactic-acid-producing bacteria are claimed to produce different antagonistic features such as bactericidal proteins, end products of metabolism, and antibiotic substances. Nevertheless, lactic acid bacteria have a narrow range of inhibitory activity, only to inhibit Gram-positive bacteria. *In vivo* studies do not give more information in this respect because the inhibitory effects of traditional bacteriocins against pathogens are restricted to the closely related species (e.g., *Lactobacilli, Bacillus,* and *Clostridium*). Nonetheless, the role of low-molecular-weight metabolites (lactic and acetic acid, hydrogen peroxide, and aroma compounds) and secondary metabolites might be of more importance since their inhibitory spectrum is broad and they are effective against *E. coli, Salmonella, Helicobacter,* and *Clostridium. L. rhamnosus* strain GG is known to produce a number of beneficial antimicrobial substances including short-chain fatty acids that have been known to retard the growth of anaerobic microbes (*Bacteroides, Clostridium, and Bifidobacterium*) and Enterobacteriaceae family *(Streptococcus, Staphylococcus,* and *Pseudomonas*) but is ineffective against other *lactobacilli*. This strain has also been known to have antagonistic properties against enteropathogenic bacteria in mice infected with *S. typhimurium* (Hudault *et al.*, 1997). *L. acidophilus* strain LB spent culture extract (SPC) has been observed to have an adverse effect on the viability of *Listeria monocytogenes, S. typhimurium, Staphylococcus aureus, Escherichia coli, Shigella flexneri, Bacillus cereus, Klebsiella pneumoniae, Pseudomonas aeruginosa,* and *Enterobacter spp. in vitro*. The antagonistic effect of this SPC toward *S. typhimurium* was confirmed by *in vivo* experiments on mice. Antibacterial compounds that do not fall in the "bacteriocin category" (unidentified but distinctive from lactic acid) were shown to be produced by *L. johnsonii* strain LJ-1 that restrain growth of a wide spectrum of Gram +ve and Gram -ve pathogenic bacteria such as *S. typhimurium, L. monocytogenes, S. aureus, S. flexneri, P. aeruginosa, Enterobacter cloacae,* and *K. pneumoniae*. Nevertheless, these studies did not establish inhibition of *bifidobacteria* and/or *lactobacilli*. An *in vivo* study confirmed inhibition of *S. typhimurium* by the LA1 strain in the mouse model (Coconnier *et al.*, 1998). Furthermore, *in vivo* and human studies have indicated antagonism of *L. acidophilus* LA1 against *H. pylori*.

11.6.4 Antimutagenicity and anticarcinogenicity

Antimutagenic and anticarcinogenic properties of probiotic microflora residing in the GIT of humans have been under investigation by researchers for many years now. The great diversity of intestinal microorganisms is responsible for the difference in metabolic activity, particularly xenobiotic biotransformation, production, and activation of carcinogens. These metabolic activities of gut microflora have a wide range of beneficial and detrimental health effects (Rowland and Gangolli, 1999). Several human and animal studies confirm the potential of probiotics to reduce the chances of carcinogenesis by affecting enzymatic activities of bacteria and antigenotoxic effects. According to Fernandez-Gine *et al.* (2005), anticarcinogenic properties of probiotics may be attributed according to one of the following actions:

(i) Binding and then degrading of (pro) carcinogens
(ii) Production of antimutagenic compounds
(iii) Modulation of procarcinogenic enzymes in the GIT
(iv) Control of tumors by a specific immune response mechanism.

LAB, utilized in dairy products, have attracted special interest with regard to their antimutagenic (especially mutagen binding) and anticarcinogenic properties (*Lactococcus, Lactobacillus, Enterococcus, Leuconostoc, and Pediococcus* species) (Zhang *et al.*, 2010). However, only some of these fall into the existing criteria for a strain to be used as a probiotic. In addition, the antimutagenic property is not solely limited to LAB or other probiotics alone; both Gram-positive and Gram-negative bacteria exhibit this characteristic. A broad spectrum of bacteria and yeasts has shown the potential to bind mutagens *in vitro*. Moreover, mutagen and carcinogen binding is also done by nonviable bacterial cells, thus reducing carcinogenicity of the intestine and enzymes, including beta-glucosidase and beta-glucuronidase and bile acid deconjugation, and improving the immunity of the host. Several reports also confirm the anticarcinogenic activity of LAB by enhancing the activity of antioxidative enzymes and by reducing oxidative stress (Kumar *et al.*, 2010). The extent of inhibition depends on the species to be used as probiotics. Pool-Zobel *et al.* (1993) concluded that *L. lactis* and *L. casei* significantly reduced mutagenicity in nitrosated beef up to 85% but *L. sake* and *L. confusus* did not show significant antimutagenic effects. Work of several other researchers also confirms the antimutagenic potential of LAB (Zhang *et al.*, 2010). Since the carcinogen binding capacity of these probiotics depends on pH, but *in vivo* trials did not show any remarkable effects on carcinogen uptake in the gut and liver due to changes in the pH. Studies also confirm the anticancer and antigenotoxic behavior of *Bifidobacterium* and *Lactobacillus* (Pool-Zobel *et al.*, 1993). In addition, reduction of fecal mutagenicity is also related to probiotic functionality. Fecal enzymes such as beta-glucuronidase and nitroreductase convert procarcinogens into carcinogens in the colon. Human trials confirm a decrease in fecal and urinary mutagenicity by the consumption of *L. acidophilus* NCFB 1748. Furthermore, consumption of the *L. casei* strain *Shirota* has been reported to decrease urinary mutagenicity (Hayatsu and Hayatsu, 1993). Similarly, consumption of *L. rhamnosus* GG, *L. gasseri* strain ADH, and *L. casei* strain Shirota was indicative of a decreased fecal enzyme content (Isolauri *et al.*, 1994). Cancer patients have benefited from consumption of *L. casei* strain Shirota, probably because it reduces the redevelopment of superficial bladder cancer and by preventing deterioration in cervical cancer patients (Okawa *et al.*, 1993). Besides, the exhibition of antimutagenicity and anticarcinogenicity cancer prevention by probiotics in humans is still litigious. Colorectal cancer (CRC) is very common in the western world and mainly associated with diet. Probiotics are also thought to control CRC in human and animal studies. Food supplementation with *L. acidophilus* can significantly reduce the chances of colon cancer in rats. It is also stated that *Lactobacillus* GG also has the potential to control the production of cancer cells in animals suffering from colon cancer (Goldin, 1999). Use of *B. longum* can also help to reduce the chances of occurrence of liver, gut, mammary, and colon tumors in rats (Orrhage *et al.*, 1994). Ingestion of *L. casei* in food is also linked with the reduced risks of bladder cancer in humans. These studies provide a strong evidence to support the anticarcinogenic efficiency of probiotics.

11.7 Disease prevention by probiotics

11.7.1 Gastrointestinal disorders

It has been established that a difference between the fecal bacterial profile of children and adults exists, with *Bifidobacteria* dominating the profile and high population of facultative anaerobes.

(a) *Alleviation of antibiotic-related diarrhea*

Diseases, such as gastroenteritis, are capable of inducing a transformation in the biochemical and bacterial profile of the gut. After use of antibiotics, there is a high frequency of dysbacteriosis in children. Probiotic foods can be used in this instance to reestablish the microflora balance in the colon and they may also be helpful in controlling the adverse effects of antibiotic treatment.

(b) *Ulcer*

Peptic ulcers are usually associated with the diet of the subject. It has been demonstrated that subjects that have a higher intake of fermented products possess lower chances of peptic ulcers. Some studies prove that *B. bifidum* and *L. acidophilus* in high amounts, up to 10^9, act as a natural therapy to decrease the incidence of gastritis. GIT ulcer activates platelets and several other inflammatory cells in the body that help in ulcer healing by producing different growth promoters and by inducing the remodeling of tissues. Different probiotic strains including *Lactobacillus*, *Bifidobacterium*, and *Streptococcus* have antitumor potential. Different polysaccharides are secreted from the cell walls of the above-mentioned microorganisms that can be used to normalize basic fibroblast and epidermal growth factors. These findings support the ability of probiotics to heal gastrointestinal ulcers and mucosa lesions. These probiotics are also known to provide high transepithelial resistance. In addition, *L. acidophilus* is also helpful to repress ileum ulcers, inhibiting the growth of various Gram-negative bacteria. *Bifidobacteria* also show antagonistic effects against different pathogens responsible for intestinal ulcers. Probiotics compete with pathogens for the same carbohydrate receptors in the intestine. In addition, various components of the cell wall and lipoteichoic acids (LTA) extracted from Gram-positive bacteria and adhering proteins of probiotics also resist pathogenic bacterial colonization.

(c) *H. pylori infection*

H. pylori infection causes inhibitory impacts on the growth of beneficial bacteria particularly *Lactobacillus* and *Bifidobacteria* and enhances the number of different pathogenic bacteria with special reference to *Enterobacteria*. This mechanism confirms the role of these good bacteria to maintain a normal immune system because deficiency of beneficial bacteria leads to *H. pylori* infection. *H. pylori* are also responsible for chronic gastritis, intestinal ulcer, and stomach carcinoma. Several studies postulate the antagonistic role of *Lactobacillus* against *H. pylori* (Aiba *et al.* 1998; Kabir *et al.* 1997). However, the antagonistic role of *Lactobacillus* is quite conflicting, as some reveal significant protection while others show no protection. Studies also demonstrated that *L. salvarius* is more productive, with regard to lactic acid production when compared to *L. casei* and *L. acidophilus*, and more effective against *H. pylori* infection (Aiba *et al.*, 1998). Researchers have established that after infection from *H. pylori*, its colonization can only be eliminated by *L. salvarius* (Kabir *et al.*, 1997). Moreover lactoferrin, a natural antibiotic found in bovine milk, has been demonstrated to pose significant activity against *H. pylori* infection.

(d) *Lactose intolerance*

Deficiency of b-galactosidase enzyme is very common all over the world and results in improper digestion and absorption of lactose. Lactose is metabolized by bacteria as a result of which different by-products are produced that cause various ailments including abdominal cramps, nausea, bloating, and diarrhea. Different strains of *Lactobacillus*, *Streptococcus*, and *Bifidobacterium* are usually added to improve digestibility in lactose-intolerant people (Gilliland and Kim 1984; Kim and Gilliland 1983). *L. acidophilus* is considered to have maximum efficiency in this regard; however, it cannot be colonized yet. These probiotic bacteria are beneficial in combating lactose intolerance by causing fermentation and by releasing lactase in the gastrointestinal tract by replication.

11.7.2 Cancer

The role of probiotics as cancer preventive agents is controversial. It has been suggested that the chances of colon cancer can be greatly reduced down by supplementation of the diet with probiotics. Lactic acid bacteria, the most commercial probiotics, are reported to possess antimutagenic and anticarcinogenic properties (Burns and Rowland 2000). A recent study was conducted to check the occurrence of anomalies in mice fed on skimmed milks fermented with *Bifidobacteria* sp. and lactic acid bacteria against uninoculated skim milk. Reductions of 49%, 61%, and 51% in the incidence of anomalies were reported, respectively

(Abdelali *et al.*, 1995). Furthermore, a reduction in cancerous cells by bacterial strains of *L. helveticus*, and a mixture of *S. thermophilus* and *Lactobacillus bulgaricus* had a significant, although variable, effect on the reduction of cancer cells (Baricault *et al.*, 1995). The suppression of carcinogens can occur due to physical binding as well as modulation of the immune system of the host by probiotic microorganisms. On the other hand, some researchers claim to have found no such effects of the *bifidobacteria* and *lactobacilli*.

(a) *Cancer studies (in vitro)*

In vitro studies indicate binding of mutagenic compounds found in meat rich with intestinal and lactic acid bacteria normally present in the gut. This explains the decrease in mutagenicity detected following the consumption of such strains (Orrhage *et al.*, 1994). This binding is proposed to take place due to physical interactions (cation exchange). Cell wall peptidoglycans and polysaccharides are postulated to be held responsible for these interactions. Nonetheless, these health claims are controversial because the *in vivo* reversibility of mutagen binding still remains unexplained. Moreover, not to our knowledge are there biologically substantial levels of mutagenic compounds and LAB. Another consideration may be the soluble compounds formed by microflora that can interact and subsequently inhibit the tumor cells *in vivo*.

(b) *Cancer studies (in vivo)*

In vivo studies on laboratory mice administered with LAB have exhibited a reduction in DNA damage, in gastric and colonic mucosa, caused by carcinogenic chemicals. Antigenotoxic behavior of *S. thermophilus, L. confusus, L. acidophilus, L. gasseri, B. longum,* and *B. breve* against N′-nitro-N-nitrosoguanidine (MNNG) has been reported (Pool-Zobel *et al.* 1993). Antigenotoxic behavior against 1,2-dimethylhydrazine (DMH) by the above-mentioned bacteria has also been reported. Particular strains of LAB have exhibited preventive effects against some preneoplastic lesions or tumors that are supposed to be produced by carcinogens. Injection of *B. infantis* and *B. adolescentis* in mice has been reported to vanquish tumors produced by 3-methylcholanthrene (Kohwi *et al.*, 1978). Dietary supplementation with *L. acidophilus* has been reported to reduce the incidence of colon cancer caused by DMH. Dietary supplementation with a lyophilized culture of *B. longum* caused a notable reduction in colon tumor occurrence, trimmed down their reproduction, and also decreased the tumor volume (Singh *et al.*, 1997). Antitumor properties of LAB have been confirmed by implanting tumor cells in animal subjects and then feeding them with LAB fermented products.

(c) *Cancer studies (human volunteers)*

Studies in volunteers show that the consumption of *lactobacilli* reduces the mutagenicity in urine and feces that occurs due to carcinogens present in cooked meat. A decrease in mutagenicity through consumption of ground beef was also reported due to the oral intake of *L. casei*.

(d) *Mechanisms of inhibition of colon cancer*

Although the precise mechanism of cancer inhibition by beneficial microbes is unknown, the following are the postulated mechanisms that are thought to be decisive for this phenomenon:

i. Enhancement of the host's immune response
ii. Binding and subsequent degradation of potential carcinogens
iii. Changes in quantity and quality of the microflora balance
iv. Manufacture of antimutagens
v. Changing the metabolic activities of the gut flora
vi. Altering the physicochemical environment of colon
vii. Affecting the physiology of the host.

11.7.3 Effect on blood lipids

Most of *Lactobacilli* possess the ability to hydrolyze conjugated bile acids found in high quantities in the intestine. Conjugated bile acids function to emulsify, digest, and absorb lipids, but the deconjugated bile acids produced by probiotics *via* hydrolysis are inefficient in doing so, as compared to conjugated bile acid. The probable justification for this may be that at a pH less than 5.5, the deconjugated bile acids are able to precipitate when coupled with cholesterol. This loss in bile acids is compensated by the liver that converts cholesterol into new bile acids. On the whole, it is still undecided whether probiotics affect cholesterol metabolism or not. However, *L. acidophilus* has been widely investigated for its cholesterol-lowering potential. Lin *et al.* (1989) used *L. acidophilus* (ATCC 4962) and *L. bulgaricus* (ATCC 33409) to study their effects on blood serum level and found a significant reduction in cholesterol in the subjects fed with these probiotics in a 16-week trial.

An investigation on effects of fermented products containing *L. paracasei* and *L. inulin* in elderly people was carried out, and a significant relationship was found to exist as the number of these strains increased in the fecal samples (Sullivan and Nord, 2002). According to this study, 45% of the subjects had $<10^6$ colony forming units per gram of stool and 55% had very high counts of $>10^6$ CFU/g stool. On the other hand, there are some studies that oppose the idea that probiotics reduce the cholesterol level (Agerholm-Larsen *et al.*, 2000). These antagonistic views result from inconsistency in usage of strains, differences in physiological conditions of subjects (healthy vs hypercholesterolemic) and inconsistent time periods for treatments.

11.8 Role of probiotics in function food development

Probiotic foods have been extensively used in Europe and Japan for many years. The United States is also focusing on promoting the development and consumption of probiotic products. However, it is important to be very confident about the biological nature, claimed functional benefits, product availability, and safety of the probiotics for researchers as well as consumers. Probiotic foods are claimed as functional foods because of their ability to provide additional health benefits beyond normal nutrition. Most of the probiotic food products comprise mainly of lactic-acid-producing bacteria, particularly *Lactobacillus* and *Bifidobacterium*. In this chapter, work of many scientists regarding the use and functionality of probiotics with special reference to meat products is reviewed. The above discussion suggests the following beneficial effects of probiotics:

- Provision of a balance of intestinal microbiota
- Blockage of pathogenic bacterial incursion in the gut
- Hindrance of colon cancer
- Therapeutic management of diarrhea
- Relief of inflammation and irritable bowel syndrome
- Hindrance of colon cancer
- Strengthening of the immune system
- Inhibition of *H. pylori* infection
- Enhancement of lactose tolerance
- Reduction of serum cholesterol level
- Improved calcium absorption.

Mechanisms associated with these benefits have also been proposed, but some of these are not yet proven. Proper viability and appropriate dose are also very important to achieve the desirable health benefits (Lin, 2003). Functional foods are found in a great variety in all food groups; however, the homogeneous distribution of these products in the market is still a big challenge for the food industry because it is very risky, costly, and complex. Furthermore, legislative requirements and consumer demands should also be taken into account while developing and marketing functional food products. The health benefits of probiotic functional foods are demonstrated through direct ingestion of living microorganisms by the host as well as consumption of different metabolites produced during fermentation. The proposed possible mechanisms of probiotics include competitive exclusion, immune response stimulation, and competition for nutrients.

Japan and Europe are considered the largest market for probiotic functional foods (Alzamora *et al.*, 2005; Ouwehand, 2007). Probiotics have been extensively utilized to provide health benefits, but viability of these beneficial microbes is still doubted in versatile environmental conditions (Charalampopoulos *et al.*, 2002, 2003). LAB and *Bifidobacteria* are the most widely used bacteria because they are also naturally present in the gut and fall in GRAS category (Salminen, 2006).

Recently, fruit juices have also been proposed as a novel medium for the development of probiotic foods because of their high consumption, but some studies have shown the development of off-flavors in probiotic juices (Luckow and Delahunty, 2004; Luckow *et al.*, 2005, 2006), which might be a limitation of the market success of these products.

11.9 Conclusion

Increasing demands of consumers for healthy and nutritious food, while meeting the required nutrition and safety criteria, has led to the development of functional foods having certain physiological roles in health maintenance while combating different immunological disorders. Probiotics contribute toward a large share of the functional food market, being extensively used in the meat, cereal, dairy, and beverage industries. Meat is a highly nutritious food and current consumer trend is toward meat products with incorporated functional ingredients. Utilization of probiotics is most promising in raw fermented meat products as these kinds of products are consumed without prior heat treatment, which would lead to the death of probiotic microorganisms. This chapter provides an overview regarding the microbial strains to be used as probiotics, technological and safety criteria for their selection, characteristics and functionality of probiotics, role of probiotics in disease prevention and product development, and probiotic meat products. The chapter concludes that several species of microorganisms, particularly lactic acid bacteria, have great potential to be used as meat probiotics. Studies also confirm the potential health benefits of these probiotic strains. However, there is a need to stimulate research work to identify some other useful strains in this regard, particularly those that can survive the high temperature of food processing and are resistant to adverse environmental conditions in the gastrointestinal tract.

REFERENCES

Abdelali, H., Cassand, P., Saussotte, B. *et al.* (1995) Antimutagenicity of components of dairy products. *Mutation Research*, **331**, 133–141.

Agerholm-Larsen, L., Raben, A., Haulrik, N., Hansen, A.S., Manders, M. and Astrup, A. (2000) Effect of 8 week intake of probiotic milk products on risk factors for cardiovascular diseases. *European Journal of Clinical Nutrition*, **54**, 288–297.

Aiba, Y., Suzuki, N., Kabir, A.M. *et al.* (1998) Lactic acid-mediated suppression of *Helicobacter pylori* by the oral administration of *Lactobacillus salivarius* as a probiotic in a gnotobiotic murine model. *American Journal of Gastroenterology*, **93**, 2097–2101.

Alander, M., Satokari, R., Korpela, R. *et al.* (1999) Persistence of colonisation of human colonic mucosa by a probiotic strain, *Lactobacillus rhamnosus* GG after oral consumption. *Applied and Environmental Microbiology*, **65**, 351–354.

Alzamora, S.M., Salvatori, D., Tapia, S.M. *et al.* (2005) Novel functional foods from vegetable matrices impregnated with biologically active compounds. *Journal of Food Engineering*, **67**, 205–214.

Ammor, M. and Mayo, B. (2007) Selection criteria for lactic acid bacteria to be used as functional starter cultures in dry sausage production: an update. *Meat Science*, **76**, 138–146.

Andersen, L. (1998) *Fermented Dry Sausages Produced with the Admixture of Probiotic Cultures*. 44th International Commitment of Meat Science and Technology – Conference Proceedings, Barcelona, Spain. Barcelona, Institut de Recerca i Tecnologia Agroalimentaries, p. 826–827.

Annuk, H., Shchepetova, J., Kullisaar, T. *et al.* (2003) Characterization of intestinal lactobacilli as putative probiotic candidates. *Journal of Applied Microbiology*, **94**, 403–412.

Arihara, K. (2006) Strategies for designing novel functional meat products. *Meat Science*, **74**, 219–229.

Arihara, K. and Itoh, M. (2000) UV-induced *Lactobacillus gasseri* mutants resisting sodium chloride and sodium nitrite for meat fermentation. *International Journal of Food Microbiology*, **56**, 227–230.

Arihara, K., Ota, H., Itoh, M. *et al.* (1998) *Lactobacillus acidophilus* group lactic acid bacteria applied to meat fermentation. *Journal of Food Science*, **63**, 544–547.

Baricault, L., Denariaz, G., Houri, J.J. *et al.* (1995) Use of HT29, a culture human colon cancer cell line, to study the effect of fermented milks on colon cancer cell growth and différentiation. *Carcinogenesis*, **2**, 245–252.

Barksby, H.E., Lea, S.R., Preshaw, P.M. and Taylor, J.J. (2007) The expanding family of interleukin-1 cytokines and their role in destructive inflammatory disorders. *Clinical and Experimental Immunology*, **149**, 217–225.

Benno, Y. and Mitsuoka, T. (1986) Development of intestinal microflora in humans and animals. *Bifidobacteria and Microflora*, **5**, 13–25.

Bertazza, L. and Mocellin, S. (2008) Tumor necrosis factor (TNF) biology and cell death. *Frontiers in Bioscience*, **13**, 2736–2743.

Burns, A.J. and Rowland, I.R. (2000) Anti-carcinogenicity of probiotics and prebiotics. *Current Issues in Intestinal Microbiology*, **1**, 13–24.

Charalampopoulos, D., Wang, R., Pandiella, S. and Webb, C. (2002) Application of cereals and cereal components in functional foods: A review. *International Journal of Food Microbiology*, **79**, 131–141.

Charteris, W.P., Kelly, P.M., Morelli, L. and Collins, J.K. (2000) Effect of conjugated bile salts on antibiotic susceptibility of bile salt-tolerant *Lactobacillus* and *Bifidobacterium* isolates. *Journal of Food Protection*, **63**, 1369–1376.

Chen, M.F. (2011) The mediating role of subjective health complaints on willingness to use selected functional foods. *Food Quality and Preference*, **22** (1), 110–118.

Coconnier, M.H., Liévin, V., Hemery, E. and Servin, A.L. (1998) Antagonistic activity against *Helicobacter* infection in vitro and in vivo by the human *Lactobacillus acidophilus* strain LB. *Applied and Environmental Microbiology*, **64**, 4573–4580.

Couper, K.N., Blount, D.G. and Riley, E.M. (2008) IL-10: the master regulator of immunity to infection. *Journal of Immunology*, **180**, 5771–5777.

Ergonul, B. and Kundakç, A. (2011) Microbiological attributes and biogenic amine content of probiotic Turkish fermented sausage. *Journal für Verbraucherschutz und Lebensmittelsicherheit*, **6**, 49–56.

Erkkila, S. and Petaja, E. (2000) Screening of commercial meat starter cultures at low pH and in the presence of bile salts for potential probiotic use. *Meat Science*, **55**, 297–300.

Erkkila, S., Peaja, E., Eerola, S. *et al.* (2001a) Flavour profiles of dry sausages fermented by selected novel meat starter cultures. *Meat Science*, **58**, 111–116.

Erkkila, S., Suihko, M.L., Eerola, S. *et al.* (2001b) Dry sausage fermented by *Lactobacillus rhamnosus* strains. *International Journal of Food Microbiology*, **64**, 205–210.

Fernandez-Gine, J.M., Fernandez-Lopez, J., Sayas-Barbera, E. and Perez-Alvarez, J.A. (2005) Meat products as functional foods: A review. *Journal of Food Science*, **70**, 37–43.

Ferreira, C.L.L.F. (2003) Grupo de bactérias lácticas – Caracterização e aplicação tecnológica de bactérias probióticas, in *Prébióticos e Probióticos: Atualização e Prospecção* (ed. C.L. de L.F. Ferreira), Editora Rubio, Viçosa.

Gabay, C. (2006) Interleukin-6 and chronic inflammation. *Arthritis Research and Therapy*, **8**, 2–3.

Ganzle, M.G., Hertel, C., van der Vossen, J.M.B.M. and Hammes, W.P. (1999) Effect of bacteriocin-producing lactobacilli on the survival of *Escherichia coli* and *Listeria* in a dynamic model of the stomach and the small intestine. *International Journal of Food Microbiology*, **48**, 21–35.

Gilliland, S.E. and Kim, H.S. (1984) Effect of viable starter culture bacteria in yogurt on lactose utilization in humans. *Journal of Dairy Science*, **67**, 1–6.

Goldin, B.R. (1999) Health benefits of probiotics. *British Journal of Nutrition*, **80**, 203–207.

Haller, D., Scherenbacher, P., Bode, C. and Hammes, W.P. (1997) Selection of potentially probiotic bacteria. *Zeitschrift fuer Ernaehrungswissenschaft*, **36**, 87.

Hammes, W.P. and Hertel, C. (1998) New developments in meat starter cultures. *Meat Science*, **49**, 125–138.

Hayatsu, H. and Hayatsu, T. (1993) Suppressing effect of *Lactobacillus casei* administration on the urinary mutagenicity arising from ingestion of fried ground beef in the human. *Cancer Letters*, **73**, 173–179.

Heenan, C.N., Adams, M.C. and Hosken, R.W. (2002) Growth medium for culturing Probiotics bacteria for applications in vegetarian food products. *LWT--Food Science and Technology*, **35**, 171–176.

Heidebach, T., Forst, P. and Kulozik, U. (2009) Microencapsulation of probiotic cells by means of rennet-gelation of milk proteins. *Food Hydrocolloids*, **23**, 1670–1677.

Heidebach, T., Forst, P. and Kulozik, U. (2010) Influence of casein-based microencapsulation on freeze drying and storage of probiotic cells. *Journal of Food Engineering*, **98**, 309–316.

Heyman, M., Crain-Denoyelle, A.M., Corthier, G. *et al.* (1986) Postnatal development of protein absorption in conventional and germ-free mice. *American Journal of Physiology*, **251**, 326–331.

Holzapfel, W.H., Haberer, P., Snel, J. *et al.* (1998) Overview of gut flora and probiotics. *International Journal of Food Microbiology*, **41**, (2), 85–101, ISSN 0168-1605.

Hudault, S., Lievin, V., Bernetcamamrd, M.F. and Servin, A.L. (1997) Antagonistic activity exerted in vivo and in vitro by *Lactobacillus casei* (strain GG) against *Salmonella typhimurium* C5 infection. *Applied and Environmental Microbiology*, **63**, 513–518.

Hutt, P., Shchepetova, J., Loivukene, K. *et al.* (2006) Antagonistic activity of probiotic lactobacilli and bifidobacteria against entero- and uropathogens. *Journal of Applied Microbiology*, **100**, 1324–1332.

Isolauri, E., Kaila, M., Mykkänen, H. *et al.* (1994) Oral bacteriotherapy for viral gastroenteritis. *Digestive Diseases and Sciences*, **39**, 2595–2600.

Izcue, A. and Powrie, F. (2008) Special regulatory T-cell review: regulatory T cells and the intestinal tract patrolling the frontier. *Immunology*, **123**, 6–10.

Kabir, A.M., Aiba, Y., Takagi, A. *et al.* (1997) Prevention of *Helicobacter pylori* infection by lactobacilli in a gnotobiotic murine model. *Gut*, **41**, 49–55.

Khan, M.I., Arshad, M.S., Anjum, F.M. *et al.* (2011) Meat as a functional food with special reference to probiotic sausages. *Food Research International*, **44**, 3125–3133.

Khan, M.I., Anjum, F.M., Sohaib, M. and Sameen, A. (2013) Tackling metabolic syndrome by functional foods. *Reviews in Endocrine and Metabolic Disorders*, **14**, 287–297.

Kidd, P. (2003) Th1/Th2 balance: the hypothesis, its limitations, and implications for health and disease. *Alternative Medicine Review*, **8**, 223–246.

Kim, H.S. and Gilliland, S.E. (1983) *Lactobacillus acidophilus* as a dietary adjunct for milk to aid lactose digestion in humans. *Journal of Dairy Science*, **66**, 959–966.

Kohwi, Y., Imai, K., Tamura, Z. and Hasimoto, Y. (1978) Antitumor effect of *Bifidobacterium infantis* in mice. *Gann*, **69**, 613–618.

Kolozyn-Krajewska, D. and Dolatowski, Z.J. (2012) Probiotics meat products and human nutrition. *Process Biochemistry*, **47**, 1761–1772.

Kołozyn-Krajewska, D. and Dolatowski, Z.J. (2012) Probiotic meat products and human nutrition. *Process Biochemistry*, **47**, 1761–1772.

Kumar, M., Kumar, A., Nagpal, R. *et al.* (2010) Cancer-preventing attributes of probiotics. *International Journal of Food Science and Nutrition*, **61**, 473–496.

Leroy, F., Verluyten, J. and Vuyst, L. (2006) Functional meat starter cultures for improved sausage fermentation. *International Journal of Food Microbiology*, **106**, 270–285.

Lilly, D.M. and Stillwell, R.H. (1965) Probiotics. Growth promoting factors produced by microorganisms. *Science*, **147**, 747–748.

Lin, D.C. (2003) Probiotics as functional foods. *Nutrition in Clinical Practice*, **18**, 497–506.

Lin, S.Y., Ayres, J.W., Winkler, W. and Sandine, W.E. (1989) *Lactobacillus* effects on cholesterol: in vitro and in vivo results. *Journal of Dairy Research*, **72**, 2885–2899.

Lucke, F.K. (2000) Utilization of microbes to process and preserve meat. *Meat Science*, **56**, 105–115.

Luckey, T.D. and Floch, M.H. (1972) Intestinal Microecology. *American Journal of Clinical Nutrition*, 204.

Luckow, T. and Delahunty, C. (2004) Consumer acceptance of orange juice containing functional ingredients. *Food Research International*, **37**, 805–814.

Luckow, T., Sheehan, V., Delahunty, C. and Fitzgerald, G. (2005) Determining the odor and flavor characteristics of probiotic, health-promoting ingredients and the effects of repeated exposure on consumer acceptance. *Journal of Food Science*, **70**, 53–59.

Luckow, T., Sheehan, V., Fitzgerald, G. and Delahunty, C. (2006) Exposure, health information and flavour-masking strategies for improving the sensory quality of probiotic juice. *Appetite*, **47**, 315–323.

Macedo, R.E.F., Pflanzer, S.B., Terra, N.N. and Freitas, R.J.S. (2005) Caracteristicas de culturas lácticas probióticas para uso em produtos cárneos fermentados: sensibilidade aos sais de cura e uso de antibióticos para contagem seletiva. *Boletim do CEPPA*, **23**, 123–134.

Macedo, R.E.F., Pflanzer, S.B., Terra, N.N. and Freitas, R.J.S. (2008) Desenvolvimento de embutido fermentado por Lactobacillus probióticos: características de qualidade. *Ciência e Tecnologia de Alimentos*, **28**, 509–519.

de-Macedo, R.E.F., Pflanzer, S.B. and Gomes, C.L. (2012) *Probiotics Meat Products in Probiotics in Animals*, Intech Open Science.

Malin, M., Verronen, P., Mykkänen, H. *et al.* (1996) Increased bacterial urease activity in feces in juvenile chronic arthritis: evidence of altered intestinal microflora. *British Journal of Rheumatology*, **35**, 689–694.

Maragkoudakis, P.A., Mountzouris, K.C., Psyrras, D. *et al.* (2009) Functional properties of novel protective lactic acid bacteria and application in raw chicken meat against *Listeria monocytogenes* and *Salmonella enteritidis*. *International Journal of Food Microbiology*, **130**, 219–226.

Marteau, P. (2002) Probiotics in clinical conditions. *Clinical Reviews in Allergy and Immunology*, **22**, 255–273.

Matteuzzi, D., Crociani, F. and Brigidi, P. (1983) Antimicrobial susceptibility of Bifidobacterium. *Annals of Microbiology*, **134A**, 339–349.

Moubareck, C., Daoud, Z., Hakime, N.I., Hamze, M., Mangeney, N., Matta, H., Mokhbat, J.E., Rohban, R., Sarkis, D.K. and Doucet-Populaire, F. (2005) Countrywide spread of community- and hospital acquired extended-spectrum beta-lactamase (CTX-M-15)-producing Enterobacteriaceae in Lebanon. *Journal of Clinical Microbiology*, **43**, 3309–3313.

Morelli, L. (2007) In vitro assessment of probiotic bacteria: from survival to functionality. *International Dairy Journal*, **17**, 1278–1283.

Muthukumarasamy, P. and Holley, R.A. (2006) Microbiological and sensory quality of dry fermented sausages containing alginate-microencapsulated *Lactobacillus reuteri*. *International Journal of Food Microbiology*, **111**, 164–169.

Okawa, T., Niibe, H., Arai, T. *et al.* (1993) Effect of LC9018 combined with radiation therapy on carcinoma of the uterine cervix. *Cancer*, **72**, 1949–1954.

Orrhage, K., Brismar, B. and Nord, C.E. (1994) Effect of supplements with *Bifidobacterium longum* and *Lactobacillus acidophilus* on the intestinal microbiota during administration of clindamycin. *Microbial Ecology in Health and Disease*, **7**, 17–25.

Ouwehand, A.C. (2007) *Success in Applying Pro- and Prebiotics in Dairy Products*. Fourth International FFNet Meeting on Functional Foods – Conference Proceedings, Budapest: Hungary, March 26–27, 2007.

Papamanoli, E., Tzanetakis, N., Litopoulou-Tzanetaki, E. and Kotzekidou, P. (2003) Characterization of lactic acid bacteria isolated from a Greek dry-fermented sausage in respect of their technological and probiotic properties. *Meat Science*, **65**, 859–867.

Pelto, L., Isolauri, E., Lilius, E.M. *et al.* (1998) Probiotic bacteria down-regulate the milk-induced inflammatory response in milk-hypersensitive subjects but have an immunostimulatory effect in healthy subjects. *Clinical and Experimental Allergy*, **28**, 1474–1479.

Penna, A.L.B. (2002) *Probioticos and Saude*, XVIII Congresso Brasileiro de Ciência e Tecnologia de Alimentos – Conference Proceedings, Porto Alegre, Brazil. Porto Alegre: SBCTA, p. 4045-4046.

Pennacchia, C., Ercolini, D., Blaiotta, G. *et al.* (2004) Selection of *Lactobacillus* strains from fermented sausages for their potential use as probiotics. *Meat Science*, **67**, 309–317.

Pool-Zobel, B.L., Bertram, B., Knoll, M. *et al.* (1993) Antigenotoxic properties of lactic acid bacteria in vivo in the gastrointestinal tract of rats. *Nutrition and Cancer*, **20**, 271–281.

Poulin, J.F., Caillard, R. and Subirade, M. (2011) Lactoglobulin tablets as a suitable vehicle for protection and intestinal delivery of probiotic bacteria. *International Journal of Pharmaceutics*, **405**, 47–54.

Rebucci, R., Sangalli, L., Fava, M. *et al.* (2007) Evaluation of functional aspects in *Lactobacillus* strains isolated from dry fermented sausages. *Journal of Food Quality*, **30**, 187–201.

Resta-Lenert, S. and Barrett, K.E. (2003) Live probiotics protect intestinal epithelial cells from the effects of infection with enteroinvasive *Escherichia coli* (EIEC). *Gut*, **52** (7), 988–997.

Rivera-Espinoza, Y. and Gallardo-Navarro, Y. (2008) Non-dairy probiotic products. *Food Microbiology*, **27**, 1–11.

Rowland, I.R. and Gangolli, S.D. (1999) Role of gastrointestinal flora in the metabolic and toxicological activities of xenobiotics, in *General and Applied Toxicology*, 2nd edn (eds B. Ballantyne, T.C. Marrs and T. Syverson), Macmillan Publishers Ltd., London, pp. 561–576.

Saarela, M., Mogensen, G., Fondén, R. *et al.* (2000) Probiotic bacteria: safety, functional and technological properties. *Journal of Biotechnology*, **84** (3), 197–215.

Salminen, S. (2006) Probiotics: basis, state of the art and future perspectives. *Functional Food Network General Meeting*, Turku, Finland, March 8–10.

Salminen, S., Isolauri, E. and Salminen, E. (1996) Clinical uses of probiotics for stabilizing the gut mucosal barrier: successful strains and future challenges. *Antonie van Leeuwenhoek*, **70**, 347–358.

Salminen, S., Bouley, C., Boutron-Ruault, M.C. *et al.* (1998) Functional food science and gastrointestinal physiology and function. *British Journal of Nutrition*, **80**, 147–171.

Salminen, S., Ouwehand, A., Benno, Y. and Lee, Y.K. (1999) Probiotics: how should they be defined? *Trends in Food Science & Technology*, **10** (3), 107–110.

Sameshima, T., Magome, C., Takeshita, K. *et al.* (1998) Effect of intestinal Lactobacillus starter cultures on the behaviour of *Staphylococcus aureus* in fermented sausage. *International Journal of Food Microbiology*, **41**, 1–7.

Santos, F.L., Ferreira, C.L.L.F. and Costa, N.M.B. (2003) Modulação da colesterolemia por meio de prebióticos e probióticos, in *Prébióticos e Probióticos: Atualização e Prospecção* (ed. C.L. de L.F. Ferreira), Viçosa, Rubio.

Shanahan, P.M., Thomson, C.J. and Amyes, S.G. (1994) Beta-lactam resistance in aerobic faecal flora from general practice patients in the UK. *European Journal of Clinical Microbiology and Infectious Diseases*, **13**, 760–763.

Singh, J., Rivenson, A., Tomita, M. *et al.* (1997) *Bifidobacterium longum*, a lactic acid-producing intestinal bacterium inhibits colon cancer and modulates the intermediate biomarkers of colon carcinogenesis. *Carcinogenesis*, **18**, 833–841.

Sperti, G.S. (1971) *Probiotics*, AVI Publishing Co. Inc., West Point, Connecticut.

Sullivan, Å. and Nord, C.E. (2002) Probiotics in human infections. *Journal of Antimicrobial Chemotherapy*, **50**, 625–627.

Taranto, M.P., Perez-Martinez, G. and Valdez, G.F. (2006) Effect of bile acid on the cell membrane functionality of lactic acid bacteria for oral administration. *Research in Microbiology*, **157**, 720–725.

Tharmaraj, N. and Shah, N.P. (2003) Selective enumeration of *Lactobacillus delbrueckii* ssp. *Bulgaricus, Streptococcus thermophilus, Lactobacillus acidophilus, Bifidobacteria, Lactobacillus casei, Lactobacillus rhamnosus and Propionibacteria. Journal of Dairy Science*, **86**, 2288–2296.

Toms, C. and Powrie, F. (2001) Control of intestinal inflammation by regulatory T cells. *Microbes and Infection*, **3**, 929–935.

Trinchieri, R. (2003) Interleukin-12 and the regulation of innate resistance and adaptive immunity. *Nature Reviews Immunology*, **3**, 133–136.

Tyopponen, S., Petaja, E. and Mattila-Sandholm, T. (2003) Bioprotectives and probiotics for dry sausages. *International Journal of Food Microbiology*, **83**, 233–244.

Vuyst, L.D., Falony, G. and Leroy, F. (2008) Probiotics in fermented sausages. *Meat Science*, **80**, 75–78.

Waldeck, F. (1990) Funktionen des Magen-Darm-Kanals, in *Physiologie des Menschen*, 24 Aufl. edn (eds R.F. Schmidt and G. Thews), Springer, Berlin, Germany.

Wu, K., Bi, Y., Sun, K. and Wang, C. (2007) IL-10-producing type 1 regulatory T cells and allergy. *Cellular and Molecular Immunology*, **4**, 269–275.

Zhang, W., Xiao, S., Samaraweera, H. *et al.* (2010) Improving functional value of meat products. *Meat Science*, **86**, 15–31.

PART III

Assessment techniques for meat quality and safety

CHAPTER 12

Rapid methods for microbial analysis of meat and meat products

Agapi Doulgeraki, Efstathios Panagou & George-John Nychas
Laboratory of Microbiology and Biotechnology of Foods, Department of Food Science and Human Nutrition, Agricultural University of Athens, Athens, Greece

12.1 Introduction

Pathogens encountered to human illness in the last decades are through consumption of undercooked or minimally processed ready-to-eat meals or fruits and vegetables (Mandal *et al.*, 2011). Although various foods can serve as sources of foodborne illness, meat and meat products are important sources of human infections with *Salmonella* spp., *Campylobacter jejuni/coli*, *Yersinia enterocolitica*, VTEC and, to some extent, *Listeria monocytogenes*. All these foodborne pathogens can be harboured in the gastrointestinal tract of food-producing animals. The most frequent chain of events leading to meatborne illness involves food animals as healthy carriers of pathogens; these organisms are subsequently transferred to humans through production, handling, and consumption of meat and meat products (Biswas *et al.*, 2008). Occurrences of *Salmonella* spp., *C. jejuni/coli*, *Y. enterocolitica* and VTEC in fresh red meat are variable, although most often are between 1% and 10%, depending on a range of factors including organisms, geographical factors, farming and/or meat production practices.

The EU has a strategically important global position in terms of beef production and consumption. Currently, the beef chain in Europe is valued at about €75 billion, which accounts for about 10% of European agricultural output. In 2003, the EU-25 region produced 8.06 million tons and consumed 8.32 million tons of beef and beef products, with an additional 0.60 and 0.045 million tons being imported and exported, respectively (Evans, 2005). Meat products are highly perishable foods, which, unless correctly stored, processed, packaged and distributed, spoil quickly and may potentially become unsafe due to microbial growth of certain pathogenic bacteria such as *L. monocytogenes*, *Escherichia coli* O157:H7 and *Salmonella* spp., despite the introduction of HACCP (Hazard Analysis and Critical Control Points) and the proliferation of food safety regulations (EFSA, 2006). In addition, the image of the European meat industry

Emerging Technologies in Meat Processing: Production, Processing and Technology, First Edition.
Edited by Enda J. Cummins and James G. Lyng.

has suffered major setbacks due to recent problems with BSE, foot and mouth disease and feeds tainted with dioxin or other chemical toxicants (see recent crisis in Ireland and Germany; http://www.efsa.europa.eu/en/efsajournal/pub/911.htm and http://ec.europa.eu/food/food/chemicalsafety/contaminants/dioxin germany_en.htm) causing huge economic losses to the industry. This loss was estimated at $4 billion per year for meat and poultry products (Evans, 2005; EFSA, 2006; Anon, 1993a,b; WHO, 2000). These setbacks are key issues not only due to short-term economic losses but also because they create an atmosphere of mistrust among meat buyers and consumers in general and this may lead to long-term problems. That potential atmosphere stresses the need for more effective meat quality and safety assurance systems (McMeekin *et al.*, 2006), which are both effective and visible to the consumer. This quality (including safety) system should not only adapt the existing technological progress and evolution of consumer lifestyle but also foresee the potential of the latest information technology as well as the future contribution of converging technologies. In addition to public health demands for safety beef chain, the industry must be competitive in a changing global market and furnish the wider desires of modern consumers.

The two most frequently reported zoonotic diseases in humans in the EU were *Campylobacter* and *Salmonella* infections with incidences of 51.6 and 38.2 cases per 100,000 population, respectively (EFSA, 2006), while in 2011 Campylobacteriosis was the most commonly reported zoonosis with 220,209 confirmed human cases (EFSA, 2013).

In regard to foodborne outbreaks, a total of 5648 cases were reported in the EU, resulting in 69,553 human cases, 7125 hospitalizations and 93 deaths. Most of the reported outbreaks were caused by *Salmonella*, bacterial toxins, *Campylobacter* and viruses; however, the outbreak with most human cases was caused by Shiga toxin-producing *E. coli*/verotoxigenic *E. coli* and associated with sprouted seeds. The most important food sources of the outbreaks were eggs and egg products, followed by mixed food and fish and fish products. Overall, 11 waterborne outbreaks were reported in 2011, caused by *Campylobacter*, calicivirus, *Cryptosporidium hominis* and verotoxigenic *E. coli*.

Thus, the microbiological examination of the food is performed for two purposes: to ascertain the absence or prevalence of pathogenic microorganisms or their hazardous by-products such as toxins to ensure the safety of foods and to enumerate total or indicator microbial to verify products' quality and monitor effectiveness of hygienic procedure (Foley and Grant, 2007; Hoorfar, 2011; Doyle, 2001). This is an ideal situation wherein rapid methods such as online monitoring systems can be useful to quickly screen large number of samples and thereby enhancing the processing efficiency (Mandal *et al.*, 2011). These purposes are increasing the demand from the food industry, the food consumer and public health authorities for more rapid methods to detect foodborne pathogens

(Foley and Grant, 2007). It is worth noted that the conventional culture methods which referred to as the reference methods for microbial analysis have been subjected to some kind of collaborative study or peer review by independent laboratories to gain a wide acceptance of proven performance (Jasson *et al.*, 2010). Thus, the availability of a rapid method alone does not mean that regulators are immediately going to take it onboard as a standard method, as the performance of the method compared with conventional cultivation methods is an important consideration for global regulators, not only in relation to their own methods but also to those used by regulators in export markets (Hoorfar, 2011).

12.2 Theory of high rapid methods

Food testing programs perform methods that evolve into laborious and time-consuming procedures. Several methods have been proposed for microbial analysis of food; however, the conventional methods take 2–3 days; in this duration, the food may have been moved from the factory and even consumed by the consumers. This arising hazard could be overcome by applying modern and rapid techniques, which have been researched, identified and developed to determine the levels of microorganisms in food. It has to be noted that rapid detection is especially critical in situations where products have already been set in motion (Hoorfar, 2011). In addition, in the case of detection of foodborne pathogens, analysis is challenging due to the complexities of the food matrix, heterogeneous distribution of low levels of pathogens, high level of normal microbiota and the presence of ingredients that can interfere with assays and bacterial stresses that may occur during food processing (Hoorfar, 2011; López-Campos *et al.*, 2012). Prior to analysis, an effective enrichment step is usually followed; however, it often extends the analysis to several days (Lazcka, Del Campo and Munoz, 2007). In addition, classical enumeration methods for the total microbiota or a specific group, requires homogenization, serial dilutions and longer incubation periods to obtain an accurate count which is labour-intensive. These traditional microbiological methods are not able to provide quick assessment regarding the safety or microbiological specification and keep up with the pace of production and distribution.

In contrast, a rapid method is generally characterized as a test giving quicker results than the standard accepted method of isolation and biochemical and/or serological identification (Ellingson *et al.*, 2004). This demand for speedy results combined with major advances in a range of technologies has led to a vast array of rapid methods being developed and investigated over the past two decades. The collaborations between workers in the fields of engineering, nanotechnology and food science have introduced new lab-on-a-chip technologies permitting development of portable, hand-held biosensors for food pathogen detection. The use of rapid methods, either in a laboratory or for on-site testing of food production premises, depends not only on the time taken to get results but also on a range of other factors such as robustness, sensitivity, specificity, accuracy, reliability, standardization, evaluation, throughput, ease of use, potential for automation, cost, convenience, validation and the throughput of the end user (Foley and Grant, 2007; Jasson *et al.*, 2010).

In several studies (de Boer and Beumer, 1999; Jasson *et al.*, 2010; Mandal *et al.*, 2011), the following requirements for alternative and rapid methods are sufficiently summarized:

(a) Accuracy – reduction or elimination of false positive and negative results
(b) Validation against standard tests
(c) Speed – giving an accurate result within hours or less than 1 day
(d) Automation and computerization
(e) Sample matrix – background biota, natural substances or debris can interfere and invalidate the results
(f) Cost
(g) Simplicity
(h) Rapid availability of reagents and supply
(i) Training, technical service, and company support is essential
(j) Space requirements.

12.2.1 Mode of action of rapid methods for microbial quality

A rapid method can be an assay that gives instant or real-time results; on the other hand, it can be also a simple modification of a procedure that reduces the assay time. The development and the use of rapid methods can clearly contribute to the rapid processing and sensitivity and should increase the likelihood that outbreak data can contribute to the overall understanding of dose response, improve the ability of risk models to estimate the probability of illness and the characterization of the prevalence of contamination (Hoorfar, 2011). Thus, rapid methods not only deal with the early detection and enumeration of microorganisms but also with the characterization of the isolates by the use of microbiological, chemical, biochemical, biophysical, molecular, biological, immunological and serological methods (Mandal *et al.*, 2011). The degree to which rapid methods and automation are accepted and used for microbiological analysis is determined by the range and type of testing required, volume throughput of samples to be tested, availability of trained laboratory staff and nature of manufacturing practices (Vasavada, 1993). Rapid methods also offer the promise of improving the risk assessment and modelling process by increasing the amount of data and information available for use in developing risk assessments, as well as in reducing the uncertainty associated with the data used (Hoorfar, 2011).

12.2.2 Advantages and disadvantages of rapid methods

The major advances in rapid methods have, in general, been in immunological and nucleic acid-based detection systems, although significant developments have been made and are being made in other areas including impedance and conductance, bacteriophages, biosensors, microscopy as well as in miniaturized, automated biochemical detection kits (Foley and Grant, 2007). However, almost all existing rapid methods are designed to detect a single target, which makes them ideal for use in quality control programs to quickly screen large number of (mostly negative) food samples for the presence of a particular pathogen or toxin (Hoorfar, 2011). This is in contrast to the cultural methods, which may simultaneously detect and isolate many pathogens by including several types of selective media in the analysis (de Boer and Beumer, 1999).

Another important challenge for detecting pathogens and toxins in foods is related to the ratio between the amount of food samples and the detection assay volume, which is particularly noteworthy in advanced technologies that only a tiny fraction of the original food samples is subjected to the actual detection assay after sample preparations (Ge and Meng, 2009). In addition, a further problem is the requirement of the ability to detect viable organisms and not those that have been destroyed by food processing. Damaging of the cells is a major disadvantage of alternative and rapid methods over cultural methods as repeat analysis using standard cultural procedures is needed in order to be able for confirmation and further characterization of the cells (de Boer and Beumer, 1999). On the other hand, enrichment serves as a major strategy to increase the concentration of target pathogens in foods and thereby diluting the effects of inhibitors and normal biota on the testing (Ge and Meng, 2009). However, enrichment is labour intensive and inevitably increases the total analysis time, stretching it to days rather than hours or even minutes. In addition to enrichment, concentration and extraction techniques are commonly applied to facilitate both pathogen and toxin detection in foods (Stevens and Jaykus, 2004). However, when food samples are homogenized and concentrated, components that are inhibitory to certain detection assays are often co-concentrated (Pillai, 2004). A major limitation with how much volume it can be taken prior to the rapid analysis for the target molecule is also arising, while the technical challenge is how to obtain an accurate representation of 25 g of food into the 100 μL sample that will go into a real-time, reverse transcriptase PCR (Jasson *et al.*, 2010; Hoorfar, 2011).

A variety of techniques have been investigated to overcome cultural enrichment, including centrifugation, filtration, aqueous polymer two-phase separation systems, ultrasound, immunomagnetic, bacteriophage and lectin separation as well as dielectrophoresis (Benoit and Donahue, 2003). None has been able to deliver 100% separation of bacteria from foods; many have improved the sensitivity of several rapid-detection methods when carried out without enrichment culture whose application provides a solution to those problems that is a challenge to the direct detection of pathogens in foods (Foley and Grant, 2007). Many of the sensors presented detect analytes in a small and relatively clean sample; however, there exist opportunities for improvement of sample preparation.

Many of the methods for rapid detection may also be used in combination, for example, immunomagnetic separation (IMS) followed by bioluminescent detection, which has led to an almost exponential increase in the number of rapid methods being reported for detecting foodborne pathogens (Foley and Grant, 2007). Thus, sometimes the use of several assays to multipathogen analyses on the food sample makes the analysis unacceptably expensive (de Boer and Beumer, 1999). Furthermore, evaluation of rapid methods shows that the same methods may perform better in some foods, which may be attributed mostly to interference by normal microbiota or inhibitors in food (Mandal *et al.*, 2011).

All of these points must be addressed in a simple, robust, rapid, generic method, and one that is amenable to automation, before the direct detection of pathogens in foods becomes a reality (Foley and Grant, 2007). In addition, regarding the selection of an appropriate method, it should be noticed that all methods have advantages and disadvantages and no method exists that is 100% sensitive and 100% specific, which can provide results without hands-on time and at low cost (Jasson *et al.*, 2010).

12.2.3 Factors affecting interaction between rapid methods and meat matrix

The complexity of the food matrices and the composition is a major obstacle to developing effective sampling, sample preparation and testing strategies (Lund, Baird-Parker and Gould, 2000; Feng, 2007). Major difficulties in detecting pathogens directly in foods arise from the diverse range of food matrices that may be encountered, the inhibitory nature of the food constituents to various detection assays and the necessity to ensure the absence of microorganisms in 25 g of food (Ellingson *et al.*, 2004). Efficient separation of pathogens or toxins from food matrices is rather difficult in meat, poultry and egg products (Ge and Meng, 2009). In addition, the analysts should be cognizant of the possibility that a method validated in one country for the analysis of a particular product may not be applicable to the same product in another country due to different production and processing practices that may result in foods containing different ingredients, pathogens or levels and types of normal bacterial biota (Hoorfar, 2011; López-Campos *et al.*, 2012).

Food matrices frequently harbour inhibitors for key reagents used in detection assays such as PCR enzymes (Ge and Meng, 2009). In addition, high amount of non-specific DNA is known to inhibit PCR sensitivity (Hoorfar, 2011). In brief, it has been reported that genetic amplification steps such as PCR and NASBA are sensitive to inhibitors and therefore require extensive sample preparation and clean-up (Nugen and Baeumner, 2008). In an earlier study, the inhibition of PCR by components food samples, microbial diagnostic assays and DNA extraction solutions was studied (Rossen *et al.*, 1992). This study reported that it is important to take into account that any component of the food sample, the growth media and DNA extraction solution may influence the effectiveness of the PCR. In the case of food, they reported as follows:

(a) The amount of homogenized food sample that could be added directly to the PCR varied considerably with the type of food product.
(b) More complex products showed a considerable inhibitory effect.
(c) The strong inhibitory effect of the cheeses on PCR may at least in part be explained by the presence of proteinases that can destroy the polymerase activity.
(d) The PCR was tolerant towards even relatively high levels of salt, carbohydrate and amino acids.
(e) Although same levels of proteins were found in ham and cheese, a further extraction step was needed for cheese because it contains 10 times the amount of fat compared to ham.

Nowadays, although many commercial PCR kits include easy-to-use sample treatments protocols that (apparently) work, independent studies show a remarkably low diagnostic sensitivity when testing naturally contaminated samples (D'Aoust *et al.*, 2007). It is also noted that this is not always an easy task with complex food systems such as peanut butter or RTE pudding that make isolation of the pathogens from the food matrix difficult; the time required to prepare the sample could possibly take far longer than the biosensor detection time (Nugen and Baeumner, 2008). The sampling of microorganisms in air and subsequent analysis by PCR was found to be an effective approach, which can circumvent the PCR inhibition problem regarding the food matrix; however, high amounts of non-specific DNA may influence detection in air depending on the quality of air sampled (Hoorfar, 2011). Nevertheless, detection of as low as 1–10 CFU and spores per air sample has been reported (Alvarez, Buttner and Stetzenbach, 1995; Makino and Cheun, 2003).

Another issue that has to be taken into account is background microbiota, as raw food naturally contains high levels of normal biota, which may interfere with the performance of a wide array of pathogen detection technologies (Ge and Meng, 2009). For instance, questions remain to be answered on how it is possible to further standardize the enrichment–enumeration techniques to minimize the impact of background microbiota, when changing from one sample type to another (Hoorfar, 2011). Furthermore, it is of great interest to establish a method that will be able to deliver 100% of bacteria, as the development of a simple generic method for food sample preparation to facilitate the direct use of rapid methods would be a huge benefit to all concerned with the microbiological safety of foods (Foley and Grant, 2007). This problem arises due to the fact that pathogens in food are generally present in low levels and frequently injured during food processing, which demands improved detection sensitivity and pre-enrichment to resuscitate the injured cells (Wu, 2008; Ge and Meng, 2009). Enrichment culture itself is an amplification technique, which allows small number of microorganisms to be detected and to grow only viable organisms. Another advantage may be enabling the repair of stressed or injured cells that may be able to cause disease but would not be detectable by direct methods. The potential for hazard is still in concern because injured microorganisms are capable of repair and toxin production (Wu, 2008). In the same study, they concluded that the incorporation of a resuscitation stage using a non-selective pre-enrichment medium or an effective recovery method improves the detection rates of rapid assays, while appropriate recovery procedures should be incorporated into current detection and enumeration methods and adopted for regulatory and quality control purposes to ensure that a true microbiological analysis is obtained.

Use of enrichment culture also dilutes the food material, thus diluting out inhibitors, and can result in microorganisms being in a more similar, simpler matrix (Foley and Grant, 2007). Although enrichment has a limitation in terms of assay speed, it is necessary for several reasons beyond increasing bacterial numbers, including diluting the effects of inhibitors, allowing the differentiation of viable from non-viable cells and allowing for resuscitation of injured cells (Dwivedi and Jaykus, 2011; Rossmanith *et al.*, 2011). In DNA-based assays, enrichment has been applied to indicate the detection of viable rather than non-viable cells (Feng, 2007). A new and fast quantitative real-time PCR approach has been reported based on the use of a simple propidium monoazide sample treatment that generated a signal from viable, but non-culturable *Campylobacter* with an intact membrane in <3 h (total assay time) (Josefsen *et al.*, 2004). In another study, it was found that the proceeding PCR step was inhibited by background biota concentrations of 6.1×10^8 CFU/swab sample but not by concentrations of 6.1×10^6 CFU/swab sample (Lofstrom *et al.*, 2011). The inhibition of background microbiota with the addition of novobiocin or other substances did not improve the detection reproducibility, while limited pre-enrichment of 8 h in buffered peptone water was unable to produce sufficient number of *Salmonella* (Hoorfar, 2011).

12.3 Rapid method tools

12.3.1 Case studies – rapid methods evaluation of foods

Advanced technologies, including convenience-based, antibody-based and nucleic acid-based assays, have revolutionized the detection methodology for microbial pathogens and toxins in food. Nevertheless, a few techniques such as PCR or ELISA have been proven sufficiently robust to be used in routine applications for the benefit of public health (Hoorfar, 2011). Several times, such assays not only identify the specific microorganism but also provide a rather comprehensive genetic or metabolic profile of the organism (Pillai, 2004). Despite the recent advances in food pathogen detection, there still exist many challenges and opportunities to improve the current technology. Techniques such as IMS and polymerase chain reaction (PCR) have paved the way for rapid and sensitive detection of foodborne pathogens, and advances in nanobiotechnology have allowed for miniaturization of devices and introduced portable, hand-held biosensors (Nugen and Baeumner, 2008).

Because of interference from the food matrix, almost all new rapid methods for food pathogen detection are performed following enrichment culture; however, a number of separation and concentration techniques to isolate pathogens or their nucleic acids from foods have been investigated in order to overcome this lengthy step (Foley and Grant, 2007). Real-time PCR quantification was applicable to *Salmonella* from pig carcasses after a short pre-enrichment phase under standardized conditions and time course (Kramer *et al.*, 2011). Thus, careful consideration should be given to the enrichment strategies for *Salmonella* cells in combination with subsequent quantitative real-time PCR analysis (Hoorfar, 2011). Similarly, Lofstrom *et al.* (2011) developed a quantitative flotation protocol based on traditional buoyant density centrifugation, for culture-independent enumeration of intact *Salmonella* in pig carcass gauze swabs ($100\,cm^2$) prior to quantitative PCR. In addition, biotracing of pathogens has been also introduced to make sound decision about possible recalls or to determine the source of contamination (Barker, Gomez and Smid, 2009). Similarly, Josefsen, Jacobsen and Hoorfar (2004) developed a semi-quantitative strategy to quantify low numbers of *Campylobacter* from chicken rinse samples.

The purpose of analysis is also affect the selection of the applied method. Briefly, in case of a competent authority, the alert can be an announcement of the presence of a hazard in a certain food, for example *E. coli* O157 outbreak in the Netherlands due to raw beef (Greenland *et al.*, 2009), but in case of a food business operator an alert can be a complaint from consumers or a critical control point (CCP) that went out of control (Jasson *et al.*, 2010). One additional consideration is the type of sampling procedures to be used, destructive versus non-destructive, which can dramatically affect microbial recovery rates, hence the detection sensitivity and overall performance (Ge and Meng, 2009).

12.3.2 Brief overview of available rapid methods

Different rapid methods had been developed in the last decades (de Boer and Beumer, 1999; Fung, 2007; Jasson *et al.*, 2010; Hoorfar, 2011; Mandal *et al.*, 2011; López-Campos *et al.*, 2012; Roda *et al.*, 2012). In this section, some of these methods are proposed in the following categories:

(i) Modified and automated conventional methods
(ii) Nucleic acid–based assays
(iii) Immunological methods
(iv) Biosensors.

12.3.2.1 Modified and automated conventional methods

Traditional culture methods use selective liquid or solid culture media to grow, isolate and enumerate the target microorganism and simultaneously prevent the growth of other microorganisms present in the food (Jasson *et al.*, 2010). In a recent study, a list of the laboratory solid media commonly used to subculture microbial foodborne pathogens is proposed (Drosinos, Paramithiotis and Andritsos, 2011). However, in the last decades, many attempts have been made to improve laboratory efficiency by making the procedures for traditional agar medium–based dilution methods more convenient and user friendly and to reduce the costs of material and labour. The continued need for rapid methods that shorten the time between receipt of a specimen or sample and the issue of a report in industrial level led to the development of instrumentation, usually known as automation, which has the additional advantage that it does not require constant attention. Modifications and automation of conventional methods in food microbiology include sample preparation, plating techniques, counting and identification test kits (de Boer and Beumer, 1999; Jasson *et al.*, 2010; Mandal *et al.*, 2011). In the same studies, the following modified conventional methods are also analysed:

- ***Spiral plater*** – It deposits a small volume onto the surface of the agar in a spiral such that there is a dilution ratio of 10 from the centre to the edge of the plate. The colonies appearing along the spiral can be counted either manually or electronically. As the volume dispensed at any point is known, this technique eliminates the need for serial dilutions before plating and less time is required for counting colonies.
- ***Chromogenic/fluorogenic culture media*** – these are microbiological growth media that contain enzyme substrates linked to a chromogen (colour reaction), fluorogen (fluorescent reaction) or a combination of both, where the target population is characterized by enzyme systems that metabolize the substrate (sugar or amino acid) to release the chromogen/fluorogen. The incorporation of such fluorogenic or chromogenic enzyme substrates into a selective medium can eliminate the need for subculture and further biochemical tests to establish the identity of certain microorganisms.
- ***Petrifilm or compact dry plates*** – alternative to agar poured plates, which consist of rehydratable nutrients that are embedded into the film along with gelling agent soluble in cold water, where the liquid sample is placed.
- ***SimPlate*** – rely on a binary detection technology for the detection and enumeration of microorganisms, where visible colour changes occur as a result of bacterial enzyme interaction with substrates present in the liquid culture medium.
- ***Tempo*** – an automated most probably number (MPN) enumeration method consists of a vial of culture medium and a card, where the target microorganism multiplies in the culture medium resulting in a signal.

- ***Soleris*** – monitors changes in the chemical characteristics of microbial liquid growth medium and detects microorganisms with pH and other sensitive reagents. The reagents change their spectral patterns as the metabolic process takes place, which can be detected photometrically.

Similarly, Collins *et al.* (2004) proposed the following automated systems:

- ***VITEK*** (*VIDAS*), an automated qualitative enzyme-linked fluorescent immune assay introduced by BioMerieux (France), which is based on the detection of microbial growth in microwells within plastic cards, that offers rapid results with an average 2- to 6-h, same-day, turnaround time.
- ***Sensititre***, the sensitive system (Trek Diagnostic Systems, Inc.), which is made up of a number of 'mix-and-match' components and provides walk-away automation for identification and susceptibility testing.
- ***Biolog***, which can detect mixed or contaminated cultures and is also able to identify environmental and fastidious organisms.
- ***Phoenix***, which is an automated, short incubation system for the rapid identification and susceptibility testing of clinically relevant bacteria, where the average time to results is 6 h for Gram-positive and 6–12 h for Gram-negative organisms.
- ***Midi microbial identification system***, a fully automated, computerized gas chromatography (GC) system that can separate, distinguish and analyse fatty acid composition that is highly conserved within a taxonomic group.

However, users of an automated system need to consider the limitations and the percentage accuracy and before the purchase and their application.

12.3.2.2 Nucleic acid–based assays

Nucleic acid–based assays allow rapid and reliable identification and typing of microorganisms, providing a clear picture of the microbial community in a particular environment. Several molecular techniques that have been developed during the last decades are applied for both identification and typing of bacteria at species or strain level; they are basically employed to overcome the drawbacks of conventional phenotypic methods. Studies of muscle (e.g. meat and fish) microbial ecology and more specific microbial meat spoilage and pathogenic microorganisms have recently benefited by the use of such tools and the most recent results described in the literature are based on the application of both culture-dependent and culture-independent molecular methods (Doulgeraki *et al.*, 2012). In the same study, a list of different techniques and their most recent applications for monitoring the bacterial communities on meat, poultry and meat products is reported. Similarly, oligonucleotides (5′–3′) and probes used as primers, for simultaneous and quantitative detection, target gene and amplicon size in pathogen detection from meat and meat products are summarized in detail in a study by Drosinos, Paramithiotis and Andritsos (2011).

12.3.2.3 PCR-based molecular typing

The PCR is a fast, accurate, sensitive and easy operating technique, which can be a powerful tool for microbial analysis. It has been reported that PCR-based method allows differentiation at the species (Welsh and McClelland, 1992) and intra-species level (Seal, Jackson and Daniels, 1992) depending on the stringency of the PCR conditions. The differences between strains can be detected by exploiting primers that are annealing in various regions of the genome, thereby producing a band pattern mainly by randomly amplified polymorphic DNA-PCR (RAPD-PCR) and repetitive extragenic palindromic-PCR (rep-PCR) (Doulgeraki *et al.*, 2012). Similarly, DNA fingerprinting methods such as restriction fragment length polymorphism (RFLP) and random amplified polymorphic DNA (RAPD) analysis are suitable for grouping of isolates in transmission studied or outbreak investigation (de Boer and Beumer, 1999; Foley and Grant, 2007). Similarly, AFLP (amplified fragment length polymorphism) is a genotyping method that is based on selective amplification of restriction fragments of digested total genomic DNA; this technique combined the reliability of RFLP with the advantages of PCR (Vos *et al.*, 1995; Foley and Grant, 2007). In addition, genetic amplification methods such as nucleic acid sequence-based amplification (NASBA) have made it possible to significantly reduce assay times while maintaining a high level of sensitivity and specificity. These methods are also able to distinguish closely related species that most antibody tests could not (Nugen and Baeumner, 2008).

PCR-based methods have some shortcomings that lead to a lower effectiveness in many cases. The major problem associated with the PCR tools is non-specific amplification, which usually arises from the specificity of primers, low reaction temperatures and amplification of unrelated PCR products. Moreover, a problem in routine use of PCR in food testing laboratories is the procedures are rather complicated and very clean environment is needed to perform the tests; however, developments in methodology make the technique more user friendly for routine use (de Boer and Beumer, 1999). Nevertheless, the drawbacks of PCR also include the inability to distinguish between live and dead cells, the presence of polymerase inhibitors in food samples and accessibility of the target organisms (de Boer and Beumer, 1999). In addition a frequently observed shortcoming of these methods is their lack of reproducibility among different and even within laboratories; the sources of variability are the primer concentration and quality, concentration and sources of polymerase, $MgCl_2$, the thermocycler and the number of thermocycles (Barrett, 1997). The reproducibility of molecular typing procedures will depend on the methods used, the organism to be typed and the experience of the analyst (de Boer and Beumer, 1999). For example, the main drawback of random PCR methods such as RAPD-PCR is their low reproducibility; however, the profiles obtained from rep-PCR analysis are specific for a species and they are also more reproducible (Cocolin, Dolci and Rantsiou, 2008).

The development of real-time PCR has been a major advance in nucleic-acid amplification technology. PCR amplicons are detected when PCR is occurring using fluorescently labelled molecules. There are many advantages of real-time PCR over conventional PCR (Fratamico and Kawasaki, 2007):

- The closed-tube detection format lowers the potential for carryover contamination and false-positive results.
- There is a shorter analytical turnaround time.
- Probe-based real-time PCR product detection results in higher assay sensitivity and specificity.
- There is no need for post-PCR processing steps to detect the products.
- Real-time PCR assays have a quantitative detection range of about 7–8 orders of magnitude compared to 2–3 logs for traditional PCR.

12.3.2.4 Typing by non-PCR-based methods

Restriction enzyme analysis coupled with pulsed-field gel electrophoresis (REA–PFGE) is a powerful method for strain characterization. Genomic DNA from different strains is digested by specific restriction enzymes and then subjected to PFGE; the result obtained is a strain-specific band pattern (Cocolin, Dolci and Rantsiou, 2008). PFGE is considered to have both good reproducibility and sufficient resolving power for epidemiological typing of bacterial isolates. Because all of the DNA extraction and restriction enzyme digestion occur within the agarose plugs, the free DNA is not pipetted; thus, the banding patterns generated are related to restriction enzyme digestion, not to non-specific mechanical shearing (Birren and Lai, 1993). A potential drawback of PFGE is that it tends to be somewhat labour intensive, sometimes requiring a few days to perform the procedure and interpret the results (Swaminathan and Barrett, 1995).

12.3.2.5 Identification and typing by PCR and/or sequencing

A valuable tool in bacterial taxonomy for determining relationships between bacterial groups and for identifying species is the sequencing of 16S rRNA gene because this gene has conserved as well as variable regions and can allow designing primers targeting all bacteria and spanning regions with species-specific sequences (Ercolini, 2004). Therefore, although 16S rRNA sequences can be used routinely to distinguish and establish relationships between genera and well-resolved species, in some cases alternative target genes are to be taken into account. Besides 16S rRNA gene, other housekeeping or functional genes might be used as targets for PCR.

Advancement of sequence-based methodologies led to the development of multi-locus sequencing typing (MLST), which has become fundamental for the characterization of foodborne microorganisms, and several genes can be used depending on the phyla of interest (Maiden *et al.*, 1998; Doulgeraki *et al.*, 2012). The MLST is based on PCR amplification and sequencing of several housekeeping genes; after sequence comparison, even one difference in the sequence is considered enough to define a different biotype (Maiden *et al.*, 1998; Foley and Grant,

2007; Doulgeraki *et al.*, 2012). The genetic relatedness of the strains is determined by comparing differences in allele profiles that make up the sequence type (Foley and Grant, 2007). In the same study, many of the genes that have been used for typing different foodborne pathogens are listed. Similarly, multiple-locus variable number tandem repeat analysis (MLVA) utilizes copy number differences in repeated sequences located throughout the bacterial genome that are used to differentiate isolates (Foley and Grant, 2007). The technique has shown promise with some of the strains that appear homogeneous with PFGE, such as *E. coli* O157:H7 isolates. MLVA is able to distinguish among many of these isolates and will prevent non-clonal strains from being called identical (Noller *et al.*, 2003).

12.3.2.6 Culture-independent methods

A small fraction of microorganisms is analysed by conventional methods, and the isolated strains often do not represent the real spectrum of microorganisms. On the other hand, culture-independent methods are believed to overcome problems associated with selective cultivation such as (1) inability to detect some bacteria on the known media, (2) lack of knowledge of the real conditions under which most of the bacteria are growing in their natural habitat and (3) difficulty to develop media for cultivation accurately resembling these conditions. The most commonly used culture-independent methods is perhaps the PCR-denaturing gradient gel electrophoresis (PCR-DGGE) of ribosomal RNA genes, which was proposed by Muyzer, De Weal and Uitterlinden (1993) and is widely applied to analyse the DNA extracted from meat and meat products (Doulgeraki *et al.*, 2012). Another culture-independent method that has been applied in food is terminal restriction fragment length polymorphism (T-RFLP) that has shown reproducibility and potential for a higher throughput than PCR-DGGE (Nieminen *et al.*, 2011). In the same study, it has to be noted that the taxonomic resolution level of the T-RFLP method when applied for the characterization of psychrotrophic bacterial communities in MAP minced meat was in between genus and species within the investigated LAB isolates and was within family and genus within the investigated Gram-negative isolates (Nieminen *et al.*, 2011).

High-throughput sequencing tools such as direct pyrosequencing of PCR amplicons came in light with the recent developments in culture-independent methods. The latter method can be an extremely valuable tool and more powerful than PCR-DGGE for the detection of microbial succession during storage of meat (Ercolini *et al.*, 2011). In fact, a wider number of species could be identified by pyrosequencing compared to PCR-DGGE and, moreover, the pyrosequencing approach can give a quantitative estimation of the abundance of a single taxon in each sample based on the number of reads of its particular 16S rRNA gene sequence (Doulgeraki *et al.*, 2012).

A method that has been also proposed for the identification of bacteria is DNA hybridization. This method refers to the detection of particular gene using complementary nucleic acid probe (Mandal *et al.*, 2011). The label probe is added after the denaturation of the target nucleic acids by high temperatures or high pH and forms a bond if the sample contains the same sequence. Fluorescent in situ hybridization (FISH) is the most common method among molecular techniques not based on PCR (López-Campos *et al.*, 2012). After hybridization, the specifically stained cells are detected using fluorescence microscopy (Wanger *et al.*, 2003).

A latest trend is the use of microarrays or gene chips that provide a miniaturized system for the simultaneous analysis of hybridization of fluorescent-labelled single-strand nucleotide chains to an array of oligonucleotide probes immobilized on a support such as glass or a synthetic membrane. PCR amplification is often used prior to hybridization to increase sensitivity of detection (Keramas *et al.*, 2003). DNA microarrays may be very useful for detecting multiple bacteria simultaneously on a single glass slide, wherein the frame of food diagnostics rather described as a tool for identification, genotyping and pathotyping (detection of appropriate virulence factors) or characterization of bacterial isolates (Jasson *et al.*, 2010). However, Roda *et al.* (2012) reported the following drawbacks of such a method: long assay time, laborious data elaboration, viable and non-viable cells are not distinguished and it is not applicable to on-site measurements.

In all above-described genomic analyses, the limitations of the information obtained must be recognized (Hoorfar, 2011). It is well known that DNA-based assays detect genetic targets (such as virulence genes), which only indicate that microorganisms with those gene sequences are present, whereas they do not indicate whether the gene is actually expressed or whether the organism is viable (de Boer and Beumer, 1999). On the other hand, current legislation requires the isolation of unintended microorganisms in situations such as food recalls, that is, cultivation methods need to be used in addition to DNA-based methods during outbreak investigations. Furthermore, DNA-based methods cannot detect proteins such as enterotoxins, which can indeed cause foodborne intoxications. Thus, proteomic techniques, including protein arrays or protein PCR (Gullberg *et al.*, 2004), may be increasingly used for food safety. In addition, when it is difficult to decide about difference probably caused by genetic changes, the use of more than one sub-typing method is advised (de Boer and Beumer, 1999).

12.3.2.7 Immunological methods

Immunologic methods rely on the binding specificity of an antibody to an antigen, usually displayed on the surface of a microorganism, or to a secreted toxin (Fratamico and Kawasaki, 2007). The selection of an appropriate antibody (monoclonal or polyclonal) is the determinant factor for the method's performance (Jasson *et al.*, 2010), while the specific binding of the antibodies/antigens is determined qualitatively and quantitatively by immunological reaction by immunoassays. Immunological assessment can be performed using various disciplines. Reflectance measurement, chemiluminescence, electrochemiluminescence, immunoelectrochemical detection, bioluminescence, time-resolved fluorescence and the most commonly applied colorimetric detection are among the techniques that have been successfully employed (Paramithiotis, Mataragas and Drosinos, 2011). Nowadays, several types of immunoassays are available in food diagnostics such as enzyme-linked immunosorbent assays (ELISAs), lateral flow devices (LFD), enzyme-linked fluorescent assays (ELFAs), latex agglutination, immunodiffusion assays, nanoparticle-labelled microfluidic immunoassays and antibody-based biosensors,3,4, which employ polyclonal or monoclonal antibodies directed against specific antigens of microorganisms (Jasson *et al.*, 2010; Roda *et al.*, 2012). ELISA, which is a biochemical technique, is widely used in food testing. This method couples an immunoassay with an enzyme assay that is usually designed as a sandwich assay (Mandal *et al.*, 2011).

A variety of immunoassay kits are available commercially, where some of them are summarized in a recent study (Roda *et al.*, 2012). However, there are limitations with the use of immunoassays including (1) cross-reactivity of the antibody with antigenically similar organisms giving rise to false-positive results, (2) quantitative differences in expression of the target antigen related to growth parameters, (3) enzyme inhibitors present in foods and (4) limited availability of antibodies against specific pathogens or specificity only at the species level or below. Generally, immunoassays yield presumptive results, which must be confirmed by biochemical and serological testing of isolates.

12.3.2.8 Biosensors

Biosensor is an analytical instrument that combines bio-specific recognition systems with physical and electrochemical signalling. Many types of biosensors have been developed such as enzymes, nucleic acids and cellular materials and the recognition signals used include electrochemical (e.g. potentiometry, voltage changes, conductance and impedance, light addressable), optical (such as UV, bioluminescence and chemiluminescence, fluorescence, laser scattering, reflection and refraction of light, surface plasmon resonance, polarized light) and miscellaneous transducers (such as piezoelectric crystals, thermistor, acoustic waves, quartz crystal) (Fung, 2007). These devices have the potential to shorten the time between sampling and results, but their future lies in reaching selectivity and sensitivities comparable to established methods at a fraction of the cost.

Biosensors are very useful in foodborne pathogens detection and their application to food testing offers several features with the goal to overcome problems associated with traditional microbiological detection techniques such as being time- and labour-intensive (Baeumner, 2003). Many of them are sensitive, portable due to the advancement in miniaturization technology and able to provide fast and real-time detection (Rasooly and Herold, 2006). Many of these detection devices are able to provide the results usually in less than

1 h (Nugen and Baeumner, 2008). In brief, ATP bioluminescence devices are commonly used to measure the effectiveness of cleaning surfaces and utensils. The main advantage to use such a device includes extreme sensitivity and speed detection, easy implementation and their ability to be used in portable field devices (Guisan, 2006); however, non-specificity of ATP assays have to be taken into account. Similarly, impedimetry methods are capable of monitoring large number of samples simultaneously, good agreement between standard viable counts and detection times and relatively short detection time (between 6 and 24 h) (Hayes, 1992); however, the influence of the food matrix on the analysis has to be recognized (de Boer and Beumer, 1999). On the other hand, piezoelectric biosensors, which are able to generate electrical potential in response to a mechanical force, are suitable for direct, label-free detection of specific nucleic-acid targets and useful for foodborne pathogen detection (Zourob, Elwary and Khademhosseini, 2010). Nowadays, newer generations of biosensors are designed to offer similar sensitivities while decreasing detection times, increasing portability or reducing the limit of detection.

12.3.3 Industry usage of rapid methods

Rapid methods of pathogen testing have been gaining increasing interest in the food industry. These methods include antibody-based assays, genetic amplification methods and newer sensor development (Nugen and Baeumner, 2008). Time and sensitivity of analysis are the most important limitation related to the usefulness of microbial testing for the food industry, which is in need for more rapid methods for the following reasons (Downes and Ito, 2006):

- To provide immediate information on the possible presence of pathogen in a raw material and finished product.
- Low numbers of pathogenic bacteria or their toxins are often present in complex biological environments.
- The presence of even a single pathogenic organism in the food may be an infectious dose.
- For monitoring of process control, cleaning and hygiene practices during manufacture.
- To reduce human errors and save time and labour cost.

It has been estimated that 38% of reported pathogen biosensors in the past 20 years were designed for the food industry (Lazcka, Del Campo and Munoz, 2007); however, the drawback of automated biosensor or immunoassay system is the initial price of the instrument. Similarly, one of the unexploited potentials of real-time PCR for the food industry is to gain information on the number of pathogens present in the food or food production environment, which could help food producers to assess the impact of hygiene measures or the implementation of an improved quality assurance program (Hoorfar, 2011).

Many rapid systems are rather directed to large capacity service labs or large in-house industry labs, as the tendency of many small and medium enterprises

(SMEs) in the food industry to outsource microbial analysis to service laboratories (Jasson *et al.*, 2010). In addition, in developing countries, many of which have made significant improvements in their national system of food control over the last decade, the use of rapid methods is limited due to offer value, the robust and reliable under the storage conditions of test kit, affordability, ready accessibility and the limited technical support and service (Hoorfar, 2011).

12.4 Future role for rapid methods in foods safety

The use of rapid, reliable and sensitive methods allow for the development of intervention strategies that could help to limit the spread of foodborne pathogens and monitor the cleanliness in HACCP programs resulting in an increased food safety. In an ideal world, the total sample preparation time should be sufficient to allow preparation, detection and reporting of multiple samples in 1 day, as the sample preparation is still a major issue affecting the speed of attainment of results (de Boer and Beumer, 1999; Foley and Grant, 2007; Ge and Meng, 2009; McGrath, Elliott and Fodey, 2012). Although the sample preparation requirements will vary from one food product to other, research into this step is required to bridge the emerging field of nanosensors with the food industry (Nugen and Baeumner, 2008).

In food testing, the development of an increasing number of monitoring devices, which are becoming more compact, mobile, digital, virtual (software-based) and personal, will result in on-site, on-line or in-line monitoring systems that obviate the need to ship samples to testing laboratories (Hoorfar, 2011). Thus, the advancements in multiplexing and the moves towards portable and integrated instrumentation should greatly improve the profile of biosensors (Ge and Meng, 2009; McGrath, Elliott and Fodey, 2012).

Another major issue is that assays will need to be developed that focus on 'overlooked' pathogens including viruses and parasites, while biosensors, at the centre of the marriage between biology and electronics, should be designed with the food matrix in mind and should be able to test larger, rather than smaller, volumes of samples (Hoorfar, 2011). In addition, alternative concentration technologies other than enrichment need to be developed to achieve the goal of real-time testing (Ge and Meng, 2009; Hoorfar, 2011). In addition, it should be able to detect pathogens in very low concentrations of the samples and must be suitable for in situ real-time monitoring as well (Velusamy *et al.*, 2010). Ideally, a non-invasive, handheld scanning device (similar to barcode scanning at a local grocery store) would be ideally used to quickly and accurately assess all aspects of a food commodity (Hoorfar, 2011). However, as usual, the clinical diagnostics is much ahead of food testing, partly because of the higher number of pathogens in clinical samples and partly because of the willingness to pay much more for a clinical testing.

References

Alvarez, A.J., Buttner, M.P. and Stetzenbach, L.D. (1995) PCR for bioaerosol monitoring: sensitivity and environmental interference. *Applied and Environmental Microbiology*, **61**, 3639–3644.

Anon (1993a) 93/43/EEC, 'Hygiene of foodstuffs', General Directive, **1993**.

Anon (1993b) Mandatory safe handling statement on labeling of raw meat and poultry products. *Federal Register*, **58** (156)9 CFR Parts 317 and 381, 52856–52873.

Baeumner, A.J. (2003) Biosensors for environmental pollutants and food contaminants. *Analytical and Bioanalytical Chemistry*, **377**, 434–445.

Barker, G.C., Gomez, N. and Smid, J.H. (2009) An introduction to biotracing in food chain systems. *Trends in Food Science and Technology*, **20**, 220–226.

Barrett, T.J. (1997) Molecular fingerprinting of foodborne bacteria: an introduction to methods, uses, and problems, in *Food Microbiological Analysis — New Technologies* (eds M.L. Tortorello and S.M. Gendel), Marcel Dekker, New York, pp. 249–264.

Benoit, P.W. and Donahue, D.W. (2003) Methods for rapid separation and concentration of bacteria in food that bypass time-consuming cultural enrichment. *Journal of Food Protection*, **66**, 1935–1948.

Birren, B. and Lai, E. (1993) *Pulsed Field Gel Electrophoresis: A Practical Guide*, Academic, San Diego.

Biswas, A.K., Kondaiah, N., Bheilegaonkar, K.N. *et al.* (2008) Microbial profiles of frozen trimmings and silver sides prepared at Indian Buffalo meat packaging plants. *Meat Science*, **80**, 418–422.

Cocolin, L., Dolci, P. and Rantsiou, K. (2008) Molecular methods for identification of microorganisms in traditional meat products, in *Meat Biotechnology* (ed. F. Toldra), Springer, New York, pp. 91–127.

Collins, C., Grange, J., Lyne, P. and Falkinham, J.O. III, (2004) *Collins and Lyne's Microbiological Methods*, 8th edn, Arnold, London, pp. 110–128.

Doulgeraki, A.I., Ercolini, D., Villani, F. and Nychas, G.-J.E. (2012) Spoilage microbiota associated to the storage of raw meat in different conditions. *International Journal of Food Microbiology*, **157**, 130–141.

D'Aoust, J.Y., Pagotto, F., Akhtar, M. *et al.* (2007) Evaluation of the BAX gel and fluorometric systems for the detection of foodborne *Salmonella*. *Journal of Food Protection*, **70**, 835–840.

de Boer, E. and Beumer, R.R. (1999) Methodology for detection and typing of foodborne microorganisms. *International Journal of Food Microbiology*, **50**, 119–130.

Downes, F.P. and Ito, K. (2006) *Compendium of Methods for the Microbiological Examination of Foods*, 4th edn, American Public Health Association, New York.

Doyle, M.P. (2001) *Food Microbiology: Fundamentals and Frontiers*, 2nd edn, ASM Press, Washington, DC.

Drosinos, E.H., Paramithiotis, S. and Andritsos, N. (2011) Microbial food-borne pathogens, in *Handbook of Analysis of Edible Animal By-Products* (eds L. Nollet and F. Toldra), CRC Press, pp. 219–238.

Dwivedi, H.P. and Jaykus, L.A. (2011) Detection of pathogens in foods: the current state-of-the-art and future directions. *Critical Reviews in Microbiology*, **37**, 40–63.

EFSA (2006) The Community Summary Report on Trends and Sources of Zoonoses, Zoonotic Agents, Antimicrobial Resistance and Foodborne Outbreaks in the European Union in 2005. *EFSA Journal*, **2006**, 94.

EFSA (2013) The European Union Summary Report on Trends and Sources of Zoonoses, Zoonotic Agents and Food-borne Outbreaks in 2011. *EFSA Journal*, **11** (4), 3129.

Ellingson, J.L., Anderson, J.L., Carlson, S.A. and Sharma, V.K. (2004) Twelve hour real-time PCR technique for the sensitive and specific detection of *Salmonella* in raw and ready-to-eat meat products. *Molecular and Cellular Probes*, **18**, 51–57.

Ercolini, D. (2004) PCR-DGGE fingerprinting: novel strategies for detection of microbes in food: a review. *Journal of Microbiological Methods*, **56**, 297–314.

Ercolini, D., Ferrocino, I., Nasi, A. *et al.* (2011) Monitoring of microbial metabolites and bacterial diversity in beef stored in different packaging conditions. *Applied and Environmental Microbiology*, **77**, 7372–7381.

Evans, D. (2005) Future trends in the European Beef Industry: a global view: The Science of Beef Quality 8th Conference of Food Industry, Bristol UK

Feng, P. (2007) Rapid methods for the detection of foodborne pathogens: current and next-generation technologies, in *Food Microbiology: Fundamentals and Frontiers* (eds M.P. Doyle and L.R. Beuchat), ASM Press, Washington, DC, pp. 911–934.

Foley, S.L. and Grant, K. (2007) Molecular techniques of detection and discrimination of foodborne pathogens and their toxins, in *Infections disease: Foodborne diseases* (ed. S. Simjee), Humana Press Inc., Totowa, NJ.

Fratamico, P.M. and Kawasaki, S. (2007) Applications of the Polymerase Chain Reaction for Detection, Identification, and Typing of Food-Borne Microorganisms, in *Microbial Food Contamination*, 2nd edn (ed. C.L. Wilson), CRC Press, pp. 213–254.

Fung, Y.C. (2007) Rapid methods for detecting microbial contaminants in foods: past, present, and future, in *Microbial Food Contamination*, 2nd edn (ed. C.L. Wilson), CRC Press, pp. 149–186.

Ge, B. and Meng, J. (2009) Advanced Technologies for pathogens and toxins detection in foods: Current applications and future detection. *Journal of the Association for Laboratory Automation*, **14**, 235–241.

Greenland, K., de Jager, C., Heuvelink, A., van der Zwaluw, K., Heck, M., Otermans, D., van Pelt, W. and Friesema, I., (2009) Nationwide outbreak of STEC O157 infection in the Netherlands, December 2008 - January 2009: continuous risk of consuming raw beef products. Euro Surveillance 14, pii 'O 19129. Available online. http://www.eurosurveillance .org/ViewArticle.aspx?ArticleId'O19129.

Guisan, J.M. (2006) *Immobilization of enzymes and cells*, Humana Press, Methods in biotechnology.

Gullberg, M., Gustafsdottir, S.M., Schallmeiner, E. *et al.* (2004) Cytokine detection by antibody-based proximity ligation. *Proceedings of the National Academy of Sciences of the United States of America*, **101**, 8420–8424.

Hayes, P.R. (1992) *Food Microbiology and Hygiene*, Chapman & Hall.

Hoorfar, J. (2011) Rapid detection, characterization and enumeration of foodborne pathogens. *APMIS*, **119**, 1–24.

Jasson, V., Jacxsens, L., Luning, P. *et al.* (2010) Alternative microbial methods: an overview and selection criteria. *Food Microbiology*, **27**, 710–730.

Josefsen, M.H., Jacobsen, N.R. and Hoorfar, J. (2004) Enrichment followed by quantitative PCR for rapid detection and as a tool for quantitative risk assessment of foodborne thermotolerant Campylobacters. *Applied and Environmental Microbiology*, **70**, 3588–3592.

Keramas, G., Bang, D.D., Lund, M. *et al.* (2003) Development of a sensitive DNA microarray suitable for rapid detection of *Campylobacter* spp. *Molecular Cell Probes*, **17**, 187–196.

Kramer, N., Lofstrom, C., Vigre, H. *et al.* (2011) A novel strategy to obtain quantitative data for modelling: combined enrichment and real-time PCR for enumeration of salmonellae from pig carcasses. *International Journal Food Microbiology*, **145**, 86–95.

Lazcka, O., Del Campo, F.J. and Munoz, F.X. (2007) Pathogen detection: A perspective of traditional methods and biosensors. *Biosensors and Bioelectronics*, **22**, 1205–1217.

Lofstrom, C., Schelin, J., Norling, B. *et al.* (2011) Culture-independent quantification of *Salmonella enterica* in carcass gauze swabs by flotation prior to real-time PCR. *International Journal of Food Microbiology*, **145**, 103–109.

López-Campos, G., Martínez-Suárez, J.V., Aguado-Urda, M. and López-Alonso, V. (2012) *Detection, Identification, and Analysis of Foodborne Pathogens. Microarray Detection and Characterization of Bacterial Foodborne Pathogens*, Springer Briefs in Food, Health, and Nutrition, pp. 13–32.

Lund, B.M., Baird-Parker, T.C. and Gould, G.W. (2000) *The Microbiological Safety and Quality of Foods*, Aspen Publishers, Gaithersburg, MD.

Maiden, M.C., Bygraves, J.A., Feil, E. *et al.* (1998) Multilocus sequence typing: a portable approach to the identification of clones within populations of pathogenic microorganisms. *Proceedings of the National Academy of Sciences of the United States of America*, **95**, 3140–3145.

Makino, S. and Cheun, H. (2003) Application of the real-time PCR for the detection of airborne microbial pathogens in reference to the anthrax spores. *Journal of Microbiology Methods*, **53**, 141–147.

Mandal, P.K., Biswas, A.K., Choi, K. and Pal, U.K. (2011) Methods for Rapid Detection of Foodborne Pathogens: An Overview. *American Journal of Food Technology*, **6**, 87–102.

McGrath, T.F., Elliott, C.T. and Fodey, T.L. (2012) Biosensors for the analysis of microbiological and chemical contaminants in food. *Analytical and Bioanalytical Chemistry*, **403**, 75–92.

McMeekin, T.A., Baranyi, J., Bowman, J. *et al.* (2006) Information systems in food safety management. *International Journal of Food Microbiology*, **112**, 181–194.

Muyzer, G., De Weal, E.C. and Uitterlinden, A. (1993) Profiling of complex microbial populations by denaturing gradient gel electrophoresis analysis of polymerase chain reaction-amplified genes coding for 16S rRNA. *Applied and Environmental Microbiology*, **59**, 695–700.

Nieminen, T.T., Vihavainen, E., Paloranta, A., Lehto, J., Paulin, L., Auvinen, P., Solismaa, M. and Bjorkroth, K.J. (2011) Characterization of psychrotrophic bacterial communities in modified atmosphere-packed meat with terminal restriction length polymorphism. *International Journal of Food Microbiology* **144**, 360–366.

Noller, A.C., McEllistrem, M.C., Stine, O.C. *et al.* (2003) Multilocus sequence typing reveals a lack of diversity among *Escherichia coli* O157:H7 isolates that are distinct by pulsed-field gel electrophoresis. *Journal of Clinical Microbiology*, **41**, 675–679.

Nugen, S.R. and Baeumner, A.J. (2008) Trends and opportunities in food pathogens detection. *Analytical and Bioanalytical Chemistry*, **391**, 451–454.

Pillai, S.D. (2004) Molecular methods for microbial detection, in *Preharvest and Postharvest Food Safety: Contemporary Issues and Future Directions* (eds R.C. Beier, S.D. Pillai, T.D. Phillips and R.L. Ziprin), Blackwell Publishing Professional, Ames, IA.

Rasooly, A. and Herold, K.E. (2006) Biosensors for Microbial Diagnostics, in *Encyclopedia of Rapid Microbiological Methods*, (ed. M.J. Miller), PDA/DHI vol. **3**, pp. 407–448.

Roda, A., Mirasoli, M., Roda, B. *et al.* (2012) Recent developments in rapid multiplexed bioanalytical methods for foodborne pathogenic bacteria detection. *Microchimica Acta*, **178**, 7–28.

Rossen, L., Norskov, P., Hoimstrom, K. and Rasmussen, O.F. (1992) Inhibition of PCR by components of food samples, microbial diagnostic assays and DNA-extraction solutions. *International Journal of Food Microbiology*, **17**, 37–45.

Rossmanith, P., Hedman, J., Radstrom, P. *et al.* (2011) Preanalytical Sample Preparation and Analyte Extraction, in *Rapid Detection, Characterization and Enumeration of Food-Borne Pathogens* (ed. J. Hoorfar), American Society for Microbiology, Washington, DC, pp. 121–135.

Seal, S.E., Jackson, L.A. and Daniels, M.J. (1992) Use of tRNA consensus primers indicate subgroups of *Pseudomonas solanacearum* by polymerase chain reaction. *Applied and Environmental Microbiology*, **58**, 3759–3761.

Stevens, K.A. and Jaykus, L.A. (2004) Bacterial separation and concentration from complex sample matrices: a review. *Critical Reviews in Microbiology*, **30**, 7–24.

Swaminathan, B. and Barrett, T.J. (1995) Amplification methods for epidemiologic investigations of infectious disease. *Journal of Microbiology Methods*, **2**, 129–139.

Paramithiotis, S., Mataragas, M. and Drosinos, E.H. (2011) Food safety assessment at the molecular level: merits and pitfalls, in *Food Quality: Control, Analysis and Consumer Concerns* (eds D.A. Medina and A.M. Laine), Nova Publishers, pp. 337–357.

Vasavada, P.C. (1993) Rapid methods and automation in dairy microbiology. *Journal of Dairy Science*, **76**, 3101–3113.

Velusamy, V., Arshak, K., Korostynska, O. *et al.* (2010) An overview of foodborne pathogen detection: In the perspective of biosensors. *Biotechnology Advances*, **28**, 232–254.

Vos, P., Hogers, R., Bleeker, M. *et al.* (1995) AFLP: a new technique for DNA fingerprinting. *Nucleic Acid Research*, **23**, 4407–4414.

Wanger, M., Horn, M. and Daims, H. (2003) Fluorescence in situ hybridisation for the identification and characterisation of prokaryotes. *Current Opinions in Microbiology*, **6**, 302–309.

Waswa, J.W., Debroy, C. and Irudayaraj, J. (2006) Rapid detection of *Salmonella enteritidis* and *Escherichia coli* using surface plasmon resonance biosensor. *Journal of Food Process Engineering*, **29**, 373–385.

Welsh, J. and McClelland, M. (1992) PCR-amplified length polymorphisms in tRNA intergenic spacers for categorizing staphylococci. *Molecular Microbiology*, **6**, 1673–1680.

WHO (2000) WHO Surveillance Programme for Control of Foodborne Infections and Intoxications in Europe 7th Report.

Wu, V.C.H. (2008) A review of microbial injury and recovery methods in food. *Food Microbiology*, **25**, 735–744.

Zourob, M., Elwary, S. and Khademhosseini, A. (2010) *Recognition Receptors in Biosensors*, Springer.

CHAPTER 13

The use of hyperspectral techniques in evaluating quality and safety of meat and meat products

Di Wu & Da-Wen Sun

Food Refrigeration and Computerised Food Technology (FRCFT), School of Biosystems and Food Engineering, University College Dublin, National University of Ireland, Agriculture and Food Science Centre, Belfield, Dublin, Ireland

13.1 Introduction

Meat and meat products are highly valued and consumed food in the human diet in not only most developed countries but also many developing countries. Many consumers choose meat as a very good source of high-quality animal protein. In purchasing meat, the main aspects that consumers are always concerned about include nutrition, appearance, taste, and freshness in addition to price and safety. As consumers have increased awareness and sophistication, the meat industry needs to guarantee product quality to be successful in the highly competitive meat market. Consumers who have the expectation of meat products of high quality prefer not only the good quality of individual meat product but also the consistency of products from a batch, so that they could continuously choose this brand. On the other hand, consumers with low income prefer meat products with low price, but these products should also not be harmful to human health. Therefore, meat products with good quality should be manufactured to meet the demands of different consumers, which encourages consumer loyalty and results in an expanding market share. This in turn has increased the need to measure the quality of meat products before distribution as it is wasteful to present high-quality meat in the low-end market and a merchant's reputation can be damaged by providing low-quality meat in the high-end market (Jackman, Sun and Allen, 2011).

Inspection of meat quality attributes should be taken throughout the entire meat process and management system to conform to consumers' expectations on meat quality and safety. Commonly used ways to assess external quality attributes such as color, texture, and visual appearance rely on sensory judgment, while laboratory chemical tests are generally used to measure chemical attributes such as water-holding capacity, intramuscular fat, and protein contents. The

Emerging Technologies in Meat Processing: Production, Processing and Technology, First Edition.
Edited by Enda J. Cummins and James G. Lyng.

shortcoming of a sensory judgment is that it is dependent to a great extent on subjective and sometimes inconsistent assessment of the panelist (Sigurgisladottir *et al.*, 1999). The processes of sensory tests are also tedious, time-consuming, and costly. Also, the designer and executor of a sensory assessment should have adequate technical ability and experience to get trustworthy outputs (Ashton, Michie and Johnston, 2010). On the other hand, chemical measurements could provide comparable results between laboratories or batches from the same laboratory, if stringent standardization is strictly controlled (Ashton, Michie and Johnston, 2010). Compared with sensory evaluation, chemical measurements are less influenced by human behavior factors so that the variation among measurements can be reduced (Casas *et al.*, 2006). However, although objective and consistent, chemical measurements cannot be used routinely by the industry, because they are time-consuming, destructive, and inefficient. Moreover, both sensory assessment and chemical analysis are selective examinations, which can only evaluate small number of samples. Therefore, it is obvious that the development of rapid and accurate quality inspection systems are important for the meat industry to ensure the safe production of meat during processing operations and the correctly labeling of products related to the quality, safety, authenticity, and compliance. For a modern meat plant, as its meat throughput is increasing as well as the tightening of quality tolerance, the employment of automatic methods for quality control and safety is necessary (Gunasekaran, 1996).

In the past few decades, a number of different techniques have been researched and applied for rapid and nondestructive evaluation and inspection for meat quality and safety. Many sensing techniques have unprecedented capabilities to measure, assess, classify, and sort meat products accurately and efficiently and have been considered as the primary instrumental ways to evaluate quality and quality-related attributes of meat products. In recent years, developing advanced optical-based techniques and instrumentation has been emphasized for online and noninvasive inspection. Optical properties include the electromagnetic spectrum in the ultraviolet (UV), visible (VIS), and near-infrared (NIR) regions from reflectance, transmittance, absorbance, or scatter of polychromatic or monochromatic radiation, usually measured by spectral instruments. A qualitative or quantitative relationship for the meat product can be established according to the correlation between spectral response (optical properties) and the target quality feature of a product, usually a chemical component or physical attribute (Park *et al.*, 2002). Hyperspectral techniques such as spectroscopy and hyperspectral imaging can obtain a spectral response with an abundance of contiguous wavelengths (normally less than 10 nm) and therefore are extremely advantageous for meat quality and safety inspection. With the advantages of simplicity, accuracy, low maintenance cost, rapidness, minimal sample preparation, and permitting an online noninvasive inspection, hyperspectral techniques are promising for industrial use since they have the potential to be used as a small handheld device or an online instrument for noncontact measurements of meat quality on a conveyer belt during production. The automatic quality and safety evaluation and inspection is a relatively new and interesting area of research, and the need for accurate and efficient methods to guarantee the quality and safety of products has increased the interest in the research and application of hyperspectral techniques in the meat industries.

13.2 Hyperspectral techniques

13.2.1 Basics of spectroscopy

In 1665, Sir Isaac Newton first described the concept of dispersion of light and the optomechanical hardware of a spectrometer, when he observed the splitting of light into colors by passing light through a prism. In general, the electromagnetic spectrum consists of several regions, including gamma rays, X-rays, ultraviolet radiation (UV), visible light (VIS), infrared radiation (IR) (near-infrared (NIR), mid-infrared (MIR), and far-infrared (FIR) radiation), microwaves, and radio waves (FM and AM). Different energies cause a specific kind of atomic or molecular transition in a corresponding region. The detailed fingerprints of the biological specimen can be provided using spectroscopic methods due to the physical characteristics of the interaction between components of specimen and their electromagnetic radiation. Qualitative and quantitative component information within either absorbed or emitted spectrum can be obtained by spectroscopic analysis using the interaction of electromagnetic radiation with atoms and molecules. Particularly, visible and near-infrared (Vis–NIR) spectra are mostly utilized for qualitative and quantitative assessment of chemical and physical features of meat and other meat products. Vis–NIR spectroscopy typically utilizes the spectral range of 400–2500 nm (25,000–4000 cm^{-1}), where the wavelength range 400–700 nm is generally defined as the visible region and 700–2500 nm range is viewed as the NIR region, which can be divided into short-wave NIR (SW-NIR) (700–1300 nm) region and common NIR region (1300–2500 nm) (Cen and He, 2007). Visible spectra mainly contain the pigment information and near-infrared spectra arise from overtones and combination bands of C—H, O—H, N—H, and S—H stretching and bending vibrations. Usually, the information of moisture, carbohydrate, and fat are rich in O–H bonds, the information of organic compounds and petroleum derivatives are rich in C—H bonds, and the information of proteins and amino acids are rich in N—H bonds, resulting in a good qualitative and quantitative determination ability of NIR spectra on the quality assessment of meat products. Because all components in meats contain O—H, C—H, N—H molecular bonds, when a meat product is under the exposure of NIR radiation, the vibration response energy of bonds of organic molecules in the meat are changed, and a complex spectrum containing the physical and chemical information of a meat product can be obtained within absorption peaks through the spectrum. Therefore, qualitative and quantitative information of various components in meat products is contained within the spectrum of absorbed energy (Carlomagno *et al.*, 2004). Indeed, the modern spectroscopic application for assessing quality traits in meat products requires a high signal-to-noise spectrometer, spectral acquisition process controllable by a computer, and the multivariate mathematical and statistical chemometric algorithms to analyze the spectral data. The process of spectral analysis includes the collection of spectra of a meat product and the prediction of the quality trait of this product, which is executed on the basis of an established calibration model. The establishment of a calibration model is a process to relate the acquired spectral data to the reference quality trait value ascertained using a standard laboratory method. When the model is established, it can be used to predict the qualities of new samples quickly. The measured spectra of new samples only need to be introduced into the established spectral model, and the corresponding qualities of meat products will be calculated quickly in a computer based on the established calibration equation. It is very important to guarantee the accuracy of the reference values of qualities determined by standard methods, and the reference values are associated with the spectral data of the samples used for calibration. Once this calibration stage is completed and good correlation exists between spectra and the corresponding traits, the application result on a new sample would be close to the value measured using an ideal analytical method (Peiris *et al.*, 1999). Indeed, spectroscopy has been intensively applied in quality evaluation of different types of meat and other foods (Armenta *et al.*, 2010; Prieto *et al.*, 2009; Bock and Connelly, 2008; Huang *et al.*, 2008; Nicolai *et al.*, 2007; Cen and He, 2007; Rodriguezotero, Hermida and Centeno, 1997). However, although spectroscopy techniques have the capability to be used in an online industrial application, due to its advantages of being rapid, low cost and noninvasive, they cannot give information on morphology and spatial distribution of meat product traits. This greatly limits the application of spectroscopy to quantify spatial-related distribution and morphology-related attributes. The use of hyperspectral imaging would be a good solution in such cases, because of its ability to acquire spectral and spatial information simultaneously.

13.2.2 Basics of hyperspectral imaging

Since meats are very heterogeneous products for chemical composition, technological, and sensory attributes, spatial information is important for meat quality inspection. Measurement of spatial features of meat products can be achieved by conventional imaging techniques, or more specifically computer vision with the minimum of human intervention. Conventional imaging systems in the form of machine or computer vision have been applied in the meat industry to replace the traditional human inspectors, due to its superior speed, accuracy, consistency, and cost-effectiveness. Computer vision with image process algorithms has become an integral part of the realization for the industry's automation. A computer vision system usually includes an image sensor, an illumination unit, a personal computer, electrical and mechanical devices (Du and Sun, 2006). In addition, image process and analysis algorithms and realization from the knowledge of mathematics, computer science, and software programming are also important for the application of computer vision. Efficiency of several seconds per object for human inspectors can be improved to several objects per second for a computer vision system. Automated visual inspection by computer-based systems with superior spatial information has been used as inspection tools for quality and safety assessment, classification, and sorting of meat products. Unfortunately, a conventional imaging system has some drawbacks that make it incompetent for certain industrial applications, such as predicting chemical components and detecting invisible defects. This is mainly because it usually operates at visible wavelengths in the forms of monochromatic or color images. It can only detect the visible features of the object, such as size, shape, color, surface texture, and external defects, while with difficulty can detect water hold capacity (WHC), tenderness, and protein content (Du and Sun, 2004).

Spectroscopy can provide information about meat quality based on light absorption of samples but fails to obtain the position/location information of meat quality. On the other hand, the position of a certain feature can be located by the naked eye or a computer vision system, but quantitative prediction for a certain component is not easy to be achieved. By assembling the spectroscopy and traditional imaging techniques, hyperspectral imaging, also called imaging spectroscopy, emerged in the past decade as a novel, rapid, and nondestructive technique for food quality evaluation (Elmasry *et al.*, 2012b; Elmasry *et al.*, 2012a; Sun, 2010; He, Wu and Sun, 2013). A typical hyperspectral imaging system consists of three main parts: a wavelength dispersion device, an area detector, and a light source (Figure 13.1). Both spectral and spatial information are acquired from an object simultaneously and then generated in a "hypercube" form. The "hypercube" is a three-dimensional data block (x, y, λ) including two spatial (x, y) and one wavelength dimension (λ), providing hundreds of contiguous wavebands for each spatial position of a target sample (Figure 13.2). In particular, for a hyperspectral image, the complete spectrum of a sample over continuous wavelengths is obtained at each point in the imaging plane. On the basis of the combination of the series of images, a three-dimensional "hyperspectral cube" can be established, which includes one dimension representing spectral information and the other two dimensions for describing spatial information. In this cube, each point (or pixel) has both spatial and spectral signature (spectrum) information. From the pixels that have the same spatial positions, the entire measured spectrum of the corresponding imaged spatial point can be recorded. The ability of the specimen in absorbing or scattering the

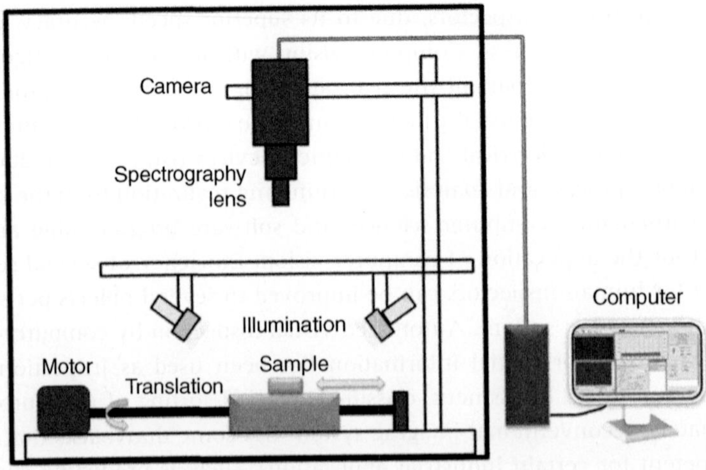

Figure 13.1 Schematic diagram of a line-scan hyperspectral imaging system.

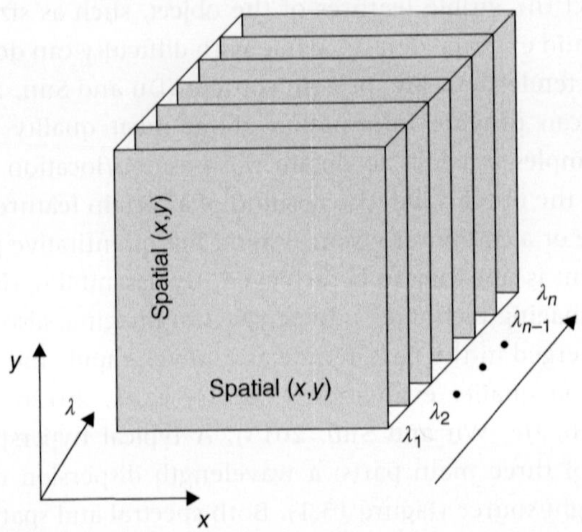

Figure 13.2 A hyperspectral image as a three-dimensional data block covering spatial (x, y) and spectral (λ) information.

exciting light can be indicated by the measured spectrum, and then through analyzing the informative cube, the inherent chemical and physical properties of the specimen can be ascertained. Recently, several approaches and applications of hyperspectral imaging techniques have been reported on quality assessment of food products, showing the usefulness to predict different qualities in beef (Wu *et al.*, 2013), pork (Barbin *et al.*, 2012a), lamb (Kamruzzaman *et al.*, 2011), ham (ElMasry *et al.*, 2011a), fish (Wu and Sun, 2013), apple (ElMasry *et al.*, 2008), citrus (Gomez-Sanchis *et al.*, 2008), melon (Tsuta, Sugiyama and Sagara, 2002), shrimp (Wu *et al.*, 2012), tomato (Cho *et al.*, 2013), wheat (Singh *et al.*, 2012), and maize (Williams *et al.*, 2012). As a comparison, a monochromatic image typically reflects the light intensity only in a single band of the spectrum, a color image usually reflects the intensity over the electromagnetic spectrum in red, green, and blue bands, and when the number of bands increased greatly, the information contained in the corresponding image greatly expands. Hyperspectral imaging techniques can obtain images containing information from many bands with a good resolution. Therefore, hyperspectral imaging techniques have the capability for quantitative prediction of components simultaneously with their spatial distribution.

13.2.2.1 Classes of spectral imaging

According to the number of images at different wavelengths for a spectral image, there are generally three classes of spectral imaging, which are multispectral imaging, hyperspectral imaging, and ultraspectral imaging. The concepts of these three imaging techniques are similar. Many researchers believe that the number of wavebands and the form of the spectrum obtained during image acquisition are the main differences to distinguish these three classes. However, no formal definition currently exists; various popular notations are used by many scientists working in this field. Generally, a system called hyperspectral imaging can acquire a spectral image with a series of contiguous wavelengths, while multispectral imaging systems only acquire very few wavelengths with large intervals. Each pixel in a hyperspectral image has the spectra at wavelengths over a contiguous wavelength range. Ultraspectral imaging usually refers to the one acquiring a spectral image with a higher spectral resolution than a hyperspectral image. Multispectral images have discrete and somewhat narrow bands. Setting band-pass filters (or interference filter) in front of a monochrome camera lens is usually the simplest method to obtain images at a discrete wavelength region (Chen, Chao and Kim, 2002). In addition, a system with only several wavebands can also be considered as a hyperspectral imaging system if it produces spectral of all pixels within a certain spectral range (VIS, NIR, SWIR, IR, etc.). For example, the visible spectrum usually covers a wavelength range of 400–700 nm (approximately 300 nm). The system within this range of only 20 wavebands of 15 nm bandwidth can also be called a hyperspectral imaging system.

Hyperspectral images preserve a substantial amount of data that provide information about the component of food products being imaged. However, hyperspectral imaging cannot be directly used for the online industry application, because the data acquisition of hyperspectral images is relatively slow compared to multispectral imaging, and more sophisticated methods are needed to process the extraordinary large data of hyperspectral images. On the contrary, although multispectral images do not produce the "spectrum" of an object, if a few optimal wavelengths are selected by means of analyzing the hyperspectral images, the high-dimensional hyperspectral images can be reduced to form multispectral images based on these selected wavelengths, and then a multispectral imaging system with the advantages of simple structure and low cost could be established to meet the speed requirement of production lines and would be incomparable for process monitoring and real-time inspection.

13.2.2.2 Hyperspectral cube

A hyperspectral image is a three-dimensional (3-D) block of data containing a stack of two-dimensional images at a series of continuous wavelengths with the physical and/or chemical information of a material under test (Cogdill *et al.*, 2004). Several congruent images containing intensities at different wavelength bands form a hyperspectral image, which is composed of voxels (vector pixels) containing spectral information (of λ wavelengths) as well as two-dimensional spatial information (of x rows and y columns). Voxel (also called pixel) in the two-dimensional image at a single wavelength is an image element in the hyperspectral cube. The spectral data of any voxel or a group of voxels in a region of interests (ROIs) at a whole wavelength region are called the spectrum. A pixel in a hyperspectral cube contains the spectrum of this pixel with its spatial position. The resulting spectrum of a certain position within the specimen can be considered as its own unique spectral fingerprint of this pixel to characterize the composition of that particular pixel.

Technically speaking, there are two main features characterized in the hyperspectral image data: first, the volume of data of a hyperspectral image is usually very large and suffers from collinearity problems, which causes difficulty to store, manage, process, and analyze hyperspectral image data. Second, although hyperspectral images are composed of high-dimensional data with a large number of spectral bands and offer access to rich information, they also represent a dilemma in themselves for image processing especially for the major purpose of using a hyperspectral imaging system in an online industrial application. For instance, if one hyperspectral image has dimensions of $512 \times 512 \times 256$, this image cube can be viewed as 262,144 spectra, each with 256 wavelength variables. Alternatively, the same hypercube can be interpreted as 256 single channel images with the size of 512×512 pixels each. This huge amount of data presents challenges to any data analysis approach and also creates new opportunities for mining detailed important information.

13.2.2.3 Acquisition of hyperspectral images

The acquisition of a hyperspectral image is generally carried out in four ways: point scanning, line scanning, area scanning, and single shot. The point scanning method (also known as the whiskbroom method) was first patented by NASA to measure the complete spectrum of a single pixel at a time using airborne visible/infrared imaging spectrometer (AVIRIS). In this method, a single point is scanned at one pixel to provide the spectrum of this point (Figure 13.3). Other points are scanned by moving either the detector or the sample in the direction

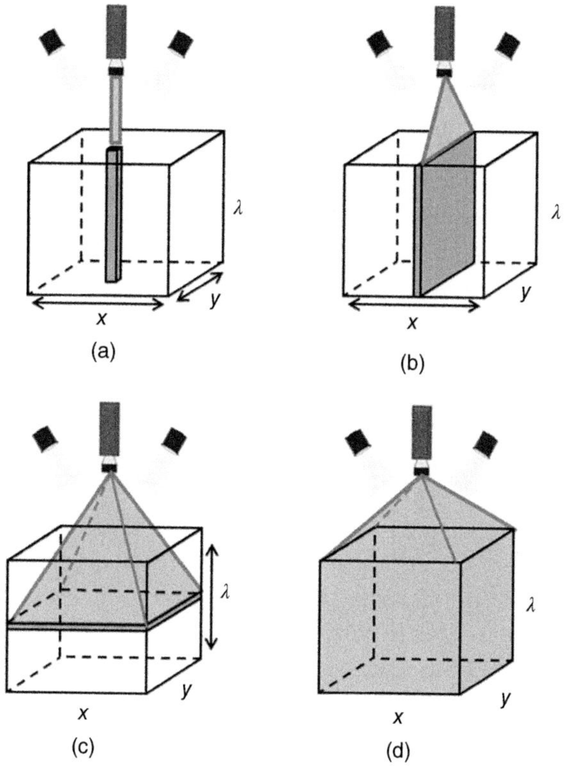

Figure 13.3 Schematic of four ways to acquire a hyperspectral image cube (I (x, y, λ)). Scanning directions are shown by arrows. (a) Point scan, (b) line scan, (c) area scan, (d) single shot.

of either the x or y dimension. Such scanning is iterative for all the target positions within the tested object to generate a hyperspectral image, which is stored as band-interleaved-by-pixel (BIP) format. A hyperspectral image equipped with a point detector is commonly used when spectra of several single points in the scene are required. The advantage of a whiskbroom scanner is that the light passes through the same path in the optical system for each spatial point measurement. To obtain a hyperspectral cube using the whiskbroom method, an exhaustive manner is required to accumulate the spectrum pixel by pixel in a two-axis motorized positioning table to finish the image acquisition. The whiskbroom method is commonly used to acquire microscopic images, where the information within the scene almost does not change during the acquisition time.

The line scanning method (also known as the pushbroom method) as illustrated in top-right corner of Figure 13.3 can be considered as an extension version of the point scanning method (whiskbroom method). Instead of scanning a single pixel at a time, this method records a whole line of an image as well as spectral information simultaneously corresponding to each spatial pixel in the line. The line scanning system uses a two-dimensional detector array and two-dimensional dispersing element. For each line scan, a 2-D image (y, λ) is acquired, in which λ represents the whole spectral dimension, and y the mean of two spatial dimensions. A complete hyperspectral cube is taken as the line is scanned along the direction of the x dimension. Such configuration is usually used when either the sensor of the specimen is moving along one direction of the spatial 2-D (Macgregor, Liu and Yu, 2007), and stores the hyperspectral cube in a band-interleaved-by-line (BIL) format. Because of its rapid and versatile nature, line scanning is particularly suitable in conveyor belt systems and is becoming the most popular method of hyperspectral imaging (Macgregor, Liu and Yu, 2007).

It is noted that both whiskbroom scanning and pushbroom scanning are spatial scanning methods. On the other hand, the area or plane scanning (also known as the band sequential method or wavelength scanning) is known as a spectral scanning method. As shown in the lower-left corner of Figure 13.3, this approach acquired a 2-D monochrome image (x, y) with full spatial information at a single wavelength at one time. To build up a hyperspectral cube, such scanning repeats at one wavelength after another over the whole wavelength range, resulting in a stack of single band images. No relative movement between the detector and the object is required in this method, and, therefore, the detector can sequentially scan a full spatial scene at every spectral wavelength/band. A rotating filter wheel containing fixed band-pass filters or an electronic tunable filter such as an acousto-optic tunable filter (AOTF) or liquid crystal tunable filter (LCTF) are commonly used filter-based configurations for the area scanning, and other used configurations include Fourier transform imaging spectrometers and by sequentially changing filters in front of the sensor (Chen, Chao and Kim, 2002). Because AOTF or LCTF filters are used, a dispersing spectrograph is not necessary in area scanning design. The acquired hyperspectral cube is stored in band sequential (BSQ) format. The area scanning approach is only applicable in multispectral or similar imaging systems, where the number of wavelengths required is limited (Garini, Young and Mcnamara, 2006).

The above-introduced three scanning methods need either spatial scanning or spectral scanning to obtain a 3-D hyperspectral cube. Different from these methods, the single shot method records both spatial and spectral information using a large area detector with one exposure to capture the images, making it much more attractive when fast hyperspectral imaging is required. However, it is still in the early stage and only a few systems with complicated fore-optics configuration and image reconstructions using computationally intensive postprocessing are currently being developed, which have limited resolutions for spatial dimension and narrow ranges for spectral dimension.

Each approach to acquire a 3-D hyperspectral image cube has the specific advantages in terms of data acquisition and processing operations and should be chosen depending on different applications. The BSQ and BIP formats acquired by area scanning and point scanning perform well for spatial and spectral accesses of the hyperspectral data, respectively, and the BIL format integrated the features of spatial and spectral accesses. If a moving sample has to be analyzed like on a conveyer belt for online scanning, one should use pushbroom imaging. If lots of samples are required to be inspected in an overview, then the band sequential method would be appropriate. Whiskbrook imaging is a typical choice to find out minor details on a small surface (i.e., impurity detection) (Macgregor, Liu and Yu, 2007). The three data storage formats mentioned above can be converted to each other using appropriate reconstruction algorithms.

13.2.3 Hyperspectral acquisition modes

There are four typical light interaction modes used for hyperspectral acquisition, namely, transmission, reflectance, transflection, and interaction (interactance) as illustrated in Figure 13.4. Positions of the light source and the optical detector (cameral, spectrograph, and lens) are different for each acquisition mode. The different arrangements of light source and detector result in a different interaction between the imaged specimen and the light beam, which is then received by the detector. The different interactions with the tested sample cause the light received by different modes to have a great influence on the signal-to-noise rate of the measured spectra, which would then be analyzed to determine chemical compositions of the sample, to predict the changes occurred in the sample because of relevant physiological disorders or diseases, and to detect morphological changes within the sample. When an object is illuminated by the radiation from the lighting system, the light is partly reflected, partly absorbed, and partly transmitted through, due to the optical properties of this object. The properties of the food products and the incident light determine information on the traits of the food products. Therefore, information on the traits of the food products can be obtained by determining the optical characteristics of the products.

For the reflectance mode, both the light source and the detector are placed on the same side of a sample. The detector captured the reflected light from the illuminated sample in a specific conformation to avoid specular reflectance. External quality features such as size, shape, color, surface texture, and external defects are typically detected using reflectance mode. The components near the surface of the sample have also been popularly measured using reflectance mode. Transmittance model has the detector located in the opposite side of the light source and captures the transmitted light through the sample, which carries more valuable information but is often very weak (Schaare and Fraser, 2000). In transmission mode, the incident light propagates through a sample and is measured by a detector located in the opposite side of the light source. The transmittance model is usually used to detect internal qualities and defects of relative transparent materials such as fish, fruits, and vegetables. Transflection and interactance modes combine the concepts of transmission and reflectance and are used for measuring liquid and solid samples, respectively. In the transflection mode, the positions of the sample, light source, and detector are the same as those in the

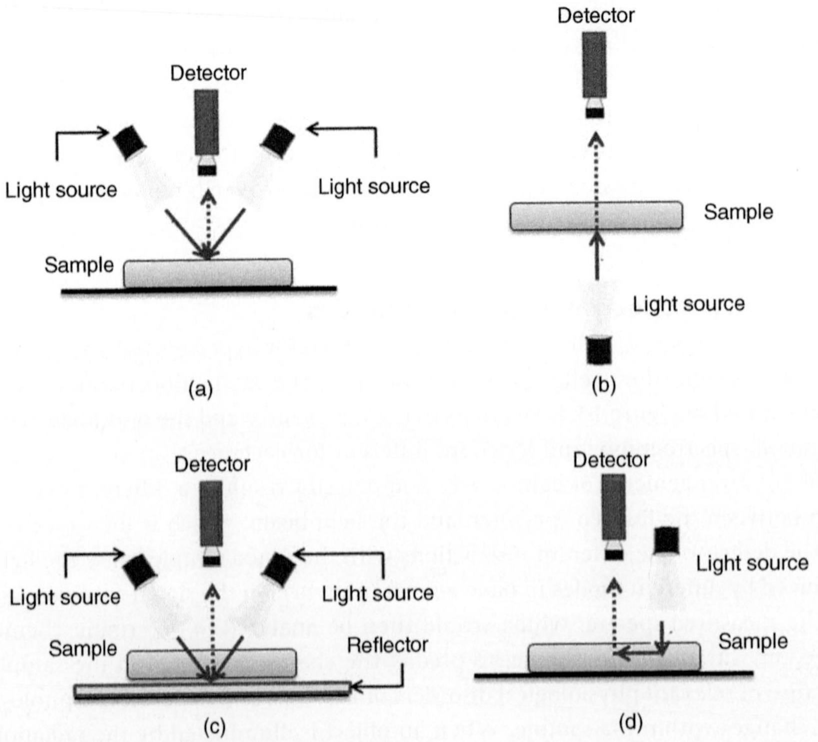

Figure 13.4 Four typical hyperspectral acquisition modes. (a) Reflectance, (b) transmittance, (c) transflection, (d) interactance.

reflectance mode. The difference is that there is a reflector located opposite to the light source on the far side of the sample in the transflection mode. In the interactance model, both light source and the detector are located in the same side and parallel to each other (ElMasry *et al.*, 2008). This mode requires a special setup to seal light in order to prevent specular reflectance and the interference light from the environment, which may affect the signal-to-noise rate of the received light (ElMasry and Wold, 2008; Nicolai *et al.*, 2007).

In practice, the mode selection depends on many factors such as sample type and installment location. The reflectance model has been widely used within hyperspectral imaging applications, while transmittance and interactance modes have also been researched. Interactance mode is more commonly used by spectroscopy rather than hyperspectral imaging, as the measurement of liquid samples does not require spatial information. Reflectance mode is usually applied for solid materials in whole wavelength of 400–2500 nm, while the utilization of transmittance mode is more specific in the wavelength range of 700–900 nm (Givens, Deboever and Deaville, 1997) or 800–1100 nm (Ariana and Lu, 2010). Compared to reflectance model, transmittance mode has potential in predicting internal component concentration and in detecting internal defects within food products (Ariana and Lu, 2008; Ariana and Lu, 2010; Schmilovitch *et al.*, 2004), because the light path has a long distance within the sample in transmittance mode, which results in more spectral information related to the internal properties. McGlone and Martinsen (2004) found that light can penetrate the deeper region of fruit (>2 mm) in transmittance mode than in reflectance mode. The shortcoming of transmittance model is the low signal level from light attenuation due to light absorption and scattering. Therefore, it requires a sensitive detector and a more powerful light source.

13.2.4 Terminology

Some familiarity with essential expressions, technical information, and definitions are useful in dealing with a hyperspectral imaging system. In this section, some basic terminologies normally used in hyperspectral imaging are explained.

13.2.4.1 Spectral range

The spectral range means the spectral wavelength regions that can be measured by the hyperspectral imaging system. Usually, a spectral range of a hyperspectral imaging system could cover the ultraviolet radiation (UV), visible light (VIS), near-infrared (NIR), and mid-infrared (MIR), based on different detectors. Among different spectral ranges, VIS and the short-wave NIR range (400–1000 nm) are mostly used for a hyperspectral imaging system with food quality inspection. Nowadays, the long-wave NIR ranges (900–1700 nm and 900–2500 nm) are available and have been applied to inspect qualities and component distribution of food products.

13.2.4.2 Spatial resolution

The spatial resolution of the hyperspectral imaging system is related to the size of the smallest object that can be seen on the surface of the sample, in other words, the size of a distinct object that can be separated from its surroundings. Higher spatial resolution means more spatial details of the object are explained. Spatial resolution can also be considered as the size of the area that is represented by a pixel in the scene. Therefore, the clarity of an image is not decided by the number of pixels within the image but its spatial resolution.

13.2.4.3 Spectral resolution

The spectral resolution means the narrowest spectral feature that can be resolved by a hyperspectral imaging system. Its magnitude is determined by the sizes of the entrance and exit apertures and the wavelength dispersion of the spectrograph. Spectral resolution can be considered as the ability to separate the monochromatic spectra of two adjacent wavelengths. The true spectral profile of an emitting light from the tested object should be reconstructed accurately as much as possible.

13.2.4.4 Band number

The number of bands is the main feature to characterize the type of spectral imaging system. A multispectral image usually has fewer than 10 bands in the electromagnetic spectrum, while there are about 100–250 spectral bands in a hyperspectral image. Both band number and bandwidth are two important criteria for choosing a hyperspectral system.

13.2.4.5 Bandwidth

The bandwidth is defined as the full width at half maximum (FWHM) of a spectral profile, describing the narrowest spectral feature that is resolved by a spectrograph. Bandwidth is different from the spectral sampling intervals.

13.2.4.6 Signal-to-noise ratio (SNR or S/N)

The signal-to-noise ratio is the ratio of the level of a desired signal to the level of background noise. In other words, it compares the signal measured to the noise generated by the detector and instrument. In hyperspectral imaging systems, the SNR decreases radiance toward longer wavelengths. Therefore, SNR values are not the same at different wavelengths. Higher SNR means less influence from the background noise.

13.3 Applications in evaluating quality and safety of meat and meat products

Meat quality is defined in terms of required nutrition, consumer appreciation of texture and flavor, and food safety (Maltin *et al.*, 2003). Due to the limitation of manual inspection, the development of rapid, stable, noncontact, and accurate inspection systems is required for assessing quality and safety of meat products under the pressure of increasing meat consumption and demand, as seen over the past two decades. In recent years, successful applications of using hyperspectral techniques for meat quality evaluation have been demonstrated by many researchers.

Tenderness is an important concern in meat and meat products because it influences consumer satisfaction. Vis–NIR spectroscopy has had some success as a less invasive and more predictive method than other traditional tenderness measurements. Consumer responses for beef tenderness were successfully predicted with a coefficient of determination (r^2) of 0.74 by visible and near-infrared (Vis–NIR) spectroscopy using 400–2498 nm region (Yancey *et al.*, 2010). In another study, online classifications of beef carcasses and pork loins for tenderness using Vis–NIR spectroscopy (350–2500 nm) have been developed and validated (Shackelford, Wheeler and Koohmaraie, 2012; Shackelford, King and Wheeler, 2011). Moreover, tenderness (shear force of 14-day aged, cooked beef) of fresh beef-ribeye steaks (*longissimus dorsi*) has been determined by using two pushbroom hyperspectral imaging systems in the wavelength range of 400–1000 nm and in the spectral range of 900–1700 nm, respectively (Naganathan *et al.*, 2008a; Naganathan *et al.*, 2008b). In the first work (Naganathan *et al.*, 2008b), PCA was applied to reduce the dimension of the original images, and gray level textural co-occurrence matrix (GLCM) analysis was executed to extract texture features. By using a discrimination model, steak samples were grouped into three tenderness categories, namely tender ($SSF \leq 205.80$ N), intermediate ($205.80\,N < SSF < 254.80$ N), and tough ($SSF \geq 254.80$ N). Results of 96.4% accuracy were obtained for the three tenderness categories, indicating that hyperspectral imaging has considerable promise for predicting beef tenderness on all tough samples. However, it is very time-consuming to use PCA for the dimensional reduction of each steak image and to calculate a huge number of textural features from PC images. In the subsequent work (Naganathan *et al.*, 2008a), PLSR was used for the reduction of high-dimensional hyperspectral images instead of PCA, and GLCM was calculated again for the extraction of texture features. The established model obtained an overall correct rate of 77.0%. Another example for the tenderness prediction was accomplished by Cluff *et al.* (2008) who used a hyperspectral imaging system (496–1036 nm) equipped with optical scattering of light. The results of the established stepwise regression model have a prediction of Warner–Bratzler shear (WBS) force with an $R = 0.67$. An accuracy of 98.4% was obtained when the categories of "tender" and "intermediate" (no "tough" samples) were considered to classify the samples as described by Naganathan *et al.* (2008a).

Moisture is a frequently used quality attribute for meat. Moisture content directly affects the economic profits of meat product because meat is always sold by weight. The oven drying method is the most common method of water analysis usually accepted at a commercial level, which is based on the mass loss after drying of the sample (Mathlouthi, 2001). However, this method is time-consuming, destructive, inefficient, and high-energy consuming, and consequently unsuitable for rapid detection of water content in meat and meat products (Wu *et al.*, 2013). Recently, moisture content in pork meat samples was determined in a rapid and nondestructive way using NIR spectroscopy (1300–2100 nm) coupled with partial least squares regression (PLSR) with a standard error of 0.5% water (m/m) (Kestens *et al.*, 2008). Vis–NIR spectroscopy (400–1700 nm) was successfully used to estimate the moisture of homogenized and minced pork sausages (Ortiz-Somovilla *et al.*, 2007) and pork dry-cured sausages (Gaitan-Jurado *et al.*, 2008) both with r^2 of 0.98. Hyperspectral imaging has also been considered to measure moisture content and evaluate its distribution in lamb (Kamruzzaman *et al.*, 2012a), pork (Barbin *et al.*, 2012b), beef (Elmasry, Sun and Allen, 2013), presliced turkey hams (Iqbal, Sun and Allen, 2013), and Spanish cooked ham (Talens *et al.*, 2013). Good water prediction abilities of hyperspectral imaging were proved with r^2 between 0.87 and 0.89 for the above meats.

Water-holding capacity (WHC), which is commonly measured as drip loss, is an important criterion used for exhibiting the ability of meat to retain all or part of its own water during the process of cutting, grinding, pressing, and heating. Destructive or time-consuming methods such as the filter paper method, bag method, tray method, EZ-DripLoss method, and centrifuge force method are the most often used traditional techniques, which are however not suitable for online inspection. Recently, Vis–NIR spectroscopy in a spectral range of about 400–2500 nm was investigated to determine WHC of beef (Ripoll *et al.*, 2008) and broiler breast meat (Samuel *et al.*, 2011), resulting in r^2 of 0.82 and 0.72, respectively. WHC/drip loss in beef, pork, and lamb meats, and their spatial distributions were also determined using a hyperspectral imaging system in the range of 900–1700 nm (Elmasry, Sun and Allen, 2011b; Kamruzzaman *et al.*, 2012b; Barbin *et al.*, 2012c). When full range spectra were considered to establish the calibration models using PLSR, coefficient of determination for cross-validation (r_{cv}^2) of 0.89, 0.88, and 0.77 for beef, pork, and lamb meats were obtained, respectively. After that, the regression coefficients resulting from the best PLSR prediction models were used to select important wavelengths for reducing the high dimensionality with redundancy among contiguous variables (wavelengths). Finally, the distribution maps of WHC/drip loss were generated by transferring the predicting models to each pixel in all portions of the samples.

Fat is an important factor significantly related to the quality grade of meat. A high fat content is greatly valued as in premium beef from Japanese Black (Wagyu) cattle. Meat trimmings are valued by fat content, the lower the fat content is, the higher the purchase value. Manual operation is commonly carried out in industrial practice today by workers based on training and experience, which is however a difficult task with large deviations from target fat content. Spectroscopy in the range of 400–2500 nm was used to measure intramuscular fat (IMF) content in minced beef and pork and intact pork samples (Prevolnik *et al.*, 2005). Remarkable prediction ability of NIR spectroscopy has been proven with coefficient of determination in cross-validation (r_{cv}^2) between 0.84 and 0.99 and standard error in cross-validation (SECV) between 0.14% and 0.53%. Hyperspectral imaging has also been used to measure fat content in beef and pork meats. Interactance imaging techniques between 760 and 1040 nm were applied online to determine fat content in inhomogeneous beef (Wold *et al.*, 2011) and pork trimmings (O'Farrell *et al.*, 2010). The root mean square error of prediction (RMSEP) decreased with increasing batch size. In particular, RMSEP was 8.7% for single trimmings, 1.33% for trimming batches from 10 to 24 kg, and 0.6% for 100 kg batches. For the evaluation of online measurement of fat content in inhomogeneous pork trimmings, the established PLSR model got RMSEP of 1.9% on 20 kg batches of inhomogeneous meat trimmings and was further tested twice with an RMSEP of 3.4% for a fat range of 8–55% in the first industrial trial and

2.82% in the second industrial trial. Recently, hyperspectral imaging techniques, working in the long-wave near-infrared spectral region, were used for assessing fat content in intact and minced pork (Barbin *et al.*, 2012b), lamb (Kamruzzaman *et al.*, 2012a) beef (Elmasry, Sun and Allen, 2013), and cooked ham (Talens *et al.*, 2013). Results of these studies show that hyperspectral imaging was feasible to determine fat content of pork, lamb, and beef with good r^2 of 0.95, 0.91, and 0.89, respectively, while failing for cooked ham ($r^2 = 0.369$). In another study, Kobayashi *et al.* (2010) utilized near-infrared (NIR) hyperspectral imaging at wavelengths of 1000–2300 nm for the determination of total fat, the proportions of total saturated fatty acid (SEA), total unsaturated fatty acid (UFA), and the main fatty acids of myristic (C14:0), palmitic (C16:0), stearic (C18:0), myristoleic (C14:1), palmitoleic (C16:1), oleic (C18:1), and linoleic (C18:2). PLSR models were developed for the prediction of fat and fatty acid content based on the mean spectrum from an area extracted from the hyperspectral image. Satisfactory results of 0.90, 0.87, and 0.89; 4.81%, 1.69%, and 3.41%; and 2.84, 2.43, and 2.84 for r^2, SEP, and RPD, respectively, were obtained for the prediction of total fat, SEA, and UFA. The determination of individual fatty acids obtained r^2 and RPD values ranging from 0.68 to 0.89 and 1.69 to 2.85, respectively. Visualization of fat content distribution was also realized using the spectrum of each pixel. As a conclusion, it is feasible to use hyperspectral techniques for the inspection of fat content of beef.

Existence of tumors in poultry carcasses can cause immediate consumer rejection of the products. Therefore, the poultry industry must deliver products free from tumors. In the past decade, several studies have focused on the detection of chicken skin tumors using hyperspectral imaging techniques with good results (Elmasry and Sun, 2010; Chao, 2010). In recent years, more efforts have been made to mine useful data in hyperspectral images, resulting in increased ability to detect tumors. Khan, Kim and Kong (2009) applied kernel independent component analysis (KICA) for the dimensionality reduction of hyperspectral images for the selection of a minimum number of bands that contain maximum information for detection in hyperspectral images. The experimental results show that KICA-based band selection has high accuracy compared to that of fast independent component analysis (fastICA)-based band selection for dimensionality reduction and analysis for hyperspectral images. In another study, a new method was presented for detecting poultry skin tumors based on a serial feature fusion in hyperspectral images (Xu, Kim and Kong, 2009). Two kinds of feature sets were extracted from the hyperspectral fluorescence images and hyperspectral reflectance images, respectively. The experimental result based on the proposed method shows the better performance in detecting tumors compared with previous works. More interestingly, an optimal design method of emission filter was presented, which provides the optimal weighted combination of emission wavelengths in terms of discriminant power, unlike the band selection method that just finds a subset of significant spectral bands (Kim, Cho and Kim, 2009). In addition, an optimized tumor detection system was developed by employing a new adaptive branch-and-bound algorithm and a support vector machine classifier, which were proved to be significantly faster than other versions of the branch-and-bound algorithm (Nakariyakul and Casasent, 2009).

Contaminants are important public health hazards in terms of meat consumption. For the poultry industry, the main interesting research area is focused on using hyperspectral techniques for detection of various contaminations on the surface of the poultry carcasses and on differentiation between wholesome and unwholesome freshly slaughtered chickens. The principle of using hyperspectral techniques for the contamination inspection of chicken and poultry products is based on the spectral dissimilarity among different compositions and contaminations within the products. Windham *et al.* (2003a) characterized the fecal contamination of chicken carcasses by using Vis–NIR spectroscopy ranging from 400 to 2500 nm. The diagnostic system with spectral profiles from hyperspectral techniques could be applied as a rapid and noncontact implementation to inspect a production line of chicken or poultry products during all key points of production. In addition, the spatial information from the hyperspectral images can be used for the location of contaminations, making the hyperspectral imaging more suitable for contamination detection than spectroscopy. Nowadays, some hyperspectral imaging systems of different sensitivities and designs have already been developed for the determination of ingesta and fecal matter (Park *et al.*, 2006b; Windham *et al.*, 2003b; Windham *et al.*, 2005a; Windham *et al.*, 2005b). Park *et al.* (2006a) developed a prototype real-time multispectral imaging system equipped with a common aperture camera with three optical trim filters (517, 565, and 802 nm), which can detect fecal and ingesta contaminants on broiler carcasses using a decision tree algorithm. The object-oriented unified modeling language (UML) was used to program the real-time imaging software for online assessment. It is a powerful programming language capable of visualizing, specifying, constructing, and documenting software systems. Both hardware and software used in this system were evaluated at a pilot-scale poultry processing plant. Effective detection results were obtained for identifying feces (from duodenum, ceca, and colon) and ingesta on the surface of poultry carcasses using the multispectral imaging system with an inspection speed of 140 birds per minute.

Microbial spoilage is a main source of foodborne outbreaks. It is caused by the growth and enzymatic activity of microorganisms. The decomposition and the formation of metabolites make the spectrum of spoiled meat different from that of fresh meat. Good quantitative relationships of r_{cv}^2 of 0.94–0.97 were found between the spectral data in 240–1200 nm range and plate count on the pork meat surface (Oto *et al.*, 2012). In another study, Gómez-De-Anda *et al.* (2012) determined *Trichinella spiralis* in pig muscles by using mid-Fourier transform infrared (MID-FTIR) spectroscopy with attenuated total reflectance (ATR) and soft independent modeling of class analogy (SIMCA), leading to the 100% recognition and rejection rates. The efforts for the rapid, accurate, and nondestructive detection of bacterially spoiled or contaminated meat have also been carried out using hyperspectral imaging techniques. Peng *et al.* (2011) exploited hyperspectral imaging over the spectral region between 400 and 1100 nm to measure biochemical changes within fresh beef. Total viable count (TVC) per gram of beef was obtained by classical microbiological plating methods. The spectral scattering profiles at each wavelength were fitted accurately with a two-parameter Lorentzian distribution function. MLR was used to establish TVC prediction models between the $\log_{10}$ (TVC) value and individual Lorentzian parameters and their combinations at different wavelengths. The best predictions of $r^2 = 0.95$ and SEP = 0.30 were obtained, which demonstrated the potential of using hyperspectral imaging techniques for nondestructive and real-time detection of bacterial spoilage in beef. In other studies, hyperspectral imaging was also used for TVC determination of pork meat (Wang, Peng and Zhang, 2010; Wang *et al.*, 2011; Tao *et al.*, 2010). Results indicated that LS-SVM obtained better results with r^2 of 0.9872 (Wang, Peng and Zhang, 2010) and 0.9236 (Wang *et al.*, 2011), while poorer results with r of 0.863 (Tao *et al.*, 2010) and r^2 of 0.7194 (Wang, Peng and Zhang, 2010) were obtained using PLSR. The performances of ANNs ($r^2 = 0.7718$) (Wang, Peng and Zhang, 2010) and MLR ($r = 0.886$) (Tao *et al.*, 2010) were also poorer than LS-SVM. This might be because there was a nonlinear relationship between TVC and reflectance spectra (Wang, Peng and Zhang, 2010). In another study, instead of directly using the extracted spectral data from hyperspectral images for model calibration, Dissing *et al.* (2013) mapped the spectrum of each pixel to identify clusters for indicating areas of spoilage using unsupervised K-means algorithm. There was a correlation found between the spoilage degree and the counted number of pixels in the images belonging to the various clusters. Then a quantitative PLSR model, with the output being spoilage degree, was established using the sizes of the clustered areas. Results showed that the TVC was successfully predicted with a standard error of prediction of 7.47%.

Color is an important external feature for meat quality evaluation by consumers. NIR spectroscopy in 1100–2498 nm was used for predicting color traits of beef, in which lightness (L^*) and redness (a^*) were well predicted with r^2 of 0.80 and 0.82, respectively, but yellowness (b^*) was not (Cecchinato *et al.*, 2011). In another study, Vis–NIR reflectance spectroscopy in the range of 400–2498 nm showed good predictability of luminosity at 0 (L^*t0) and 60 min (L^*t60) of beef ($r^2 = 0.85$ and 0.82; SECV = 1.16 and 1.36, respectively) (Andres *et al.*, 2008). pH, another major factor affecting meat quality, its initial and ultimate values can influence the extent of protein denaturation and fresh meat quality attributes such as color and water-holding capacity (Bidner *et al.*, 2004). Andres *et al.* (2008) used Vis–NIR reflectance spectroscopy to measure the pH of beef, achieving a high r^2 of 0.97. When hyperspectral imaging was considered, several works have been conducted for color parameters (L^*, a^*, b^*) and pH prediction by analyzing hyperspectral images in the wavelength range of 900–1700 nm. These studies include beef (Elmasry, Sun and Allen, 2012c), pork (Barbin *et al.*, 2012c), lamb analysis (Kamruzzaman *et al.*, 2012b), and sliced turkey hams (Iqbal, Sun and Allen, 2013). All of these attempts relied on determining color and pH non-invasively by extracting spectral data from the regions of interest of meat and then relating these spectral data with the reference values of color and pH determined by colorimeter and a pH meter, respectively. When feature wavelengths were selected, good predictions of L^* and pH were achieved with r_{cv}^2 of 0.88 and 0.75 for beef, 0.93 and 0.87 for pork, 0.91 (L^*) for lamb, and 0.81 (pH) for sliced turkey hams, respectively.

Protein plays an important role in providing nutrients for consumers and its content is often used as an index for quality evaluation. The protein content in beef was successfully predicted using NIR spectroscopy (Ripoll *et al.*, 2008), resulting in r^2 of 0.91. Recently, in a series of works using NIR hyperspectral imaging (900–1700 nm) in tandem with PLSR modeling for determining protein content in meats, good prediction outputs were obtained with r^2 of 0.85, 0.91, 0.91, and 0.855 for lamb (Kamruzzaman *et al.*, 2012a), pork (Barbin *et al.*, 2012b), beef (Elmasry, Sun and Allen, 2013), and Spanish cooked ham (Talens *et al.*, 2013), respectively.

Sensory evaluation plays an important role in quality characterization of meat. Sensory analysis is a method for describing the eating quality of meat based on human senses (sight, smell, taste, touch, and sound). The attributes usually cover appearance, odor, flavor, and texture including tenderness and juiciness (Bejerholm and Aaslyng, 2004). Andres *et al.* (2007) used the NIR reflectance spectroscopy in the 400–2098 nm region to predict the sensory characteristics of lamb meat samples such as flavor, abnormal flavor, juiciness, and texture. However, the output results were not good ($r^2 < 0.4$). In another study, Ripoll *et al.* (2008) used the same spectral range to predict tenderness, juiciness, and overall appraisal in beef. Tenderness was the best-predicted variable ($r^2 = 0.98$), while juiciness and overall appraisal had the r^2 of 0.588 and 0.589, respectively. In recent years, the feasibility of using hyperspectral imaging for determining sensory attributes of meats have also been investigated (Kamruzzaman *et al.*, 2013; Barbin *et al.*, 2012c). The results show that some attributes such as juiciness ($r^2 = 0.49$) and tenderness ($r^2 = 0.54$ for pork and 0.69 for lamb) could be assessed reasonably, while the predictions of other attributes such as flavor and overall acceptability were not successful. The results of the above work show that hyperspectral techniques are not very suitable for predicting sensory variables and should be further investigated.

An important aspect for the meat industry, to meet the needs of consumers, is to realize the discrimination and classification of meat products for its pricing and authentication. Good quality pork is generally red, firm, and nonexudative (RFN). Other quality grades of pork meat include RSE (red, soft, and exudative), PFN (pale, firm, and nonexudative), and PSE (pale, soft, and exudative), which are determined by various combinations of color, texture, and drip loss. The commonly used method of inspecting pork quality is through the direct assessment of the exposed loin eye at the 10th/11th rib interface of the longissimus muscle. It is difficult to classify all the four quality grades (RFN, RSE, PFN, and PSE) by using only the visible spectral information. Qiao *et al.* (2007) used hyperspectral imaging techniques for the classification of 40 pork samples of different quality grades (RFN, PSE, PFN, and RSE). The used hyperspectral imaging system equipped with pushbroom structure in the spectral range of 400–1000 nm consisted of a CMOS camera, a spectrograph device, a fiber-optic line light source, and a conveyor belt controlled by a computer. The appropriate conveyor speed was set to avoid distortion on spatial resolution, image size, and fitting the camera exposure time. The main spectrum was extracted from the region of interest for each sample after the image calibration. The smoothing of 10-point mean filter followed with second derivatives executed as the spectral preprocessing for each spectrum to correct multiplicative scatter and correct the baseline. PCA was then performed for the reduction of dimensionality of the extracted spectral data. Different numbers of PCs obtained from the PCA process were used for the establishment of a classification model by using cluster analysis and ANN, respectively. The classification results of 75–80% overall accuracy were obtained using cluster analysis and the correct answer rates of 69% and 85% were calculated by using the ANN model with the first 5 and 10 PCs, respectively. Later, similar work has been conducted based on hyperspectral imaging with the wavelength range extended into 900–1700 nm (Barbin *et al.*, 2012a). Pork cuts from three quality grades (PSE, RFN, and DFD) were classified according to the selected spectra of lonely loin eye area. A better result with an overall accuracy of 96% was obtained in this work, which is better than the results of previous work considering hyperspectral imaging in 400–1000 nm. In another study, Kamruzzaman *et al.* (2011) evaluated the feasibility of using a NIR hyperspectral imaging system for the discrimination of three types of lamb muscles. A pushbroom hyperspectral imaging system with a spectral range of 900–1700 nm was used for the image acquisition of samples from *semitendinosus* (ST), *Longissimus dorsi* (LD), and *Psoas Major* (PM) of *Charollais* breed. Muscles were cut into slices of 1 in. in thickness by a scalpel and cutting machine. There were 105 lamb samples including PM (35), ST (35), and LD (35) used for further research. PCA was performed to reduce high dimensionality, select optimal wavelengths, and visualize chemical images. The eigenvector plot of PCA was used to select optimal wavelengths, which were 934, 974, 1074, 1141, 1211, and 1308 nm in this study. An LDA calibration model was established with these six optimum wavelengths and overall accuracy of 100% was obtained. Based on the optimum wavelengths, a multispectral imaging system can be set up and installed over the conveyor belt with minimum modifications of the existing industrial system. A high overall accuracy in a pixel-wise scale was also obtained for visualizing classification results using an image processing algorithm. Details of hyperspectral data analysis is illustrated in Figure 13.5, which

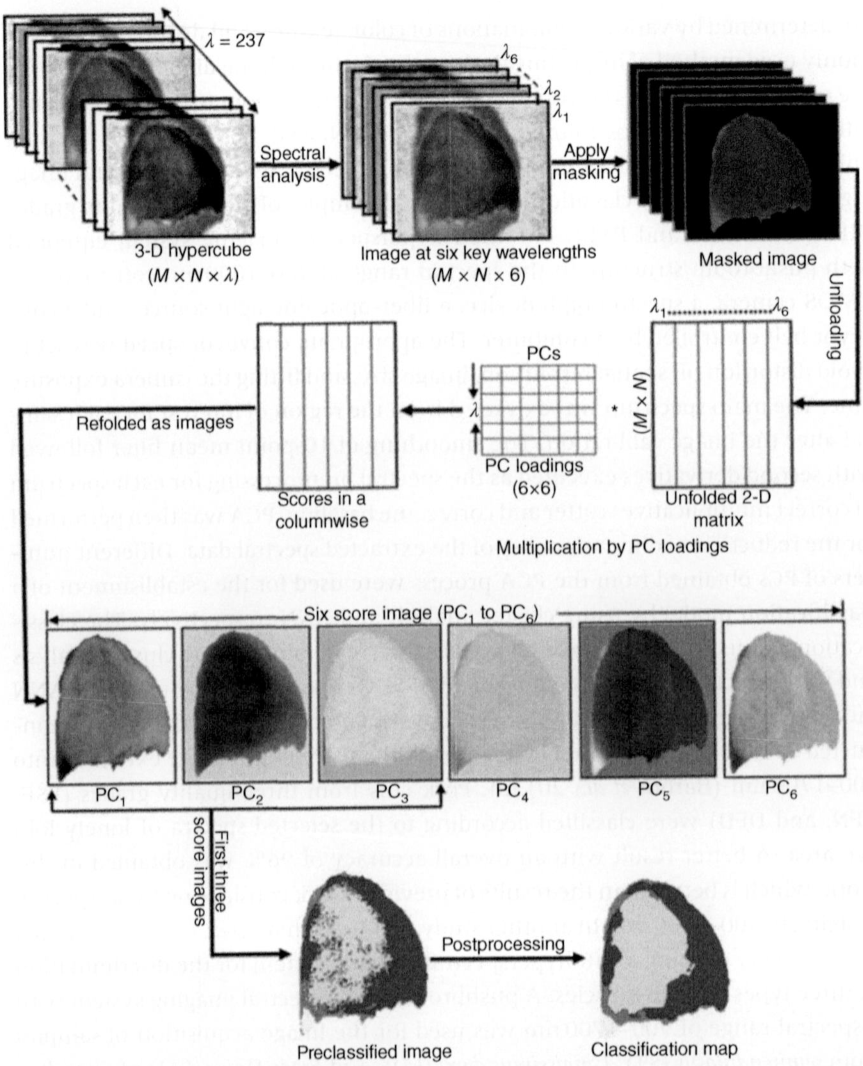

Figure 13.5 Key steps involved in building classification map, which starts from 3-D hypercube of an ST muscle and ends with its classification map. The final classification map shows that the outer borders of ST muscle are misclassified as PM (light grey boundary) and LD (dark grey, mainly on right hand side, and around border) and adjacent fat of muscle is classed as fat (white, front left component).

shows the step-by-step procedure for creating a classification map. The results of this study show the great potential of using NIR hyperspectral imaging together with chemometrics as an efficient and reliable tool for the discrimination of lamb muscles. Moreover, to prove the advantages of using hyperspectral imaging systems over other systems, its classification results were compared with those of computer vision and Minolta chroma meter, which only had the classification accuracy of 81.9% and 89.5%, respectively.

13.4 Advantages and disadvantages of hyperspectral techniques in meat applications

Hyperspectral techniques are able to achieve accurate, constant, and reliable quality and safety analysis of meat and meat products compared to sensory assessment and chemical methods. Therefore, they have a great potential to replace manual routine inspection and random inspection using chemical methods and can increase the efficiency of the inspection process. However, hyperspectral techniques also have some shortcomings that need to be overcome or at least noted.

13.4.1 Common advantages of spectroscopy and hyperspectral imaging

Both spectroscopy and hyperspectral imaging are noninvasive, nondestructive, and chemical-free assessment methods requiring little/no sample preparation. Compared with traditional sensory and chemical methods, they save on measurement time, reagents, and labor costs. Based on the established and validated calibration models, the quality classification or prediction of new samples could be extremely simple and expeditious using hyperspectral techniques. Both techniques can determine several quality and safety attributes simultaneously for the same sample. Such determination allows different labeling and pricing to be available in sorting meat products according to consumer preference, market requirements, and/or product specifications.

13.4.2 Common disadvantages of spectroscopy and hyperspectral imaging

Hyperspectral techniques are indirect inspection techniques, which require reference quality values for model calibration and would not have higher prediction accuracy than the reference methods. On the other hand, because hyperspectral techniques do not need to touch the sample, but only rely on the optical characteristics of samples, their detection limits are usually higher than those of chemical-based analytical methods. Moreover, hyperspectral data have the well-known problem of multicollinearity. Two ways are generally applied to reduce this problem, namely multivariate analysis and variable selection.

13.4.3 Advantages of hyperspectral imaging over spectroscopy

Spectroscopy has the capability to characterize homogenous materials but fails to assess quality attributes of meat with high spatial variations because its spectral measurement focuses only on a round area of the sample being analyzed. The average spectra of the measured area can only be used to predict the average contents of components. However, when a sample is inhomogeneous, a misleading interpretation of the result could occur on using a single measurement to represent the bulk value of the entire specimen. Therefore, the heterogeneity in the specimen can only be entirely and accurately determined when the spectral measurement is systematically repeated many times. In other words, due to the inhomogeneous distribution of the qualities within the specimen, the spectral acquisition might be necessary at different positions on the specimen, which is obviously inefficient for an industrial application (Nicolai *et al.*, 2006). Therefore, spectroscopy is not suitable for the situation when spatial distribution of quality parameters is needed.

Somewhat different from spectroscopy, which can only collect a single spectrum of one spot on a sample at once, the spectrum of each pixel within the sample and background can be acquired by using hyperspectral imaging. Therefore, hyperspectral imaging allows the visualization of the distribution of different components within a sample based on their spectral signatures. The bulk composition can be replaced by the visualization of the spatial distribution and concentration of the components in the product. Even after image acquisition, hyperspectral imaging can choose any ROI within the image. Moreover, if the spectrum of an ROI or a pixel presents very obvious spectral characteristics, it can be considered as a spectral signature and be saved in a spectral library.

It should be noted that, in many applications of hyperspectral imaging for meat analysis, only the spectral information of the acquired hyperspectral images was used for the analysis, which makes almost no difference when compared directly to using spectroscopy techniques. The only improvement in using hyperspectral imaging is that the spectral data were extracted from different regions of interests with representative pixels, while spectroscopy can only consider a small area of the sample surface where the probe is located.

13.4.4 Limitations of hyperspectral imaging

Although hyperspectral imaging has the aforementioned advantages, it does have some constraints to limit its application. One of the main shortcomings of hyperspectral imaging is that hyperspectral images contain much more redundant data than spectroscopic data. The acquisition and calculation of such a high amount of raw hyperspectral images bring arduous tasks for the hardware and software designers. Before the development of hyperspectral imaging systems for the rapid acquisition and analysis of the huge hyperspectral data cube, many efforts have been made in the laboratories to do feature extraction and wavelength selection, so that optimized multispectral imaging systems could be developed based on feature wavelengths to meet the requirement of speed and efficiency for online inspection of meat and meat products. Another problem of hyperspectral imaging is that the reference values of quality attributes of each pixel within a sample cannot be measured. Therefore, the prediction model can only be calibrated using the region within the sample, where reference values of this region could be measured using reference physical or chemical methods and the mean spectral profile of this region could be also extracted. In addition, the analysis of liquids or homogenous samples does not need hyperspectral imaging but only spectroscopy, as spatial information is useless for this kind of sample.

13.5 Conclusion

This chapter covers the basic principles of two typical hyperspectral techniques, namely spectroscopy and hyperspectral imaging, and their recent applications in meat quality and safety evaluation. As competent ways for the rapid and nondestructive evaluation of meat and meat products, hyperspectral techniques have outstanding characteristic identification capabilities and rich physical and chemical information and are very suitable for numerous meat applications. There are many research endeavors on the applications of off-line and online quality inspection and documentation of meat and meat products using hyperspectral techniques. Hyperspectral techniques can replace sensory evaluation and chemical methods when rapid and noninvasive inspection of the quality and safety of meat and meat products is required. In addition, hyperspectral techniques can also be used as the primary evaluation of large batches of products, followed by elaborative inspection of some particular subsets of samples that need further analysis using manual or chemical methods. Although hyperspectral techniques face a problem of dimensionality as high spectral resolution is used, useful information could be extracted using data mining and multivariate processing techniques, which provides practical feasibility for both qualitative classification and quantitative prediction of quality and safety attributes of meat and meat products. The successful application of hyperspectral techniques in meat inspection could help the meat processing industry to better understand and control meat quality and safety, leading to increased competitiveness in international markets.

Acknowledgments

The authors would like to acknowledge the financial support provided by the Irish Research Council for Science, Engineering and Technology under the Government of Ireland Postdoctoral Fellowship scheme.

References

Andres, S., Murray, I., Navajas, E.A. *et al.* (2007) Prediction of sensory characteristics of lamb meat samples by near infrared reflectance spectroscopy. *Meat Science*, **76**, 509–516.

Andres, S., Silva, A., Soares-Pereira, A.L. *et al.* (2008) The use of visible and near infrared reflectance spectroscopy to predict beef M-longissimus thoracic et lumborum quality attributes. *Meat Science*, **78**, 217–224.

Ariana, D.P. and Lu, R. (2008) Detection of internal defect in pickling cucumbers using hyperspectral transmittance imaging. *Transactions of the ASABE*, **51**, 705–713.

Ariana, D.P. and Lu, R.F. (2010) Evaluation of internal defect and surface color of whole pickles using hyperspectral imaging. *Journal of Food Engineering*, **96**, 583–590.

Armenta, S., Moros, J., Garrigues, S. and De La Guardia, M. (2010) The Use of Near-Infrared Spectrometry in the Olive Oil Industry. *Critical Reviews in Food Science and Nutrition*, **50**, 567–582.

Ashton, T.J., Michie, I. and Johnston, I.A. (2010) A Novel Tensile Test Method to Assess Texture and Gaping in Salmon Fillets. *Journal of Food Science*, **75**, S182–S190.

Barbin, D.F., Elmasry, G., Sun, D.-W. and Allen, P. (2012a) Near-infrared hyperspectral imaging for grading and classification of pork. *Meat Science*, **90**, 259–268.

Barbin, D.F., Elmasry, G., Sun, D.-W. and Allen, P. (2012b) Non-destructive determination of chemical composition in intact and minced pork using near-infrared hyperspectral imaging. *Food Chemistry*, **138**, 1162–1171.

Barbin, D.F., Elmasry, G., Sun, D.-W. and Allen, P. (2012c) Predicting quality and sensory attributes of pork using near-infrared hyperspectral imaging. *Analytica Chimica Acta*, **719**, 30–42.

Bejerholm, C. and Aaslyng, M.D. (2004) The influence of cooking technique and core temperature on results of a sensory analysis of pork—depending on the raw meat quality. *Food Quality and Preference*, **15**, 19–30.

Bidner, B., Ellis, M., Brewer, M. *et al.* (2004) Effect of ultimate pH on the quality characteristics of pork. *Journal of Muscle Foods*, **15**, 139–154.

Bock, L.E. and Connelly, R.K. (2008) Innovative uses of near-infrared spectroscopy in food processing. *Journal of Food Science*, **73**, R91–R98.

Carlomagno, G., Capozzo, L., Attolico, G. and Distante, A. (2004) Non-destructive grading of peaches by near-infrared spectrometry. *Infrared Physics & Technology*, **46**, 23–29.

Casas, C., Martinez, O., Guillen, M.D. *et al.* (2006) Textural properties of raw Atlantic salmon (*Salmo salar*) at three points along the fillet, determined by different methods. *Food Control*, **17**, 511–515.

Cecchinato, A., De Marchi, M., Penasa, M. *et al.* (2011) Near-infrared reflectance spectroscopy predictions as indicator traits in breeding programs for enhanced beef quality. *Journal of Animal Science*, **89**, 2687–2695.

Cen, H.Y. and He, Y. (2007) Theory and application of near infrared reflectance spectroscopy in determination of food quality. *Trends in Food Science & Technology*, **18**, 72–83.

Chao, K. (2010) Automated poultry carcass inspection by a hyperspectral-multispectral line-scan imaging system, in *Hyperspectral Imaging for Food Quality Analysis and Control*, 1st edn (ed. D.-W. Sun), Academic Press/Elsevier, San Diego, California, USA.

Chen, Y.R., Chao, K.L. and Kim, M.S. (2002) Machine vision technology for agricultural applications. *Computers and Electronics in Agriculture*, **36**, 173–191.

Cho, B.-K., Kim, M.S., Baek, I.-S. *et al.* (2013) Detection of cuticle defects on cherry tomatoes using hyperspectral fluorescence imagery. *Postharvest Biology and Technology*, **76**, 40–49.

Cluff, K., Naganathan, G.K., Subbiah, J. *et al.* (2008) Optical scattering in beef steak to predict tenderness using hyperspectral imaging in the Vis–NIR region. *Sensing and Instrumentation for Food Quality and Safety*, **2**, 189–196.

Cogdill, R.P., Hurburgh, C.R., Rippke, G.R. *et al.* (2004) Single-kernel maize analysis by near-infrared hyperspectral imaging. *Transactions of the ASAE*, **47**, 331–320.

Dissing, B., Papadopoulou, O., Tassou, C. *et al.* (2013) Using multispectral imaging for spoilage detection of pork meat. *Food and Bioprocess Technology*, **6**, 2268–2279.

Du, C.J. and Sun, D.-W. (2004) Recent developments in the applications of image processing techniques for food quality evaluation. *Trends in Food Science & Technology*, **15**, 230–249.

Du, C.J. and Sun, D.-W. (2006) Learning techniques used in computer vision for food quality evaluation: a review. *Journal of Food Engineering*, **72**, 39–55.

Elmasry, G., Barbin, D.F., Sun, D.-W. and Allen, P. (2012a) Meat Quality Evaluation by Hyperspectral Imaging Technique: An Overview. *Critical Reviews in Food Science and Nutrition*, **52**, 689–711.

Elmasry, G., Iqbal, A., Sun, D.-W. *et al.* (2011a) Quality classification of cooked, sliced turkey hams using NIR hyperspectral imaging system. *Journal of Food Engineering*, **103**, 333–344.

Elmasry, G., Kamruzzaman, M., Sun, D.W. and Allen, P. (2012b) Principles and applications of hyperspectral imaging in quality evaluation of agro-food products: A review. *Critical Reviews in Food Science and Nutrition*, **52**, 999–1023.

Elmasry, G. and Sun, D.-W. (2010) Meat quality assessment using a hyperspectral imaging system, in *Hyperspectral Imaging for Food Quality Analysis and Control*, 1st edn (ed. D.-W. Sun), Academic Press/Elsevier, San Diego, California, USA.

Elmasry, G., Sun, D.-W. and Allen, P. (2011b) Non-destructive determination of water-holding capacity in fresh beef by using NIR hyperspectral imaging. *Food Research International*, **44**, 2624–2633.

Elmasry, G., Sun, D.-W. and Allen, P. (2012c) Near-infrared hyperspectral imaging for predicting colour, pH and tenderness of fresh beef. *Journal of Food Engineering*, **110**, 127–140.

Elmasry, G., Sun, D.-W. and Allen, P. (2013) Chemical-free assessment and mapping of major constituents in beef using hyperspectral imaging. *Journal of Food Engineering*, **117** (2), 235–246.

Elmasry, G., Wang, N., Vigneault, C. *et al.* (2008) Early detection of apple bruises on different background colors using hyperspectral imaging. *LWT--Food Science and Technology*, **41**, 337–345.

Elmasry, G. and Wold, J.P. (2008) High-speed assessment of fat and water content distribution in fish fillets using online imaging spectroscopy. *Journal of Agricultural and Food Chemistry*, **56**, 7672–7677.

Gómez-De-Anda, F., Dorantes-Álvarez, L., Gallardo-Vel Zquez, T. *et al.* (2012) Determination of *Trichinella spiralis* in pig muscles using Mid-Fourier Transform Infrared Spectroscopy (MID-FTIR) with Attenuated Total Reflectance (ATR) and Soft Independent Modeling of Class Analogy (SIMCA). *Meat Science*, **91**, 240–246.

Gaitan-Jurado, A.J., Ortiz-Somovilla, V., Espana-Espana, F. *et al.* (2008) Quantitative analysis of pork dry-cured sausages to quality control by NIR spectroscopy. *Meat Science*, **78**, 391–399.

Garini, Y., Young, I.T. and Mcnamara, G. (2006) Spectral imaging: Principles and applications. *Cytometry. Part A*, **69A**, 735–747.

Givens, D.I., Deboever, J.L. and Deaville, E.R. (1997) The principles, practices and some future applications of near infrared spectroscopy for predicting the nutritive value of foods for animals and humans. *Nutrition Research Reviews*, **10**, 83–114.

Gomez-Sanchis, J., Gomez-Chova, L., Aleixos, N. *et al.* (2008) Hyperspectral system for early detection of rottenness caused by *Penicillium digitatum* in mandarins. *Journal of Food Engineering*, **89**, 80–86.

Gunasekaran, S. (1996) Computer vision technology for food quality assurance. *Trends in Food Science & Technology*, **7**, 245–256.

He, H.-J., Wu, D. and Sun, D.-W. (2013) Non-destructive spectroscopic and imaging techniques for quality evaluation and assessment of fish and fish products. *Critical Reviews in Food Science and Nutrition*.

Huang, H.B., Yu, H.Y., Xu, H.R. and Ying, Y.B. (2008) Near infrared spectroscopy for on/in-line monitoring of quality in foods and beverages: A review. *Journal of Food Engineering*, **87**, 303–313.

Iqbal, A., Sun, D.-W. and Allen, P. (2013) Prediction of moisture, colour and pH in cooked, pre-sliced turkey hams by NIR hyperspectral imaging system. *Journal of Food Engineering*, **117** (1), 42–51.

Jackman, P., Sun, D.-W. and Allen, P. (2011) Recent advances in the use of computer vision technology in the quality assessment of fresh meats. *Trends in Food Science & Technology*, **22**, 185–197.

Kamruzzaman, M., Elmasry, G., Sun, D.-W. and Allen, P. (2011) Application of NIR hyperspectral imaging for discrimination of lamb muscles. *Journal of Food Engineering*, **104**, 332–340.

Kamruzzaman, M., Elmasry, G., Sun, D.-W. and Allen, P. (2012a) Non-destructive prediction and visualization of chemical composition in lamb meat using NIR hyperspectral imaging and multivariate regression. *Innovative Food Science & Emerging Technologies*, **16**, 218–226.

Kamruzzaman, M., Elmasry, G., Sun, D.-W. and Allen, P. (2012b) Prediction of some quality attributes of lamb meat using near-infrared hyperspectral imaging and multivariate analysis. *Analytica Chimica Acta*, **714**, 57–67.

Kamruzzaman, M., Elmasry, G., Sun, D.-W. and Allen, P. (2013) Non-destructive assessment of instrumental and sensory tenderness of lamb meat using NIR hyperspectral imaging. *Food Chemistry*, **141**, 389–396.

Kestens, V., Charoud-Got, J., Bau, A. *et al.* (2008) Online measurement of water content in candidate reference materials by acousto-optical tuneable filter near-infrared spectrometry (AOTF-NIR) using pork meat calibrants controlled by Karl Fischer titration. *Food Chemistry*, **106**, 1359–1365.

Khan, A., Kim, I. and Kong, S.G. (2009) Dimensionality reduction of hyperspectral images using kernel ICA. Proc. SPIE 7315. *Sensing for Agriculture and Food Quality and Safety*, 731510.

Kim, T., Cho, B.-K. and Kim, M. S. (2009) *Optical filter design of fluorescence emission to detect poultry skin tumors*. ASAE Annual International Meeting. Reno, Nevada.

Kobayashi, K., Matsui, Y., Maebuchi, Y. *et al.* (2010) Near infrared spectroscopy and hyperspectral imaging for prediction and visualisation of fat and fatty acid content in intact raw beef cuts. *Journal of Near Infrared Spectroscopy*, **18**, 301–315.

Macgregor, J.F., Liu, Z. and Yu, H.L. (2007) Standardization of line-scan NIR imaging systems. *Journal of Chemometrics*, **21**, 88–95.

Maltin, C., Balcerzak, D., Tilley, R. and Delday, M. (2003) Determinants of meat quality: tenderness. *Proceedings of the Nutrition Society*, **62**, 337–347.

Mathlouthi, M. (2001) Water content, water activity, water structure and the stability of foodstuffs. *Food Control*, **12**, 409–417.

Mcglone, V.A. and Martinsen, P.J. (2004) Transmission measurements on intact apples moving at high speed. *Journal of Near Infrared Spectroscopy*, **12**, 37–43.

Naganathan, G.K., Grimes, L.M., Subbiah, J. *et al.* (2008a) Partial least squares analysis of near-infrared hyperspectral images for beef tenderness prediction. *Sensing and Instrumentation for Food Quality and Safety*, **2**, 178–188.

Naganathan, G.K., Grimes, L.M., Subbiah, J. *et al.* (2008b) Visible/near-infrared hyperspectral imaging for beef tenderness prediction. *Computers and Electronics in Agriculture*, **64**, 225–233.

Nakariyakul, S. and Casasent, D.P. (2009) Fast feature selection algorithm for poultry skin tumor detection in hyperspectral data. *Journal of Food Engineering*, **94**, 358–365.

Nicolai, B.M., Beullens, K., Bobelyn, E. *et al.* (2007) Nondestructive measurement of fruit and vegetable quality by means of NIR spectroscopy: A review. *Postharvest Biology and Technology*, **46**, 99–118.

Nicolai, B.M., Lotze, E., Peirs, A. *et al.* (2006) Non-destructive measurement of bitter pit in apple fruit using NIR hyperspectral imaging. *Postharvest Biology and Technology*, **40**, 1–6.

O'Farrell, M., Wold, J.P., Hoy, M. *et al.* (2010) On-line fat content classification of inhomogeneous pork trimmings using multispectral near infrared interactance imaging. *Journal of Near Infrared Spectroscopy*, **18**, 135–146.

Ortiz-Somovilla, V., Espana-Espa, A.F., Gaitan-Jurado, A. *et al.* (2007) Proximate analysis of homogenized and minced mass of pork sausages by NIRS. *Food Chemistry*, **101**, 1031–1040.

Oto, N., Oshita, S., Kawagishi, S. *et al.* (2012) Non-destructive estimation of ATP contents and plate count on pork meat surface by UV–Vis reflectance spectrum analysis. *Journal of Food Engineering*, **110**, 9–17.

Park, B., Abbott, A.J., Lee, K.J., Choi, C.H. and Choi, K.H. (2002) *Near-infrared spectroscopy to predict soluble solids and firmness in apples*. ASAE/CIGR Annual International Meeting, Chicago, Illinois, USA.

Park, B., Kise, M., Lawrence, K. C., Windham, W. R., Smith, D. P. and Thai, C. N. (2006a) Real-time multispectral imaging system for online poultry fecal inspection using UML, in: *Optics for Natural Resources, Agriculture, and Foods. Proceeding of SPIE* (eds Y.-R. Chen, G. E. Meyer and S.-I. Tu), pp. 63810W-1-63810W-12.

Park, B., Lawrence, K.C., Windham, W.R. and Smith, D.P. (2006b) Performance of hyperspectral imaging system for poultry surface fecal contaminant detection. *Journal of Food Engineering*, **75**, 340–348.

Peiris, K.H.S., Dull, G.G., Leffler, R.G. and Kays, S.J. (1999) Spatial variability of soluble solids or dry-matter content within individual fruits, bulbs, or tubers: Implications for the development and use of NIR spectrometric techniques. *Hortscience*, **34**, 114–118.

Peng, Y.K., Zhang, J., Wang, W. *et al.* (2011) Potential prediction of the microbial spoilage of beef using spatially resolved hyperspectral scattering profiles. *Journal of Food Engineering*, **102**, 163–169.

Prevolnik, M., Candek-Potokar, M., Skorjanc, D. *et al.* (2005) Predicting intramuscular fat content in pork and beef by near infrared spectroscopy. *Journal of Near Infrared Spectroscopy*, **13**, 77–85.

Prieto, N., Roehe, R., Lavin, P. *et al.* (2009) Application of near infrared reflectance spectroscopy to predict meat and meat products quality: A review. *Meat Science*, **83**, 175–186.

Qiao, J., Ngadi, M.O., Wang, N. *et al.* (2007) Pork quality and marbling level assessment using a hyperspectral imaging system. *Journal of Food Engineering*, **83**, 10–16.

Ripoll, G., Alberti, P., Panea, B. *et al.* (2008) Near-infrared reflectance spectroscopy for predicting chemical, instrumental and sensory quality of beef. *Meat Science*, **80**, 697–702.

Rodriguezotero, J.L., Hermida, M. and Centeno, J. (1997) Analysis of dairy products by near-infrared spectroscopy: A review. *Journal of Agricultural and Food Chemistry*, **45**, 2815–2819.

Samuel, D., Park, B., Sohn, M. and Wicker, L. (2011) Visible–near-infrared spectroscopy to predict water-holding capacity in normal and pale broiler breast meat. *Poultry Science*, **90**, 914–921.

Schaare, P.N. and Fraser, D.G. (2000) Comparison of reflectance, interactance and transmission modes of visible-near infrared spectroscopy for measuring internal properties of kiwifruit (*Actinidia chinensis*). *Postharvest Biology and Technology*, **20**, 175–184.

Schmilovitch, Z., Shenderey, C., Shmulevich, I., Alchanatis, V., Egozi, H. and Hoffman, A. (2004) *NIRS detection of mouldy core in apples*. 2004 CIGR International Conference, Beijing, China.

Shackelford, S., King, D. and Wheeler, T. (2011) Development of a system for classification of pork loins for tenderness using visible and near-infrared reflectance spectroscopy. *Journal of Animal Science*, **89**, 3803–3808.

Shackelford, S., Wheeler, T. and Koohmaraie, M. (2012) Technical note: Validation of a model for online classification of US Select beef carcasses for longissimus tenderness using visible and near-infrared reflectance spectroscopy. *Journal of Animal Science*, **90**, 973–977.

Sigurgisladottir, S., Hafsteinsson, H., Jonsson, A. *et al.* (1999) Textural properties of raw salmon fillets as related to sampling method. *Journal of Food Science*, **64**, 99–104.

Singh, C., Jayas, D., Paliwal, J. and White, N. (2012) Fungal damage detection in wheat using short-wave near-infrared hyperspectral and digital colour imaging. *International Journal of Food Properties*, **15**, 11–24.

Sun, D.-W. (2010) *Hyperspectral Imaging for Food Quality Analysis and Control*, Academic Press/Elsevier, San Diego, California, USA.

Talens, P., Mora, L., Morsy, N. *et al.* (2013) Prediction of water and protein contents and quality classification of Spanish cooked ham using NIR hyperspectral imaging. *Journal of Food Engineering*, **117**, 272–280.

Tao, F.F., Wang, W., Li, Y.Y. *et al.* (2010) A rapid nondestructive measurement method for assessing the total plate count on chilled pork surface. *Spectroscopy and Spectral Analysis*, **30**, 3405–3409.

Tsuta, M., Sugiyama, J. and Sagara, Y. (2002) Near-infrared imaging spectroscopy based on sugar absorption band for melons. *Journal of Agricultural and Food Chemistry*, **50**, 48–52.

Wang, W., Peng, Y., Huang, H. and Wu, J. (2011) Application of hyper-spectral imaging technique for the detection of total viable bacteria count in pork. *Sensor Letters*, **9**, 1024–1030.

Wang, W., Peng, Y.K. and Zhang, X.L. (2010) Study on modeling method of total viable count of fresh pork meat based on hyperspectral imaging system. *Spectroscopy and Spectral Analysis*, **30**, 411–415.

Williams, P.J., Geladi, P., Britz, T.J. and Manley, M. (2012) Investigation of fungal development in maize kernels using NIR hyperspectral imaging and multivariate data analysis. *Journal of Cereal Science*, **55**, 272–278.

Windham, W., Lawrence, K., Park, B. and Buhr, R. (2003a) Visible/NIR spectroscopy for characterizing fecal contamination of chicken carcasses. *Transactions-American Society of Agricultural Engineers*, **46**, 747–754.

Windham, W.R., Heitschmidt, G.W., Smith, D.P. and Berrang, M.E. (2005a) Detection of ingesta on pre-chilled broiler carcasses by hyperspectral imaging. *International Journal of Poultry Science*, **4**, 959–964.

Windham, W.R., Smith, D.P., Berrang, M.E. *et al.* (2005b) Effectiveness of hyperspectral imaging system for detecting cecal contaminated broiler carcasses. *International Journal of Poultry Science*, **4**, 657–662.

Windham, W.R., Smith, D.P., Park, B. *et al.* (2003b) Algorithm development with visible/near-infrared spectra for detection of poultry feces and ingesta. *Transactions of the ASAE*, **46**, 1733–1738.

Wold, J.P., O'Farrell, M., Hoy, M. and Tschudi, J. (2011) On-line determination and control of fat content in batches of beef trimmings by NIR imaging spectroscopy. *Meat Science*, **89**, 317–324.

Wu, D., Shi, H., Wang, S. *et al.* (2012) Rapid prediction of moisture content of dehydrated prawns using online hyperspectral imaging system. *Analytica Chimica Acta*, **726**, 57–66.

Wu, D. and Sun, D.-W. (2013) Potential of time series-hyperspectral imaging (TS-HSI) for non-invasive determination of microbial spoilage of salmon flesh. *Talanta*, **111**, 39–46.

Wu, D., Wang, S., Wang, N. *et al.* (2013) Application of time series-hyperspectral imaging (TS-HSI) for determining water distribution within beef and spectral kinetic analysis during dehydration. *Food and Bioprocess Technology*, **6**, 2943–2958.

Xu, C., Kim, I. and Kong, S. G. (2009) *Feature level fusion for hyperspectral images*. Proc. SPIE 7315, Sensing for Agriculture and Food Quality and Safety.

Yancey, J., Apple, J., Meullenet, J.-F. and Sawyer, J. (2010) Consumer responses for tenderness and overall impression can be predicted by visible and near-infrared spectroscopy, Meullenet–Owens razor shear, and Warner–Bratzler shear force. *Meat Science*, **85**, 487–492.

CHAPTER 14

Online meat quality and compositional assessment techniques

Kumari Shikha Ojha[1,2], Brijesh K. Tiwari[1], Joseph P. Kerry[2] & Patrick J. Cullen[3,4]

[1] *Department of Food Biosciences, Teagasc Food Research Centre, Dublin, Ireland*
[2] *School of Food and Nutritional Sciences, University College Cork, Co Cork, Ireland*
[3] *School of Chemical Engineering, New South Wales University, Sydney, Australia*
[4] *School of Food Science & Environmental Health, Dublin Institute of Technology, Dublin, Ireland*

14.1 Summary

The meat industry is constantly seeking novel in vivo and ex vitro meat evaluation techniques for rapid, real-time assessment of livestock, carcasses and meat products to assure meat quality. With the advancement in sensor technology and computational power, it is possible for online determination of meat quality traits both in vivo and in vitro. Various meat evaluation techniques with particular emphasis on non-invasive methods that provide spatial information (ultrasound, X-ray computed tomography (CT) and nuclear magnetic resonance (NMR)/magnetic resonance imaging (MRI)) are discussed in this chapter. Methods available for both the prediction of body and carcass composition are also detailed in this chapter.

14.2 Introduction

Carcass evaluation is generally the basis for judging the commercial value of livestock and is consequently one of the most common quality control tests performed in the meat industry. Coupled with weight analysis, carcass evaluation typically forms the basis for the financial transactions between livestock producers and meat processors. Carcasses must be evaluated based on both quality and compositional attributes, all the while considering the ever-increasing consumer demand for healthier and consistent meat products. Carcass quality attributes include tenderness, cut size, fat cover, marbling, meat and fat colour, whereas,

Emerging Technologies in Meat Processing: Production, Processing and Technology, First Edition.
Edited by Enda J. Cummins and James G. Lyng.

composition attributes include saleable meat yield or proportions of fat, lean and bone (Harrington and Kempster, 1989). Carcass evaluation is a methodology to describe the quality of livestock in terms of their suitability and commercial value for various end usages including retail cut and processed meat. Currently, a number of approaches are available for prediction of carcass composition and quality, which may also allow grading of carcass into various categories. For example, in the United States, beef carcasses are evaluated based on established Standards for Grades of Slaughter Cattle and Standards for Grades of Carcass Beef. According to this quality metric, grades are determined by marbling and overall maturity. There are eight quality grade designations: Prime, Choice, Select, Standard, Commercial, Utility, Cutter and Canner. Prime, Choice, Select and Standard are classified as young beef (maturity levels A and B) and must be less than 42 months of age. Commercial, Utility, Cutter and Canner are cow grades from carcasses greater than 42 months. Similarly, in the European Union, adult bovine carcasses are classified according to the EUROP grid system, which is based on visual assessment scores according to the defined standards implemented by the European Community regulations 1208/81 and 1026/91 (Craigie *et al.*, 2012). The EUROP classification scheme includes carcass conformation scores on a 15-point scale with five main classes, E (excellent conformation), U, R, O and P (poor conformation), and 10 subclasses, and five main fatness scores (1 (low fatness), 2, 3, 4 and 5 (excessive fatness)) also with 10 subclasses (Craigie *et al.*, 2013; Fisher, 2007). Carcass evaluation is performed manually by trained graders using photographic references. Various carcass classification schemes adopted worldwide have been criticised due to the subjective nature of the process, with a high degree of inconsistencies in such manual grade assessments (Gupta *et al.*, 2013). With growing concern over the qualitative value of carcasses and the possibility to improve the consistency and subjectivity of assessment, instrumented carcass evaluation techniques are gaining importance in the meat industry. This chapter focuses on various in vivo and ex vivo techniques for the prediction of body and carcass composition. These techniques include ultrasound, X-ray computed tomography (CT), NMR, MRI, total body electrical conductivity (TOBEC), video image analysis (VIA), CT scanning and resistance sensors. Some of these technologies can be used in vivo (i.e. ultrasound and computed tomography), others can be used online (i.e. visual image analysis, near-infrared spectroscopy) with most feasible for post-mortem evaluation. Feasibility, cost, speed, precision and current adoption by the meat industry of such technologies are outlined in this chapter.

14.3 In vivo methods of carcass evaluation

In vivo methods for carcass evaluation that do no harm animal performance are increasingly popular because they are non-destructive and have no impact on live animal performance. It can be employed for the selection of breeding stock and/or estimation of the market value of animals (Stanford, Jones and Price, 1998). The use of advanced technologies for more accurate measure of carcass traits in live animals has increased the rate of genetic improvement in body composition traits (Kvame and Vangen, 2006). Various non-destructive methods including ultrasound, X-ray computed tomography and NMR have been proposed and tested for the prediction and evaluation of various compositional traits in live animals. Computer tomography and magnetic resonance image analysis are commonly used for in vivo 3D structure images that can be reconstructed to obtain volumetric measurements of target area. Table 14.1 shows the comparative assessment of selected carcass evaluation techniques.

14.3.1 X-ray-based technologies

X-ray-based technologies are well recognised as a useful sensing technology for detection, monitoring and diagnosis in medical applications. However, in recent years, X-ray has emerged as a promising non-destructive scanning technique for application in agrifood sector (Barigou and Douaire, 2013; Blazek and Gilbert, 2011; Chen *et al.*, 2013; Mery *et al.*, 2011; Nielsen, Christensen and Feidenhans'l, 2014). Within the meat sector, X-ray-based imaging techniques, mainly computed tomography (CT) and dual-energy X-ray absorptiometry (DEXA), have been successfully employed to study the changes in the body composition of live animals and correlated with carcass quality (Karamichou *et al.*, 2006; Marcoux, Bernier and Pomar, 2003; Marcoux, Faucitano and Pomar, 2005; Mitchell *et al.*, 1998b). Both technologies are non-invasive imaging techniques, which can be employed on live animals and on carcasses or primal cuts.

CT along with image analysis methodologies can classify and quantify the fat content of livestock. This technique has been used in the selection of breeding animals. In general, CT scanning involves food withdrawal 24 h prior to scanning to reduce gut fill, followed by general anaesthesia to minimise animal stress and movement during scanning. Figure 14.1 shows the application of CT scanning of anaesthetised pig and 3D reconstruction of the carcass relevant parts of the pig (Lambe *et al.*, 2012).

DEXA is a quantitative body composition analysis method based on the tissue-specific attenuation of X-rays (Pietrobelli *et al.*, 1998). DEXA has been successfully used to measure the chemical carcass composition of livestock in vivo with a high accuracy for prediction of body composition compared with chemical analysis or dissection (Bernau *et al.*, 2015; Kremer, Förster and Scholz, 2013; Marcoux, Faucitano and Pomar, 2005; Mitchell, Scholz and Conway, 1998a; Mitchell, Scholz and Pursel, 2003; Mitchell, Solomon and Rumsey, 1997; Scholz *et al.*, 2007; Suster *et al.*, 2003). This technique quickly and accurately provides reliable information about mass and the percentage of lean tissue and fat, bone mineral content and bone mineral density (Kremer *et al.*, 2012;

Table 14.1 Comparison of various carcass evaluation techniques.

Parameters	CT scan (computed (axial) tomography)	MRI (magnetic resonance imaging)	X-ray (X-radiation or Rontgen radiation)	Ultrasound
Principle used for imaging	Uses X-rays for imaging	Uses large external field, RF pulse and three different gradient fields	X-rays attenuated (blocked) by denser tissue creates a shadow on the image	High-frequency sound waves (ultrasound) are used for imaging
Radiation exposure	The effective radiation dose from CT ranges from 2 to 10 mSv	No ionising radiation	Exposure to ionising radiation	No radiation
Cost	Scan cost is lower compared to MRI and ultrasound	MRI scan is expensive compared to CT, X-ray and/or ultrasound	X-ray is relatively cheaper than CT, MRI and/or ultrasound	Relatively higher than CT and MRI
Scan time	Scan time < 30 s	Depending on the need usually few minutes	Few seconds	Approximately 10–15 min
Effects on the animal body	May pose risk of ionising radiation	No risk	Risk of exposure to ionising radiation	Cavitation (non-thermal effect), temperature rise in tissues
Ability to change the imaging plane without moving the animal	Yes, multiplanar reformation function is possible	MRI machines can produce images in any plane 3D isotropic imaging can also be produced Multiplanar reformation	Does not have this ability	Possible depending on the system

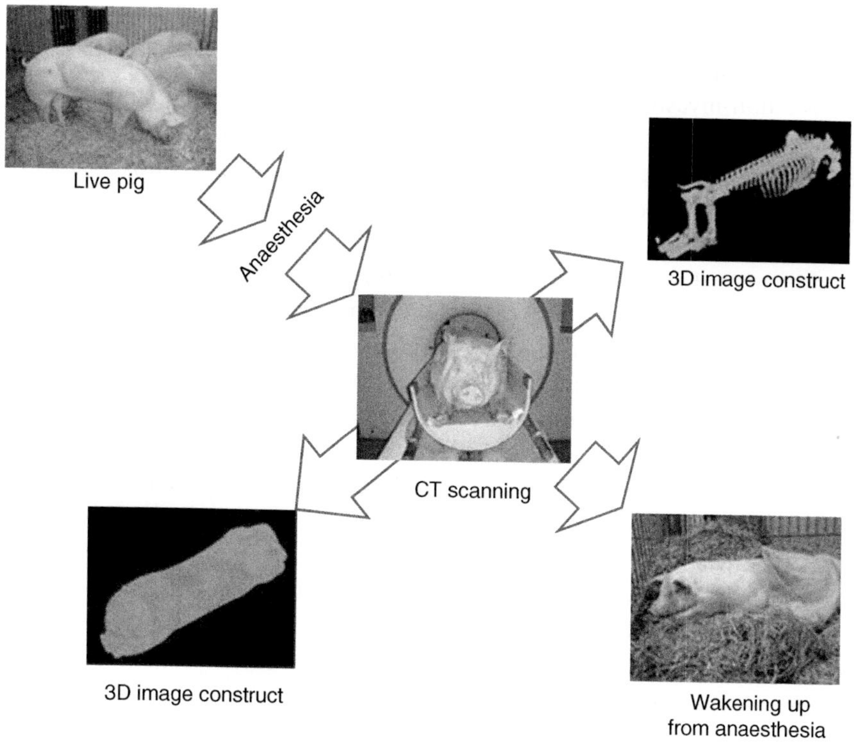

Figure 14.1 Application of CT scanning in pigs and 3D reconstruction of the carcass relevant parts of the pig (Modified from Lambe *et al.* (2012)).

Marcoux, Faucitano and Pomar, 2005; Suster *et al.*, 2003). Scholz *et al.* (2007) employed DXEA for analysis of pigs at two different positions using two different instruments yielding a close relationship between both in vivo and carcass DXEA fat and lean measurements and the reference data (R^2 values ranged from 0.545 to 0.839).

X-ray computed tomography (CT) has been demonstrated to be an accurate and reliable non-destructive method to analyse carcass traits in live animals. In vivo measurements via CT for body composition has enhanced the accuracy of selection and the rate of genetic improvement in body composition traits for livestock (Damez and Clerjon, 2013). X-ray computed tomography (CT), which rotates around the sample facilitating the gathering of information from different planes, has been used to predict salt, water and fat content in bone decisions regarding carcass quality and, more recently, some components of meat quality (Bünger *et al.*, 2011). Karamichou *et al.* (2006) employed CT for the estimation of genetic parameters for carcass composition and meat quality traits in Scottish sheep and demonstrated the feasibility of genetic improvement of carcass composition and meat quality traits using X-ray tomography.

14.3.2 Nuclear magnetic resonance and magnetic resonance imaging

MRI is a non-invasive imaging technique that uses strong magnetic fields combined with the application of radiofrequency waves to produce images. Originally developed for medical diagnostics, MRI is widely recognised as one of the most powerful diagnostic tools that can generate cross-sectional images of a body with a high degree of contrast between various soft tissues (Baulain, 1997). The basic principle of MRI is the magnetic resonant properties of protons associated with molecules (e.g. water and lipids) of tissues when positioned in a strong magnetic field, which results in a range of signal intensities (Mitchell *et al.*, 2001; Szabo *et al.*, 1999). The data obtained is reconstructed to obtain 2D or 3D images of the object that finds applications in medical diagnostics and assessment of clinical conditions. The basic difference between MRI and CAT scan is that MRI uses a magnetic field, whereas CAT scan uses X-rays, consequently MRI has the distinct advantage of having no radiation exposure compared to X-rays. Additionally, NMR has capabilities to evaluate muscle metabolism and prediction of carcass quality attributes including water-holding capacity (Lazzaroni, Gigli and Gabiña, 2007). NMR allows exceptional contrast among muscle, fat and connective tissue structures present in meats. The 3D nature of the NMR allows not only the measurement of superficial fat but also volumetric content of fat (Ballerini *et al.*, 2000). An obvious disadvantage of NMR over CAT is the operational cost for livestock species. MRI has been used successfully for the determination of in vivo body composition of various livestock including boar, pig, sheep, chicken and turkey (Baulain, 1997; Bernau *et al.*, 2015; Kremer, Förster and Scholz, 2013; Mitchell *et al.*, 1991). A patent was granted for an MRI-based animal sorting and grading system that can compare, rank, sort and grade livestock or carcasses (Ellis, 2005). Studies to date involving the use of MRI as an in vivo technique have focused on the prediction of body composition. For example, Mitchell (REF) employed in vivo MRI to measure the volume of several organs and tissues of 111 pigs (males and females) and compared tissue or organ weights obtained by dissection. The authors observed a strong correlation between MRI volume and dissected weight as shown in Figure 14.2. Similarly, Monziols *et al.* (2006) employed MRI to predict the tissue composition of 24 pig carcasses and cuts. A good prediction of muscle content for meat cuts and carcasses (R^2 ranging from 0.970 to 0.997) was reported. Collewet *et al.* (2005) carried out MRI measurements (left side) and automatic image analysis (right side) of 120 pig carcasses and reported a precision of 586 g and 1.10% for lean meat weight and lean meat percentage, respectively. Such results point to the feasibility of MRI to predict the fat and lean components of a carcass. Figure 14.3 shows MRI images from a ham and a belly after MRI acquisition (original images) and after image processing to separate the cuts from the background and segmenting pixels into classes corresponding to muscle, subcutaneous fat and intermuscular fat.

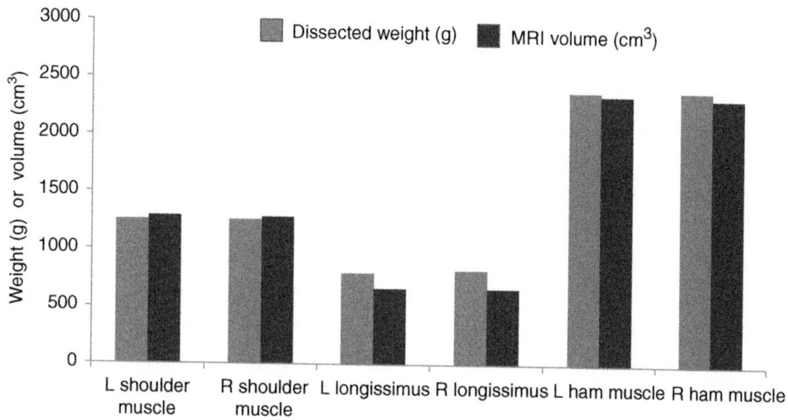

Figure 14.2 Relationship between magnetic resonance imaging (MRI) volume analysis and dissected weights of various tissues (L: left and R: right) from male and female pigs (Adapted from Mitchell *et al.*, 2001).

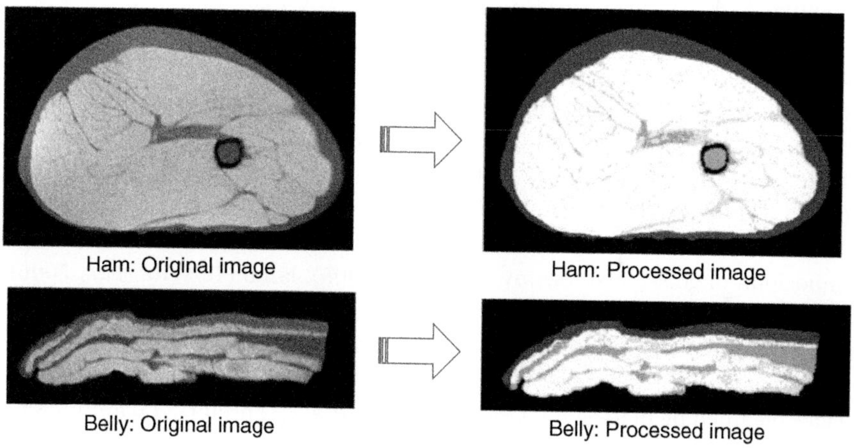

Figure 14.3 Original and processed MRI images of a ham and a belly after MRI acquisition and after image processing separating the cuts from the background and segmenting pixels into classes corresponding to muscle (1), subcutaneous fat (2), intermuscular fat (3), background (4), intermuscular fat/muscle partial volumes (5), and subcutaneous fat/muscle partial volumes (6), respectively (Modified from Monziols *et al.* (2006).

14.3.3 Ultrasound

The history of scientific advances and discovery of ultrasound are rooted in the study of sound, with Sir Isaac Newton first proposing his theory of sound waves in 1687 (Mason, 1998). Ultrasonic waves are waves having frequencies above the audible range (>20 kHz) and less than microwave frequencies (up to 10 MHz). The sound ranges employed can be divided into high-frequency, low-energy diagnostic ultrasound and low-frequency, high-energy power ultrasound. Ultrasonic can also be broadly classified as low-intensity (<1 W/cm^2) and high-intensity (10–1000 W/cm^2) ultrasound. Low-intensity ultrasound is typically used as a non-destructive analytical technique for quality assurance and process control, especially in the case of physicochemical properties such as composition, structure and physical state of matter, whereas high-intensity ultrasound is employed for extraction and processing applications. Ultrasound scanning technology is a non-invasive method to predict carcass characteristics that utilises ultrasonic waves to generate ultrasonic images that are employed for visualisation of various body parts, with real-time tomographic imaging feasible. Ultrasonography is a viable, reliable and cost-effective technique to provide precise and accurate determination of carcass composition (Houghton and Turlington, 1992; Lambe *et al.*, 2010; Peña *et al.*, 2014). The basic principle of diagnostic ultrasound is that when low–power, high-frequency sound waves, in the range of 1–15 MHz, propagate by longitudinal motion through the body tissues, some waves are reflected when an interface between two tissues is encountered. The reflected signals are detected and are then transcribed and presented as a 2D image or cross section of muscle mass (Szabo *et al.*, 1999). Two-dimensional images are constructed based on the reflection of ultrasonic waves, for example, fluids like blood will have minimal reflection, whereas bone will have higher reflection. Ultrasound as a diagnostic tool may cause heating due to thermal effects of ultrasound. Formation and collapse of cavitation bubble may introduce mechanical effects leading to tissue damage (Fowlkes and Holland, 2000).

Ultrasound has been reported for estimating carcass quality traits and body composition of goat (Teixeira, Joy and Delfa, 2008), lamb (Teixeira *et al.*, 2006), sheep (Silva *et al.*, 2006), pig (Kiefer *et al.*, 2014) and cattle (Nunes *et al.*, 2015). Table 14.2 shows some applications of ultrasound for in vivo estimation of quality and compositional factors. Carcass traits are estimated based on ultrasonic images obtained by placing ultrasonic transducers at the preferred scanning site.

Table 14.2 Selected examples demonstrating the use of ultrasound for in vivo estimation.

Animal	Technique	Objective	Reference
Beef bulls of the Slovakian Pied breed	Ultrasound combined with body dimensions	In vivo estimates of slaughter value	Polak *et al.* (2001)
Charollais lambs	Ultrasound alone	Traits of carcass quality	Milerski (2001)
Chicken	Ultrasound	Amount and yield of breast meat	Remignon, Seigneurin and Moati (2000)
Austrian meat sheep	Ultrasound scanning and subjective muscle scoring system	Prediction of carcass quality and composition	Grill *et al.* (2015)

Ultrasonic images can be obtained in live animals over skin without shaving or clipping hair. In most cases, edible oil/vegetable oil is applied over the skin as a coupling agent to allow better acoustic contact between the probe and the skin (Peña *et al.*, 2014). Ultrasonic images are processed typically using specialised software. Peña *et al.* (2014) employed real-time ultrasound to predict carcass traits in 300 lean beef cattle from three breeds, and they observed that the predictions improve with the proximity to slaughter and ultrasonic measurements were positively correlated with the corresponding carcass attributes. Ultrasound has been used for fat and muscle depth measurements, which along with the live weight can be good predictors of carcass composition (Teixeira *et al.*, 2006). Polak *et al.* (2001) investigated the feasibility of using ultrasound combined with body dimensions for in vivo estimates of slaughter value of 101 Slovakian Pied beef bulls. The authors observed that the combined use of the sonographic technique with body dimensions was suitable for the estimation of slaughter value in live bull beef.

14.4 Post-mortem compositional analysis

Ex vivo compositional analysis is an essential part of carcass evaluation. Compositional analysis has traditionally been carried out by physical dissection of animals or carcass, which is destructive, time consuming and expensive (Stanford, Jones and Price, 1998) for meat carcass quality evaluation. Post-mortem compositional analysis is generally carried out on relatively small sample sizes in the laboratory. With advancements in analytical technologies, several non-destructive methods have been proposed for compositional analysis including TOBEC, VIA and optical probes.

14.4.1 Total body electrical conductivity

TOBEC is a non-invasive and rapid technique for determination of body composition. TOBEC technology works on the principle that lean tissue and body fat differ considerably in their electrical properties. Lean tissue, because of its higher content of water and electrolytes, is a good electrical conductor, whereas fat or bone is relatively anhydrous and impedes electrical current (Keim *et al.*, 1988). The signal from the TOBEC is therefore strongly related to the lean content of carcass. During TOBEC measurement, the subject is placed in a chamber generating a radiofrequency electromagnetic field. In the chamber, the target absorbs energy from the field in proportion to its conductivity that is measured by detectors (Hyánková and Szebestová, 2010). TOBEC has proven to be an accurate means of prediction for carcass composition under controlled laboratory conditions. TOBEC can be applied to warm carcasses or chilled carcass sides with nearly equal prediction. The first commercial application of this technology in the meat industry was to measure the lean content of boxed boneless beef in Australia (Allen and McGeehin, 2001). Berg, Forrest and Fisher (1994) showed high correlations between carcass characteristics and TOBEC measurements. Similarly, Gu *et al.*, (1992) reported high R^2 values, in the range from 0.64 to 0.93 for different carcass measurements in a large sample size of 181 barrows from five genetic populations. Shelton *et al.* (2003) also estimated carcass composition of 24 barrows with high accuracy using TOBEC (Simeonova, Todorov and Schinckel, 2012). Several reports have shown that TOBEC can predict beef carcass composition with a reasonable degree of accuracy, accounting for 85–90% of the variation in lean content of beef quarters a primal from a homogeneous population of steer carcasses (Gwartney *et al.*, 1994).

14.4.2 Video image analysis

VIA is a rapid, automated and objective method of carcass evaluation, which has gained increasing attention in the past few decades. VIA follows the principle that areas of different light intensity received by a video camera's photosensitive element generate different voltages so that areas of light (fat) can be quantitatively differentiated from areas of dark (lean) areas (Stanford, Jones and Price, 1998; Wood *et al.*, 1991). VIA-based systems generally consist of stationary image-capture devices with a standardised lighting source; image analysing and processing software; and a conveyor to move carcasses through the VIA's field of view. VIA can be applied either on the whole or half carcass before chilling, or it can be applied to the quartered carcass after chilling. VIA has been applied for evaluation of different traits of carcass and meat eating quality (Craigie *et al.*, 2012). Previous studies have shown the successful use of VIA for improved carcass classification based on the levels of intramuscular fat at the 12/13th rib, estimation of carcass shape and composition, prediction of tenderness from the meat surface, measurement of meat colour, evaluation of water-holding capacity and quantification of the amount of intramuscular fat in beef and lean meat yield percentage (LMY%) (Craigie *et al.*, 2012). VIA-based systems that assess whole sides of beef can be installed online in abattoirs to operate autonomously at line speeds up to 800 carcasses per hour (Craigie *et al.*, 2012; Rius-Vilarrasa *et al.*, 2009). The latter approach has been developed for use at the 12/13th rib interface and is used extensively in the United States and Canada. For markets where there was a need to automate visual assessment of whole carcasses, the whole-side approach to VIA is preferred. VIA systems tend to be able to account for more variation in SMY% with greater repeatability than with visual assessment methods. Other features that make VIA appealing are that it is objective, fast and, in the case of whole-side systems, totally automated. In the European Union, where whole-side VIA is used to predict carcass conformation and fat class, VIA machines must be validated against a panel of five trained assessors before they can replace visual classification in a member state. The use of the expert panel for validating VIA for carcass classification on the EUROP grid is not representative of everyday operating conditions where a single classifier is used. Regression equations can be used to determine the ability of VIA to predict SMY%, but the accuracy and precision of predictions are heavily dependent on the SMY% variation in the sample carcasses as well as the level and variability in fat trimming. The use of CT scanning to determine carcass composition (e.g. LMY%, fat%, muscle distribution and muscle-to-bone ratio) may be a more appropriate method for validating VIA. CT scanning is an accurate, precise and non-destructive tool that can determine carcass composition in an easily standardised manner. Individual plants could then determine how to calculate SMY% from CT-predicted LMY% in accordance with their individual market requirements. Three different brands of commercially available whole-side VIA systems are used in Europe to classify beef carcasses according to the EUROP conformation and fat class grid, including the BCC-2, VBS 2000 and the Normaclass Machine à classer (MAC) ((Craigie *et al.*, 2013) (Figure 14.4).

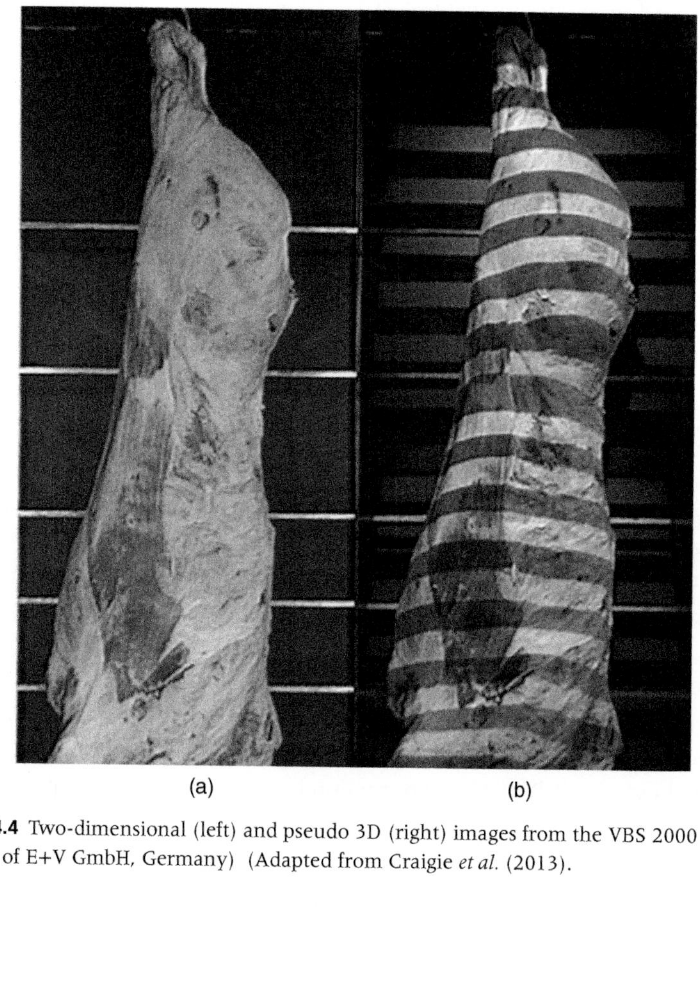

Figure 14.4 Two-dimensional (left) and pseudo 3D (right) images from the VBS 2000 (Image courtesy of E+V GmbH, Germany) (Adapted from Craigie *et al.* (2013).

14.4.3 Optical probes

Optical probes objectively measure fat and muscle depths and are routinely used to predict lean meat percentage in many countries (Price, 1995). Optical probe consists of a light-emitting diode that illuminates the meat and two photodetector diodes, which measure the reflected light. Optical probes are based on the principle that fat usually has a much higher reflectance than muscle, enabling the estimation of fat and proportional muscle yield. The device is placed on the carcass, and the probe is pushed into and out of the carcass. As the probe is withdrawn from the carcass, the amount of light detected by the sensor changes when the optical window passes from muscle to fat. A number of optical probes including Hennessy Grading Probe (HGP), the AUS-Meat Sheep Probe (SASTEK, Hamilton, Queensland, Australia) and the Swedish FTC Lamb Probe (Upplands, Vasby, Sweden) have been developed for grading and predicting the lean yield of carcasses. Such probes can operate at 9–10 carcasses per minute (Kirton *et al.*, 1995; Swatland, 1995a). The HGP has been used for measurement of fat depth in beef carcases (Hopkins, 1989; Phillips, Herrod and Schafer, 1987), pork carcases (Goenaga, Lloveras and Améndola, 2008) and lamb carcases (Garrett *et al.*, 1992; Kongsro *et al.*, 2009). Schinckel *et al.* (2010) reported the suitability of three optical probes (HGP4 Hennessey Grading Probe, Destron-Feering PG-100 probe and Giraldo OPTO-Electronic PG-200 probe) to predict the carcass percentage of five alternative measures of carcass composition (fat-tissue-free lean, lipid-free soft tissue, lipid-free lean, total fat tissue and soft tissue lipid) in barrows and gilts from different genetic populations. Normally, these optical probes are poorly suited for the PSE (pale, soft, exudative) meat because of the diminished reflectance difference between muscle and fat. This problem can be addressed by using automated multiple probe sites coupled with neural networks (Swatland, 1995b).

14.5 Conclusions

While the majority of carcass evaluation techniques were originally developed for clinical applications, many have emerged as powerful non-destructive measurement techniques for the meat sector. Research to date has demonstrated that numerous opportunities exist for their potential application in the meat industry. Carcass evaluation techniques have been successfully proposed and undergone various degrees of development and application in predicting in vivo and ex vivo carcass quality traits and body composition in livestock. Some of the applications discussed have been successfully adopted by the industry while others of these are still at the experimental stages. Technology advances are resulting in more sensitive, reliable, user-friendly and cost-effective solutions with a greater focus on food application. Given these attributes and the potential applications discussed, increased adoption of these approaches by the meat industry for improved efficiency is likely.

References

Allen, P. and McGeehin, B. (2001) Measuring the lean content of carcasses using TOBEC: Teagasc. ISBN 1 84170 235 8

Ballerini, L., Hogberg, A., Borgefors, G., Bylund, A.-C., Lindgard, A., Lundstrom, K., Rakotonirainy, O. and Soussi, B. (2000) Testing MRI and image analysis techniques for fat quantification in meat science. In *Nuclear Science Symposium Conference Record, 2000 IEEE*, vol. 3 (pp. 18/136-118/140 vol. 133): IEEE.

Barigou, M. and Douaire, M. (2013) 9 - X-ray micro-computed tomography for resolving food microstructures. In V. J. M. Groves (Ed.), Food Microstructures, (pp. 246-272): Woodhead Publishing.

Baulain, U. (1997) Magnetic resonance imaging for the in vivo determination of body composition in animal science. *Computers and Electronics in Agriculture*, **17** (2), 189–203.

Berg, E., Forrest, J. and Fisher, J. (1994) Electromagnetic scanning of pork carcasses in an on-line industrial configuration. *Journal of Animal Science*, **72** (10), 2642–2652.

Bernau, M., Kremer, P.V., Lauterbach, E. *et al.* (2015) Evaluation of carcass composition of intact boars using linear measurements from performance testing, dissection, dual energy X-ray absorptiometry (DXA) and magnetic resonance imaging (MRI). *Meat Science*, **104**, 58–66.

Blazek, J. and Gilbert, E.P. (2011) Application of small-angle X-ray and neutron scattering techniques to the characterisation of starch structure: A review. *Carbohydrate Polymers*, **85** (2), 281–293.

Bünger, L., Glasbey, C., Simm, G. *et al.* (2011) *Use of X-ray Computed Tomography (CT) in UK Sheep Production and Breeding*, INTECH Open Access Publisher.

Chen, Q., Zhang, C., Zhao, J. and Ouyang, Q. (2013) Recent advances in emerging imaging techniques for non-destructive detection of food quality and safety. *TrAC Trends in Analytical Chemistry*, **52**, 261–274.

Collewet, G., Bogner, P., Allen, P. *et al.* (2005) Determination of the lean meat percentage of pig carcasses using magnetic resonance imaging. *Meat Science*, **70** (4), 563–572.

Craigie, C.R., Navajas, E.A., Purchas, R.W. *et al.* (2012) A review of the development and use of video image analysis (VIA) for beef carcass evaluation as an alternative to the current EUROP system and other subjective systems. *Meat Science*, **92** (4), 307–318.

Craigie, C.R., Ross, D.W., Maltin, C.A. *et al.* (2013) The relationship between video image analysis (VIA), visual classification, and saleable meat yield of sirloin and fillet cuts of beef carcasses differing in breed and gender. *Livestock Science*, **158** (1–3), 169–178.

Damez, J.-L. and Clerjon, S. (2013) Quantifying and predicting meat and meat products quality attributes using electromagnetic waves: An overview. *Meat Science*, **95** (4), 879–896.

Ellis, J. S. (2005) Animal sorting and grading system using MRI to predict maximum value. In): US Patent 6,877,460.

Fisher, A. (2007) Beef carcass classification in the EU: An historical perspective. *Publication-European Association for Animal Production*, **123**, 19.

Fowlkes, J.B. and Holland, C.K. (2000) Section 4: Bioeffects in tissues with gas bodies. *Journal of Ultrasound in Medicine*, **19** (2), 97–108.

Garrett, R., Edwards, J., Savell, J. and Tatum, J. (1992) Evaluation of the Hennessy grading probe to predict yields of lamb carcasses fabricated to multiple end points. *Journal of Animal Science*, **70** (4), 1146–1152.

Goenaga, P., Lloveras, M. and Améndola, C. (2008) Prediction of lean meat content in pork carcasses using the Hennessy Grading Probe and the Fat-O-Meater in Argentina. *Meat Science*, **79** (3), 611–613.

Grill, L., Ringdorfer, F., Baumung, R. and Fuerst-Waltl, B. (2015) Evaluation of ultrasound scanning to predict carcass composition of Austrian meat sheep. *Small Ruminant Research*, **123** (2–3), 260–268.

Gu, Y., Schinckel, A., Martin, T. *et al.* (1992) Genotype and treatment biases in estimation of carcass lean of swine. *Journal of Animal Science*, **70**, 1708–1718.

Gupta, S., Kumar, A., Kumar, S. *et al.* (2013) Recent trends in carcass evaluation techniques-a review. *Journal of Meat Science and Technology*, **1** (2), 50–55.

Gwartney, B.L., Calkins, C.R., Lin, R.S. *et al.* (1994) Electromagnetic scanning of beef quarters to predict carcass and primal lean content. *Journal of Animal Science*, **72** (11), 2836–2842.

Harrington, G. and Kempster, A. (1989) Improving lamb carcass composition to meet modern consumer demand, in *Reproduction, Growth and Nutrition in Sheep* (eds O.R. Dyrmundsson and S. Thorgeirsson), pp. 79–90.

Hopkins, D. (1989) An evaluation of the Hennessy Grading Probe for measuring fat depth in beef carcasses. *Animal Production Science*, **29** (6), 781–784.

Houghton, P. and Turlington, L. (1992) Application of ultrasound for feeding and finishing animals: a review. *Journal of Animal Science*, **70** (3), 930–941.

Hyánková, L. and Szebestová, Z. (2010) Prediction of body composition by total body electrical conductivity technique is affected by fat reserves of Japanese quail. *Poultry Science*, **89** (11), 2506–2513.

Karamichou, E., Richardson, R.I., Nute, G.R. *et al.* (2006) Genetic analyses of carcass composition, as assessed by X-ray computer tomography, and meat quality traits in Scottish Blackface sheep. *Animal Science*, **82**, 151–162.

Keim, N.L., Mayclin, P.L., Taylor, S.J. and Brown, D.L. (1988) Total-body electrical conductivity method for estimating body composition: validation by direct carcass analysis of pigs. *The American Journal of Clinical Nutrition*, **47** (2), 180–185.

Kiefer, C., Suguisawa, L., Gonçalves, L. *et al.* (2014) Assess of carcass of immuno-castrated pigs by real time ultrasound technique. *Revista Agrarian*, **7** (23), 118–123.

Kirton, A., Mercer, G., Duganzich, D. and Uljee, A. (1995) Use of electronic probes for classifying lamb carcasses. *Meat Science*, **39** (2), 167–176.

Kongsro, J., Røe, M., Kvaal, K. *et al.* (2009) Prediction of fat, muscle and value in Norwegian lamb carcasses using EUROP classification, carcass shape and length measurements, visible light reflectance and computer tomography (CT). *Meat Science*, **81** (1), 102–107.

Kremer, P., Fernandez-Figares, I., Förster, M. and Scholz, A. (2012) In vivo body composition in autochthonous and conventional pig breeding groups by dual-energy X-ray absorptiometry and magnetic resonance imaging under special consideration of Cerdo Iberico. *Animal*, **6** (12), 2041–2047.

Kremer, P., Förster, M. and Scholz, A. (2013) Use of magnetic resonance imaging to predict the body composition of pigs in vivo. animal, 7(06), 879-884.

Kvame, T. and Vangen, O. (2006) In-vivo composition of carcass regions in lambs of two genetic lines, and selection of CT positions for estimation of each region. *Small Ruminant Research*, **66** (1), 201–208.

Lambe, N., Ross, D., Navajas, E. *et al.* (2010) The prediction of carcass composition and tissue distribution in beef cattle using ultrasound scanning at the start and/or end of the finishing period. *Livestock Science*, **131** (2), 193–202.

Lambe, N., Wood, J.D., McLean, K.A., Walling, G.A., Whitney, H., Jagger, S., Fullarton, P., Cesaro, G., Maltin, C.A., Bayntun, J., Glasbey, C.A. and Bünger, L. (2012) Use of computed tomography (CT) in a longitudinal body composition study in pigs fed different diets

In *FARM ANIMAL IMAGING DUBLIN* (Eds C. Maltin, C. Craigie and L. Bünger). EU Cost Action report available online at http://www.cost-faim.eu/FA2487%20FAIM%20Report_%20LoRes%20cm.pdf

Lazzaroni, C., Gigli, S. and Gabiña, D. (2007) *Evaluation of Carcass and Meat Quality in Cattle and Sheep*, Wageningen Academic Publishers.

Marcoux, M., Bernier, J.F. and Pomar, C. (2003) Estimation of Canadian and European lean yields and composition of pig carcasses by dual-energy X-ray absorptiometry. *Meat Science*, **63** (3), 359–365.

Marcoux, M., Faucitano, L. and Pomar, C. (2005) The accuracy of predicting carcass composition of three different pig genetic lines by dual-energy X-ray absorptiometry. *Meat Science*, **70** (4), 655–663.

Mason, T.J. (1998) Power ultrasound in food processing – the way forward, in *Ultrasound in Food Processing* (eds M.J.W. Povey and T.J. Mason), Thomson Science, London, pp. 105–126.

Mery, D., Lillo, I., Loebel, H. *et al.* (2011) Automated fish bone detection using X-ray imaging. *Journal of Food Engineering*, **105** (3), 485–492.

Milerski, M. (2001) In vivo assessment of meatiness and fattiness of Charollais ram-lambs. *Czech Journal of Animal Science*, **46** (6), 275–280.

Mitchell, A., Scholz, A., Wang, P. and Song, H. (2001) Body composition analysis of the pig by magnetic resonance imaging. *Journal of Animal Science*, **79** (7), 1800–1813.

Mitchell, A., Wang, P., Rosebrough, R. *et al.* (1991) Assessment of body composition of poultry by nuclear magnetic resonance imaging and spectroscopy. *Poultry Science*, **70** (12), 2494–2500.

Mitchell, A.D., Scholz, A.M. and Conway, J.M. (1998a) Body composition analysis of pigs from 5 to 97 kg by dual-energy X-ray absorptiometry. *Applied Radiation and Isotopes*, **49** (5–6), 521–523.

Mitchell, A.D., Scholz, A.M. and Pursel, V.G. (2003) Prediction of pork carcass composition based on cross-sectional region analysis of dual energy X-ray absorptiometry (DXA) scans★. *Meat Science*, **63** (2), 265–271.

Mitchell, A.D., Scholz, A.M., Pursel, V.G. and Evock-Clover, C.M. (1998b) Composition analysis of pork carcasses by dual-energy X-ray absorptiometry. *Journal of Animal Science*, **76** (8), 2104–2114.

Mitchell, A.D., Solomon, M.B. and Rumsey, T.S. (1997) Composition analysis of beef rib sections by dual-energy X-ray absorptiometry. *Meat Science*, **47** (1–2), 115–124.

Monziols, M., Collewet, G., Bonneau, M. *et al.* (2006) Quantification of muscle, subcutaneous fat and intermuscular fat in pig carcasses and cuts by magnetic resonance imaging. *Meat Science*, **72** (1), 146–154.

Nielsen, M.S., Christensen, L.B. and Feidenhans'l, R. (2014) Frozen and defrosted fruit revealed with X-ray dark-field radiography. *Food Control*, **39**, 222–226.

Nunes, J., Piquerez, M., Pujadas, L. *et al.* (2015) Beef quality parameters estimation using ultrasound and color images. *BMC Bioinformatics*, **16** (Suppl 4), S6.

Peña, F., Molina, A., Juárez, M. *et al.* (2014) Use of serial ultrasound measures in the study of growth- and breed-related changes of ultrasonic measurements and relationship with carcass measurements in lean cattle breeds. *Meat Science*, **96** (1), 247–255.

Phillips, D., Herrod, W. and Schafer, R. (1987) The measurement of subcutaneous fat depth on hot beef carcasses with the Hennessy grading probe. *Animal Production Science*, **27** (3), 335–338.

Pietrobelli, A., Gallagher, D., Baumgartner, R. *et al.* (1998) Lean R value for DXA two-component soft-tissue model: Influence of age and tissue or organ type. *Applied Radiation and Isotopes*, **49** (5–6), 743–744.

Polak, P., Sloniewski, K., Sakowski, T. *et al.* (2001) In vivo estimates of slaughter value of bulls using ultrasound and body dimensions. *Czech Journal of Animal Science*, **46** (4), 159–164.

Price, M. (1995) Development of carcass grading and classification systems, in *Quality and Grading of Carcasses of Meat Animals* (ed. S.D. Morgan Jones), CRC Press, Boca Raton, 27 p.

Remignon, H., Seigneurin, F. and Moati, F. (2000) In vivo assessment of the quantity of breast muscle by sonography in broilers. *Meat Science*, **56** (2), 133–138.

Rius-Vilarrasa, E., Bünger, L., Maltin, C. *et al.* (2009) Evaluation of Video Image Analysis (VIA) technology to predict meat yield of sheep carcasses on-line under UK abattoir conditions. *Meat Science*, **82** (1), 94–100.

Schinckel, A., Wagner, J., Forrest, J. and Einstein, M. (2010) Evaluation of the prediction of alternative measures of pork carcass composition by three optical probes. *Journal of Animal Science*, **88** (2), 767–794.

Scholz, A.M., Mitchell, A.D., Förster, M. and Pursel, V.G. (2007) Two-site evaluation of the relationship between in vivo and carcass dual energy X-ray absorptiometry (DXA) in pigs. *Livestock Science*, **110** (1–2), 1–11.

Shelton, J., Southern, L., Bidner, T. *et al.* (2003) Effect of microbial phytase on energy availability, and lipid and protein deposition in growing swine. *Journal of Animal Science*, **81** (8), 2053–2062.

Silva, S., Afonso, J., Santos, V. *et al.* (2006) In vivo estimation of sheep carcass composition using real-time ultrasound with two probes of 5 and 7.5 MHz and image analysis. *Journal of Animal Science*, **84** (12), 3433–3439.

Simeonova, M., Todorov, N. and Schinckel, P. (2012) In vivo-methods for quantitative measurement of protein deposition rate in animals with emphasize on swine. *Bulgarian Journal of Agricultural Science*, **18**, 455–481.

Stanford, K., Jones, S. and Price, M. (1998) Methods of predicting lamb carcass composition: A review. *Small Ruminant Research*, **29** (3), 241–254.

Suster, D., Leury, B.J., Ostrowska, E. *et al.* (2003) Accuracy of dual energy X-ray absorptiometry (DXA), weight and P2 back fat to predict whole body and carcass composition in pigs within and across experiments. *Livestock Production Science*, **84** (3), 231–242.

Swatland, H. (1995a) *On-line Evaluation of Meat*, CRC Press.

Swatland, H.J. (1995b) Objective assessment of meat yield and quality. *Trends in Food Science & Technology*, **6** (4), 117–120.

Szabo, C., Babinszky, L., Verstegen, M.W.A. *et al.* (1999) The application of digital imaging techniques in the in vivo estimation of the body composition of pigs: a review. *Livestock Production Science*, **60** (1), 1–11.

Teixeira, A., Joy, M. and Delfa, R. (2008) In vivo estimation of goat carcass composition and body fat partition by real-time ultrasonography. *Journal of Animal Science*, **86** (9), 2369–2376.

Teixeira, A., Matos, S., Rodrigues, S. *et al.* (2006) In vivo estimation of lamb carcass composition by real-time ultrasonography. *Meat Science*, **74** (2), 289–295.

Wood, J., Newman, P., Miles, C. and Fisher, A. (1991) Video image analysis: Comparisons with other novel techniques for carcass assessment. In *Proceedings of a symposium on the electronic evaluation of meat in support of value-based marketing*, (pp. 145-169).

CHAPTER 15
Meat authenticity

Yan Zhao[1,2]

[1] *Institute of Quality Standard & Testing Technology for Agro-Products, Key Laboratory of Agro-product Quality and Safety, Chinese Academy of Agricultural Sciences, Beijing, China*

[2] *Key Laboratory of Agro-product Quality and Safety, Ministry of Agriculture, Beijing, China*

15.1 Introduction

Meat authenticity is a major concern not only for consumers but also for farmers, retailers, and government authorities. Meat authenticity, including traceability, is of particular importance for consumer protection due to meat incidents such as BSE (bovine spongiform encephalopathy), FMD (foot-and-mouth disease), and avian influenza. New regulations, and in particular the newly issued European regulation 178/2002, applicable on January 1, 2005, imposed the requirement for the authenticity of agroproducts (Kelly *et al.*, 2005). Meat is a global product, which is traded between regions, countries, and continents. In order to ensure consumers receive a high-quality product that is free from contamination, diseases, and fraudulent geographic origin, many techniques need to be developed and applied in the production, transportation, and retailer stages. In this chapter, research and applications of innovative technologies to meat products are introduced.

15.2 Theory of authenticity in the meat industry

Meat authenticity has focused on several areas: identification of geographical origin of meat, affirmation of animal species, and the feeding regime differentiation. There exist two categories of techniques to achieve these aims. One main technique to authenticate the meat is electronic informative coding techniques. It records the information of animal source and species, slaughter management, process management, distribution management, and sales management (Zhao *et al.*, 2012). All of the information from field to table are recorded and coded into tags by which the meat product can be traced (Yan *et al.*, 2013). However, traceability methods based only on batch codes or paper documents cannot always be trusted. Another effective technique of authenticate meat is analytical

Emerging Technologies in Meat Processing: Production, Processing and Technology, First Edition.
Edited by Enda J. Cummins and James G. Lyng.

technologies including physical, chemical, and biological approaches. The merit of these analytical methods is that the information of producing, processing, and storage can be obtained without knowing the background of the meat product (Osorio *et al.*, 2011a,b; Zhao *et al.*, 2013a; Zhao *et al.*, 2014).

15.2.1 Significance of the application of authenticity techniques on meat

In the last few years, meat authenticity issues have increased in importance due to the consumers' interest to meat quality and safety. The consumer's lack of confidence, especially about meat of animal origin, is due to many reasons including economic changes and the influence of food safety incidents. BSE has been one of the most serious food safety incidents of the past number of years, leading to destructive damage to the reputation of the beef industry (Osorio *et al.*, 2011a,2011b). This was followed by the avian influenza scare in poultry. The more recent horse meat event has brought to the fore the issue regarding traceability techniques and authenticity systems (Zhang *et al.*, 2012). Traceability is considered an important program that supports and builds a strong foundation for meat safety management programs. The requirement for traceability has been enforced by EU law and demands food manufacturers and retailers implement the system as part of their food safety procedure (Jin and Zhou, 2014). In China, the traceability system of agroproduct quality needs to be implemented since the law on quality and safety of agricultural products enacted in October 2006 (Zhao *et al.*, 2013b). In Japan, the traceability of beef is mandatory since December 2004 to improve consumer confidence in meat safety. The Law for Special Measures Concerning the Management and Relay of Information for Individual Identification of Cattle was enacted in 2004 (Jin and Zhou, 2014). Meanwhile, meat traceability systems can reduce the anxiety of consumers regarding safety by serving as a form of quality assurance (Zhang *et al.*, 2012). A number of researches have gathered data regarding consumers' willingness to buy traceable meat (Zhang *et al.*, 2012; Bu *et al.*, 2013; Jin and Zhou, 2014). A lot of studies have looked at the aspects of information the consumers are interested in. Date of production and production certification have a high ratio of respondents among those investigated. The government also encourages the company to establish food traceability systems that can protect their own brands. For meat, the species and the geographical origination are important and essential for government administration, farmers, and consumers.

15.2.2 Content of authenticity of meat

15.2.2.1 Geographical origin of meat

Analytical method

With regard to meat, food incidents including BSE and FMD indicate that geographic origin is an important risk factor and a factor in influencing consumer purchasing interests. Researchers developed several techniques (stable isotope (Boner and Forstel, 2004; Franke *et al.*, 2008a; Perini *et al.*, 2009; Guo *et al.*, 2010), multi-element (Kelly *et al.*, 2005; Guo *et al.*, 2007; Sun *et al.*, 2011a,b), near infrared (Sun *et al.*, 2011a,b)) to identify the geographical origin of meat. In this section, we concentrate on the stable isotope technique, which is a newly emerged method and can be carried out rapidly and effectively.

The first application of stable isotope analysis was investigated in European beef due to the BSE crisis. The single $\delta^{18}O°$ value of tissue water can be clearly separated from beef samples in Germany ($n = 175$) and Argentina ($n = 42$). However, $\delta^{18}O°$ values did not provide resolution to discriminate the United Kingdom samples from German and Argentina beef (Hegerding *et al.*, 2002). Furthermore, the beef samples collected from three different locations on farms in German could be geographically differentiated by using the stable isotopes of nitrogen and sulfur of defatted protein. Nakashita *et al.* (2008) detected the carbon, nitrogen, and oxygen isotope signatures of beef defatted protein from Australia, Japan, and the United States. Defatted dry matter of beef from the United States was characterized by higher $\delta^{13}C$ values than that from Japan and Australia. Samples from Australia were characterized by higher $\delta^{18}O$ than that from Japan and the United States (Nakashita *et al.*, 2008). An experiment to compare beef samples originating from the major cattle producing regions of the world has been conducted by the British Institute of Food Research. They determined the C and N isotope composition of defatted dry mass and H and O isotope composition of lipids. The results showed that ^{13}C content of beef produced in Brazil and the United States was significantly higher than that of British beef, for the reason that intensive maize and C_4 pasture were fed to the cattle from Brazil and the United States while the British beef are fed predominantly on C_3 pasture and fodder. The mean $\delta^{2}H°$ and $\delta^{18}O°$ values of beef lipid correlated well with the latitude of production regions and the relationship between the H and O isotopic contents were found to parallel the Meteoric Water Line (Heaton *et al.*, 2008). In the recent report, beef samples from Korea and some of the main beef exporting countries (the United States, Mexico, Australia, and New Zealand) were examined for carbon, nitrogen, and hydrogen isotope ratios of the defatted dried mass. The carbon isotope composition of the samples showed that New Zealand beef had the most depleted values, followed by Australian beef samples; Korean samples ranged from −19.5° to −15.5°, Mexican from −15.5° to −13.5°, and American beef showed the most enriched values. The hydrogen isotope composition slightly overlapped for all investigated countries apart from Australia (Horacek and Min, 2010). Examination of variation in carbon and nitrogen stable isotope ratios in cattle tissues from different provinces in China was done by Chinese scientists (Guo *et al.*, 2010; Zhao *et al.* 2013).

Multi-element (C, N, H, S) isotope analysis of beef muscle was used as an analytical tool to verify the geographical origin from several European and non-European countries. The mean isotopic values from different countries had highly significant differences, and 84.9% of the samples were correctly assigned to the country of origin by discriminate analysis (Osorio *et al.*, 2011a,2011b). The four stable isotope ratios (C, N, H, S) were also used as an analytical tool to verify the preslaughter diet of beef (Osorio *et al.*, 2011a,2011b). The ratios of stable isotope provided useful information to confirm lamb meat origin. Multi-element (H, C, N, O, S) stable isotope characteristics of lamb meat from different Italian regions were researched (Perini *et al.*, 2009). δ^2H and $\delta^{18}O$ values of the defatted dry matter were significantly correlated with the corresponding values in meteoric waters, thus it enabled a trace of the variability of geoclimatic factors. Geographical origin of Apulian lamb meat by stable isotope was done by Sacco in 2005(Sacco *et al.*, 2005). Twenty-five lamb meat samples from three areas located in Apulia were analyzed for stable isotope ratios, major elements, and trace metals. ^1H NMR spectra were also applied to the samples. $\delta^{13}C$ values were significantly different between five sample groups divided according to the lamb type and area. The utilization of multivariate statistical analysis to data containing metals and NMR spectral data combining with isotope ratios could predict correct geographical origin. The authenticity and traceability research of pork and poultry are less than those of beef and lamb. Recently there were some reports about the application of stable isotopes to trace poultry. In Japan, $\delta^{13}C$ of poultry did not differ among muscle tissue samples from different body parts, but the values clearly varied among samples from China and the United States (Sakamoto *et al.*, 2002). Franke *et al.* distinguished the poultry country origin by $\delta^{18}O$; however, $^{87}Sr/^{86}Sr$ ratio did not give a sufficient differentiation by geographic origin of poultry (Franke *et al.* 2008).

Tag method

In order to meet the demands of government supervision of meat production safety, tag methods including radio frequency identification (RFID), intelligent PDA mobile inquiring and reading, GPRS, Internet, and other information technologies were developed (Zhao *et al.*, 2009; Yan *et al.*, 2013). With a wide range of applications of information technology, the popularity of RFID technology in tracking the geographical origin of meat products is increasing. RFID tags usually record key traceability information in the supply chain from breeding, slaughter, processing, and transportation. The platform meets with the requirements of tracking, traceability, recall, and monitoring of products in the meat supply chain with the participation of consumers, enterprises, and the government. So the geographical origin can be traced by the system. Consumers are the main users and they can get information from the product that enhances the desirability and confidence in the products. A RFID-based traceability system for beef product was developed in China (Feng *et al.*, 2013). The system provided full information in the supply chain of beef with real-time, accurate data on the process. Apart from beef, recording of pork and poultry meat supply chains were also proposed by RFID (Ma *et al.*, 2008; Zhao *et al.*, 2009; Chen, 2012; McMillin *et al.*, 2012).

15.2.2.2 Feeding regime

Feedstuff source

The carbon isotope composition provides information about the animal feed. The cattle from the Jilin Province of China were mainly fed on maize, which is a C_4 plant, and the C stable isotope value was higher than that of other provinces where cattle were mainly fed on C_3 pasture grass (Guo *et al.*, 2010). Bahar *et al.* investigated the effect of substituting grass silage with maize silage on the carbon and nitrogen stable isotope composition of beef. They found that the $\delta^{13}C$ value of cattle fed with maize silage resulted in a less negative value. So ratio analysis of the carbon stable isotope of beef could be used to quantify C_3/C_4 dietary constituents (Bahar *et al.*, 2005). Muscle and tail hair samples were collected from animals fed either pasture, a barley-based concentrate, silage followed by pasture, or silage followed by pasture with concentrate for 1 year. The $^{13}C/^{12}C$ and $^{34}S/^{32}S$ were the best indicators for authentication of beef from animals consuming the different diets. Analysis of $^{13}C/^{12}C$ and $^{15}N/^{14}N$ in tail hair sections provided an archival record of changes to the diet of the cattle for periods of over 1-year preslaughter (Osorio *et al.*, 2011a,b). Piasentier *et al.* evaluated the stable isotope ratios (C and N) in lamb meat as a method of origin authentication. Analyses were developed on meat from 12 lamb types and divided in three groups by feeding regime: suckled milk only, pasture and supplementation containing maize grain. The results showed that $^{13}C/^{12}C$ of crude fat and protein mainly reflected the feeding regime, and $^{15}N/^{14}N$ showed significant differences between lamb types (Piasentier *et al.*, 2003). Another study focused on the feed regime of lambs (Moreno-Rojas *et al.*, 2008). Lambs were divided into three groups: 14 lambs were fed vetch, 14 lambs received a barley-based concentrate, and 6 lambs were fed maize-based concentrate. The results showed that for lambs that were fed herbage, the muscle $\delta^{13}C$ values were higher and $\delta^{15}N$ values were lower than those of the lambs receiving concentrates. The $\delta^{15}N$ and $\delta^{13}C$ values in the muscle and $\delta^{13}C$ values in the adipose tissue showed correct discrimination between the lambs fed three different diets.

Organic and conventional feed regime

Organic meat is gradually increasing in popularity. With the development of the economy, organic pork products have received increasing attention in both developed and developing countries. Organic animal rearing system is on the basis of the standards that absolutely avoid genetically modified organisms, animal by-products, and chemical fertilizer. The system should also offer enough space indoors with a lower intensity and free access to the outdoors. Therefore, organic animal products are considered to be healthier and safer products compared to the traditional intensive products because the organic pork are environmentally friendly products.

$\delta^{13}C°$ value above −20° was suggested as a limit for differentiating between conventional and organically beef in Germany (Boner and Forstel, 2004). Bahar *et al.* investigated the influence of season variation in the C, N, and S stable isotope composition of organic and conventional Irish beef. They found that the $\delta^{13}C$ values for conventional beef gradually became less negative from December to June; the $\delta^{13}C$ value of organic beef was only slightly less negative from March to May. The data revealed that $\delta^{13}C$ values of conventional and organic beef showed significant differences during winter, spring, and summer. Observing $\delta^{15}N$ data by season showed that $\delta^{15}N$ values of organic and conventional samples differed during spring, summer, and autumn. So the authors suggested that there was a need to consider possible seasonal variation when applying stable isotopes for the authentication of organic beef (Bahar *et al.*, 2008). Nitrogen isotopic composition of animal tissue also reflects cattle diet. From the data of $\delta^{15}N$ values, we can expect enriched $^{15}N/^{14}N$ in the organic meat, as the nitrogen isotope composition of manure is higher in the $^{15}N/^{14}N$ ratio compared to mineral fertilizer (Boner and Forstel, 2004). Apart from that, for common beef, a low $\delta^{15}N$ value is speculated about feed containing clover (or other leguminous plant feed), which has a very low $\delta^{15}N$ ratio of about 0° (Horacek and Min, 2010).

15.2.2.3 Species identification

Recently, an incident in the British meat industry, with horse adulteration in beef, resulted in a serious problem in the meat industry. In China, fox, mink, and rat meat was sold as lamb after processing with chemical additives. These food incidents put an important issue back in focus with regard to safety of meat species. Nowadays, many techniques based on morphological classification, protein, and DNA analysis have been established to enhance consumers' confidence in meat safety. However, the method based on the morphology is not suitable due to the counterfeiting technology of adding pigments, water, and glutin. Techniques on the basis of protein including electrophoresis, enzyme-linked immunosorbent assay (ELISA), high-performance liquid chromatography-mass spectrometry (HPLC-MS) cannot differentiate the species when the meat is processed by the high temperature (Rodriguez-Ramirez *et al.*, 2011). Nucleic acid-based molecular techniques for the differentiation of the meat species have attracted the increased attention of researchers. Currently, techniques based on the polymerase chain reaction (PCR) are proposed for the identification of meat species (Zarringhabaie *et al.*, 2011; Rogberg-Munoz *et al.*, 2013). Early methods based on PCR techniques combined with gel electrophoresis (Mafra *et al.*, 2008; Rodriguez-Ramirez *et al.*, 2011), as random amplified polymorphic DNA (RAPD) (Koh *et al.*, 1998; Mane *et al.*, 2008), single sequence repeats (SSRs) (Jaayid, 2013), multiplex PCR (Md. Eaqub *et al.*, 2014; Nejad *et al.*, 2014), and so on. Recently, the PCR method was improved by the introduction of real-time PCR for DNA quantification and species differentiation (Bajzik *et al.*, 2011; Farrokhi and Joozani, 2011). Real-time PCR does not need the use of gel electrophoresis; meanwhile, species-mixed meat and heavily processed products can also be analyzed (Soares *et al.*, 2013). A common adulteration of meat products is the addition of pork to beef, which is more expensive. In order to protect the consumers who refuse pork product because of the medical or halal verification, real-time PCR approach was utilized. A multiplex real-time PCR method was investigated to quantify beef and pork in minced meat (Iwobi *et al.*, 2015). The multiplex real-time PCR method also realized the discrimination of deer, cattle, goat, duck, horse, donkey, pig, turkey, and chicken (Kesmen *et al.*, 2012; Cheng *et al.*, 2014; You *et al.*, 2014). The method shows that the limit of detection for adulterated meat is at the nanogram level, or even lower, implying that it is sensitive and specific for meat species identification and can be used for high-throughput demand for meat authentication. However, the results of measurement uncertainty are within a wide range from 1.8–124% (Koeppel *et al.*, 2013). In order to improve the measurement uncertainty, further study needs to be done.

15.2.3 Factors affecting accuracy

15.2.3.1 Tag methods

RFID technology applies tags paste to animal or organ/tissue to record real-time information to trace the meat production and supply chain. An antenna, interrogator, and transponder make up the whole system. The tags contain all the information about the animals' health, feed source, slaughter plant, and distribution site (Sun *et al.*, 2007). However, some drawbacks exist when using the system. The RFID technology increases the tag cost because a special reading device is needed to read the information, which may increase the product's price. Compared to paper barcode, RFID can store large amounts of information, which lead to higher costs (Chrysochou *et al.*, 2009). It was reported that the cost was 40 CNY for tracing one ox or cow. Most of consumers complained that this was too expensive (Feng *et al.*, 2013). In addition, the tag has the chance to be damaged due to the environmental changes: metals and water interfere with the label. Whether the reading device has the authorization to obtain the tags' information is another problem that may result in privacy and legal issues.

15.2.3.2 Analytical method

There are many analytical techniques to trace the geographical origin and feed regime of animals. One of the main factors to affect accuracy is the amount of samples. In order to identify the origin of meat, a large number of samples as possible should be gathered to set up a reliable database. Then we can get useful information after analyzing the samples, and key parameters can be screened by the mathematical statistics (Perini *et al.*, 2009; Osorio *et al.*, 2011a,b; Zhao *et al.*,

2013b). Geographical range is another factor to influence the accuracy rate of distinguishing the meat. Meanwhile, environmental factors involving soil type, climate, altitude, and latitude should be taken into consideration when using stable isotope and multi-element analysis (Zhao *et al.*, 2014). In addition, tissue types, collection time, processing, and packing methods of meat products should also be characterized (Franke *et al.*, 2005). When tracing meat origin, animals fed with similar feeds can be classified as same geographical site, this can especially happen with pork traceability. Another disadvantage of analyzing methods is the expensive cost of analytical equipment and high requirement for data manipulation.

15.3 Authenticity methods

15.3.1 Case studies – authenticity applications in meat

Currently, pork production is mainly carried out on an intensive indoor system. Organic pork draws consumers' attention because organic pig-rearing farms do not allow genetically modified organisms or chemical fertilizer. In order to classify the two kinds of pork, we collected 18 samples from supermarket and did the chemical analysis including four kinds of stable isotope (C, N, H, and O) and 17 kinds of mineral elements. The detection method was according to the reference method of our research group (Zhao *et al.*, 2013b). The statistical analysis of the data was performed using the SPSS 22.0 version. In order to reduce the dimensionality of the data set and to describe all the variability of the system using a smaller number of variables, principal component analysis (PCA) was used. The results of stable isotope and multi-element are shown in Tables 15.1 and 15.2. It was shown that four isotope values of two kinds of pork were different due to the feeds. Based on the combination of mineral element and stable isotope data, the PCA results showed that a better distinction between organic and conventional meat can be obtained (Figure 15.1).

15.3.2 Brief overview of authenticity equipment

15.3.2.1 Near-infrared spectroscopy (NIR)

Near infrared (NIR) is a spectroscopy technique. Its radiation (750–2500 nm) is between the visible and mid-infrared spectrum. NIR spectroscopy has several advantages over other analytical techniques: rapidity of analysis, no use of chemicals, minimal or no sample preparation, easily applicable in different work environments (on/in/at line applications). Certain molecular bonds (such as O—H, N—H, and C—H) respond to NIR radiation, and a fingerprint spectrum can be generated. Based on previous data results on the authentication of meat products, instances of meat fraud can be detected after mathematically processing spectral data by chemometric tools. NIR spectroscopy has some advantages

Table 15.1 The stable isotope results for organic and conventional pork.

Sample	Organic pork				Conventional pork			
	C (‰)	N (‰)	H (‰)	O (‰)	C (‰)	N (‰)	H (‰)	O (‰)
No. 1	−17.00	3.20	−117.41	9.49	−15.43	2.62	−118.32	8.92
No. 2	−17.06	3.31	−120.10	9.53	−21.05	2.15	−119.42	7.44
No. 3	−14.89	4.64	−122.41	10.34	−16.53	2.64	−118.48	8.59
No. 4	−17.64	3.39	−120.06	11.06	−15.99	3.22	−114.61	9.47
No. 5	−16.86	3.30	−120.63	10.72	−18.79	2.17	−111.29	10.14
No. 6	−16.50	3.68	−123.58	9.43	−17.63	1.20	−117.07	8.89
No. 7	−16.77	3.17	−126.17	10.57	−18.73	2.11	−119.74	8.83
No. 8	−16.62	3.43	−122.49	10.51	−17.74	2.79	−114.88	10.91
No. 9	−17.24	3.56	−125.49	11.90	−20.39	2.32	−118.93	10.36
Average	−16.73	3.52	−122.04	10.39	−18.03	2.36	−116.97	9.28

Table 15.2 The multi-element result of organic and conventional pork.

Conventional pork	Average (ppb)	Organic pork	Average (ppb)
^{23}Na	1,679.69	^{23}Na	1,798.52
^{24}Mg	1,162.67	^{24}Mg	1,157
^{27}Al	7.78	^{27}Al	8.57
^{39}K	12,601.3	^{39}K	14,425.56
^{44}Ca	1,074.98	^{44}Ca	1,133.7
^{51}V	189.81	^{51}V	208.52
^{55}Mn	326.06	^{55}Mn	356.41
^{57}Fe	27,985.19	^{57}Fe	33,070.37
^{65}Cu	1,779.94	^{65}Cu	2,119.44
^{66}Zn	31,548.33	^{66}Zn	42,057.41
^{69}Ga	25.87	^{69}Ga	43.56
^{75}As	1,247.22	^{75}As	1,222.67
^{82}Se	528.07	^{82}Se	823.57
^{88}Sr	515.63	^{88}Sr	547.44
^{90}Zr	11.44	^{90}Zr	12.31
^{95}Mo	1,945.93	^{95}Mo	1,946.11
^{118}Sn	154.56	^{118}Sn	141.39

compared to other analytical methods: fast speed of detection, nondestruction of samples, and is easily applicable by a technician. NIR is helpful to trace the geographical origin of meat. It was used to identify the origin of lamb meat from five areas of China. Linear discriminant analysis based on two methods including principal component analysis and partial least squares was investigated. Both of the statistics methods could classify the lamb meat from pasturing and farming area with a 100% correct classified rate (Sun *et al.*, 2011a,b; Sun

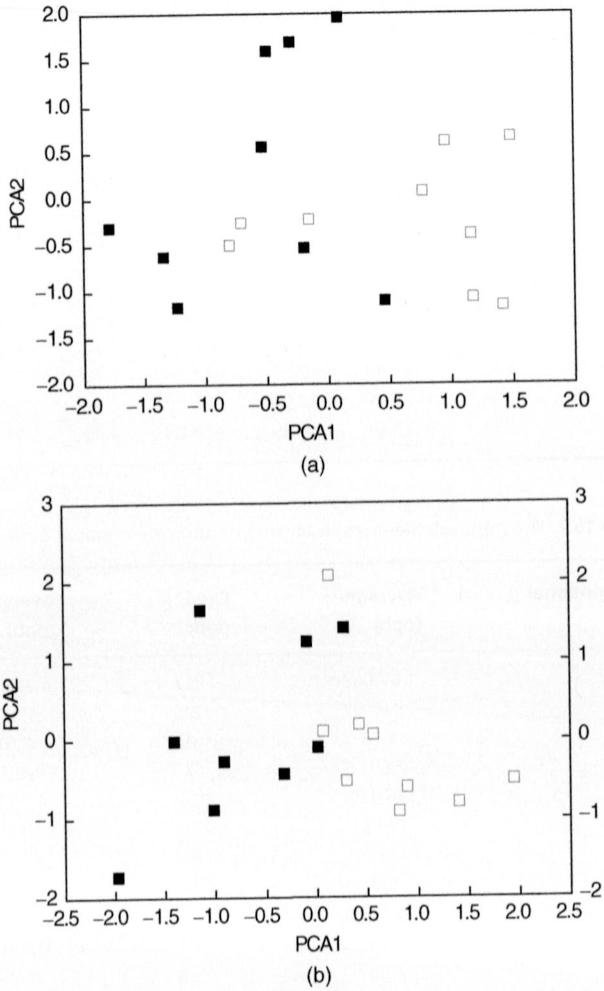

Figure 15.1 Principal component analysis of (a) mineral element in pork from organic and conventional farms. (b) Stable isotope in pork from organic and conventional farms. (c) Combine mineral element and stable isotope from organic and conventional farms. □ Organic pork, ■ Conventional pork. (*See color plate section for the color representation of this figure.*)

et al., 2012a,b). The successful application of NIR to trace the dried beef originating from different countries was realized with 78% correct classification rate (Franke *et al.*, 2006b). On the other hand, NIR is more applicable in the authenticity of meat, involving species identification and feed regime differentiation. A total of 230 samples from five kinds of meat (chicken, turkey, pork, beef, and lamb) were collected and scanned by visible and near-infrared spectra. It was difficult to distinguish chicken and turkey; however, 85–100% correct identification was achieved by three chemometric methods (McElhinney *et al.*, 1999).

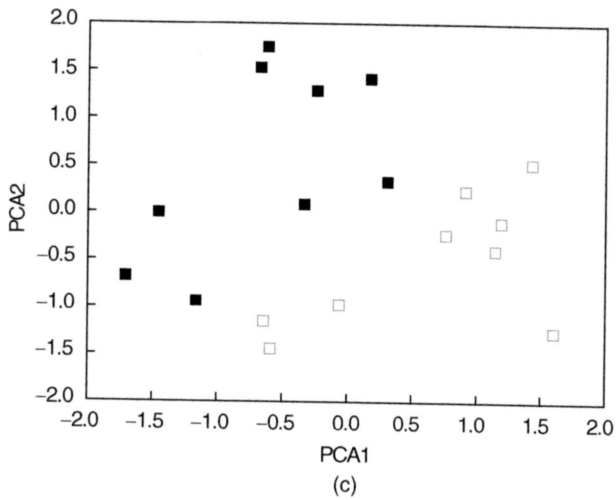

Figure 15.1 (*Continued*)

Species adulteration also happens in seafood due to economic benefits. Gayo and Hale established a method based on visible and NIR to detect and quantify species authenticity in crabmeat samples. The addition of 10% blue swimmer crabmeat into high-value Atlantic blue crabmeat could be determined (Gayo and Hale, 2007). However, some disadvantages hindered the development of NIR for meat authenticity: for example, low capacity to predict a compound at low content, high demands to treat the huge amount of data generated, and difficulty to transfer the results among instruments from different brands and labs.

15.3.2.2 Nuclear magnetic resonance (NMR)

Nuclear magnetic resonance spectroscopy is one of the instruments used by chemists to identify compound molecular structures (Marcone *et al.*, 2013). Currently, NMR spectroscopy has been used to obtain metabolite data of a series of food including fruit juice, wine, milk, and beer (Gall and Colquhoun, 2003). The NMR data combined with multivariate analysis can provide a nontargeted method to differentiate food origin. Concerning meat, ^{1}H high resolution magic angel spinning nuclear magnetic resonance (HR-MAS NMR) was utilized to identify meat geographical origin. Twenty-three raw beef samples from five places (Australia, Brazil, Canada, Switzerland, and the United States) were selected. The results showed that fat and specific metabolite could be indicators of origin (Shintu *et al.*, 2007). Two-dimensional NMR HSQC, HMBC, and TOCSY coupled with ^{1}H NMR spectra were applied as nontargeted and targeted methods to trace beef origin from four countries: Australia, Korea, New Zealand, and the United States. ^{1}H NMR spectra combined with PCA and orthogonal projection to latent structure-discriminant analysis (OPLS-DA) showed clear separation between the countries. Two-dimensional NMR spectra identified the special compounds as various amino acids. So NMR methods can provide characteristic fingerprints of meat, and biomarkers can also be verified to discriminate the geographical origin (Jung *et al.*, 2010). The application of NMR in cattle production systems was developed by Osorio *et al.* (2012). Urine and muscle samples of cattle raised on outdoor pasture, a barley-based concentrate indoor, silage followed by pasture outdoor, and silage followed by pasture outdoor with concentrate over 1 year were selected. The results showed that it was possible to separate the production system according to the NMR, and potential markers were identified. NMR techniques were also used in fish differentiation (Melis *et al.*, 2014). Lipid extract of farmer and wild-gilthead sea breams were screened by NMR. The results show that polyunsaturated fatty acids, especially n-3 fatty acids, play an important role in differentiating the farmed fish from wild counterparts.

15.3.2.3 Inductively coupled plasma mass spectrometry (ICP-MS)

There are some instruments used for multi-element analysis: flame atomic absorption spectrometry (FAAS), inductively coupled plasma atomic emission spectrometry (ICP-AES), inductively coupled plasma mass spectrometry (ICP-MS), and so on. Among several techniques for element analysis, ICP-MS possesses higher sensitivity compared to common methods (Tokalioglu, 2012). Franke *et al.* applied element analysis to trace beef and poultry from different countries. Twenty-five poultry breast samples from 6 countries and 23 beef samples from 5 countries were collected for analyzing 66 and 46 elements, respectively. The results showed that four kinds of elements (As, Na, Rb, and Tl) in poultry and 18 kinds of elements were significantly different among the samples (Franke *et al.*, 2006a). Stable isotope analysis coupled with mineral elements is a more useful tool to identify the meat origin. The multi-element fingerprints data could provide the key factors after multivariate analysis (Heaton *et al.*, 2008). Element fingerprint combined with multivariate statistical analysis was effective in evaluating the geographical traceability of mutton in China. The results showed 100% accuracy to differentiate the mutton samples from agricultural regions and pastoral regions. It indicated that a multi-element fingerprint method is of high potential use to verify the feed source of animals (Sun *et al.*, 2010; Sun *et al.n* 2012a,b).

15.4 Future role for authenticity in food

In the past number of years, numerous food safety incidents happened throughout the world. In order to improve the situation, government departments have made policies to guarantee food safety. Meanwhile, researchers developed many relevant techniques to resolve the problems. The demand for good quality food overrides basic feeding needs. Protected designation of origin (PDO) and protected geographical indication (PGI) are used to protect the reputation of products and economic benefits to farmers. Organic food and ecological food are also emerged in the food markets; however, adulteration of food with food of false geographical origin and feed regime disturb normal economic activities. Effective, reliable, low-cost techniques for food authenticity are required. On the one hand, tag method is a basic management tool to trace food origin and considers geographical origin and quality traits. On the other hand, analytical methods are complementary and essential when no background information can be provided. Furthermore, certain techniques that can combine the tag method together with chemical analysis will be a new research direction and of high potential. Also the development of Internet tools will be critical, and information technology will play a more important role in the food authenticity.

Reference

Bahar, B., Monahan, F.J. *et al.* (2005) Alteration of the carbon and nitrogen stable isotope composition of beef by substitution of grass silage with maize silage. *Rapid Communications in Mass Spectrometry*, **19** (14), 1937–1942.

Bahar, B., Schmidt, O. *et al.* (2008) Seasonal variation in the C, N and S stable isotope composition of retail organic and conventional Irish beef. *Food Chemistry*, **106** (3), 1299–1305.

Bajzik, P., Golian, J. *et al.* (2011) Authentication of food of animal origin using real-time PCR method. *Folia Veterinaria*, **55** (Supplement 2), 7–9.

Boner, M. and Forstel, H. (2004) Stable isotope variation as a tool to trace the authenticity of beef. *Analytical and Bioanalytical Chemistry*, **378** (2), 301–310.

Bu, F., Zhu, D. *et al.* (2013) Research on the consumers' willingness to buy traceable pork with different quality information: a case study of consumers in Weifang, Shandong province. *Asian Journal of Agricultural Research*, **5** (5), 121–124.

Chen, G. (2012) Radio frequency identification technology in applications of traceable poultry food safety. *Food Research and Development*, **33** (11), 218–220.

Cheng, X., He, W. *et al.* (2014) Multiplex real-time PCR for the identification and quantification of DNA from duck, pig and chicken in Chinese blood curds. *Food Research International*, **60**, 30–37.

Chrysochou, P., Chryssochoidis, G. *et al.* (2009) Traceability information carriers. The technology backgrounds and consumers' perceptions of the technological solutions. *Appetite*, **53** (3), 322–331.

Farrokhi, R. and Joozani, R.J. (2011) Identification of pork genome in commercial meat extracts for Halal authentication by SYBR green I real-time PCR. *International Journal of Food Science and Technology*, **46** (5), 951–955.

Feng, J., Fu, Z. *et al.* (2013) Development and evaluation on a RFID-based traceability system for cattle/beef quality safety in China. *Food Control*, **31** (2), 314–325.

Franke, B.M., Gremaud, G. *et al.* (2005) Geographic origin of meat - elements of an analytical approach to its authentication. *European Food Research and Technology*, **221** (3–4), 493–503.

Franke, B.M., Haldimann, M. *et al.* (2006a) Indications for the applicability of element signature analysis for the determination of the geographic origin of dried beef and poultry meat. *European Food Research and Technology*, **225** (3–4), 501–509.

Franke, B.M., Koslitz, S. *et al.* (2008a) Tracing the geographic origin of poultry meat and dried beef with oxygen and strontium isotope ratios. *European Food Research and Technology*, **226** (4), 761–769.

Franke, B.M., Ziolko, T. *et al.* (2006b) Investigation of the determination of the geographic origin of dried beef using NIR spectroscopy. *Mitteilungen aus Lebensmitteluntersuchung und Hygiene*, **97** (5), 345–347.

Franke, K., Schoene, F. *et al.* (2008b) Influence of dietary iodine on the iodine content of pork and the distribution of the trace element in the body. *European Journal of Nutrition*, **47** (1), 40–46.

Gall, G. l. and I. J. Colquhoun (2003). NMR spectroscopy in food authentication.

Gayo, J. and Hale, S.A. (2007) Detection and quantification of species authenticity and adulteration in crabmeat using visible and near-infrared spectroscopy. *Journal of Agricultural and Food Chemistry*, **55** (3), 585–592.

Guo, B.L., Wei, Y.M. *et al.* (2007) Determination of beef geographical origin based on multi-element analysis. *Scientia Agricultura Sinica*, **40**, 2842–2847, (in Chinese).

Guo, B.L., Wei, Y.M. *et al.* (2010) Stable C and N isotope ratio analysis for regional geographical traceability of cattle in China. *Food Chemistry*, **118** (4), 915–920.

Heaton, K., Kelly, S.D. *et al.* (2008) Verifying the geographical origin of beef: The application of multi-element isotope and trace element analysis. *Food Chemistry*, **107** (1), 506–515.

Hegerding, L., Seidler, D., Danneel, H.J. *et al.* (2002) Oxygen isotope-ratio-analysis for the determination of the origin of beef. *Fleischwirtschaft*, **82** (4), 95–100.

Horacek, M. and Min, J.-S. (2010) Discrimination of Korean beef from beef of other origin by stable isotope measurements. *Food Chemistry*, **121** (2), 517–520.

Iwobi, A., Sebah, D. *et al.* (2015) A multiplex real-time PCR method for the quantification of beef and pork fractions in minced meat. *Food Chemistry*, **169**, 305–313.

Jaayid, T.A. (2013) Rapid and sensitive identification of horse and donkey meat in Iraqi markets using SSR and PCR-RFLP based on mitochondrial DNA cytochrome B gene. *Journal of Agricultural Science and Technology A*, **3** (11), 896–903.

Jin, S. and Zhou, L. (2014) Consumer interest in information provided by food traceability systems in Japan. *Food Quality and Preference*, **36**, 144–152.

Jung, Y., Lee, J. *et al.* (2010) Discrimination of the geographical origin of beef by H-1 NMR-based metabolomics. *Journal of Agricultural and Food Chemistry*, **58** (19), 10458–10466.

Kelly, S., Heaton, K. *et al.* (2005) Tracing the geographical origin of food: The application of multi-element and multi-isotope analysis. *Trends in Food Science & Technology*, **16** (12), 555–567.

Kesmen, Z., Yetiman, A.E. *et al.* (2012) Detection of chicken and turkey meat in meat mixtures by using real-time PCR assays. *Journal of Food Science*, **77** (2), C167–C173.

Koeppel, R., Daniels, M. *et al.* (2013) Multiplex real-time PCR for the detection and quantification of DNA from duck, goose, chicken, turkey and pork. *European Food Research and Technology*, **236** (6), 1093–1098.

Koh, M.C., Lim, C.H. *et al.* (1998) Random amplified polymorphic DNA (RAPD) fingerprints for identification of red meat animal species. *Meat Science*, **48** (3–4), 275–285.

Ma, C., Zhao, D.a. *et al.* (2008) Overall monitoring of pork industrialization production and development of traceability system. *Transactions of the Chinese Society of Agricultural Engineering*, **24** (9), 121–125.

Mafra, I., Ferreira, I.M.P.L.V.O. *et al.* (2008) Food authentication by PCR-based methods. *European Food Research and Technology*, **227** (3), 649–665.

Mane, B.G., Tanwar, V.K. *et al.* (2008) RAPD markers for differentiation of meat species. *The Indian Journal of Veterinary Research*, **17** (2), 9–13.

Marcone, M.F., Wang, S. *et al.* (2013) Diverse food-based applications of nuclear magnetic resonance (NMR) technology. *Food Research International*, **51** (2), 729–747.

McElhinney, J., Downey, G. *et al.* (1999) Chemometric processing of visible and near infrared reflectance spectra for species identification in selected raw homogenised meats. *Journal of Near Infrared Spectroscopy*, **7** (3), 145–154.

McMillin, K.W., Lampila, L.E. *et al.* (2012) Advances in Meat, Poultry and Seafood Packaging, A volume in Woodhead Publishing Series in Food Science, Technology and Nutrition, 2012, Pages 565–595.

Md. Eaqub, A., Md. Abdur, R. *et al.* (2014) Multiplex PCR in species authentication: Probability and prospects - a review. *Food Analytical Methods*, **7** (10), 1933–1949.

Melis, R., Cappuccinelli, R. *et al.* (2014) Addressing marketplace gilthead sea bream (*Sparus aurata* L.) differentiation by H-1 NMR-based lipid fingerprinting. *Food Research International*, **63**, 258–264.

Moreno-Rojas, J.M., Vasta, V. *et al.* (2008) Stable isotopes to discriminate lambs fed herbage or concentrate both obtained from C(3) plants. *Rapid Communications in Mass Spectrometry*, **22** (23), 3701–3705.

Nakashita, R., Suzuki, Y. *et al.* (2008) Stable carbon, nitrogen, and oxygen isotope analysis as a potential tool for verifying geographical origin of beef. *Analytica Chimica Acta*, **617** (1–2), 148–152.

Nejad, F.P., Tafvizi, F. *et al.* (2014) Optimization of multiplex PCR for the identification of animal species using mitochondrial genes in sausages. *European Food Research and Technology*, **239** (3), 533–541.

Osorio, M.T., Moloney, A.P. *et al.* (2012) Authentication of beef production systems using a metabolomic-based approach. *Animal*, **6** (1), 167–172.

Osorio, M.T., Moloney, A.P. *et al.* (2011a) Beef authentication and retrospective dietary verification using stable isotope ratio analysis of bovine muscle and tail hair. *Journal of Agricultural and Food Chemistry*, **59** (7), 3295–3305.

Osorio, M.T., Moloney, A.P. *et al.* (2011b) Multielement isotope analysis of bovine muscle for determination of international geographical origin of meat. *Journal of Agricultural and Food Chemistry*, **59** (7), 3285–3294.

Perini, M., Camin, F. *et al.* (2009) Multielement (H, C, N, O, S) stable isotope characteristics of lamb meat from different Italian regions. *Rapid Communications in Mass Spectrometry*, **23** (16), 2573–2585.

Piasentier, E., Valusso, R. *et al.* (2003) Stable isotope ratio analysis for authentication of lamb meat. *Meat Science*, **64** (3), 239–247.

Rodriguez-Ramirez, R., Gonzalez-Cordova, A.F. *et al.* (2011) Review: Authentication and traceability of foods from animal origin by polymerase chain reaction-based capillary electrophoresis. *Analytica Chimica Acta*, **685** (2), 120–126.

Rogberg-Munoz, A., Posik, D.M. *et al.* (2013) Recent patents for detecting the species of origin in animal feedstuff, and raw and processed meat products. *Recent Patents on Food, Nutrition & Agriculture*, **5** (1), 3–8.

Sacco, D., Brescia, M.A. *et al.* (2005) Geographical origin and breed discrimination of Apulian lamb meat samples by means of analytical and spectroscopic determinations. *Meat Science*, **71** (3), 542–548.

Sakamoto, N., Ishida, T. *et al.* (2002) Concentrations of radiocarbon and isotope compositions of stable carbon in food. *Journal of Nuclear Science and Technology*, **39** (4), 323–328.

Shintu, L., Caldarelli, S. *et al.* (2007) Pre-selection of potential molecular markers for the geographic origin of dried beef by HR-MAS NMR spectroscopy. *Meat Science*, **76** (4), 700–707.

Soares, S., Amaral, J.S. *et al.* (2013) A SYBR Green real-time PCR assay to detect and quantify pork meat in processed poultry meat products. *Meat Science*, **94** (1), 115–120.

Sun, C., Ji, Z. *et al.* (2007) A traceability system for beef products based on radio frequency identification technology in China. *New Zealand Journal of Agricultural Research*, **50** (5), 1269–1275.

Sun, S.-m., Guo, B.-l. *et al.* (2011a) Application of near infrared spectral fingerprint technique in lamb meat origin traceability. *Spectroscopy and Spectral Analysis,* **31** (4), 937–941.

Sun, S., Guo, B. *et al.* (2010) Multi-element analysis for determining the geographical origin of mutton from different regions of China. *Food Chemistry,* **124** (3), 1151–1156.

Sun, S., Guo, B. *et al.* (2011b) Multi-element analysis for determining the geographical origin of mutton from different regions of China. *Food Chemistry,* **124** (3), 1151–1156.

Sun, S., Guo, B. *et al.* (2012a) Classification of geographical origins and prediction of delta C-13 and delta N-15 values of lamb meat by near infrared reflectance spectroscopy. *Food Chemistry,* **135** (2), 508–514.

Sun, S., Guo, B. *et al.* (2012b) Geographical origin traceability of lamb based on mineral element fingerprints. *Transactions of the Chinese Society of Agricultural Engineering,* **28** (17), 237–243.

Tokalioglu, S. (2012) Determination of trace elements in commonly consumed medicinal herbs by ICP-MS and multivariate analysis. *Food Chemistry,* **134** (4), 2504–2508.

Yan, B., Shi, P. *et al.* (2013) Development of traceability system of aquatic foods supply chain based on RFID and EPC internet of things. *Transactions of the Chinese Society of Agricultural Engineering,* **29** (15), 172–183.

You, J., Huang, L. *et al.* (2014) Species-specific multiplex real-time PCR assay for identification of deer and common domestic animals. *Food Science and Biotechnology,* **23** (1), 133–139.

Zarringhabaie, G.E., Pirany, N. *et al.* (2011) Molecular traceability of the species origin of meats using multiplex PCR. *African Journal of Biotechnology,* **10** (73), 16461–16465.

Zhang, C., Bai, J. *et al.* (2012) Consumers' willingness to pay for traceable pork, milk, and cooking oil in Nanjing, China. *Food Control,* **27** (1), 21–28.

Zhao, D.-a., Teng, C.-f. *et al.* (2009) *Design of Traceability System for Pork Safety Production Based on RFID.* Second International Conference on Intelligent Computation Technology and Automation.

Zhao, M., O'Donnell, C.P. *et al.* (2013a) Detection of offal adulteration in beefburgers using near infrared reflectance spectroscopy and multivariate modelling. *Journal of Near Infrared Spectroscopy,* **21** (4), 237–248.

Zhao, Q., Wang, Y. *et al.* (2012) Application prospects of organic RFID tags for animal food tracing. *Transactions of the Chinese Society of Agricultural Engineering,* **28** (8), 154–158.

Zhao, Y., Zhang, B. *et al.* (2013b) Tracing the geographic origin of beef in China on the basis of the combination of stable isotopes and multielement analysis. *Journal of Agricultural and Food Chemistry,* **61**, 7055–7060.

Zhao, Y., Zhang, B. *et al.* (2014) Recent developments in application of stable isotope analysis on agro-product authenticity and traceability. *Food Chemistry,* **145**, 300–305.

CHAPTER 16

Regulation and legislative issues

Fiona Lalor & Patrick Wall
School of Public Health, Physiotherapy and Population Science, University College Dublin, Dublin, Ireland

16.1 Introduction

In the Western and developed world, we have the luxury of an abundant well-regulated safe food supply. In many countries, food safety is not a burning issue, rather it is the overconsumption of food leading to diet-related diseases and obesity that are concerning. The foods people consume today are varied and very different to those eaten 50–100 years ago. International food trade has existed for thousands of years but until recently food was mainly produced, sold and consumed locally. Due to population shifts and changes in consumer demands and exposure to new tastes and flavours, the food chain has become much more complex than it used to be. The idea of 'seasonal' foods has changed, and consumers now expect all types of foods to be available all year around. Satisfying the demands of modern consumers has presented massive challenges to the industry in terms of transportation and refrigeration and also to the regulators in their efforts to oversee the food chain. Over the last century, the amount of food traded internationally has grown exponentially, and a quantity and variety of food never before possible, travels the globe today.

The consumer has been led to believe that the food chain is a simple straight line and the phrase 'farm to fork' is mentioned in the legislation in many jurisdictions. However, the food chain is a lot more complex and is more like a maze that a straight line. At farm level, feed stuffs, vitamins and minerals, agrochemicals and animals remedies are globally sourced. In the food chain, there is often co-mingling of products from many processors often from different countries and even from different continents. Even if the meat in a product is from one country, spices, herbs and other additives or preservatives are globally sourced. This creates extreme difficulty when it is necessary to recall a contaminated ingredient that may have ended up in a processed product. Tracing primal cuts of meat precisely is possible, but once meat enters into secondary processing precise traceability becomes extremely difficult. The more the steps in the food chain, the more opportunities that exist for something to go wrong, and the more people

Emerging Technologies in Meat Processing: Production, Processing and Technology, First Edition.
Edited by Enda J. Cummins and James G. Lyng.

that are involved in the food chain the increasing likelihood that one may be a bad workman or worse still a criminal.

In many jurisdictions, harmonised food legislation exists; however, the same resources for enforcement do not exist in all these jurisdictions. Legislation that is not enforced does not provide the intended consumer protection, and with increasing cutbacks in the public sector in many countries due to challenging economic situations, policing the food chain and surveillance is compromised because of inadequate resources.

Food fraud is becoming increasingly prevalent and the processed meat sector has not escaped. In 2013, a pan-EU scandal was uncovered with beef being replaced with horsemeat in many products distributed across the European Union.

Increasingly, more and more of the food consumed today is eaten outside of the home (Poti and Popkin, 2011; Angulo and Jones, 2006). This food is eaten in restaurants and cafes, is purchased in supermarkets and grocery shops and is also consumed in institutional settings such as schools, workplace canteens, hospitals and so on. The location of this food consumption is as varied as the foodstuffs themselves (Burke *et al.*, 2007). As a result of this lengthening of the food chain, more and more people become involved in the preparation of our meals. The traditional home-cooked meal day after day with all components cooked from scratch is not a reality in many homes in the Western world anymore. This development and change in our eating habits has served to increase the importance and responsibility towards the regulatory environment within the food production industry.

16.2 Overview of principles of food regulation

The global distribution of food has necessitated international rules and regulation to enable trade to occur safely. There are many regulatory and standard setting agencies but increasingly they are collaborating to share risk assessments, risk management methodologies and risk communication strategies.

The Food and Agriculture Organisation (FAO) of the United Nations is an autonomous agency within the United Nations (UN) system. It was established in 1943, and as of 2001, it had a membership of 180 countries worldwide. The FAO was created to 'raise levels of nutrition and to improve production and distribution of food and agricultural products'. The FAO governing body (known as The Conference) meets every 2 years and is a major forum for the discussion of food and agriculture issues. The FAO created the Codex Alimentarius Commission in 1961 and was joined in its operation by the World Health Organisation (WHO) in 1962. The Commission develops and maintains the Codex Alimentarius, which is a collection of internationally recognised standards, codes of practice, guidelines and other recommendations relating to foods, food production and food safety, and it has issued over 200 international food standards (Casadei and Albert, 2003) under the Joint FAO/WHO Food Standards Programme. The aims of this programme are to protect consumer health, ensure fair trade and promote coordination of all food standards as developed under intergovernmental or non-governmental organisations.

16.2.1 Risk assessment should underpin risk management decisions

EU Regulations state that risk assessments shall be based on the best available scientific evidence and undertaken in an independent, objective and transparent manner. They also formally establish the precautionary principle as an option open to risk managers when decisions have to be made to protect public health but scientific information concerning the risk is inconclusive or incomplete.

The precautionary principle is relevant in circumstances where risk managers have found reasonable grounds for concern that an unacceptable level of risk to health exists, but the supporting evidence is insufficient for a comprehensive and accurate risk assessment.

When scientific evidence is scarce, risk managers may use the precautionary principle and take appropriate action to protect the public until more information on the nature of the risk becomes available. Such measures are provisional and have to comply with the normal principles of non-discrimination and proportionality.

No country is fully self-sufficient in terms of food production, and some degree of food importation is inevitable. Technical regulations and product standards may vary from country to country. Having many different regulations and standards makes operations difficult for producers and exporters. If regulations are set arbitrarily and without solid foundation, they could be used as an excuse for protectionism.

The World Trade Organisation has established international trade agreements which call for, and aim to establish, fair and free food trade between all countries (Lupien, 2002). Specifically, the sanitary and phytosanitary measures (SPS) and on technical barriers to trade (TBT) have been agreed to by over 140 countries, whereby the aim is to promote free and free trade of good quality and safe food between all countries.

16.2.2 Sanitary and phytosanitary measures (SPS) and technical barriers to trade (TBT)

The measures that the WTO apply can be classified as sanitary (human or animal life or health) or phytosanitary (plant life or health).

The SPS agreement is focused on two things – *health* and *international trade.* It was designed so that, while recognising that the trade of food is global, countries must be in a position to protect their citizens from health-threatening risks. The *health* aspect means that WTO members can protect human, animal or plant life or health, by applying measures to manage the risks associated with imports. The measures usually take the form of quarantine or food safety requirements.

The Agreement on TBT tries to ensure that regulations, standards, testing and certification procedures do not create unnecessary obstacles while providing members with the right to implement measures to achieve legitimate policy objectives, such as the protection of human health and safety, or the environment.

The *International Trade* aspect of the SPS agreement means that in seeking to protect health, WTO members must not use SPS measures that are as follows:

- Unnecessary
- Not science-based
- Arbitrary
- Constitute a disguised restriction on international trade.

16.2.3 International laws

Given the global movement of goods in general and food in particular, food safety has wide-ranging implications with both international trade and also public health. How countries perceive and handle food safety risks is complicated and partly depends on their access to and use of science, detection technology and mitigation methods (Buzby, 2001).

The food chain is very complex, and in most jurisdictions different government agencies have responsibility for overseeing elements of the chain including (1) Ministries of Agriculture, (2) Trade, (3) Health, (4) Transport, (5) Customs and Excise, (6) Municipal authorities, (7) local and regional authorities and so on. Often these government agencies act in isolation, and there is little communication between the regulator stakeholders. Not all risks along the food chain receive equal attention, and there are often gaps in the continuum of enforcement from farm to fork. The ideal national situation is a seamless inspectorate from farms through to final sale to consumers with a clear handover between agencies and a sharing of inspection information and surveillance data. Due to the global nature of food distribution, there needs to be global cooperation in overseeing the distribution network and good collaboration and cooperation between jurisdictions. The development of food safety agencies with the primary objective to protect consumers has created the conduit for collaboration across country boundaries.

A. **Food safety systems in the United States**

The Food and Drugs Act and the Meat Inspection Act were published in 1906 and were the first step by the US government to protect consumers and their food supply. Additionally, since 1979, the Food and Drug Administration, as part of the Department of Health and Human Services, has been responsible for protecting the public from unsafe foods, drugs, medical devices and cosmetics.

Today, there are a number of Federal Agencies with a role in food safety (Longely, 2010; United States Federal Agencies for Food Safety).

- The Food Safety and Inspection Service (FSIS)
  FSIS is the public health agency in the US Department of Agriculture responsible for ensuring that the nation's commercial supply of meat, poultry and egg products is safe, wholesome and correctly labelled and packaged.
  The USDA's regulatory authority comes from the Federal Meat Inspection Act, the Poultry Products Inspection Act, the Egg Products Inspection Act and the Humane Methods of Livestock Slaughter Act.
- US Food and Drug Administration (FDA)
  The FDA is charged with protecting consumers against impure, unsafe and fraudulently labelled products. FDA, through its Centre for Food Safety and Applied Nutrition (CFSAN), regulates foods other than the meat, poultry and egg products that are regulated by FSIS. FDA is also responsible for the safety of drugs, medical devices, biologics, animal feed and drugs, cosmetics, and radiation-emitting devices.
  The FDA is authorised to conduct its work by the federal Food, Drug and Cosmetic Act and the Public Service Act. New regulations giving the FDA the authority to inspect large commercial egg farms took effect in July 2010.
- Centers for Disease Control and Prevention (CDC)
  CDC leads federal efforts to gather data on foodborne illnesses, investigate foodborne illnesses and outbreaks, and monitor the effectiveness or prevention and control efforts in reducing foodborne illnesses. CDC also plays a key role in building state and local health department epidemiology, laboratory and environmental health capacity to support foodborne disease surveillance and outbreak response.

B. **Food safety systems in China (Bai *et al.*, 2007)**

The food safety system in China was very fragmented. The following Government agencies each had a role:

- Ministry of Agriculture
- Ministry of Health and family planning
- State Administration for Industry and Commerce
- Administration of Quality Supervision, Inspection and Quarantine (AQSIQ) – is responsible for national quality, metrology, entry/exit commodity inspection, entry/exit health quarantine, entry/exit animal and plant quarantine, import–export food safety, certification and accreditation standardisation as well as administrative law enforcement.
- State Food and Drug Administration (SFDA) (http://eng.sfda.gov.cn/WS03/CL0756/) – established in 2003. Among many other things, SFDA's mission is to formulate policies and programmes on food safety at consumption stage. The organisation was created to take control of food hygiene licensing and food safety supervision at consumption stage. This

includes the investigation and punishment of illegal activities of food safety at consumption stage.

- State Drug Administration
- Ministry of Commerce
- Ministry of Science and Technology
- National Institute of Nutrition and Food Safety.

In order to address the fragmented nature of regulation in China, the China Food and Drug Administration (CFDA) was founded in March 2013. It replaced a large group of overlapping regulators and created an entity similar to the Food and Drug Administration of the United States, streamlining regulation processes for food and drug safety. The China Food and Drug Administration is directly under the State Council of the People's Republic of China, which is in charge of comprehensive supervision on the safety management of food, health food and cosmetics and is the competent authority of drug regulation in mainland China.

C. **Food safety systems in Latin America**

Latin America includes the following countries: Brazil, Mexico, Columbia, Argentina, Peru, Venezuela, Chile, Ecuador, Cuba, Bolivia, Dominican Republic, Paraguay, Costa Rica, Uruguay, El Salvador, Honduras and Suriname.

Most countries in Latin America have a fragmented responsibility for food safety. Usually, there are a number of different ministerial offices involved which, like elsewhere in the world, makes coordination difficult. However, these countries are making efforts to align their own regulatory frameworks with the requirements of the WTO SPS/TBT agreements because currently in Latin America most countries apply SPS risk management measures that are forced by International market demands for entering food trade or for maintaining their markets. SPS measures only become important when they are related to trade and economic development, but they can result in double standards – one for internal trade and one for external trade/the export market.

These countries have recognised a need to continuously build capacity/expertise in terms of risk assessment and risk analysis as well as increase their laboratory capacity and networks in order to deal with and prevent food safety issues.

D. **Food safety systems in Australia/New Zealand**

Food Standards Australia New Zealand is a binational Government agency whose principal responsibility is to develop and administer the Australia New Zealand Food Standards Code (the Code), which lists requirements for foods such as additives, food safety, labelling and GM foods. Enforcement and interpretation of the Code is the responsibility of State/Territory departments and food agencies within Australia and New Zealand.

In Australia, FSANZ has broader coverage and prepares standards across the food supply chain. They develop standards for primary production and processing and for food hygiene. They also set residue limits for and veterinary products. In New Zealand, these activities are undertaken by the New Zealand Ministry for Primary Industries.

16.3 Food safety regulation within the European Union

While attempts are always made by Regulators to be ahead of the technical advances of industry in terms of legislation, in reality, this is often not the case. Oftentimes legislation evolves in response to industrial developments or new production practices. For example, in 1997, evidence emerged that Bovine Spongiform Encephalopathy (BSE), a neurological condition of cows, might be transmissible to humans. This rocked the foundations of the legislative system of food law not only in the United Kingdom where the disease initially arose, but across Europe and the world (O'Rourke, 2001). Until this point, the regulation of the food supply in reality put the needs of the industry ahead of that of the consumer. The need to support the concept of the Single Market via 'Free movement of Goods' and the Theory of Mutual Recognition were of paramount importance and drove all legislative developments. Post-BSE, however, things were very different and protecting consumers' health, and consumer's confidence began to be considered before commercial concerns. In April 1997, the EU Commission published a Green Paper on Food Law in order to encourage discussion on the topic and this established the following goals of EU food law:

- Ensuring a high level of protection of public health for the consumer
- Ensuring the free movement of goods within the internal market
- Ensuring that legislation is primarily based on scientific evidence
- Ensuring the competitiveness of the European food industry
- Placing the primary responsibility for safe food on industry producers and suppliers
- Ensuring that food law is coherent and user friendly.

Following the publication of this Green Paper, the EU Commission published the White Paper on Food Safety in 1999. The major objective of this paper was the attainment of the highest possible standards for food safety, and it was envisaged that the creation of a single EU-wide food agency would be best placed to achieve this. Building on these objectives and principles, the Commission published EU Regulation 178/2002 (European Parliament and Council, 2002), which was to become the bedrock of all EU food law. This Regulation (1) lays down the principles of food law, (2) establishes the European Food Safety Authority and (3) lays down procedures in matters of food safety. This regulation ensures the quality of foodstuffs intended for human consumption and animal feed. It guarantees the free circulation of safe and secure food and feed in the internal market and has formed the basis for all legislative developments in food and feed safety.

A. **EU Regulation 178/2002 – the General Food Law Regulation**

The EU integrated approach to food safety aims to assure a high level of food safety, animal health, animal welfare and plant health within the European Union through coherent farm to table measures and adequate monitoring, while ensuring the effective functioning of the internal market.

In order to implement this approach, a variety of legislative steps have been taken, in addition to other actions:

- To assure effective control systems and evaluate compliance with EU standards in the food safety and quality, animal health, animal welfare, animal nutrition and plant health sectors within the European Union and in third countries in relation to their exports to the European Union;
- To manage international relations with third countries and international organisations concerning food safety, animal health, animal welfare, animal nutrition and plant health;
- To manage international relations with third countries and international organisations concerning food safety, animal health, animal welfare, animal nutrition and plant health;
- To manage relations with the European Food Safety Authority (EFSA) and ensure science-based risk management.

In all Member States and in many third countries, the overarching principles concerning food safety and consumer protection are established in national legislation. However, at the EU level, food legislation has evolved without some of these basic principles having been established in an overarching legal instrument. This problem was addressed with the publication of the General Food Law Regulation (178/2002) (European Parliament and Council, 2002).

In addition to what was outlined earlier, the aim of the General Food Law Regulation is to provide a framework to ensure a coherent approach in the development of food legislation. At the same time, it provides the general framework for those areas not covered by specific harmonised rules but where the functioning of the internal market is ensured by mutual recognition. It lays down definitions, principles and obligations covering all stages of food/feed production and distribution (European Commission General Food Law).

One of the key concepts in EU Food Law is that Food Business Operators (FBOs), at all stages of production, processing and distribution within a business 'are responsible for ensuring the fulfillment of food law which are relevant to their activities' (Pisanello, 2010). This aspect of the legislation represented a major shift in the focus of food law – making those who are 'doing', also responsible for ensuring they are 'doing within the law' and moving resources away from the Regulatory system.

B. **Official control of foodstuffs**

The General Food Law Regulation (178/2002) makes it clear that Member States must maintain a system of official controls and other activities as appropriate to the circumstances, including public communication on food and feed safety and risk, food and feed safety surveillance and other monitoring activities covering all stages of production, processing and distribution (European Parliament and Council 2002). The general requirements for official controls on foodstuffs are addressed by Regulation (EC) No. 882/2004 of the European Parliament and of the Council of 29 April 2004 on official controls performed to ensure the verification of compliance with food and feed law, animal health and welfare rules and by Regulation 854/2004 of the European Parliament and of the Council of 29 April 2004 laying down specific rules for the organisation of official controls on products of animal origin intended for human consumption.

C. **The hygiene package**

In April 2004, the EU adopted three regulations forming the core of what has become known as the 'Food Hygiene Package'. The regulations are as follows:

- Regulation No. 852/2004 on the hygiene of foodstuffs
- Regulation No. 853/2004 laying down specific hygiene rules for food of animal origin in order to guarantee a high level of food safety and public health
- Regulation No. 854/2004 putting in place a Community framework of official controls on products of animal origin intended for human consumption.

D. **Food processing**

In terms of legislation and food production and processing techniques, the rules in general are horizontal rather than specific and the principle is that the food should be safe for human consumption. With the exception of irradiation, which upon its introduction was very controversial, most legislation merely declares that the food must be safe for human consumption and so are not process specific.

Foodstuffs should not contain micro-organisms or their toxins or metabolites in quantities that present an unacceptable risk for human health. In terms of all food safety legislation with the European Union, the safety of foods is mainly ensured by a preventive approach, such as implementation of good hygiene practice and application of procedures based on hazard analysis and critical control point (HACCP) principles. Testing against the microbiological criteria set down in EU Regulation 2703/2005 (European Commission, 2005) should be undertaken when validating or verifying the correct and appropriate functioning of systems in place in any given processing facility.

E. **Hazard analysis and critical control points (HACCP)**

HACCP is a systematic approach to identifying physical, biological and chemical hazards in a food processing facility and then identifying systems to reduce these risks to a level that is not unsafe for human consumption. HACCP was first conceived in the 1960s when the US National Aeronautics and Space Administration (NASA) was trying to identify a system to ensure the safety of food supply during space flights. It is now in widespread use across the cosmetic, pharmaceutical and food industries. It is very different to any previous systems in that, by its very design, it aims to identify hazards and remove them in order to reduce human risk. Previous systems merely identified faulty or 'below par' products at the end of the system and isolated them. Ultimately, HACCP, as a preventive system, requires control over the raw materials, the process, the environment, personnel, storage and distribution, beginning as early in the system as possible (Pearson and Dutson, 1995).

The seven principles of HACCP are as follows:

1. Identify the hazards
2. Determine the critical control points (CCPs)
3. Establish critical limits
4. Establish as system to monitor control of the CCP
5. Establish the corrective action to be taken when monitoring indicates that a particular CCP is not under control
6. Establish procedures for verification to confirm the HACCP system is working effectively
7. Establish documentation concerning all procedures and records appropriate to these principles and their application.

With the exception of food irradiation, there is no specific legislation for individual food processing techniques in the European Union. However, the EU have established some microbiological limits and maximum limits for chemicals and pesticide residues. European Food law in general is a long and complicated list of published legislation and the Food Safety Authority of Ireland (Food Safety Authority of Ireland, 2011) categorises it as outlined in Table 16.1: Food Legislation - general areas.

16.4 Meat inspection

In most jurisdictions, there are specific controls on products of animal origin intended for human consumption, and inspection tasks are prescribed including (1) ante-mortem inspection, (2) animal welfare, (3) post-mortem inspection, (4) addressing specified risk material and other by-products and (5) laboratory testing. The purpose of the inspection is to assess if the meat is fit for human consumption in general and to address a number of specific hazards, in particular the following issues: transmissible spongiform encephalopathies (only ruminants), cysticercosis, trichinosis, glanders (only solipeds), tuberculosis, brucellosis,

Table 16.1 Food legislation – general areas.

Aquaculture – bivalve molluscs	Foods for special nutritional purposes (PARNUTS)	Meat – meat products
Contamination of foodstuffs	Game	Meat – minced meat, meat preparations and mechanically separated meat
Eggs	Fruit and vegetables	Meat – poultry meat
Fish and fishery products	General product safety	Milk and milk products
Flavourings	General principles of food safety	Novel foods
Food additives	Genetically modified organisms (GMOs)	Official control of foodstuffs
Food enzymes	Imports and exports	Organic foods
Food fortification	Internal carriage of perishable foodstuffs	Pesticide residues in foodstuffs
Food hygiene	Labelling, presentation and advertising of foodstuffs	Veterinary medicines, animal remedies, control of illegal substances and poisons
Food products with compositional requirements	Manufacturing and processing methods	Water (included bottled water)
Food products	Materials and articles in contact with foodstuffs	Zoonoses
Food standards	Meat – fresh meat	

contaminants (e.g. heavy metals), residues of veterinary drugs and unauthorised substances or products.

There is usually a continuous presence of inspectors either lay or veterinary qualified in abattoirs and all meat is inspected; however, this situation is under review as it is considered that historic practices of visual inspection and incising glands and organs fail to address the current zoonotic risks and the risk of chemical contamination with environmental contaminants in feed contaminants or animal remedies. The official controls are likely to be modernised to address the risks to public health on the basis of reviews of the past practices to deliver a more risk-based inspection regimen (European Food Safety Authority, 2013).

Unlike the permanent presence of inspectors in abattoirs, the intensity of inspections varies greatly in meat processing facilities. There has been a move from undertaking inspection of premises to audits of the business's food safety management systems in an attempt to get greater return for regulatory resources.

Many businesses participate in third-party audited quality assurance schemes where they comply with additional private standards over and above the regulatory requirements. The hope by companies is that this will be recognised by the regulators and may result in less frequent official Government inspections.

16.4.1 General standards for meat processing

Recommendations for handling all meat products are to keep them clean, cold and covered in order to maintain quality and protect against food poisoning and disease. Generally, contamination occurs when the products come into contact with dirty hands, clothing equipment or facilities. Abattoir's are not operating theatres and meat can never be assumed to be pathogen free but if the product is kept clean, these will be minimal additional contamination by micro-organisms (FAO, 1991).

16.4.2 Microbiological limits for meat products

The total viable count of bacteria (TVC), expressed as organisms/cm^2 or as organisms/g on fresh meat or meat products, sets a limit to its shelf life. Meat will 'spoil' with TVC at $10^6/cm^2$ because of off-odours. Slime and discolouration appear at $10^8/cm^2$. The main factors determining the time it takes for the TVC to reach these levels are the initial count due to contamination during slaughtering and processing, further contamination during storage, and the temperature, pH and relative humidity (FAO, 1991).

16.4.3 Novel processing techniques

A. **Irradiation of foods in Europe**

Since 1999, there have been two pieces of legislation governing the irradiation of foodstuffs within the European Union.

Directive 1999/2 which covers general and technical aspects for carrying out the process, labelling of irradiated foods and conditions for authorising food irradiation (European Commission, 1999).

Directive 1999/3 which provides a list of products that are authorised for irradiation. In Europe, this list only contains 'dried aromatic herbs, spices and vegetable seasonings' (European Commission Food Irradiation).

B. **High-pressure processing for meat and meat products**

Addressing the growing demand for minimally processed foods that are safe and with superior sensory and nutritional features, the food industry has employed HPP to develop products that have the quality of fresh foods but have an extended shelf life, without the use of preservatives (Marco, 2010). Industrial applications have developed rapidly, especially in the stabilisation of ready-to-eat meats and cured products, satisfying the demands of the USDA-FSIS.

Areas of experimentation for industrial applications in the meat sector include the following:

- Optimisation of HPP conditions to in activate target micro-organisms for each product and commercial presentation
- New packaging systems and combination with natural antimicrobial substances to enhance the shelf life extension.
- Development of new meat products based on cold gelification of starches,

- HPP-thermal coagulation of proteins, selective enzymatic inactivation and meat separation.

In 2003, the US Department of Agriculture issued a LNO (letter of no objection) to the use of HPP as a post-packaging intervention in controlling *Listeria monocytogenes* in RTE meat and poultry products in US companies (Marco, 2010).

Since then, a number of applications have been made to the USDA for the use of HPP to prevent contamination of RTE foods (Santa Maria Foods).

In the European Union, HPP foods are classified as novel foods, which means that either they have to go through a system of approval prior to use or else they need to be designated as substantially equivalent to other products on the marketplace (Garriga *et al.*, 2004). In the United States, they are viewed quite differently and there is no equivalent to the novel foods regulation. Instead, to these foods, traditional sanitary regulations are applied.

C. **Novel foods regulation**

The EU Commission considers food and food ingredients that have not been used for human consumption to a significant degree in the European Union before 15 May 1997 as 'novel foods' and novel food ingredients, and EU Regulation 258/97 (European Commission, 1997) provides for their regulation. To place a novel food or ingredient on the market in the European Union, in accordance with this legislation, authorisation must be sought. The granting of authorisation involves presenting scientific information and a safety assessment report to the Commission for consideration. The Commission seeks the opinion of the Standing Committee on Food Chain and Animal Health, which reviews the application in terms of the following:

1. Proposed conditions of use
2. Designation of novel food or novel food ingredient
3. Specification and labelling requirements.

An alternative procedure exists which enables companies go to market with a process known as 'substantial equivalence'. If you can demonstrate that your product is 'substantially equivalent' to one that is already on the market, then the extensive authorisation process is no longer a requirement.

If an EU country considers a novel food or ingredient a risk to human or environmental health because of new information, it may suspend or temporarily restrict the marketing and use of any novel food or ingredient on their territory. The country informs the Commission that either extends the national measures to all EU countries or asks that they are repealed.

16.4.3.1 Regulation EC 258/97 – article 2

The Regulation on novel foods does not apply to foods and ingredients for which an alternative approval procedure exists:

- **Food additives** within **Directive 89/107/EEC** – approximating EU countries' laws on food additives

- **Flavourings** for use in foods within **Directive 88/388/EEC** – approximating EU countries' laws on flavourings
- **Extraction solvents** used in the production of foods within **Directive 88/344/EEC** – approximating EU countries' laws
- **GMOs** for food and feed – **Regulation EC 1829/2003**
- If foods and/or food ingredients were used exclusively in food supplements, new uses in other foods require authorisation under the Novel Food Regulation, for example food fortification require authorisation.

16.4.3.2 Food fraud

With the increasing complexity of the food chain and the global distribution of product, the opportunity exists for unscrupulous operators to engage in food fraud substituting fake or inferior product for more expensive ingredients. Many different product categories have been involved from wine to olive oil to fish to processed meats. The food safety inspectorate in most jurisdictions usually focus on hygiene-related issues and are not experienced nor trained to deal with food fraud and criminal activity.

In 2013, a pan European fraud was uncovered, triggered by the investigations of the Irish Food Safety Authority of Ireland, which revealed horse DNA in frozen beef burgers manufactured in Ireland. It subsequently emerged that horsemeat was substituted for beef in a range of processed meat products identified in several countries including Ireland, the United Kingdom, France, Norway, Austria, Switzerland, Holland, Sweden and Germany. Confidence in the integrity of the processed beef supply chain in the honesty of labels and in the word 'traceability' was destroyed.

Major companies, who believed that they had robust controls, had horse DNA identified in products they were selling. Several major British retailers had to undertake costly recalls and were the subject of substantial adverse publicity. Similarly, multinational companies and brands, for example, Findus, Nestle, Burger King and Birds Eye discovered that the controls in their supply chains, and contracted manufacturers' facilities, were not as good as they might have been.

Ironically while analytical chemists were testing down to levels of parts per billion and microbiologists were seeking minute levels of bacterial contamination, truckloads of horsemeat were moving around the European Union masquerading as beef. Most of the companies selling mislabelled products were using private standards, such as the British Retail Consortium (BRC), for enhanced quality assurance believing them to be more stringent than the regulatory requirements. However, both the routine regulatory oversight and the private standards' audits failed to identify this fraud. DNA testing was not a routine food safety test and, because of the cost, will not become one. After initial intensive testing, in the immediate aftermath of the crisis, verifiable paper traceability and random checking are used as they are more economically sustainable.

One burger manufacturer in Ireland substituted cheap Polish beef for the retailer's specification for Irish and British beef and this fundamental breach of contract was compounded when it was revealed that the 'Polish beef' contained horsemeat. Staff are a company's greatest asset, but careless or dishonest staff can be a phenomenal liability. Those charged with overseeing procurement and policing quality control are the frontline guardians of a company's reputation and should omit nothing to ensure that all purchasing specifications are being met.

The role of meat traders, and middlemen, demonstrated that from 'farm to fork' proved to be a longer distance than many believed. The big lesson from this debacle for the retailers and the major brands is that a company's reputation is only as secure as the standards of their weakest supplier. Therefore, shorter supply chains, with long-term strategic relationships between the stakeholders, are now becoming normalised. Retailers and big brands forcing prices down was only incentivising some players to engage in unlawful practices. The distribution of margin along the chain to enable every player remain commercially viable is the only sustainable approach.

The impact of the social media was huge in this scandal and compounded brand and reputational damage. The tone in the social media went from humour to rising surprise and shock, then anger and finally recriminations as the crisis continued to escalate. Consumer confidence has been severely damaged in the processed meat sector and in the industry players that consumers have trusted to look after their interests. Unimplicated companies in the sector suffered as sensational generalisations made for great headlines. An additional lesson for companies is that the emphasis has to be on prevention, rather than crisis management, as with the proliferation of the social networks it is now impossible to control the media (FoodRisC).

16.5 Marketing challenges: reports of adverse health effects

A series of publications highlighting an association between both red meat and processed meats and cancer and/or cardiovascular disease is presenting challenges for those marketing these products. Heretofore, those in the meat industry considered quality parameters and price as the key factors in influencing customers purchasing behaviour. However, the industry now must address some of the health issues raised by the scientists and deal with the associated adverse publicity.

Several reputable scientific publications support a link between colorectal cancer and red meat consumption, especially processed (cured) red meat (hot dogs, bologna, sausages, bacon, ham, lunch meats) (Aune *et al.*, 2013). A dose–response relationship was demonstrated by Chan and colleagues, who reported that the risk for colorectal increased by 21% for every 50 g/day of processed red meat intake and by 29% for every 100 g/day consumed (Chan *et al.*, 2011).

Pro-cancer factors in red meat might be excess fat, protein or iron or heat-induced mutagens (Corpet, 2011). Red meat contains high levels of heme iron, which has a catalytic effect on the endogenous formation of carcinogenic *N*-nitroso compounds and on the formation of cytotoxic and genotoxic aldehydes by lipoperoxidation (Bastide *et al.*, 2011). Sodium nitrate ($NaNO_3$) and sodium nitrite ($NaNO_2$) are found in processed meats, and both are currently authorised as food additives within the European Union (European Commission, 2008). They function as preservatives in food and are used extensively to both enhance the colour and extend the shelf life of processed meats (Butler *et al.*, 2003). Nitrite is considered the active curing ingredient responsible for the preservation of the meat in combination with other ingredients such as salt. Nitrate, when added to meat, converts to nitrite before exerting a preservative function. Nitrite prevents the growth of a harmful bacterium called *Clostridium botulinum*, and it may also have preservation effects on other harmful and spoilage bacteria. In addition, nitrite develops cured meat flavour and colour and retards the development of rancidity and off-odours and off-flavours during storage of cured meats. Nitrite is responsible for the characteristic pink colour of cured meats. In the EU and in many other jurisdictions, there is a requirement to include nitrate and nitrite on the list of ingredients on the label of pre-packaged meats (European Union, 2000; European Parliament and Council, 2012). Sodium nitrite reacts with stomach acid and other chemicals in the stomach to produce nitrosamines, which have been shown to cause cancer in animals when consumed in large quantities. However, there is not much sodium nitrate/nitrite in meats, and we consume sodium nitrate/nitrite from other foods as well, so it is not clear that they are harmful in the quantities we get from meats. Cooking meat at high temperatures or on barbeques produces heterocyclic amines (HCAs) and polycyclic aromatic hydrocarbons (PAHs), which are potent carcinogens (Heddle *et al.*, 2001).

Explaining what all these findings mean to the public is not easy as, to date, none of the proposed mechanisms have been able to convincingly explain the link between red meat intake and cancer risk. For example, in a study that demonstrated a dose-dependent effect of red meat intake on colon cancer, the etiologically relevant compound in cooked meat was thought to be HCAs (Butler *et al.*, 2003). However, grilled and fried chicken contain much higher levels of HCAs than beef, but the intake of poultry is not related to cancer risk.

The EU EPIC study demonstrated that rather than being associated with zero meat consumption, the lowest mortality rates were found among those who consumed about 10–20 g of processed meat per day (equivalent to eating processed meat once or twice a week). This is consistent with previous studies showing that vegetarians and non-vegetarians who eat low amounts of meat have similar mortality rates (Rohrmann *et al.*, 2013).

The American Institute for Cancer Research (AICR) has concluded the need to advise consumers to limit their intake of red meat and processed meats. Their recommendation is to 'limit consumption of red meats (such as beef, pork and lamb) and avoid processed meats'. To reduce the risk for cancer, the AICR suggests eating no more than 18 oz (510 g) (cooked weight) per week of red meat (beef, pork, lamb) and avoid processed red meats (ham, bacon, salami, hot dogs and sausages) (American Institute for Cancer Research. Recommendations for Cancer Prevention).

There is always a difficulty in balancing risk and benefits. The overarching nutritional benefit of meat consumption questions the necessity of some of these drastic recommendations. A greater understanding of the mechanism by which cancer originates and the gene environment interactions will be needed before a definitive link can be conclusively proven. The meat industry needs to address the concerns of scientists and the associated consumers and, be seen to do so, to assist in establishing the true relationship between meat and cancer.

Historically, animal nutritionists considered formulating rations to improve food conversion efficiency and weight gain in the animals, but now it is possible to modify the composition of the meat to make it healthier by adjusting the animal diet. Strategies such as raising the levels of omega 3 fatty acids and reducing the levels of saturated fats may prove to be effective approaches to address consumer concerns about adverse health effects associated with meat.

16.6 Conclusion

There is a strong positive relationship between the level of income and the consumption of animal protein. With the emerging middle and upper classes in several parts of the world including China, India and many Latin American countries, demand for animal protein will continue to rise. Annual meat production is predicted to rise from a level of 218 million tonnes recorded in 1999 to 376 million tonnes by 2030 (World Health Organisation, 2003). With increasing affluence also comes a demand for more convenience food and the processed meat sector is responding to this demand. Worldwide meat production has tripled over the last four decades and increased 20% in just the last 10 years (World Watch).

Large-scale meat production also has serious implications for the world's climate. Animal waste releases methane and nitrous oxide and greenhouse gases that are 25 and 300 times more potent than carbon dioxide, respectively (Environmental Protection Agency, 2014). The economic crisis faced by some countries of the world, the public health threat of obesity, and the increasing reports of cancer associated with the consumption of processed meat and environmental impacts of meat production are the factors that act as constraints for this industry.

References

American Institute for Cancer Research. *Recommendations for Cancer Prevention* (2014) http://preventcancer.aicr.org/site/PageServer?pagename=recommendations_05_red_meat.

Angulo, F.J. and Jones, T.F. (2006) Eating in restaurants: a risk factor for foodborne disease? *Clinical Infectious Diseases*, **43** (10), 1324–1328.

Aune, D., Chan, D.S. and Viera, A.R. (2013) Red and processed meat intake and risk of colorectal adenomas: a systematic review and meta-analysis of epidemiological studies. *Cancer Causes and Control*, **24**, 611–627.

Bai, L. *et al.* (2007) Food safety assurance systems in China. *Food Control*, **18** (5), 480–484.

Bastide, N.M., Pierre, F.H. and Corpet, D.E. (2011) Heme iron from meat and risk of colorectal cancer: a meta-analysis and a review of the mechanisms involved. *Cancer Prevention Research*, **4**, 177–184.

Burke, S. *et al.* (2007) An examination of the influence of eating location on the diets of Irish children. *Public Health Nutrition*, **10** (6), 599–607.

Butler, L.M., Sinha, R. and Millikan, R.C. (2003) Heterocyclic amines, meat intake and association with colon cancer in a population-based study. *American Journal of Epidemiology*, **157**, 434–445.

Buzby, J.C. (2001) Effects of food-safety perceptions (Economic Research Services/USSA) on food demand and global trade, in *Changing Structure of Global Food Consumption and Trade*, China General Administration of Quality Supervision, Inspection and Quarantine of PRC http://english.aqsiq.gov.cn pp. 55–66.

Casadei, E. and Albert, J. (2003) Food and agriculture organization of the united nations, in *Encyclopedia of Food Sciences and Nutrition*, Second edn (ed. C. Benjamin), Academic Press, Oxford, pp. 2587–2593.

Chan DS, et al., (2011) Red and processed meat and colorectal cancer incidence: meta-analysis of prospective studies. *PLoS*, 2011. **6**(6): e20456.

China: Administration of Quality Supervision Inspection and Quarantine, http://english.aqsiq.gov.cn/.

Corpet, D. (2011) Red meat and colon cancer: should we become vegetarians or can we make meat safer? *Meat Science*, **89**, 310–316.

Environmental Protection Agency (2014) [cited January 2014], epa.gov/climatechange/ghgemissions/gases/ch4.html (last accessed 27 April 2016).

European Commission (1997) *EU Regulation on novel foods and novel food ingredients*, (OJ L 43 p.1 14/02/1997).

European Commission (1999) Directive 1999/2 on food irradiation, (OJ L66 p.16 13/3/1999).

European Commission (2005) *EU Regulation 2703/2005 on microbiological criteria for foodstuffs*, (OJ L 338, p.1 22/12/2005)

European Commission (2008) EC Regulation 1335/2008 on food additives, (OJ L354 p.16 31/12/2008).

European Commission Food Irradiation, http://ec.europa.eu/food/food/biosafety/irradiation/comm_legisl_en.htm (last accessed 27 April 2016).

European Commission General Food Law, http://ec.europa.eu/food/food/foodlaw/index_en.htm(last accessed 27 April 2016).

European Food Safety Authority, *Panel on Biological Hazards: Scientific Opinion on the public hazards to be covered by inspection of meat.* 11, 2013. 6: p. 3263.

European Parliament and Council (2002) EU Regulation 178/2002 on General Food Law, (OJ L31 p.1–24 1/2/2002).

European Parliament and Council (2012) *Regulation 1169/2011 on the provision of food information to consumers.*

European Parliament and Council, *Regulation 178 of 2002 of the European Parliament and of the Council laying down the general principles and requirements of food law, establishing the European Food Safety Authority and laying down procedures in matters of food safety*, in *OJ L 031, 1.2.2002*2002.

European Union, *EU Directive 2000/13 on labelling, presentation and advertising of foodstuffs*, European Union, Editor 2000: Brussels.

FAO (1991) *Guidelines for slaughtering meat, cutting and further processing*.

Food Safety Authority of Ireland (2011) *Food Legislation.*, www.fsai.ie/legislation/food_legislation.html.

Food Standards Australia New Zealand, (xxxx) http://www.foodstandards.gov.au/scienceandeducation/aboutfsanz/whatwedoanddontdo.cfm (last accessed 27 April 2016).

FoodRisC, (2016) www.foodrisc.org (last accessed 27 April 2016).

Garriga, M. *et al.* (2004) Microbial inactivation after high-pressure processing at 600 MPa in commercial meat products over its shelf life. *Innovative Food Science & Emerging Technologies*, **5** (4), 451–457.

Heddle, J.A. *et al.* (2001) *A test of the mutagenicity of cooked meats in vivo. Mutagenesis*, **16**, 103–107.

Longely, R. (2010) *The US Food Safety System: A Case of shared government responsibilities* [cited 2013], http://usgovinfo.about.com/od/consumerawareness/a/The-Us-Food-Safety-System.htm (last accessed 27 April 2016).

Lupien, J.R. (2002) The precautionary principles and other non-tariff barriers to free and fair international food trade. *Critical Reviews in Food Science and Nutrition*, **42** (4), 403–15.

Marco, C. (2010) High Pressure Processing of meat, meat products and seafood. *Food Engineering Reviews*, **2**, 256–273.

O'Rourke, R. (2001) *European Food Law*, Second edn, Palladian Law Publishing Ltd., Bembridge, Isle of Wight.

Pearson, A.M. and Dutson, T.R. (1995) *HACCP in Meat, Poultry and Fish Processing*, Aspen, New York.

Pisanello, D. (xxxx) *Shortcomings in Enforcing EU Food Law. What Does EU Food Law Stand for?* Lex Alimentaria Food Law, Editor, Taviano, Italy.

Poti, J.M. and Popkin, B.M. (2011) Trends in energy intake among US children by eating location and food source, 1977–2006. *Journal of the American Dietetic Association*, **111** (8), 1156–1164.

Rohrmann, S., Overvad, K. and Burno-de-Mesquito, H.B. (2013) Meat consumption and mortality results from the European Prospective Investigation into Cancer and Nutrition. *BMC*, **11**, 63.

Santa Maria Foods (2010) *Petition for New Uses of High Pressure Processing Technology, a USDA-FSIS approved technology*. 200? http://www.fsis.usda.gov/PDF/Petition_Professional_%20Research_Organization.pdf (last accessed 27 April 2016).

United States Federal Agencies for Food Safety (xxxx) www.foodsafety.gov/about/federal/index.html (last accessed 27 April 2016).

World Health Organisation (2003) Diet, Nutrition and Prevention of chronic diseases, in *Global and Regional Food Consumption Patterns*. WHO Technical Report Series No. 916.

World Watch, www.worldwatch.org/global-meat-production-and-consumption-continue-to-rise-1.

Index

Emerging Technologies in Meat Processing: Production, Processing and Technology, First Edition.
Edited by Enda J. Cummins and James G. Lyng.

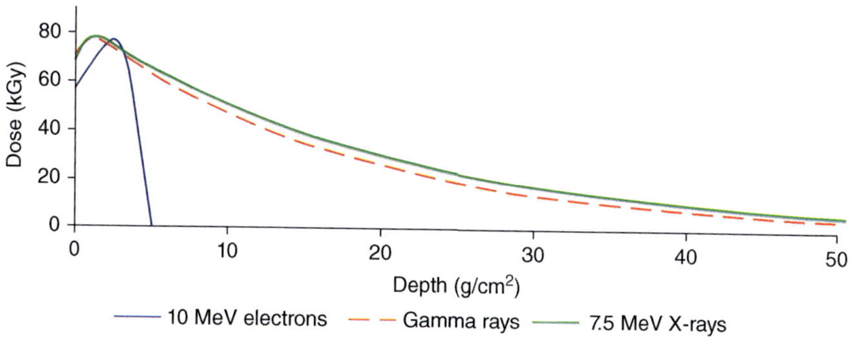

Figure 2.10 Electron beam, gamma ray, and X-ray penetration. (From IAEA (2011). Reproduced with permission of IAEA.)

Emerging Technologies in Meat Processing: Production, Processing and Technology, First Edition.
Edited by Enda J. Cummins and James G. Lyng.

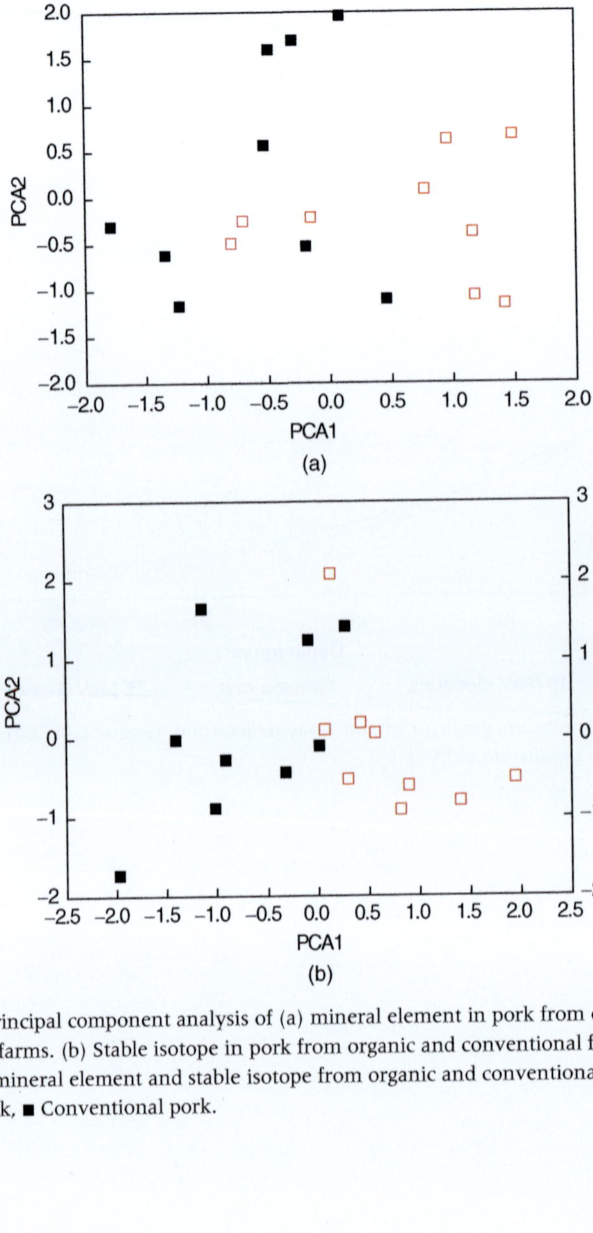

Figure 15.1 Principal component analysis of (a) mineral element in pork from organic and conventional farms. (b) Stable isotope in pork from organic and conventional farms. (c) Combine mineral element and stable isotope from organic and conventional farms. □ Organic pork, ■ Conventional pork.

Printed and bound by CPI Group (UK) Ltd, Croydon, CR0 4YY

17/09/2026

14964301-0001